AF410812

Biosensors for Diagnosis and Monitoring

Biosensors for Diagnosis and Monitoring

Editor

Mònica Mir

MDPI • Basel • Beijing • Wuhan • Barcelona • Belgrade • Manchester • Tokyo • Cluj • Tianjin

Editor
Mònica Mir
Institute for Bioengineering of Catalonia (IBEC)
Spain

Editorial Office
MDPI
St. Alban-Anlage 66
4052 Basel, Switzerland

This is a reprint of articles from the Special Issue published online in the open access journal *Biosensors* (ISSN 2079-6374) (available at: https://www.mdpi.com/journal/biosensors/special_issues/biosens_diagno_monitor).

For citation purposes, cite each article independently as indicated on the article page online and as indicated below:

LastName, A.A.; LastName, B.B.; LastName, C.C. Article Title. *Journal Name* **Year**, *Volume Number*, Page Range.

ISBN 978-3-0365-3635-4 (Hbk)
ISBN 978-3-0365-3636-1 (PDF)

Contents

About the Editor

Mònica Mir graduated with a degree in chemistry from the Rovira i Virgili University (Spain) in 1998, and in 2006, she obtained her doctorate in biotechnology from the same University. She carried out predoctoral stays in Greece and the United Kingdom and a postdoctoral stay at the Max Planck Institute for Polymer Research (Germany). In 2008, she joined the Institute of Bioengineering of Catalonia (IBEC) (Spain) as a senior researcher at the Center for Biomedical Research Network on Bioengineering (CIBER-bbn) while also teaching as an Associate Professor at the University of Barcelona. Her main research motivations are related to the areas of electrochemical biosensors and implantable sensors for biomedical applications, integrated in microfluidic for lab-on-a-chip, point-of-care, and organ-on-a-chip platforms. She is the co-author of more than 50 publications with an h index of 15 and has 1242 citations (Scopus, 2021).

Preface to "Biosensors for Diagnosis and Monitoring"

In recent decades, biosensor technologies have received great interest, and especially in recent years due to the health alert caused by the COVID-19 pandemic. Increasingly sensitive and selective biosensors are required with a fast and economical response in our daily life, increasing the applications and market of this technology.

The book "Biosensors for Diagnosis and Monitoring" has been published thanks to contributions from its authors and is dedicated to all areas of research related to biosensor technology. Reviews, perspective articles and research articles in different areas of biodetection such as portable sensors, point-of-care platforms, pathogen detection for biomedical applications as well as for environmental monitoring will introduce the reader to these relevant topics in the field.

Mònica Mir
Editor

Perspective

Point-of-Care PCR Assays for COVID-19 Detection

Niharika Gupta [1], Shine Augustine [1], Tarun Narayan [2], Alan O'Riordan [2], Asmita Das [1], D. Kumar [3], John H. T. Luong [4,*] and Bansi D. Malhotra [1,*]

[1] Department of Biotechnology, Delhi Technological University, Shahbad Daulatpur, Delhi 110042, India; niharika.gupta990@gmail.com (N.G.); shine2089@gmail.com (S.A.); asmita1710@gmail.com (A.D.)

[2] Nanotechnology Group, Tyndall National Institute, University College Cork, T12 K8AF Cork, Ireland; tarun.narayan@tyndall.ie (T.N.); alan.oriordan@tyndall.ie (A.O.)

[3] Department of Applied Chemistry, Delhi Technological University, Shahbad Daulatpur, New Delhi 110042, India; dkumar@dce.ac.in

[4] School of Chemistry, University College Cork, T12 K8AF Cork, Ireland

* Correspondence: j.luong@ucc.ie (J.H.T.L.); bansi.malhotra@gmail.com (B.D.M.)

Abstract: Molecular diagnostics has been the front runner in the world's response to the COVID-19 pandemic. Particularly, reverse transcriptase-polymerase chain reaction (RT-PCR) and the quantitative variant (qRT-PCR) have been the gold standard for COVID-19 diagnosis. However, faster antigen tests and other point-of-care (POC) devices have also played a significant role in containing the spread of SARS-CoV-2 by facilitating mass screening and delivering results in less time. Thus, despite the higher sensitivity and specificity of the RT-PCR assays, the impact of POC tests cannot be ignored. As a consequence, there has been an increased interest in the development of miniaturized, high-throughput, and automated PCR systems, many of which can be used at point-of-care. This review summarizes the recent advances in the development of miniaturized PCR systems with an emphasis on COVID-19 detection. The distinct features of digital PCR and electrochemical PCR are detailed along with the challenges. The potential of CRISPR/Cas technology for POC diagnostics is also highlighted. Commercial RT–PCR POC systems approved by various agencies for COVID-19 detection are discussed.

Keywords: polymerase chain reaction; COVID-19; electrochemical; digital PCR; point-of-care

Citation: Gupta, N.; Augustine, S.; Narayan, T.; O'Riordan, A.; Das, A.; Kumar, D.; Luong, J.H.T.; Malhotra, B.D. Point-of-Care PCR Assays for COVID-19 Detection. *Biosensors* **2021**, *11*, 141. https://doi.org/10.3390/bios11050141

Received: 16 March 2021
Accepted: 28 April 2021
Published: 1 May 2021

Publisher's Note: MDPI stays neutral with regard to jurisdictional claims in published maps and institutional affiliations.

1. Introduction

The coronavirus disease 2019 (COVID-19) outbreak crisis has changed the shape of our world since its first report in December 2019. While some countries seem to be recovering from the crisis and are reporting fewer cases, others are still witnessing an increasing number of cases [1]. Clinical diagnosis has been the forerunner in controlling the COVID-19 pandemic. Molecular nucleic acid amplification tests (NAATs) were the first to be developed for detecting SARS-CoV-2 RNA in patient samples. Particularly, reverse transcriptase-polymerase chain reaction (RT-PCR) and its quantitative variant (qRT-PCR) have been the keystone for diagnosis of SARS-CoV-2 with the capacity to detect target nucleic acids (<100 copies/mL) with remarkable sensitivity [2]. However, the analysis proved time-intensive, requiring up to a few hours, and could only be performed in a centralized laboratory. The high false-negative rates with some RT-PCR assays also raised concern. Thus, attention shifted to faster, cheaper, and equally sensitive (if not more) point-of-care (POC) biosensing devices that could be deployed for mass screening.

Therein began a major shift in the clinical diagnostic industry, with point-of-care testing (POCT) becoming the focus of attention almost overnight. Lateral flow assays (LFAs), chemiluminescence, and nanoparticle-based colorimetric detection were developed for detecting SARS-CoV-2-related antigens and antibodies produced in response to its infection [3–8]. Faster, miniaturized isothermal amplification tests emerged that could detect the virus within a few minutes and with sensitivity at par with RT-PCR assays [5,9,10].

Although different types of POCT devices have been authorized in various countries for emergency use, many novel biosensing strategies and designs still seek validation and are currently subject to academic inquiry.

These devices have shorter response times and have cost-effectively enabled population-wide mass screening. However, evidence suggests that the analytic performance (sensitivity, specificity, positive and negative predictive values, etc.) of current antigen diagnostic tests is not at par with that of RT-PCR and other NAATs [11]. Thus, while rapid antigen tests and other POCT are being widely used for COVID-19 screening, it is still uncertain whether such tests will be regularized and used in routine diagnostic procedures. In attempt to synergize the sensitivity of NAATs and the ease of use of POCT assays, miniaturized NAAT-based POCT devices and assays were devised for faster screening and diagnosis of COVID-19. One of the first such devices was the Abbott ID Now, which integrates isothermal amplification with colorimetric detection to yield results within 5 min. However, questions were soon raised about its utility as a singular diagnostic test due to its low positive predictive value (PPA) and high false-negative rates, especially in samples with low viral load [10]. More rapid devices based on isothermal amplification with improved performance were devised. Thus, although the integration of isothermal amplification in POC devices has gained some success, they are not as successful as RT-PCR for COVID-19 detection. In general, the high temperature requirements of RT-PCR prevent non-specific amplification, which is more common in isothermal amplification techniques. Conversely, these temperature requirements somewhat complicate the development of PCR-based rapid devices.

Nonetheless, efforts have been directed toward miniaturizing PCR to make it an automated, high-throughput device that can be applied at point-of-use. In this review, we summarize studies related to the development of miniaturized, high-throughput PCR biosensors for COVID-19 detection. The distinct features, limitations, and advantages of various types of PCR biosensors and chips are discussed. The advantages and limitations of PCR chips over biosensors based on other amplification assays are listed. The potential of biosensing formats to be integrated with RT-PCR is explored, along with the path-breaking integration of CRISPR/Cas technology with amplification assays toward the development of faster, miniaturized devices and chips.

2. RT-PCR: The Gold Standard

RT-PCR is the first molecular diagnostic test to be employed for detecting SARS-CoV-2 RNA in patient samples and is currently considered the gold standard for COVID-19 diagnosis. Different RT-PCR assays have been designed for detecting SARS-CoV-2 virus RNA in different body fluids, such as nasopharyngeal swabs, lower respiratory tract fluid, sputum, saliva, etc. [12–14]. However, RT-PCR is prone to false-negative results that reduce the overall sensitivity of the diagnosis. This may be because of various reasons such as low viral load in the pharyngeal, nasal, and sputum samples; storage and transport of samples; and improper handling [15,16]. Moreover, any mismatches between the primers and probe–target regions compromise the assay performance, leading to false-negative results [15,17]. Another major challenge faced by RT-PCR is that it can yield false-positive results by amplifying RNA from dead, noninfectious viruses as well [18]. Thus, recovered patients that no longer hold the threat of transmitting the disease may be positive per RT-PCR tests.

The current challenges of the qRT-PCR method include the use of fluorescent label binding to the source signal produced by the amplified DNA, which not only increases the cost of the instrument, but also the complexities. This technology is less appealing to developing nations or remote locations with limited resources. Commercial RT-PCR kits have not been subject to rigorous quality control. Personnel skills and good laboratory practice play an important role in Biosafety Level 3. Optimum sample types and timing for peak viral load remain to be fully investigated as sputum or nasal swabs are the most accurate sample for diagnosis of COVID-19, but not throat swabs.

Despite these limitations, RT-PCR remains the gold standard for confirming the diagnosis of COVID-19. There have been multiple attempts to develop portable PCR systems since the inception of the pandemic. Lab-in-tube systems incorporating lysis, reverse transcription, amplification, and detection in a single tube within 36 min were demonstrated in May 2020 [19]. A lab-on-chip device, CovidNudge, can be used to perform sample processing and real-time RT-PCR outside of a laboratory setting [20] (Figure 1). The chip consists of detection arrays for seven SARS-CoV-2 genes and one host gene as a sample adequacy control. This device detects the virus in 90 min and reduces the collection-to-result turnaround time significantly by eliminating the requirement of sample transport from the site of collection to a centralized lab. The sensitivity of this POC test (94%) is comparable to that of lab-based tests in clinical settings. As of September 2020, over 5 M CovidNudge kits had been deployed in the U.K. for COVID-19 testing. Of note is a portable RT-PCR workstation for COVID-19 detection in under-served and remote areas [21]. This workstation is a chip-based, battery-operated qRT-PCR system with the capability of network data transfer and automated reporting. Almost 3.8% (2.7 million) of the total tests conducted in India were performed on these workstations (as of September 2020). The average cost of an RT-PCR varies in different parts of the world. In India, for example, the cost of a conventional RT-PCR test currently varies from INR 400 (~USD 5.30) to INR 950 (~USD 12.6), and POC rapid antigen tests are free. While the CovidNudge test (Table 1) deployed in the U.K. costs around GBP 10 (per test) (equivalent to ~USD 13.80), which is almost 10 times cheaper than the average cost (~GBP 100) of a conventional RT-PCR test in the country.

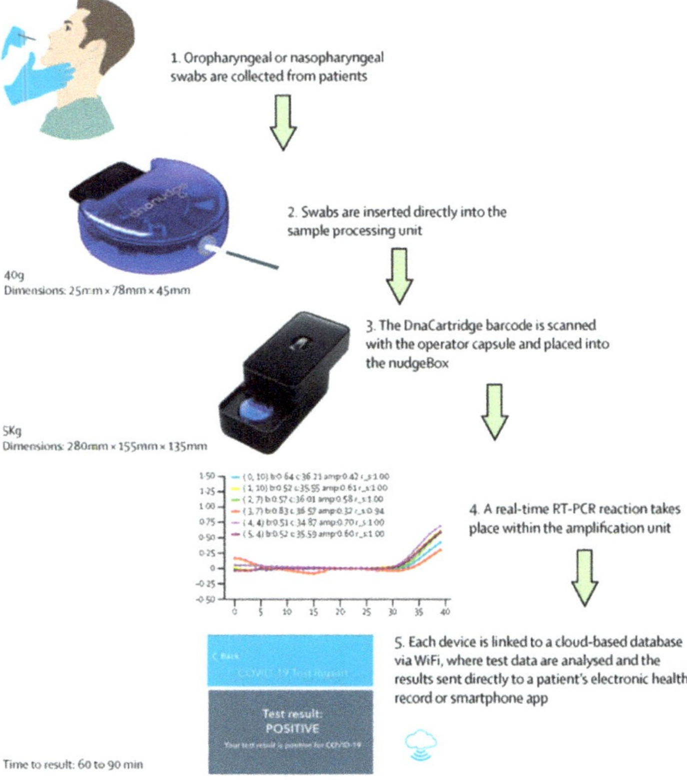

Figure 1. Schematic diagram depicting the various steps performed by the CovidNudge assay for automated detection of SARS-CoV-2 RNA (Reprinted with permission from Ref. [20]).

Table 1. List of commercial, automated RT-PCR systems authorized under emergency use.

Name of the Kit	Target Genes	Type	Sample Preparation	No. of Tests	Time	LOD	Sensitivity	Specificity	Cost (Per Test)	Reference
CovidNudge	rdrp1, rdrp2, E gene, N gene, n1, n2, and n3	RT-PCR	Automated	NA	~90 min	5 copies/µL	>94%	100%	GBP 10	[20]
Accula SARS-CoV-2 Test	N gene	RT-PCR	Automated	NA	~30 min	NA	100%	100%	USD 20	[22]
Cepheid Xpert Xpress SARS-CoV-2 assay	N2 and E	RT-PCR (real time)	Automated	10 per kit		0.02 PFU/mL			USD 19.8	[23]
FastPlex Triplex SARS-CoV-2 Detection Kit	ORF1ab, N, RPP30	RT-dPCR	Manual	96 test per kit	90 min	285.7 copies/mL	>95%	95.7%	USD 1152	[23]
Gnomegen COVID-19 RT-Digital PCR Detection Kit	N1, N2	RT-dPCR	Manual	48 samples per day	180 min	2.5 copies per reaction	>95%	99%	NA	[23]
Bio-Rad SARS-CoV-2 ddPCR Test	N1, N2	RT-dPCR	Manual	96 samples	NA	400 copies/mL			NA	[23]
ePlexSARS-CoV-2 Test	N gene	End-point RT-PCR with electrochemical Detection	Automated	12 tests/kit	NA	1×10^3 copies/mL	99.02%	98.41%	NA	[23]

There have been several other innovations related to the fabrication of PCR chips and biosensors for COVID-19 detection. The following sections cover some of these studies and discuss the potential and challenges faced by such devices in emerging as viable commercial products.

3. RT-PCR Biosensors

3.1. Digital RT-PCR

The concept of digital PCR (dPCR) was pioneered by Vogelstein and Kinzler in 1999 [24]. The principle of dPCR is to partition the reaction mixture into many sub-reactions before amplification; the original numbers are determined by counting the partition showing negative and positive reactions [25] (Figure 2). It does not require a standard curve or reference genes and is more resistant to interference factors such as specific template amplification inhibitors [26,27]. The quantification results are analyzed from Poisson's distribution and can achieve an accurate estimation of low concentrations of nucleic acid samples [26]. Therefore, a method like dPCR offers high sensitivity, higher precision, and resistance to inhibitors, which are required for an accurate SARS-CoV-2 diagnosis. The dPCR method can be classified into three types based on liquid separation: droplet-based (ddPCR), chip-based (cdPCR), and microfluidic digital PCR (mdPCR). The primary difference between these three types of digital PCR is the design of the sample partitioning system in the detection platform: ddPCR combines several millions partitioning of the PCR test into individual droplets in a water-in-oil emulsion [26,28], whereas cdPCR uses an active partitioning approach. It has two chip halves with two arrays of microwells. The chambers are aligned so that the opposite halves form continuous channels [28,29]. In mdPCR, microfluidic chambers are used to split the samples. These chambers are fluidically designed such that each sample can be partitioned into tens of thousands of wells [30].

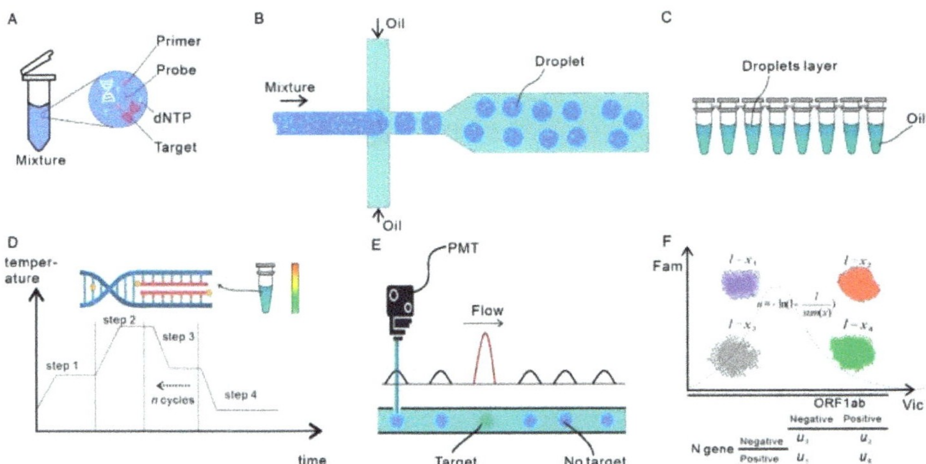

Figure 2. Schematic depicting workflow of a ddPCR system: (**A**) preparation for amplification, (**B**) generation of water-in-oil droplets using a microfluidic flow system, (**C**) collection of the droplets in PCR tubes, (**D**) PCR amplification, (**E**) analysis of fluorescence in the droplets after amplification, and (**F**) fitting to Poisson distribution to determine the absolute copy numbers of the target molecules (Reprinted from Ref. [28]).

dPCR can be used for the quantification of a low viral load, monitoring of the virus in the environment, evaluation of anti-SARS-CoV-2 drugs [28], and the detection of viral mutations [31]. Many types of clinical samples can be used for COVID-19 testing using dPCR, including blood, urine, sputum, stool, nasal swabs, and throat swabs. Studies have compared RT-PCR with RT-dPCR for the presence of SARS-CoV-2 in pharyngeal swab samples and found RT-dPCR to be more sensitive and accurate than RT-PCR [32,33].

Lu and group showed that RT-dPCR has a detection limit ten-fold lower than that of RT-PCR [34]. They compared the RT-dPCR and RT-PCR of 36 COVID-19 patients with 108 specimens, including blood, pharyngeal swab, and stool, in which four pharyngeal samples yielding negative results in RT-PCR were positive per RT-dPCR. Another study demonstrated that suspected patients who tested negative by RT-PCR were found to be positive by ddPCR [35]. The results of ddPCR were validated by the serological testing of anti-COVID-19 antibodies in the samples. The ddPCR can yield better and more precise quantitation of viral loads of SARS-CoV-2 [36–38]. However, most of the reported ddPCR procedures included an RNA extraction and purification step, which can lead to potential amplification errors [38]. Moreover, direct quantification by ddPCR targeting the envelope (E) gene [39], ORF1ab gene [40], and nucleocapsid (N) [41] region have also been reported. The viral load can be quantified in throat swabs, sputum, nasal swabs, blood, and urine [37]. Droplet-based dPCR was also used to detect SARS-CoV-2 RNA in airborne aerosols [42], in which the viral load in the toilets used by some medical personnel and patients was found to be high. This study indicated the significance of sanitization and room ventilation for limiting COVID-19 spread. The primary advantage of dPCR is its good sensitivity and high-throughput analysis, which has been the key requirement for COVID-19 detection. Currently, there are three commercial dPCR tests authorized for emergency use by the USFDA (Table 1).

However, a few challenges require the utmost attention before dPCR can be used in routine diagnostics. Particularly, much like conventional PCR tests, dPCR also requires expensive instruments, reagents, and professional experts to operate the system. The fabrication of the dPCR chips requires complex steps, making it a costly operation. Moreover, much like other POC tests, strict standards and guidelines need to be followed to assure the quality of results obtained from dPCR systems.

3.2. Electrochemical PCR: Unexplored Potential

The integration of electrochemistry with RT-PCR aims to provide a rapid, miniaturized, hand-held instrument. Electrochemical biosensors work by modification of a working electrode with a biomolecule that interacts with a specific target analyte present in an aqueous electrolyte and generates an electrical signal corresponding to its concentration. In the case of an electrochemical PCR, there is an electroactive species whose oxidation or reduction signal is correlated to the amount of PCR amplified product. A more challenging approach is the use of nanomaterials to tag the DNA primers used in the PCR amplification step, such as gold nanoparticles (AuNPs) or semiconductor quantum dots (QDs). The labeled amplified products are then further quantified via the generation of electrochemical signals.

Electrochemical systems offer the benefits of being seamlessly implemented into compact and intelligent systems, enabling high versatility and real-time detection. Moreover, electrochemically active labels (such as metal-complex, organic molecules, etc.) are more durable than fluorescent dyes (Cy5, FAM, etc.) and are a notable factor toward the commercial applications of electrochemical-RT-PCR (EPCR). The power and sample volume requirements are lower for electrochemical biosensors compared with RT-PCR. Despite the considerable interest, electrochemical biosensors have garnered in the context of COVID-19 detection, the clinical industry appears reluctant to adopt this technology for practical and commercial use.

The pre-COVID era witnessed the emergence of PCR-free electrochemical assays for detecting different nucleic acid targets, including microRNA, viral RNA and DNA, and cancer-related genes [43–45]. Perhaps the research community has been confident that electrochemical assays can compete with the existing PCR technology in terms of sensitivity and turnaround times and eliminate the use of costly reagents and dyes [46]. There have been some studies on PCR-integrated electrochemical biosensors in the last 5 years. Some of the recent studies have demonstrated innovative PCR-free electrochemical sensors for

SARS-CoV-2 RNA detection with remarkable detection limits [47,48]; however, none has yet achieved a commercial or authorized status.

Integrating PCR with electrochemical transducers poses various challenges; the primary challenge includes the capability of the sensing surface to withstand the harsh temperature changes and salt concentrations required during PCR [49]. Isothermal amplification techniques are preferred over PCR for integration with electrochemical sensors. A rapid electrochemical detection system based on rolling circle amplification (RCA) was demonstrated for multiplex detection of the S and N genes of SARS-CoV-2 [50] (Figure 3). Sandwich hybridization was employed in this study, with oligonucleotide probes consisting of redox-active labels (methylene blue-and-acridine orange) for electrochemical detection using differential pulse voltammetry. This assay detects the N or S viral gene at a concentration as low as 1 copy/μL within 2 h with high selectivity and sensitivity.

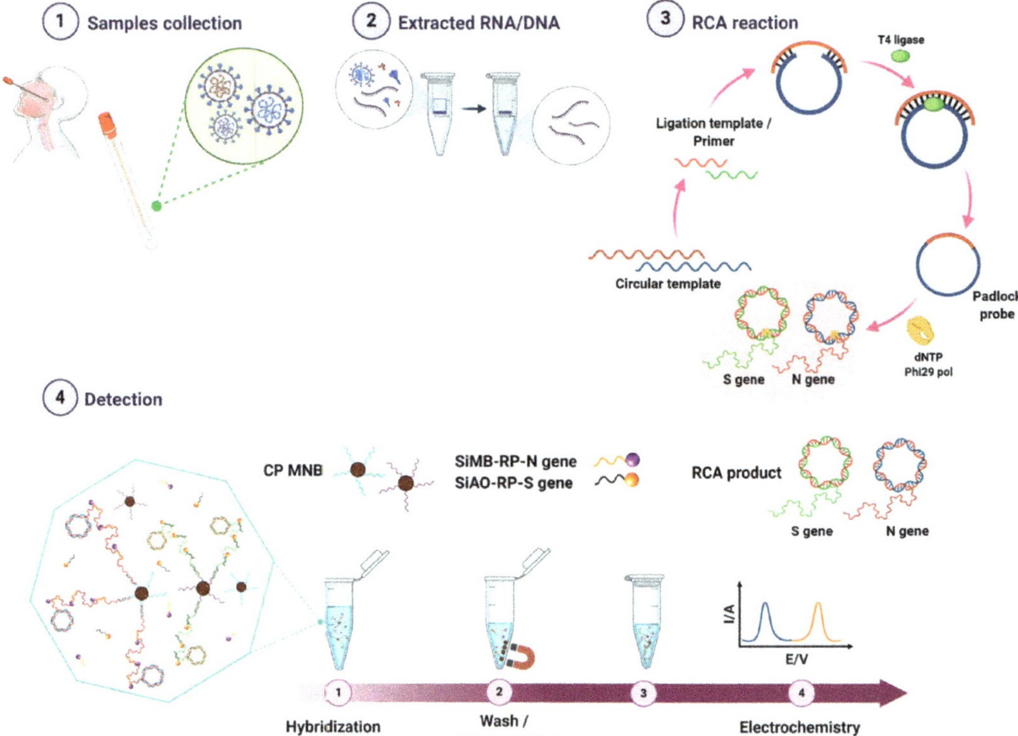

Figure 3. Workflow of the RCA-based electrochemical sensor for SARS-CoV-2 detection (Reprinted from Ref. [50]).

The recent advances in microfluidics technology have enabled the integration of electrochemical electrodes with miniaturized reaction chambers (or chips) designed for PCR. The USFDA recently approved the GenMark ePlex® SARS-CoV-2 test, which automates RNA extraction and amplification, and then further integrates it into competitive hybridization-based electrochemical detection [51]. This system uses the principle of electrowetting (digital microfluidics) to manipulate the movement of samples and reagents on a printed circuit board (PCB) (Table 1).

4. CRISPR/Cas-Based Sensors: The New Alternative

CRISPR stands for clustered regularly interspaced short palindromic repeat, which utilizes genetic information of bacterial species as a part of an antiviral process. CRISPR/Cas is a genetic editing technology whose precise and specific DNA and RNA cleavage ability makes it a useful tool in nucleic acid diagnostics. CRISPR/Cas-based sensors mainly utilize single guide RNA in conjunction with the Cas system to bind to a target sequence or cleave target DNA and RNA, resulting in signal generation. Owing to their high specificity, they are an attractive alternative to POC RT-PCR devices. CRISPR/Cas-based diagnostics circumvents the issue of long turnaround times and enhances the assay specificity [52]. Recently, Hou et al. developed a rapid assay known as CRISPR–COVID for detecting SARS-CoV-2 with less turnaround time (~40 min) compared with RT-PCR and metagenomics sequencing [53]. Another advantage of using CRISPR/Cas systems is the exclusion of RNA isolation and amplification, making it a faster analysis method. An ultrasensitive RT-RPA CRISPR–fluorescence detection system (FDS) assay can eliminate the need for RNA isolation for SARS-CoV-2 detection [54]. It uses a saliva sample that is subject to a mix of chemicals that amplify the viral RNA, which is then subjected to CRISPR/Cas12a-based fluorescence signal amplification. The linear range of this handheld CRISPR-based test was found to be 1 to 10^5 copies/mL with a limit of detection of 0.38 copies/mL, which is consistent with the result obtained using qRT-PCR. In another approach, the need for SARS-CoV-2 RNA pre-amplification was eliminated with the use of CRISPR-Cas13a, which aids the detection of SARS-CoV-2 RNA from nasal swabs [55]. The main highlight of this study was the use of different sets of crRNAs to increase the sensitivity by activation of a greater number of Cas13a per target RNA. Additionally, the study reported the ability to directly translate the fluorescent signal into viral loads, thus resulting in remarkable sensitivity compared with other CRISPR-based assays for COVID detection.

5. Future Outlook

COVID-19 diagnostics has evolved significantly since its first appearance. The range and types of diagnostic devices that have emerged in the past year are immensely diverse. Several earlier diagnostic devices and assays were only the subject of academic interest and research but are now commercially available for use. However, since most of the POC devices for COVID-19 detection have been authorized under emergency use, caution should be taken when extrapolating the use of such devices for the diagnosis of other diseases.

Despite the advances, there are limitations associated with RT-PCR POC devices and biosensors concerning sample preparation in ePCR, false negatives and positives, and reagent evaporation in dPCR. Efforts to identify the limitations in current PCR devices for COVID-19 detection can soon help in the design of improved diagnostic devices. Additionally, different detection strategies and platforms can be integrated to develop new, hybrid devices for improved performance. For example, electrokinetic focusing on microfluidic chips was used to automate the process of nucleic acid purification and amplification with a reduction in non-specific amplification [56]. A recent study used isotachophoresis (ITP), an ionic focusing technique, on a microfluidic chip to automate SARS-CoV-2 RNA purification and subsequent detection by CRISPR-based technique within 35 min [57]. This on-chip device uses a smaller volume of reagents (<100 times lower) and automates sample preparation and subsequent detection. Reduction in bubble generation and reagent evaporation in dPCR systems was also demonstrated by creating a vertical polymeric barrier leading to ultrafast PCR amplification [58].

Centrifugal microfluidic platforms (or lab-on-a-disc) for automated sample preparation and subsequent RT-PCR can also be conceived. These devices use different layers of polymeric substrates to integrate multiple steps involving complex fluid flow. These centrifugal systems were shown to improve reaction rates using efficient mixing, thus enabling high sensitivity and reduced hybridization times [59]. Paperfluidic devices that involve the creation of microfluidic channels on paper can also be realized for SARS-CoV-2 RNA detection. Apart from being inexpensive, paperfluidic devices do not require any additional step to render the channels hydrophilic for fluid flow; the intrinsic hydrophilicity of paper allows fluid flow via capillary action, thus eliminating the need for external pumps. This allows their use in resource-limited, POC settings. These devices, much like LFAs, can be batch fabricated at minimal cost and can thus be used in mass screening operations in resource-limited settings. A paper-based assay, FnCas9 editor-linked uniform detection assay (FELUDA), was developed in India, which enables detection of single nucleotide variants [60]. This test uses RT-PCR followed by CRISPR-based detection in a lateral flow format. Similarly, paperfluidic devices that can integrate RNA extraction, amplification, and subsequent detection can be realized [61].

COVID-19 diagnostics has provided new opportunities and advances in the clinical diagnostic sector. It will be interesting to see how these developments affect the overall diagnostics landscape over time.

6. Conclusions

Molecular diagnostics has been the cornerstone in controlling the ongoing COVID-19 pandemic. RT-PCR is currently the primary gold standard for COVID-19 diagnosis. Simultaneously, this crisis has brought us to realize the importance of low-cost, sensitive, and high-throughput devices that can be deployed in POC settings. On-site analysis that is fast, reliable, and helps to reduce the economic costs of infection transmission and potential quarantine is required. Different rapid POC tests have been authorized and deployed for mass screening and diagnostic purposes. Yet, RT-PCR has remained the primary and the only method for COVID-19 confirmation. Miniaturized PCR and PCR biosensors, devices that integrate PCR with different detection modalities, have emerged as tools that can address the issue of the low sensitivity of the current rapid POC tests and simultaneous analysis of samples in a high-throughput manner outside of a centralized lab. Digital PCR has emerged as an efficient high-throughput system. However, it does not eliminate the use of expensive reagents and often requires professional involvement in its operation. Electrochemical PCR is also a viable option for faster, cost-effective, and sensitive COVID-19 detection. However, the difficulty of the integration of PCR with electrochemical systems still creates formidable challenges in realizing a commercially adaptable system. CRISPR/Cas-based systems have further created a scope for diagnostic devices that do not require RNA extraction and amplification before detection.

The active transition from routine diagnostic laboratories to the realm of high sensitivity molecular diagnostics can significantly increase the efficiency and responsiveness of POCTs and facilitate the management of outbreaks in difficult settings. Devices such as those mentioned above can readily aid healthcare professionals in making faster medical decisions. However, there are still limitations to be addressed in such systems. Sample preparation errors and false positives and negatives need to be addressed before these assays can eventually be used for other diagnostic applications as well. Although different formats of POC RT-PCR assays have emerged, there is still scope for the development of hybrid, integrated systems that have better performance in terms of specificity and response time. Rigorous validation protocols and a high sampling rate would determine whether these devices are capable of use in the long run.

Author Contributions: Conceptualization, B.D.M., J.H.T.L., and N.G.; writing–original draft writing, N.G., S.A., and T.N.; writing–review and editing, B.D.M., J.H.T.L., A.D., D.K., and A.O. All authors have read and agreed to the published version of the manuscript.

Funding: This research received no external funding.

Institutional Review Board Statement: Not applicable.

Informed Consent Statement: Not applicable.

Data Availability Statement: Not applicable.

Acknowledgments: N.G., S.A., A.D., D.K., and B.D.M. acknowledge Yogesh Singh, Vice-Chancellor, Delhi Technological University, Delhi, India, for providing necessary facilities. N.G. and S.A. thank Delhi Technological University, Delhi, India; and the Council of Scientific and Industrial Research (CSIR; 08/133/(0013)/2018-EMR-I), India, respectively, for a fellowship award. T.N. is thankful for funding from the European Union's Horizon 2020 Research & Innovation Programme under the

Biosensors **2021**, 11, 141

Marie Sklodowska-Curie grant agreement no.: H2020-MSCA-ITN-813680. B.D.M. thanks the Science & Engineering Research Board (SERB), Govt. of India, for the award of a Distinguished Fellowship (SB/DF/011/2019).

Conflicts of Interest: The authors declare no conflict of interest.

References

1. *COVID-19 Weekly Epidemiological Update, 9 March 2021*; World Health Organization: Geneva, Switzerland, 2021.
2. Arnaout, R.; Lee, R.A.; Lee, G.R.; Callahan, C.; Yen, C.F.; Smith, K.P.; Arora, R.; Kirby, J.E. SARS-CoV2 testing: The limit of detection matters. *bioRxiv* **2020**. [CrossRef]
3. Grant, B.D.; Anderson, C.E.; Williford, J.R.; Alonzo, L.F.; Glukhova, V.A.; Boyle, D.S.; Weigl, B.H.; Nichols, K.P. SARS-CoV-2 coronavirus nucleocapsid antigen-detecting half-strip lateral flow assay toward the development of point of care tests using commercially available reagents. *Anal. Chem.* **2020**, *92*, 11305–11309. [CrossRef]
4. Ragnesola, B.; Jin, D.; Lamb, C.C.; Shaz, B.H.; Hillyer, C.D.; Luchsinger, L.L. COVID19 antibody detection using lateral flow assay tests in a cohort of convalescent plasma donors. *BMC Res. Notes* **2020**, *13*, 1–7. [CrossRef]
5. Zhu, X.; Wang, X.; Han, L.; Chen, T.; Wang, L.; Li, H.; Li, S.; He, L.; Fu, X.; Chen, S. Multiplex reverse transcription loop-mediated isothermal amplification combined with nanoparticle-based lateral flow biosensor for the diagnosis of COVID-19. *Biosens. Bioelectron.* **2020**, *166*, 112437. [CrossRef]
6. Huang, C.; Wen, T.; Shi, F.-J.; Zeng, X.-Y.; Jiao, Y.-J. Rapid detection of IgM antibodies against the SARS-CoV-2 virus via colloidal gold nanoparticle-based lateral-flow assay. *ACS Omega* **2020**, *5*, 12550–12556. [CrossRef]
7. Cai, X.-f.; Chen, J.; Hu, J.-l.; Long, Q.-x.; Deng, H.-j.; Liu, P.; Fan, K.; Liao, P.; Liu, B.-z.; Wu, G.-c. A peptide-based magnetic chemiluminescence enzyme immunoassay for serological diagnosis of coronavirus disease 2019 (COVID-19). *J. Infect. Dis.* **2020**, *222*, 189–193. [CrossRef]
8. Padoan, A.; Cosma, C.; Sciacovelli, L.; Faggian, D.; Plebani, M. Analytical performances of a chemiluminescence immunoassay for SARS-CoV-2 IgM/IgG and antibody kinetics. *Clin. Chem. Lab. Med.* **2020**, *58*, 1081–1088. [CrossRef] [PubMed]
9. Yu, L.; Wu, S.; Hao, X.; Dong, X.; Mao, L.; Pelechano, V.; Chen, W.-H.; Yin, X. Rapid detection of COVID-19 coronavirus using a reverse transcriptional loop-mediated isothermal amplification (RT-LAMP) diagnostic platform. *Clin. Chem.* **2020**, *66*, 975–977. [CrossRef] [PubMed]
10. Basu, A.; Zinger, T.; Inglima, K.; Woo, K.-m.; Atie, O.; Yurasits, L.; See, B.; Aguero-Rosenfeld, M.E. Performance of Abbott ID Now COVID-19 rapid nucleic acid amplification test using nasopharyngeal swabs transported in viral transport media and dry nasal swabs in a New York City academic institution. *J. Clin. Microbiol.* **2020**, *58*. [CrossRef] [PubMed]
11. Pray, I.W. *Performance of an Antigen-Based Test for Asymptomatic and Symptomatic SARS-CoV-2 Testing at Two University Campuses— Wisconsin, September–October 2020*; Centers for Disease Control and Prevention: Atlanta, GA, USA, 2021.
12. Wang, X.; Yao, H.; Xu, X.; Zhang, P.; Zhang, M.; Shao, J.; Xiao, Y.; Wang, H. Limits of detection of 6 approved RT–PCR kits for the novel SARS-coronavirus-2 (SARS-CoV-2). *Clin. Chem.* **2020**, *66*, 977–979. [CrossRef] [PubMed]
13. Babady, N.E.; McMillen, T.; Jani, K.; Viale, A.; Robilotti, E.V.; Aslam, A.; Diver, M.; Sokoli, D.; Mason, G.; Shah, M.K. Performance of severe acute respiratory syndrome coronavirus 2 real-time RT-PCR tests on oral rinses and saliva samples. *J. Mol. Diagn.* **2021**, *23*, 3–9. [CrossRef] [PubMed]
14. Patel, M.R.; Carroll, D.; Ussery, E.; Whitham, H.; Elkins, C.A.; Noble-Wang, J.; Rasheed, J.K.; Lu, X.; Lindstrom, S.; Bowen, V. Performance of Oropharyngeal Swab Testing Compared with Nasopharyngeal Swab Testing for Diagnosis of Coronavirus Disease 2019—United States, January 2020–February 2020. *Clin. Infect. Dis.* **2021**, *72*, 482–485. [CrossRef] [PubMed]
15. Zhou, Y.; Pei, F.; Ji, M.; Wang, L.; Zhao, H.; Li, H.; Yang, W.; Wang, Q.; Zhao, Q.; Wang, Y. Sensitivity evaluation of 2019 novel coronavirus (SARS-CoV-2) RT-PCR detection kits and strategy to reduce false negative. *PLoS ONE* **2020**, *15*, e0241469. [CrossRef] [PubMed]
16. Arevalo-Rodriguez, I.; Buitrago-Garcia, D.; Simancas-Racines, D.; Zambrano-Achig, P.; Del Campo, R.; Ciapponi, A.; Sued, O.; Martinez-Garcia, L.; Rutjes, A.W.; Low, N. False-negative results of initial RT-PCR assays for COVID-19: A systematic review. *PLoS ONE* **2020**, *15*, e0242958. [CrossRef] [PubMed]
17. Tahamtan, A.; Ardebili, A. Real-time RT-PCR in COVID-19 detection: Issues affecting the results. *Expert Rev. Mol. Diagn.* **2020**, *20*, 453–454. [CrossRef] [PubMed]
18. Singanayagam, A.; Patel, M.; Charlett, A.; Bernal, J.L.; Saliba, V.; Ellis, J.; Ladhani, S.; Zambon, M.; Gopal, R. Duration of infectiousness and correlation with RT-PCR cycle threshold values in cases of COVID-19, England, January to May 2020. *Euro Surveill.* **2020**, *25*, 2001483. [CrossRef]
19. Wee, S.K.; Sivalingam, S.P.; Yap, E.P.H. Rapid direct nucleic acid amplification test without RNA extraction for SARS-CoV-2 using a portable PCR thermocycler. *Genes* **2020**, *11*, 664. [CrossRef]
20. Gibani, M.M.; Toumazou, C.; Sohbati, M.; Sahoo, R.; Karvela, M.; Hon, T.-K.; De Mateo, S.; Burdett, A.; Leung, K.F.; Barnett, J. Assessing a novel, lab-free, point-of-care test for SARS-CoV-2 (CovidNudge): A diagnostic accuracy study. *Lancet Microbe* **2020**, *1*, e300–e307. [CrossRef]
21. Gupta, N.; Rana, S.; Singh, H. Innovative point-of-care molecular diagnostic test for COVID-19 in India. *Lancet Microbe* **2020**, *1*, e277. [CrossRef]
22. *Accula SARS-CoV-2 Test-Letter of Authorization*; The U.S. Food and Drug Administration: Silver Spring, MD, USA, 2021.

23. In Vitro Diagnostics EUAs. Available online: https://www.fda.gov/medical-devices/coronavirus-disease-2019-covid-19-emergency-use-authorizations-medical-devices/vitro-diagnostics-euas (accessed on 10 February 2021).

24. Vogelstein, B.; Kinzler, K.W. Digital PCR. *Proc. Natl. Acad. Sci. USA* **1999**, *96*, 9236–9241. [CrossRef]

25. Quan, P.-L.; Sauzade, M.; Brouzes, E. dPCR: A technology review. *Sensors* **2018**, *18*, 1271. [CrossRef] [PubMed]

26. Hindson, B.J.; Ness, K.D.; Masquelier, D.A.; Belgrader, P.; Heredia, N.J.; Makarewicz, A.J.; Bright, I.J.; Lucero, M.Y.; Hiddessen, A.L.; Legler, T.C. High-throughput droplet digital PCR system for absolute quantitation of DNA copy number. *Anal. Chem.* **2011**, *83*, 8604–8610. [CrossRef] [PubMed]

27. White, R.A.; Blainey, P.C.; Fan, H.C.; Quake, S.R. Digital PCR provides sensitive and absolute calibration for high throughput sequencing. *BMC Genomics* **2009**, *10*, 1–12.

28. Tan, C.; Fan, D.; Wang, N.; Wang, F.; Wang, B.; Zhu, L.; Guo, Y. Applications of digital PCR in COVID-19 pandemic. *View* **2021**, *2*. [CrossRef]

29. Nykel, A.; Kaszkowiak, M.; Fendler, W.; Gach, A. Chip-based digital PCR approach provides a sensitive and cost-effective single-day screening tool for common fetal aneuploidies—A proof of concept study. *Int. J. Mol. Sci.* **2019**, *20*, 5486. [CrossRef] [PubMed]

30. Dueck, M.E.; Lin, R.; Zayac, A.; Gallagher, S.; Chao, A.K.; Jiang, L.; Datwani, S.S.; Hung, P.; Stieglitz, E. Precision cancer monitoring using a novel, fully integrated, microfluidic array partitioning digital PCR platform. *Sci. Rep.* **2019**, *9*, 1–9. [CrossRef]

31. Wong, Y.C.; Lau, S.Y.; Wang To, K.K.; Mok, B.W.Y.; Li, X.; Wang, P.; Deng, S.; Woo, K.F.; Du, Z.; Li, C. Natural transmission of bat-like SARS-CoV-2$_{\Delta PRRA}$ variants in COVID-19 patients. *Clin. Infect. Dis.* **2020**. [CrossRef]

32. Suo, T.; Liu, X.; Feng, J.; Guo, M.; Hu, W.; Guo, D.; Ullah, H.; Yang, Y.; Zhang, Q.; Wang, X. ddPCR: A more accurate tool for SARS-CoV-2 detection in low viral load specimens. *Emerg. Microbes Infect.* **2020**, *9*, 1259–1268. [CrossRef] [PubMed]

33. Dong, L.; Zhou, J.; Niu, C.; Wang, Q.; Pan, Y.; Sheng, S.; Wang, X.; Zhang, Y.; Yang, J.; Liu, M. Highly accurate and sensitive diagnostic detection of SARS-CoV-2 by digital PCR. *Talanta* **2021**, *224*, 121726. [CrossRef] [PubMed]

34. Lu, R.; Wang, J.; Li, M.; Wang, Y.; Dong, J.; Cai, W. SARS-CoV-2 detection using digital PCR for COVID-19 diagnosis, treatment monitoring and criteria for discharge. *MedRxiv* **2020**. [CrossRef]

35. Alteri, C.; Cento, V.; Antonello, M.; Colagrossi, L.; Merli, M.; Ughi, N.; Renica, S.; Matarazzo, E.; Di Ruscio, F.; Tartaglione, L. Detection and quantification of SARS-CoV-2 by droplet digital PCR in real-time PCR negative nasopharyngeal swabs from suspected COVID-19 patients. *PLoS ONE* **2020**, *15*, e0236311. [CrossRef]

36. Liu, X.; Feng, J.; Zhang, Q.; Guo, D.; Zhang, L.; Suo, T.; Hu, W.; Guo, M.; Wang, X.; Huang, Z. Analytical comparisons of SARS-COV-2 detection by qRT-PCR and ddPCR with multiple primer/probe sets. *Emerg. Microbes Infect.* **2020**, *9*, 1175–1179. [CrossRef]

37. Yu, F.; Yan, L.; Wang, N.; Yang, S.; Wang, L.; Tang, Y.; Gao, G.; Wang, S.; Ma, C.; Xie, R. Quantitative detection and viral load analysis of SARS-CoV-2 in infected patients. *Clin. Infect. Dis.* **2020**, *71*, 793–798. [CrossRef]

38. Lv, J.; Yang, J.; Xue, J.; Zhu, P.; Liu, L.; Li, S. Detection of SARS-CoV-2 RNA residue on object surfaces in nucleic acid testing laboratory using droplet digital PCR. *Sci. Total Environ.* **2020**, *742*, 140370. [CrossRef]

39. Mio, C.; Cifù, A.; Marzinotto, S.; Bergamin, N.; Caldana, C.; Cattarossi, S.; Cmet, S.; Cussigh, A.; Martinella, R.; Zucco, J.; et al. A streamlined approach to rapidly detect SARS-CoV-2 infection avoiding RNA extraction: Workflow validation. *Dis. Markers* **2020**, *2020*. [CrossRef] [PubMed]

40. Ternovoi, V.; Lutkovsky, R.Y.; Ponomareva, E.; Gladysheva, A.; Chub, E.; Tupota, N.; Smirnova, A.; Nazarenko, A.; Loktev, V.; Gavrilova, E. Detection of SARS-CoV-2 RNA in nasopharyngeal swabs from COVID-19 patients and asymptomatic cases of infection by real-time and digital PCR. *Klin. Lab. Diagn.* **2020**, *65*, 785–792. [CrossRef]

41. Deiana, M.; Mori, A.; Piubelli, C.; Scarso, S.; Favarato, M.; Pomari, E. Assessment of the direct quantitation of SARS-CoV-2 by droplet digital PCR. *Sci. Rep.* **2020**, *10*, 1–7. [CrossRef]

42. Liu, Y.; Ning, Z.; Chen, Y.; Guo, M.; Liu, Y.; Gali, N.K.; Sun, L.; Duan, Y.; Cai, J.; Westerdahl, D. Aerodynamic analysis of SARS-CoV-2 in two Wuhan hospitals. *Nature* **2020**, *582*, 557–560. [CrossRef] [PubMed]

43. Chen, Y.-X.; Zhang, W.-J.; Huang, K.-J.; Zheng, M.; Mao, Y.-C. An electrochemical microRNA sensing platform based on tungsten diselenide nanosheets and competitive RNA–RNA hybridization. *Analyst* **2017**, *142*, 4843–4851. [CrossRef]

44. Lynch III, C.A.; Foguel, M.V.; Reed, A.J.; Balcarcel, A.M.; Calvo-Marzal, P.; Gerasimova, Y.V.; Chumbimuni-Torres, K.Y. Selective Determination of Isothermally Amplified Zika Virus RNA Using a Universal DNA-Hairpin Probe in Less than 1 Hour. *Anal. Chem.* **2019**, *91*, 13458–13464. [CrossRef] [PubMed]

45. Feng, D.; Su, J.; He, G.; Xu, Y.; Wang, C.; Zheng, M.; Qian, Q.; Mi, X. Electrochemical DNA Sensor for Sensitive BRCA1 Detection Based on DNA Tetrahedral-Structured Probe and Poly-Adenine Mediated Gold Nanoparticles. *Biosensors* **2020**, *10*, 78. [CrossRef]

46. Santhanam, M.; Algov, I.; Alfonta, L. DNA/RNA electrochemical biosensing devices a future replacement of PCR methods for a fast epidemic containment. *Sensors* **2020**, *20*, 4648. [CrossRef]

47. Zhao, H.; Liu, F.; Xie, W.; Zhou, T.-C.; OuYang, J.; Jin, L.; Li, H.; Zhao, C.-Y.; Zhang, L.; Wei, J. Ultrasensitive supersandwich-type electrochemical sensor for SARS-CoV-2 from the infected COVID-19 patients using a smartphone. *Sens. Actuators B Chem.* **2021**, *327*, 128899. [CrossRef] [PubMed]

48. Alafeef, M.; Dighe, K.; Moitra, P.; Pan, D. Rapid, ultrasensitive, and quantitative detection of SARS-CoV-2 using antisense oligonucleotides directed electrochemical biosensor chip. *ACS Nano* **2020**, *14*, 17028–17045. [CrossRef]

49. Patterson, A.S.; Hsieh, K.; Soh, H.T.; Plaxco, K.W. Electrochemical real-time nucleic acid amplification: Towards point-of-care quantification of pathogens. *Trends Biotechnol.* **2013**, *31*, 704–712. [CrossRef]
50. Chaibun, T.; Puenpa, J.; Ngamdee, T.; Boonapatcharoen, N.; Athamanolap, P.; O'Mullane, A.P.; Vongpunsawad, S.; Poovorawan, Y.; Lee, S.Y.; Lertanantawong, B. Rapid electrochemical detection of coronavirus SARS-CoV-2. *Nat. Commun.* **2021**, *12*, 1–10. [CrossRef]
51. *ePlex®SARS-CoV-2 Test Assay Manual*; The United States Food and Drug Administration: Silver Spring, MD, USA, 2020.
52. Kumar, P.; Malik, Y.S.; Ganesh, B.; Rahangdale, S.; Saurabh, S.; Natesan, S.; Srivastava, A.; Sharun, K.; Yatoo, M.I.; Tiwari, R. CRISPR-Cas system: An approach with potentials for COVID-19 diagnosis and therapeutics. *Front. Cell Infect. Microbiol.* **2020**, *10*, 576875. [CrossRef] [PubMed]
53. Hou, T.; Zeng, W.; Yang, M.; Chen, W.; Ren, L.; Ai, J.; Wu, J.; Liao, Y.; Gou, X.; Li, Y. Development and evaluation of a rapid CRISPR-based diagnostic for COVID-19. *PLoS Pathog.* **2020**, *16*, e1008705. [CrossRef]
54. Ning, B.; Yu, T.; Zhang, S.; Huang, Z.; Tian, D.; Lin, Z.; Niu, A.; Golden, N.; Hensley, K.; Threeton, B. A smartphone-read ultrasensitive and quantitative saliva test for COVID-19. *Sci. Adv.* **2021**, *7*, eabe3703. [CrossRef]
55. Fozouni, P.; Son, S.; de León Derby, M.D.; Knott, G.J.; Gray, C.N.; D'Ambrosio, M.V.; Zhao, C.; Switz, N.A.; Kumar, G.R.; Stephens, S.I. Amplification-free detection of SARS-CoV-2 with CRISPR-Cas13a and mobile phone microscopy. *Cell* **2021**, *184*, 323–333.e329. [CrossRef] [PubMed]
56. Ouyang, W.; Han, J. One-step nucleic acid purification and noise-resistant polymerase chain reaction by electrokinetic concentration for ultralow-abundance nucleic acid detection. *Ang. Chem.* **2020**, *132*, 11074–11081. [CrossRef]
57. Ramachandran, A.; Huyke, D.A.; Sharma, E.; Sahoo, M.K.; Huang, C.; Banaei, N.; Pinsky, B.A.; Santiago, J.G. Electric field-driven microfluidics for rapid CRISPR-based diagnostics and its application to detection of SARS-CoV-2. *Proc. Natl. Acad. Sci. USA* **2020**, *117*, 29518–29525. [CrossRef] [PubMed]
58. Lee, S.H.; Song, J.; Cho, B.; Hong, S.; Hoxha, O.; Kang, T.; Kim, D.; Lee, L.P. Bubble-free rapid microfluidic PCR. *Biosens. Bioelectron.* **2019**, *126*, 725–733. [CrossRef]
59. McArdle, H.; Jimenez-Mateos, E.M.; Raoof, R.; Carthy, E.; Boyle, D.; ElNaggar, H.; Delanty, N.; Hamer, H.; Dogan, M.; Huchtemann, T.; et al. "TORNADO"–Theranostic One-Step RNA Detector; microfluidic disc for the direct detection of microRNA-134 in plasma and cerebrospinal fluid. *Sci. Rep.* **2017**, *7*, 1–11. [CrossRef] [PubMed]
60. Azhar, M.; Phutela, R.; Ansari, A.H.; Sinha, D.; Sharma, N.; Kumar, M.; Aich, M.; Sharma, S.; Singhal, K.; Lad, H.; et al. Rapid, field-deployable nucleobase detection and identification using FnCas9. *bioRxiv* **2020**. [CrossRef]
61. Deng, H.; Zhou, X.; Liu, Q.; Li, B.; Liu, H.; Huang, R.; Xing, D. Paperfluidic chip device for small RNA extraction, amplification, and multiplexed analysis. *ACS Appl. Mater. Interfaces* **2017**, *9*, 41151–41158. [CrossRef]

 *biosensors*

Review

Wearable Biosensors for Non-Invasive Sweat Diagnostics

Jing Xu [1,2,†], Yunsheng Fang [2,†] and Jun Chen [2,*]

1   School of Electrical & Electronic Engineering, North China Electric Power University, Beijing 102206, China;
    joycexu@ucla.edu
2   Department of Bioengineering, University of California, Los Angeles, Los Angeles, CA 90095, USA;
    fys131415@g.ucla.edu
*   Correspondence: jun.chen@ucla.edu
†   These authors contributed equally.

Abstract: Recent advances in microfluidics, microelectronics, and electrochemical sensing methods have steered the way for the development of novel and potential wearable biosensors for healthcare monitoring. Wearable bioelectronics has received tremendous attention worldwide due to its great a potential for predictive medical modeling and allowing for personalized point-of-care-testing (POCT). They possess many appealing characteristics, for example, lightweight, flexibility, good stretchability, conformability, and low cost. These characteristics make wearable bioelectronics a promising platform for personalized devices. In this paper, we review recent progress in flexible and wearable sensors for non-invasive biomonitoring using sweat as the bio-fluid. Real-time and molecular-level monitoring of personal health states can be achieved with sweat-based or perspiration-based wearable biosensors. The suitability of sweat and its potential in healthcare monitoring, sweat extraction, and the challenges encountered in sweat-based analysis are summarized. The paper also discusses challenges that still hinder the full-fledged development of sweat-based wearables and presents the areas of future research.

Keywords: point-of-care; biomonitoring; personalized healthcare; sweat; biosensors

Citation: Xu, J.; Fang, Y.; Chen, J. Wearable Biosensors for Non-Invasive Sweat Diagnostics. *Biosensors* **2021**, *11*, 245. https:// doi.org/10.3390/bios11080245

Received: 6 July 2021
Accepted: 20 July 2021
Published: 23 July 2021

1. Introduction

The first wave of medical diagnostics saw the beginning of technological development in which people produced instruments that could measure almost any analyte of interest through the collection and transfer of samples to a separate lab. The second technological wave of POCT has even brought the lab into the hands of doctors, nurses, and even patients. More recently, a third technological wave is coming where patients can take the lab with them through wearable bio-marker monitoring [1–3]. Wearable technology can be an innovative solution to current medical problems with the ability to continuously monitor both physiological and biochemical markers and physical activities and behaviors [4–8]. Vital signs such as heart rate, body temperature, and blood pressure are measured [9,10]. Through the use of these signs, including electrocardiogram, electromyogram and biological fluid oxygen saturation and physical activity, the personal health condition can be characterized and monitored [11–20]. This paper reviews the method of chemical monitoring using sweat as the biological fluid [21–23]. Generally, personal items such as shoes, glasses, clothes, gloves, and watches can be used to deploy wearable devices [24–27]. As techniques move forward, emerging chemical wearable devices have evolved into devices that can be attached to the skin for further improving the accuracy of measurement. Furthermore, data collected by wearable devices and transmitted to a remote server through smartphones will be used to generate a personalized medical model through combination with analysis algorithms [28–37].

The need for wearable bioelectronics is rising daily and it has the potential to revolutionize the healthcare industry [38–40]. Conventional medicine is a reactive model, in

13

which people wait for symptoms to manifest themselves in an individual before proceeding to diagnosis and eventual remedy. Due to this inherent nature of traditional medicine, it is quite often referred to as "Sick-care". It is therefore desirable to move towards a preemptive model that can help people diagnose or take action to treat diseases at the earliest stages, well before the visible symptoms of the disease even become apparent. To move towards such a model, we need to understand how the body functions at a chemical or molecular level. This can be enabled through continuous non-invasive wearable biomonitoring [41–43]. The current system of a one-drug fits all approach is not accurate because people have varying body types and lifestyles. Personalized medicine can be modified according to each person's characteristics that are acquired through the help of wearable biomonitoring [44–53]. As a result, our therapies will be more accurate since they rely on biomarkers rather than apparent symptoms [54–57]. Moreover, wearable bioelectronics can replace expensive and time-consuming lab tests with wearable diagnostic alternatives [58,59]. In traditional medical therapies or testing, many processes are involved, such as sample collection, preservation, and storage [60]. According to reports from the National Institutes of Health and the National Cancer Institute, adding preservatives or additives can potentially impact the protein makeup of plasma or any other bio-fluid involved in a particular test [61]. Storage temperature may also have a significant impact on sample quality. Moreover, temperature control issues may occur while transporting samples between facilities [62]. Altogether, even small errors can make a great impact on the test accuracy because all of them stack up and what reaches the lab is the summation of all of them. On top of these issues, these methods are expensive and time-consuming to complete the tests. Wearable testing measures over-ride this convoluted path by offering point-of-care-diagnostics (POCD) to the concerned individual at any time and any place while also saving time and cost.

Although there are many advantages of various wearable bioelectronics, non-invasive bioelectronics have their unique strengths and can also encourage users to take multiple readings in a day, as these tests are nondestructive and painless [63–68]. The large number of daily tests enabled by non-invasive wearable biomonitoring allows users to collect many data points across the population that can be used to gain a better understanding of diseases and will help us develop preemptive medical models to monitor healthcare before disease occurrence [69]. Great efforts have been made in wearable technologies as the acknowledgment of their uses on molecular biomonitoring has increased. These techniques are useful for sampling and analyzing the heavy metals, metabolites, toxic gases, and main electrolytes within body fluids [70,71]. Non-invasive detection of human body fluids generally includes the detection of tear, interstitial fluid (ISF), exhaustion breath, saliva, wound exudate, and sweat [72,73]. For example, in human tears, because biological fluids accumulate in the eyes, they contain certain salts, enzymes, proteins, and lipids. Consequently, eye conditions and diseases can be revealed through the analysis of the chemical composition of tears. Typically, increased levels of proline-rich proteins are considered biomarkers for the diagnosis of dry eye. In addition, tears have been used in the treatment of diabetes mellitus since the glucose concentration in tears is strongly correlated with the blood glucose concentration [74]. Figure 1a shows some of the tear-based flexible sensors developed using PDMS (Polydimethylsiloxane) or soft-lens [75]. Although blood biomarkers detection can provide more accurate reports, non-invasive detection is not possible for most of them. To circumvent this sampling problem, ISF monitoring is used to obtain information of chemical concentrations in blood alternatively. The composition of ISF is very similar to that of blood regarding the concentration of salt, protein, glucose, ethanol, and other small molecules [76,77]. A process called reverse iontophoresis is used on the skin to extract ISF. GlucoWatch, the most famous glucose sensing device approved by the United States Food and Drug Administration, uses this technology to carry ISF through the skin to an external sensor. It is an integrated wristwatch device with reverse iontophoresis and biosensor functions [74]. Figure 1b shows a wearable sensor for non-invasive monitoring of the ISF [49]. As for breathing, human breathing is a mixture of gases and vapors that we exhale through our nose or mouth. The composition of respiration is complex: it includes a mixture of nitrogen, oxygen, carbon dioxide, and water vapor [78]. In addition, respiration includes up to 500 different compounds, both endogenous and exogenous. The potential of respiratory measurement is huge because of its completely non-invasive and inherent safety. Figure 1c shows a wearable sensor for the non-invasive monitoring of human breath. Saliva is another attractive diagnostic bio-fluid that contains various disease signal biomarkers including hormones, enzymes, antibodies, antimicrobial agents, which can accurately reflect the status of humans [79]. These biochemical substances from the blood travel through spaces between cells and enter saliva. Thus, the main compounds found in the blood can also be detected in saliva, as shown in Figure 1d. Therefore, saliva is functionally analogous to the serum in that it involves changes in mood, hormones, nutrition, and metabolism. Salivary cortisol and salivary alpha-amylase, for example, are considered to be important biomarkers of physical and psychological stress. Proper monitoring of biofluids can also have benefits in the wound healing process. The formation or release of some compounds or exudates is crucial to wound healing. The pH value of the exudates, the concentration of uric acid and C-reactive protein can reflect wound healing progress and infection risk [74]. Figure 1e depicts research in a study in which wearable smart bandages were fabricated for wound monitoring. Among these bio-fluids, human sweat is crucial because it contains abundant physiological state information [80]. Figure 1f,g shows a sweat-based biosensor and its therapeutic system for wearable diabetes monitoring. The optical image of the device array on human skin with perspiration was shown in Figure 1h. There are a few advantages associated with sweat that make it a predominant candidate for most wearable biomonitoring research.

Since the distribution of sweat glands in the human body is rich (>100 glands/cm^2) and the sweat contains abundant biochemical compounds, human sweat has become a promising bio-fluid for non-invasive biosensing [81]. Since nearly every portion of human skin has eccrine glands, sweat is readily available without the use of needles or other invasive devices. Iontophoresis sweat can be extracted from anywhere which is not possible in any other case of bio-fluids [69]. Moreover, analytes including ions [76], metabolites [82], acids [64,83], hormones [84–87], and small proteins [88,89] and peptides are partitioned into the sweat. Sweat also contains various electrolytes (such as potassium, sodium, chloride, and calcium), nitrogen-containing compounds (such as urea and amino acids), as well as metabolites such as glucose, lactic acid, and uric acid, along with xenobiotics such as drugs and ethanol [90–92]. This creates a huge opportunity for research and biomonitoring. Sweat chloride analysis has been used as the gold standard for the diagnosis of cystic fibrosis [93]. Sweat also has excellent sampling and detection efficiency without foreign contamination during testing and its composition does not affect the analytes to get degraded [81].

This review focuses on the recent progress made in the field of wearable and flexible bioelectronics for non-invasive health monitoring through in-situ sweat analysis. In the following sections, we will first discuss sweat partitioning and its relation to human health. Also, we will cover the most common sweat extraction techniques and fabrication methods for sweat-based bio-sensors. Mainly, we will review some of the recent researches and applications of continuous and non-invasive sweat-based biomarker monitoring, including health monitoring and disease detection, exercise monitoring, drug metabolism monitoring, and ethanol level measurement. Finally, we will go over the overall challenges and future scope of wearable sweat-based biosensors towards personalized health monitoring.

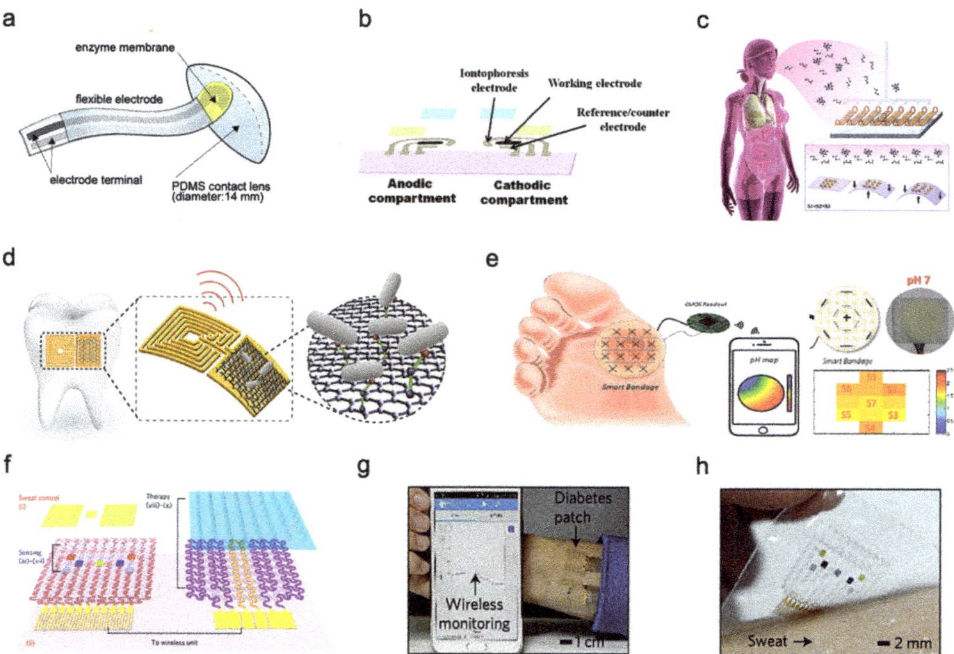

Figure 1. Wearable and flexible sensors for continuous biofluids analysis. (**a**) Soft PDMS contact lens with a glucose-sensing strip attached. Reproduced with permission from ref. [75], Copyright from 2020, Elsevier B.V. (**b**) All-printed tattoo-based ISF glucose sensor. Reproduced with permission from ref. [49], Copyright 2015, American Chemical Society. (**c**) Dynamic nanoparticle-based breath sensor. Reproduced with permission from ref. [78], Copyright from 2015, American Chemical Society. (**d**) Bacteria sensing on tooth enamel with graphene-based nanosensors. Reproduced with permission from ref. [79], Copyright from 2012, Nature Publishing Group. (**e**) Smart bandage for chemical sensing of wound pH using pH-sensitive threads. CMOS wireless readout and 2D mapping of pH levels were incorporated Reproduced with permission from ref. [74], Copyright 2017 IEEE. (**f**) Diabetes patch is composed of sweat-control (i, ii), sensing (iii–vii) and therapy (viii–x) components. (**g**) Integrated wearable diabetes monitoring and therapy system. (**h**) The electrochemical device on the human skin with perspiration. Reproduced with permission from ref. [80], Copyright from 2016, Nature Publishing Group.

2. Sweat as a Bio-Fluid for Biomonitoring

Sweat can provide abundant biomarker measurements continuously and noninvasively of ions, drugs, metabolites and biomolecules, including potassium, sodium, calcium, chlorine, lactic acid, glucose, ammonia, ethanol, urea, cortisol, and various neuropeptides and cytokines. Table 1 also summarizes these key analytes in sweat and the detection methods. In addition to the abundant biochemical components in sweat, sweat glands are widely distributed across the human body. Consequently, sweat has become an ideal platform for noninvasive biosensing, which is feasible and safe as well. Sweat from the eccrine sweat gland can be noninvasive and easily obtained. This kind of sweat contains water and various electrolytes and can be directly excreted to the skin surface. Unusual health conditions (such as electrolyte imbalances and stress) and disease are usually reflected by changes in the concentration of existing sweat components or the emergence of new components in sweat. For example, the concentration of alcohol in sweat is highly correlated with the concentration of alcohol in the blood. The increase of urea concentration in sweat may be related to kidney failure. Moreover, because the concentration of chlorine in the sweat of patients who have cystic fibrosis (CF) is abnormally high, the analysis of chlorine in sweat has been regarded as a widely used method for the diagnosis of CF [74].

Table 1. Key analytes in sweat and the related detection methods.

Target Analyte		Concentration in Sweat	Recognition Element	Sensing Modality	Ref.
Ions	Na^+	10–100 mM	Na+	Potentiometry	[44,45,93]
	Cl^-	10–100 mM	Ag/AgCl	Potentiometry	[93,94]
	K^+	1–18.5 mM	K+	Potentiometry	[44,94]
	Ca^{2+}	0.41–12.4 mM	Ca2+	Potentiometry	[95]
	pH	3–8	Polyaniline	Potentiometry	[96]
	NH^{4+}	0.1–1 mM	NH4+	Potentiometry	[97]
	Zn^{2+}	100–1560 µg L^{-1}	Bi	Square wave stripping voltammetry	[84,98]
	Cd^{2+}	<100 µg L^{-1}	Bi	Square wave stripping voltammetry	[98]
	Pb^{2+}	<100 µg L^{-1}	Bi, Au	Square wave stripping voltammetry	[98]
	Cu^{2+}	100–1000 µg L^{-1}	Au	Square wave stripping voltammetry	[98]
	Hg^+	<100 µg L^{-1}	Au	Square wave stripping voltammetry	[98]
Drugs	Levodopa	<10 µM	Au	Chronoamperometry	[99]
	Caffeine	<40 µM	Carbon	Chronoamperometry	[100]
	Alcohol	2.5–22.5 mM	Carbon	Chronoamperometry	[81,101]
Metabolites	Glucose	10–200 µM	Glucose oxidase	Chronoamperometry	[44,93,102]
	Lactate	5–20 mM	Lactate oxidase	Chronoamperometry	[96]
	Uric acid	2–10 mM	Carbon	Cyclic voltammetry	[83]
	Cortisol	8–140 µg L^{-1}	ZnO, MoS2	Electrochemical impedance spectroscopy	[85,86]
	Ascorbic acid	10–50 µM	Carbon	Chronoamperometry	[83,87]
Biomolecules	Peptides	0.1 pM–0.1 µM	Au	Chronoamperometry	[103]
	Antimicrobial peptides	-	Carbon	Resistance	[79]

2.1. Sweat Partitioning

Bio-marker partitioning in human sweat is important to study because the current gold standard for measuring biomarker concentration, for most biomarkers, is its concentration in blood. Therefore, correlating sweat biomarker concentration with the respective concentrations in the blood can help us draw meaningful inferences about the health of a person. In certain cases, independent concentrations in sweat can also have significant value. For example, chloride concentrations for patients with CF are very high in sweat independently and have been used in literature to diagnose CF [93]. The most easily obtained sweat comes from eccrine glands composed of coils and dermal ducts. These eccrine glands are the first place where sweat is produced, and the sweat is transported to the skin surface through dermal ducts [81], as shown in Figure 2a. During this process, analytes such as ions, metabolites, acids, hormones, small proteins, and peptides travel into human sweat. Sodium (Na^+) and chloride (Cl^-) ions have the highest concentration in sweat. Since they can stimulate hydration, they play a role in maintaining electrolyte balance in the human body. Ions such as Calcium (Ca^{2+}) and Potassium (K^+) also partition into human sweat from blood and are present in the mM (M for Molar) range [69]. K^+ concentration can be used to predict muscle activity [104]. During exercises, electrical activity in the exercising muscles can cause an increase in K^+ concentration. Weak acids or bases can also diffuse into the sweat gland and ionize because of the high pH of sweat. Other bio-marks, such as lactate and urea at mM levels, can come from human blood or be produced locally during the metabolic activity of the sweat gland. The lactate concentrations of human sweat are closely approximate to those of plasma [96]. Thus, the concentrations of lactate can indicate physical exertion and exercise intensity [105,106]. Sweat also contains larger molecules such as glucose, neuropeptides, and hormones, which are present in nM or pM traces. These important bio-marks in sweat carry valuable information about the human body. Take glucose as an example, monitoring the concentrations of glucose in sweat can provide continuous glycemic monitoring [102]. Apart from analytes that naturally be generated, drugs, heavy metals and alcohol can be discharged and detected in sweat as well when the human body tries to eject toxins. These biomarkers can be used for multi-purpose biomonitoring.

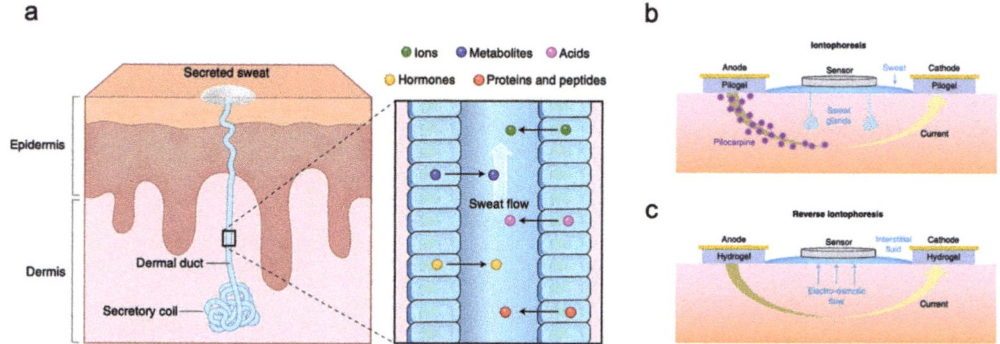

Figure 2. Sweat gland structure and biomarker partitioning. (**a**) Illustration of the sweat production process and metabolites that passed along with sweat. (**b**) Depiction of Iontophoresis to stimulate local sweat secretion at a selected site. (**c**) Reverse iontophoresis drives interstitial fluid through the epidermis to the skin surface. Reproduced with permission from ref. [69], Copyright from 2018, Nature Publishing Group.

2.2. Sweat Extraction

Although it is a viable option to test sensors by sweat from exercise, realizing continuous sweat monitoring might not be practical. Under such circumstances, alternative methods of sweat extraction are needed. Iontophoresis is the most advanced non-invasive method to induce sweat excretion at a selected location such as the wrist [74]. It depends on local sweat stimulation through the application of topical current [69]. Figure 2b shows that the topical current is applied between pilogels, or hydrogels containing the sweat stimulant-charged drug, called pilocarpine. By applying a potential drop across the two sides of the test area, pilocarpine will be driven by a small current under the skin surface and trigger nearby glands to secrete sweat for wearable sensors to collect enough samples. This technology can gain unprecedented insight into the process of the sweat secretion, facilitating a wider range of sweat-based sensing applications such as health monitoring and POCD [93].

A process similar to iontophoresis is called reverse iontophoresis, which can be used to extract ISF by attaching two electrodes to the skin and applying a potential difference between them, as shown in Figure 2c. Ions, such as Na^+, can be extracted outside of the skin. Since the net charge of the skin is negative, an electroosmotic flow will be generated to drive the interstitial fluid (ISF) to the skin surface through the epidermis without involving any drugs. The hydrogel isolates electrodes for applying current to the skin to prevent irritation. ISF-based wearable sensors can detect analytes, such as glucose, along with interstitial fluid transported to the skin surface by advection.

2.3. Sweat-Based Bio-Sensor Fabrication

There are certain fabrication methods usually involved in printing or making these sweat-based sensors into wearable devices. To make these flexible wearable sensors suitable for biomonitoring, several requirements should be satisfied. Firstly, the sensors must be highly sensitive, bio-compatible, and selective. Secondly, related circuitry should be in micro or nano size. Popular methods for this are photolithography and screen printing. Photolithography provides excellent resolution at the nanoscale by using an electron beam to write patterns. However, the cost of manufacturing is extremely high due to the equipment cost and clean room requirements. Low-cost screen printing can be used to produce large numbers of electrodes on various flexible substrates and can provide as high as one device per second throughput when roll to roll printing is used. However, the spatial resolution of the screen printing is not as good as that of lithography [106]. Other methods, such as epidermal elastomeric stamping, stamp transferring, and ink or aerosol jet printing, can also be used for fabrication. For example, stamp transfer and inkjet printing (combined with electroplating) showed great potential in high-resolution printing (down to about 2 µm). At the same time, it can provide manufacturing on non-planar substrates. Therefore, the appropriate selection and combination of these methods is ideal for the manufacture of sweat-based wearable bioelectronics.

3. Sweat-Based Wearable Bioelectronics

Wearable sweat-based sensors have been studied extensively recently for detecting analytes as they relate to human diseases and conditions. A few studies have also developed wearable devices with multi-analyte sensing and circuitry for in-situ analysis and calibration. In this section, we present some of these studies that illustrate the application directions of wearable technology, including health monitoring and disease detection, exercise monitoring, drug metabolism monitoring, and ethanol level measurement.

3.1. Health Monitoring and Disease Detection

The molecular-level view of health is extremely valuable for health and disease monitoring applications. Sweat-based biosensors provide an effective way to achieve health monitoring and disease detection. For example, diabetes is directly related to the metabolite profile of glucose, which can be easily detected through sweat [44,98]. A sweat-based glucose monitoring system was developed using an electrochemical monolithic glucose sensor with pH and temperature correction functions [53]. The device can sense glucose levels in sweat in real time and even includes a drug that uses microneedles to regulate blood glucose. Although continuous blood glucose monitoring is becoming available, the potential of sweat sampling blood glucose monitors can not only reduce the size of current equipment but also achieve painless treatment of diabetes. Figure 3a shows a wearable molecular level health monitoring tool developed by researchers at Caltech which is self-powered by human motion [107]. The developed device was powered by a flexible triboelectric nanogenerator (TENG) [108–114]. It measures Na^+ and pH of sweat using a potentiometric sensing technique, performs signal processing, and transmits this data to a mobile user interface using Bluetooth to track real-time personal health state. Figure 3b shows the schematic of a biosensor array containing both pH and sodium ion sensors patterned on a flexible plythylene terephthalate (PET) substrate [95]. The whole design is achieved in a wearable format by integrating it onto a flexible printed circuit board (PCB) that can be worn on the arm or the side of the human torso, as shown in Figure 3c. To realize the self-powering device, the authors developed a flexible TENG that can easily be integrated with the rest of the design. TENG bases on the principle of contact electrification and develops a potential difference between plates due to relative sliding between the copper and polytetrafluoroethylene plates. This helps the device power itself by harvesting energy from biomechanical energy induced by human motion. The developed TENG in this device manages to achieve a maximum power output of 0.94 mW for a 4.7 MΩ load, which is very suitable for wearable applications (Figure 3d,e). These results suggest that the device developed can be successfully used in molecular-level pH and Na^+ monitoring [94,115]. Additionally, using a TENG to power the device itself enables it to achieve a very small form factor as it doesn't use bulky batteries. It paves the way for developing innovative self-powering approaches for wearable devices used in human health monitoring.

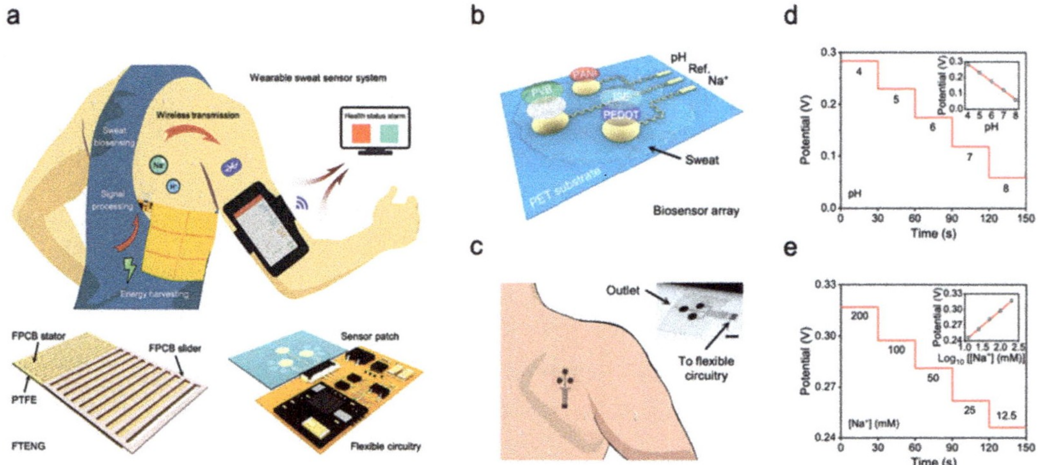

Figure 3. Sweat-based sensors for continuous health monitoring. (**a**) Describes the working of the device and shows the developed FTENG to power device. (**b**) Schematic of the sensor array. (**c**) Schematic of microfluidic sensor patch. (**d**,**e**) Open-circuit potential responses of the pH sensor in standard McIlvaine's buffer solutions (**d**) and a sodium ion sensor in NaCl solutions (**e**). Reproduced with permission from ref. [107], Copyright from 2020, the American Association for the Advancement of Science.

Beyond health monitoring, sweat-based sensors can also provide low-cost disease detection and diagnosis. A wearable device was reported for the diagnosis of CF [93]. CF is a kind of hereditary disease which can cause severe damage to the human lungs, digestive system, and other organs. It makes secretions sticky by affecting cells that produce mucus, sweat and digestive juice. These viscous secretions no longer act as lubricants of the passage, but block the tube, catheter, and passageway, especially in the lungs and pancreas, causing serious damage to the human body. This leads to symptoms such as damaged airways, chronic infections, and in serious cases, even respiratory failure. According to the Cystic Fibrosis Foundation Patient Registry, more than 70,000 people worldwide are living with CF. Therefore, CF is a very serious disease that needs to be regularly monitored. The system-level implementation of the developed system is shown in Figure 4a,b. The working process of the device is described as follows—iontophoresis is used to induce sweat with various secretion profiles as depicted in Mode 1, which is then used for real-time sensing by the front-end electronics depicted in Mode 2 (Figure 4c,d). The processed signal is then transmitted to the communication module of the circuit, which sends the concentration data to the phone and displays it in a format that is easily understood by the user. The flow for signal processing is shown in Figure 4e. In this way, the system can measure the levels of sodium ion and chloride ion in the sweat of CF patients stimulated by iontophoresis in real-time (Figure 4f). Figure 4g shows the comparison results of sweat electrolyte levels achieved by the paper between six healthy subjects and three CF patients.

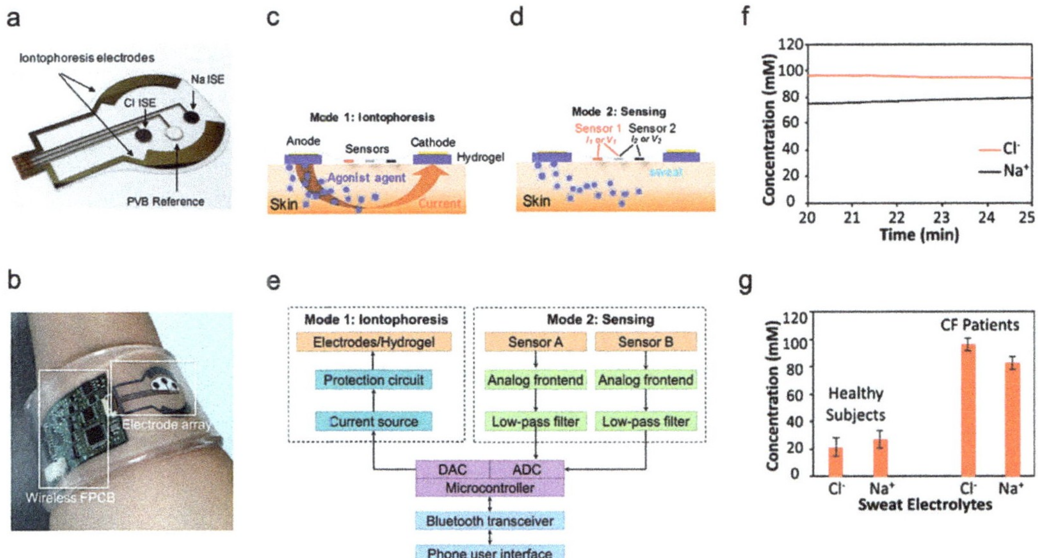

Figure 4. Overview of the developed CF monitoring device. (**a**) Electrodes used for iontophoresis and sensing. (**b**) Flexible wearable device on user's wrist. (**c,d**) Description of the working modes of system (**c**) mode 1: Iontophoresis (**d**) mode 2: Sensing. (**e**) Block-level diagram of the developed system. (**f**) Real-time on-body measurement of sweat sodium ion and chloride ion levels of a CF patient after iontophoresis-based sweat stimulation. (**g**) Comparison of sweat electrolyte levels between six healthy subjects and three CF patients. Reproduced with permission from ref. [93], Copyright 2017, National Academy of Sciences USA.

The results from the above study indicate its potential for use as a low-cost diagnosis of CF. Additionally, the demonstration of successfully integrating the design on a flexible substrate opens a plethora of possibilities for the development of wearable bioelectronics for disease diagnosis and monitoring.

3.2. Exercise Monitoring

The most common application of wearable sweat biosensors is exercise monitoring. In the study of Gao et al., a wearable sensor array was introduced where several analytes could be monitored at the same time. The sensor used a flexible integrated sensing array (FISA) and the signal was processed by a flexible PCB (FPCB) [44]. The integrated FISA and FPCB are displayed and function on a subject's head and wrist, as shown in Figure 5a,b. These sensors need to bear the stresses of everyday wearing and physical exercise. Upon bending FPCB at 1.5 cm and 3 cm bending radii, minimal change in output was observed in FISA response.

Figure 5c illustrates the structural design and working mechanism of the sensor. The working principle of Amperometric glucose and lactate sensors is immobilizing glucose oxidase and lactate oxidase inside a permeable membrane. Both the shared reference electrode and the counter electrode of the two sensors use Ag/AgCl as the electrode. Current signals proportional to the corresponding metabolites will be generated through these enzymatic sensors automatically and transported between the working electrode and the Ag/AgCl electrode. Ion-selective electrodes (ISEs) were used to measure the Na$^+$ and K$^+$ levels. The ion-selective electrode is a type of potentiometric device that combines with a reference electrode coated with polyvinyl butyral (PVB) to stabilize the potential in solutions with different ionic strengths. In addition, a resistance-based temperature sensor is achieved through the fabrication of Cr/Au microwires.

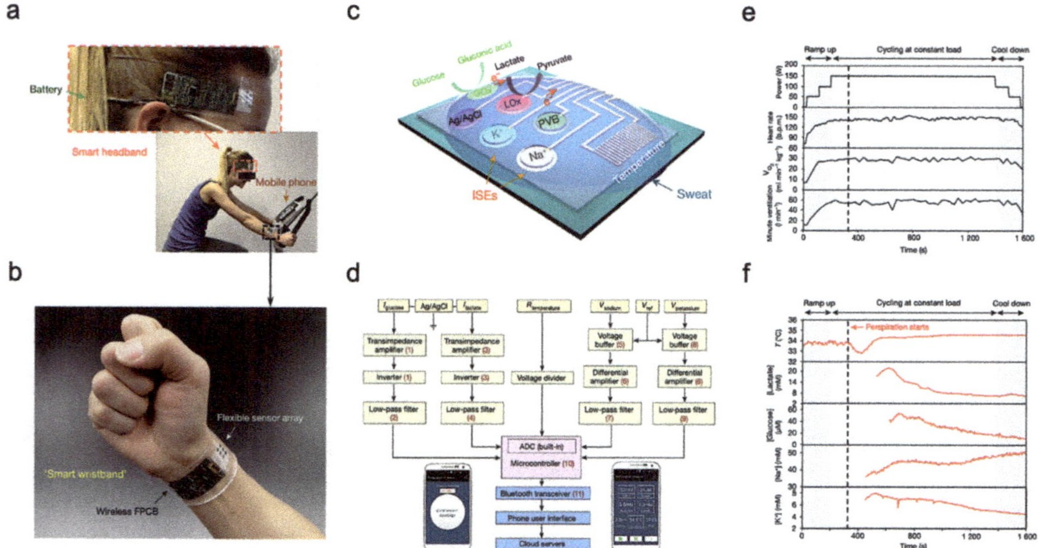

Figure 5. Sweat-based sensors for exercise monitoring. (**a**) Subject wearing forehead and wrist sensors undergoing stationary exercise. (**b**) Flexible integrated sensor array shown on a subject's wrist as part of a wireless FPCB. (**c**) Schematic of the sensor array. (**d**) System-level block diagram illustrating the flow of information. (**e**) Real-time sweat analysis results of the FISA worn on a subject's forehead. (**f**) Constant-load exercise at 150 W: power output, heart rate (in beats per minute, b.p.m.), oxygen consumption and pulmonary minute ventilation, as measured by external monitoring systems. Reproduced with permission from ref. [44], Copyright from 2016, Nature Publishing Group.

Figure 5d shows the measured potentiometric data is transmitted through an amplifier, an inverter, an analog-to-digital converter, and a Bluetooth module to be displayed on a cellphone. The sensor was tested on a subject undergoing a stationary cycling exercise as shown in Figure 5a. Specifically, the exercise program includes 3 min of accelerated cycling, 20 min of fixed 150 W cycling, and 3 min of cooling down. During exercise, heart rate, oxygen consumption, and pulmonary ventilation were measured. It showed that the values of the measured data increased in proportion with the increase of output power. Figure 5e illustrates the real-time measurement of sweat on a subject's forehead using FISA. It shows that the skin temperature increases at about 400 s and stays almost unchanged after that with continuous perspiration. Meanwhile, lactic acid and glucose concentrations in sweat gradually decrease, as shown in Figure 5f. This decrease in lactic acid and glucose can be attributed to the dilution effect caused by the increased per-spiration rate.

Unlike typical use of polymeric substrates, a textile-based platform can increase overall permeability to the affected skin area, which cultivates natural sweating and evaporative cooling [116]. This sensor is constructed from different layers of materials, as shown in Figure 6a. A commercial adhesive bandage is used as the base for the sensor, then a hydrophobic adhesive film is placed above to retain sweat samples within the patch. Additionally, specialized sensing threads are implemented in a parallel pattern to the previous layer. An absorbent gauze is applied to further enhance the collection of the sweat sample. Eventually, an adhesive film is placed on top to keep the entire patch intact.

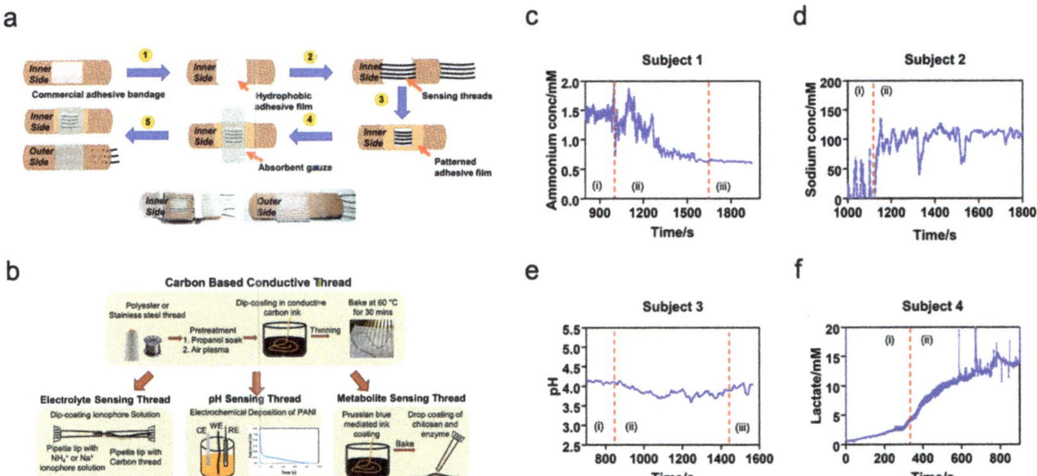

Figure 6. Sweat-based multimodal sensor patch for exercise monitoring. (**a**) Step-by-step process of fabrication and complete sensor patch prototype. (**b**) Fabrication of specialized sensing threads. (**c–f**) Measurement results of different biomarkers. Reproduced with permission from ref. [116], Copyright from 2020, Nature Publishing Group.

The thread bundle incorporated within the sensor patch contains two specialized thread types for analyzing the biomarkers within the sweat sample. These specialized threads are commonly available threads that have undergone additional treatments and processes; the specific fabrication details of the threads are shown in Figure 6b. The first thread type in discussion is a carbon-based conductive thread that can sense the electrolyte, pH, and metabolite levels within the sweat sample. These measurements directly relate to biomarkers such as sodium, lactate, and ammonium which can represent real-time physiological status of an individual undergoing strenuous physical activity. The second thread type is a silver-silver chloride thread that acts as a solid-state reference for comparison during analysis. The four graphs shown in Figure 6c–f displayed the sensor patch performance when applied onto a participant's arm while exercising on a stationary bike. Each graph is sectioned by Roman numerals with "1" as equilibrium state, "2" as real-time measurement, and "3" as cool-down period. The results demonstrate the consistency and stability of the sensor patch in collecting data from the biomarkers within the sweat sample.

3.3. Drug Metabolism Monitoring

In addition to exercise intensity monitoring, wearable sweat-based sensors for biomarks monitoring can also be applied as drug metabolism monitoring sensors. For example, Levodopa is used to treat patients suffering from Parkinson's disease. Various factors can affect an individual's response to levodopa and therefore it is important to monitor the concentration in blood. Since blood-based monitoring is usually invasive, sweat-based levodopa sensing was investigated using a sensor packaged into a sweatband (S-band) [99]. The sensor has a standard three-electrode (working electrode, reference electrode and counter electrode) design and is fabricated on a PET substrate (Figure 7a). Experiments were performed on healthy subjects after consuming fava beans as they happen to contain levodopa. In this way, the function of the S-band can be extensively tested on non-vulnerable subjects, and sweat was elicited from the subjects using iontophoresis. The concentration of levodopa in sweat was continuously monitored after consumption of fava beans and followed by iontophoresis. The S-band can detect the level of levodopa in sweat continuously, which is similar to the level in blood.

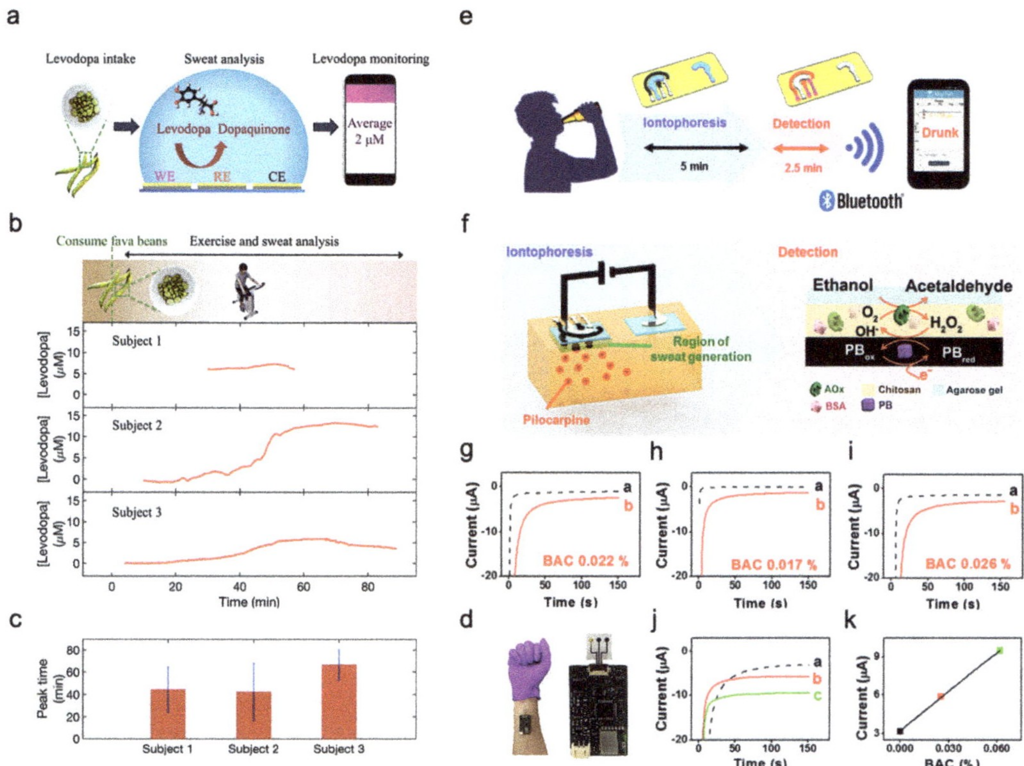

Figure 7. Drug sensing mechanism and ethanol levels measurement. (**a**) Interaction of the sensor with sweat. (**b**) Cycling and sweat analysis. Examples of sweat levodopa concentrations for three different subjects after they consume 450 g of fava beans. (**c**) Averaged time of peak levodopa concentration for three different subjects across multiple exercise trials. (**d**) Optical image of the S-band worn on a subject's wrist. Reproduced with permission from ref. [99], Copyright from 2019, American Chemical Society. (**e**) Sweat ethanol sensor used to send alerts to a smart device. (**f**) Schematic diagram of constituents in the iontophoretic system (left) and of the reagent layer and processes involved in the amperometric sensing of ethanol on the working electrode (right). (**g–i**) Experiments with consumption of 12 oz of beer measured from three different human subjects before (plot "a") and after drinking alcohol beverage (plot "b"). (**j**) Chronoamperograms obtained from a: BAC 0%, b: BAC 0.025% and c: BAC 0.062%. (**k**) Correlation between current response and the BAC level. Reproduced with permission from ref. [101], Copyright from 2016, American Chemical Society.

Sweating caused by iontophoresis lasts only for a very limited time period, while that caused by exercise generally lasts longer. This is helpful in the experimentation and validation of these sensors. Figure 7b shows the response observed from three subjects exercising on a stationary ergometer. In each trial, a subject would consume 450 g of fava beans and undergo multiple exercise trials. Cumulative results are shown in Figure 7c, depicting the average time of peak concentration. By analyzing the sweat induced by physical activities and iontophoresis, it is possible to optimize the dose of drugs by monitoring drug metabolism. Figure 7d shows the sensor apparatus mounted on a subject's wrist. The future development directions include the study of pharmacodynamics between drugs, prolonging the duration of iontophoresis sweating, and improving the lifetime of the electrode. As a result, S-band can be used to study the inherent complex drug profiles, optimize drug dosages for people with Parkinson's disease, and be incorporated into drug delivery systems.

3.4. Ethanol Levels Measurement

Sweat ethanol levels are an indicator of blood alcohol concentration. Ethanol levels were measured using a sweat-based sensor in tattoo form [101]. Figure 7e presents the illustration of a system that uses a sweat ethanol sensor to send alerts to a smart device. A tattoo-like patch based on an enzyme amperometric sweat ethanol sensor was developed with a pilocarpine iontophoresis drug delivery system and connected to intelligent devices through a Bluetooth module. The external current of iontophoresis (0.6 mA) was optimized between the effective delivery of drugs and the comfort of subjects because high current can easily cause skin irritation. The device is more reliable than the commonly used breath meter because it avoids potential errors induced by environmental factors like water vapor, or consumer products like mouthwash. Compared to other transcutaneous devices, it is a faster blood alcohol concentration measurement method because it only takes about 10 min compared with 0.5–2 h on traditional devices. Therefore, it is suggested that ethanol sensors can be used to detect illegal levels of alcohol consumption in car drivers.

This sensor system uses iontophoresis technology with constant current to induce sweat by delivery of the drug pilocarpine through the skin to perform sweat ethanol sensing. The sensor uses an alcohol oxidase enzyme electrode and a printed Prussian blue (PB) electrode transducer. All the electrodes on the wearable temporary tattoo paper are produced by screen printing for mass production and can be removed from the skin easily. Figure 7f illustrates these processes starting with iontophoresis followed by amperometric detection. Experiments were performed on three different human subjects using this sensor to validate the response of the system. Figure 7g–i shows plots that illustrate the responses of the subjects upon consumption of equal amounts of alcohol. Curve a is the amperometric response before drinking, and curve b is after drinking an alcoholic beverage. The blood alcohol concentrations of these subjects are different due to varying metabolism rates. However, after consuming alcohol, the current change from the sensor is quite apparent. Three control experiments were conducted to make sure there are no false positives. Figure 7g shows that zero blood alcohol content (BAC) results in no change in sensor response. And the sensor response after a certain amount of time had elapsed without alcohol consumption. This proves the sensor current response shown was caused by alcohol consumption. Figure 7h shows the response if no enzyme is immobilized in the sensor, showing that it is highly specific to sweat ethanol levels. Figure 7i shows the sensor response with and without iontophoresis, showing that the method of sweat extraction does not affect the sensor response. Thus, the sensor is highly specific to sweat ethanol levels and does not produce a response with 0 BAC. Figure 7j showed the sensor response at different BAC, with (a) being at 0%, (b) at 0.025%, and (c) at 0.062%. Figure 7k showed current changes at different BAC.

Additional control experiments were performed to validate the sensor. It has been shown to provide reliable information in real-world settings, which can provide a highly useful instrument to monitor alcohol for road safety. A more interesting implementation could be to fit these devices in such a way that the measured BAC determines whether a vehicle can be started or not. Future systems would involve calibrating the device, ensuring data security, and safeguarding privacy.

3.5. Biomolecules Monitoring

Compared to the three biomarkers mentioned, biomolecules including proteins, cytokines, nucleic acids, or neuropeptides are also important indicators reflecting a subject's health or infection status. Although they are always present at relatively low concentrations in physiological fluids, they are of great interest for monitoring the chronic wound healing process as well as future diagnosis or management of diseases such as wound healing, Parkinson's disease, and depression. As an example, a stretchable electrochemical immunosensor was used to detect the protein of the TNF-α antibody to monitor wound healing (Figure 8a) [103]. TNF-α was immobilized on the working electrode and detected using a voltammetric technique of differential pulse voltammetry method. Without the addition of TNF-α, the Faraday current was recorded at the redox potential that came from the redox of ferricyanide (Figure 8b). When TNF-α is added, a barrier layer is formed on the surface of the working electrode to inhibit electron transfer to reduce the recorded current. The immunosensor shows decent sensing performance both in buffer solutions with clinical concentration ranges (0.1 pM–0.1 µM) and human serum. Additionally, it can adapt to strains up to 30% due to 3D micro-patterned elastomers as a potential for body-attachable immunosensing (Figure 8c).

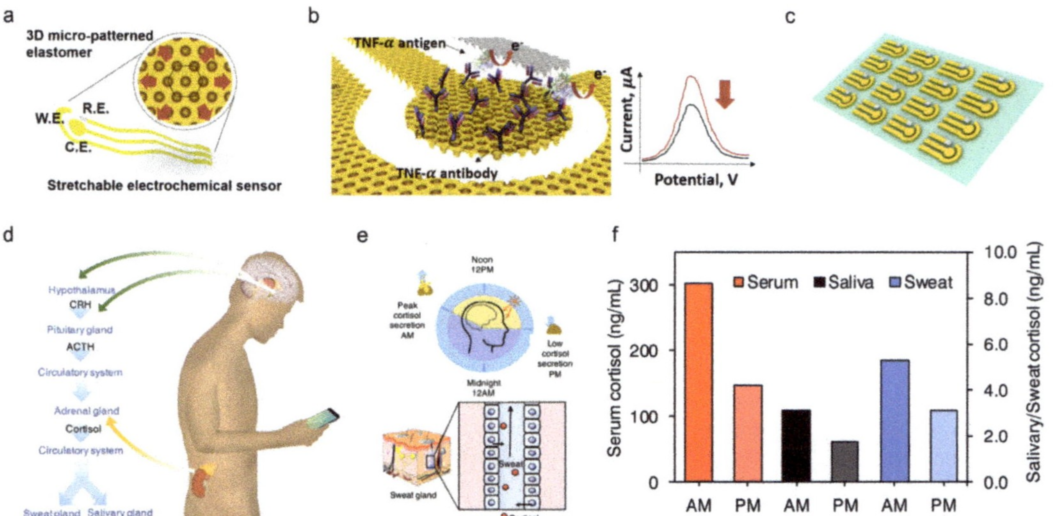

Figure 8. Sweat-based sensors for biomolecules monitoring. (**a**) Schematic showing a stretchable chemical immunosensor for TNF-α cytokine sensing. (**b**) Schematic illustration of the TNF-α cytokine proteins immobilized on the working electrode for electrochemical sensing. (**c**) Image of fabricated device arrays. Reproduced with permission from ref. [103], Copyright from 2019, Elsevier. (**d**) Schematic illustration of cortisol in sweat and saliva. CRH, corticotropin-releasing hormone; ACTH, adrenocorticotropic hormone. (**e**) Schematic showing the light and dark cycle for regulating circadian rhythm and controlling of cortisol transport to sweat. (**f**) Cortisol levels in serum, saliva, and sweat at different times from a healthy subject. Reproduced with permission from ref. [27], Copyright from 2020, Elsevier.

Besides, timely and accurate detection of stress is essential for monitoring and managing mental health. Considering that the current questionnaire and other methods are very subjective, a wearable chemical sensor was proposed with a highly sensitive, selective, and miniaturized mHealth device based on a laser flexible graphene sensor to non-invasively monitor the level of the stress hormone of cortisol (Figure 8d). It shows a strong correlation between sweat and circulating cortisol and demonstrates that changes in sweat cortisol respond quickly to acute stress stimuli. In addition, it showed a diurnal cycle and sweat cortisol pressure response curve, revealing the potential for dynamic pressure monitoring achieved by the mHealth sensor system (Figure 8e). Trends in ante meridiem/post meridiem (AM/PM) cortisol variability modulated by circadian rhythm are observed from a subject with the ratios ranging from 1.35 to 2.00 (Figure 8f). This platform provides a rapid, reliable, and decentralized health care vigilance at the metabolic level, thereby guiding an accurate snapshot of our physical, mental, and behavioral changes. Other sweat based wearable bioelectronics are also developed to detect pathogenic contamination [117] and antimicrobial peptides [79], and are compatible with the customized display interface for more convenient monitoring of health performance.

4. Conclusions and Future Scope

In the new age of wearable bioelectronics, sweat-based wearable bioelectronics has emerged as a leading technology that can measure biomarkers from a source that is boundless and easily extractable. The acquisition of biomarkers is generally noninvasive, will not degrade analytes, and has high sampling and detection efficiency. Recent advances in sweat-based wearables have also demonstrated a strong indication that this technology can be utilized in: health monitoring and disease detection, exercise intensity monitoring, drug metabolism monitoring, and ethanol level measurement. To further promote the practicality of sweat-based wearable bioelectronics, the following challenges need further attention:

(1) Improvement of biomarkers availability in sweat. Although sweat contains many biomarkers, the concentration of these biomarkers varies widely. In general, the concentration of biomarkers in sweat is significantly lower than that in other body fluid samples with a similar volume. Compared with human plasma, the proportion of sodium, potassium, lactic acid, and glucose in sweat is small. The main reason behind this huge difference is the filtration of extracellular matrix tight junctions, which limits the size of molecules that can pass through the skin. Therefore, improving the availability of biomarkers in human sweat is one of the key problems to be solved in sweat-based wearable bio-electronic devices.

(2) Sustained sample source and stable quantity. It is well known that the amount of skin perspiration varies from individual to environment. Therefore, the ideal wearable sensor must be able to accommodate the variations of different individuals and provide accurate sweat monitoring. Continuous monitoring of sweating is difficult to achieve because a person cannot sweat for a long time without external stimulation. This special problem greatly reduces the efficiency of an independent sweat-based wearable device for continuous monitoring throughout the day. Although we can use some methods to induce perspiration for providing sample sources for wearable sweat sensors, the differences in composition between heat-induced perspiration and chemical-induced perspiration are still questionable. Currently, it seems very convenient and time-saving to use chemicals such as pilocarpine to artificially produce sweat, and it can provide sweat samples continuously and indefinitely. However, it is not yet clear whether this sweating stimulation will affect the individual's common sweating function and induce health concerns of long-term use in the human body.

(3) Improvement in sample quality. The quality of sweat samples is also susceptible to various external factors which may lead to inaccurate measurement data. For example, sweat produced during strenuous physical activity is usually used to cool the body temperature. Thus, the sweat tends to evaporate on the surface of the skin to carry away heat for cooling purposes. Consequently, the concentration of biomarkers in the initial sweat sample changes during evaporation. In addition, the sweat excreted by the human body is easily contaminated by pollutants on the surface of the skin. In research settings, a protective conductive layer is added between the sensor and the skin area where the sweat is artificially induced to prevent the generation of pollutants. However, the dead volume between surfaces may cause a delay in the time from perspiration to sensing, which reduces the accuracy and increases the latency of the collected data. To improve the measurement accuracy of the sensor, more attention and effort should be paid to this problem.

Although with numerous challenges yet to be overcome, sweat-based sensors for biomonitoring are a nascent and promising field of technology. Several technological challenges will need to be resolved, as listed above, before a commercial implementation of wearable sweat-based devices can be developed. Some future directions for further research and development are also discussed here:

(1) Exploring efficient power supply methods. The great progress of wearable biosensor technology and the growing demand for multi-task processing on wearable platforms

to promote the development of advanced power supplies. To realize non-invasive wearable bioelectronics, the power supply should be efficient and sustainable and have good flexibility to meet skin contour and mechanical stress. Although great efforts have been made in noninvasive flexible fuel cells and biofuel cells, the current technology is far from the requirements to provide stable and reliable power support for most of the existing wearable bioelectronics. Besides further boosting the power output from the flexible fuel cells, another possible solution is to manufacture microsensors that consume less power. Additionally, harnessing energy from multiple sources, such as biomechanical energy and solar energy, could also be a promising solution.

(2) Developing suitable data processing and system integration methods. To obtain informative results gathered from wearable bioelectronics, it is necessary to perform appropriate post-processing on the electrical signal of the sensor, including amplification, filtering, and analog-digital conversion. Then, the processed signal is transmitted to the upper computer for analysis and display. Therefore, in this process, the electrical signal gathered by the sensor needs to be sampled by the processor and converted into a recordable value. The sampled raw data may suffer from inherent or environmental noise. Appropriate signal processing methods can reduce the influence of these noises, which is conducive to extracting useful signals from the sensor. The processed data is then transmitted to an external platform, such as a computer or mobile phone, for displaying and analysis. Here, the major role of data processing is to reduce noise and provide a user-friendly display of the recorded data. For applications requiring big data storage space and complex calculations, the data needs to be preprocessed before transmission. At present, the most popular technologies used for real-time data streaming and analysis in wearable sensing devices are low-energy Bluetooth and near field communication (NFC). However, both technologies have obvious transmission drawbacks. For example, NFC needs to be close to the receiver electronics for functioning. A transmission system that achieves the ideal connection has yet to be developed.

(3) Reducing the delay of data collection and analysis. Since it is a complex process to analyze and process the data collected by the sweat sensor, it generally takes a long time to complete. However, during this process, the evaporation of sweat on the skin surface will cause changes in the concentration of biomarkers in the sweat sample, which will pose a major obstacle. Reducing the delay between sweat collection and analysis is an important research area, which may be solved by developing low-power and high-performance microprocessors. Moreover, advanced big-data processing methods based on machine learning or deep learning algorithms can be further integrated into the system to realize the rapid and accurate extraction of the collected data.

Author Contributions: Supervision, J.C.; conceptualization, J.C. and J.X.; visualization, J.C. and J.X.; writing—original draft, J.X. and Y.F.; writing—review & editing, J.X., Y.F. and J.C.; funding acquisition, J.C. All authors have read and agreed to the published version of the manuscript.

Funding: Startup support from the University of California, Los Angeles, the 2020 Okawa Foundation Research Grant, and the 2021 Hellman Fellows Grant.

Institutional Review Board Statement: Not applicable.

Informed Consent Statement: Informed consent was obtained from all subjects involved in the study.

Data Availability Statement: Data of our study are available upon request.

Acknowledgments: We acknowledge the Henry Samueli School of Engineering & Applied Science and the Department of Bioengineering at the University of California, Los Angeles for the startup support. J.C. also acknowledges the 2020 Okawa Foundation Research Grant.

Conflicts of Interest: The authors declare no competing interest.

References

1. St John, A.; Price, C.P. Existing and Emerging Technologies for Point-of-Care Testing. *Clin. Biochem. Rev.* **2014**, *35*, 155–167.
2. Yang, Y.; Song, Y.; Bo, X.; Min, J.; Pak, O.S.; Zhu, L.; Wang, M.; Tu, J.; Kogan, A.; Zhang, H.; et al. A Laser-Engraved Wearable Sensor for Sensitive Detection of Uric Acid and Tyrosine in Sweat. *Nat. Biotechnol.* **2020**, *38*, 217–224. [CrossRef]
3. Fang, Y.; Zhao, X.; Tat, T.; Xiao, X.; Chen, G.; Xu, J.; Chen, J. All-in-One Conformal Epidermal Patch for Multimodal Biosensing. *Matter* **2021**, *4*, 1102–1105. [CrossRef]
4. Zheng, Y.L.; Ding, X.R.; Poon, C.C.; Lo, B.P.; Zhang, H.; Zhou, X.L.; Yang, G.Z.; Zhao, N.; Zhang, Y.T. Unobtrusive Sensing and Wearable Devices for Health Informatics. *IEEE Trans. Biomed. Eng.* **2014**, *61*, 1538–1554. [CrossRef] [PubMed]
5. Xiao, X.; Xiao, X.; Lan, Y.; Chen, J. Learning from Nature for Healthcare, Energy, and Environment. *The Innovation* **2021**, *2*, 100135. [CrossRef]
6. Conta, G.; Libanori, A.; Tat, T.; Chen, G.; Chen, J. Triboelectric Nanogenerators for Therapeutic Electrical Stimulation. *Adv. Mater.* **2021**, *33*, e2007502. [CrossRef]
7. Li, X.; Tat, T.; Chen, J. Triboelectric Nanogenerators for Self-Powered Drug Delivery. *Trends Chem.* **2021**. [CrossRef]
8. Xu, Q.; Fang, Y.; Jing, Q.; Hu, N.; Lin, K.; Pan, Y.; Xu, L.; Gao, H.; Yuan, M.; Chu, L.; et al. A Portable Triboelectric Spirometer for Wireless Pulmonary Function Monitoring. *Biosens. Bioelectron.* **2021**, *187*, 113329. [CrossRef]
9. Zhao, X.; Askari, H.; Chen, J. Nanogenerators for Smart Cities in the Era of 5g and Internet of Things. *Joule* **2021**, *5*, 1391–1431. [CrossRef]
10. Xiao, Y.; Chen, J.; Wang, C.; Ding, J.; Tao, W. Editorial: Emerging Micro- and Nanotechnologies for Medical and Pharmacological Applications. *Front. Pharmacol.* **2021**, *12*, 648749. [CrossRef] [PubMed]
11. Chen, G.; Au, C.; Chen, J. Textile Triboelectric Nanogenerators for Wearable Pulse Wave Monitoring. *Trends Biotechnol.* **2021**. [CrossRef] [PubMed]
12. Shen, S.; Xiao, X.; Xiao, X.; Chen, J. Wearable Triboelectric Nanogenerators for Heart Rate Monitoring. *Chem. Commun.* **2021**, *57*, 5871–5879. [CrossRef]
13. Lama, J.; Yau, A.; Chen, G.; Sivakumar, A.; Zhao, X.; Chen, J. Textile Triboelectric Nanogenerators for Self-Powered Biomonitoring. *J. Mater. Chem. A* **2021**. [CrossRef]
14. Tat, T.; Libanori, A.; Au, C.; Yau, A.; Chen, J. Advances in Triboelectric Nanogenerators for Biomedical Sensing. *Biosens. Bioelectron.* **2021**, *171*, 112714. [CrossRef] [PubMed]
15. Zhang, S.; Bick, M.; Xiao, X.; Chen, G.; Nashalian, A.; Chen, J. Leveraging Triboelectric Nanogenerators for Bioengineering. *Matter* **2021**, *4*, 845–887. [CrossRef]
16. Su, Y.; Wang, J.; Wang, B.; Yang, T.; Yang, B.; Xie, G.; Zhou, Y.; Zhang, S.; Tai, H.; Cai, Z.; et al. Alveolus-Inspired Active Membrane Sensors for Self-Powered Wearable Chemical Sensing and Breath Analysis. *ACS Nano* **2020**, *14*, 6067–6075. [CrossRef]
17. Zhou, Z.; Padgett, S.; Cai, Z.; Conta, G.; Wu, Y.; He, Q.; Zhang, S.; Sun, C.; Liu, J.; Fan, E.; et al. Single-Layered Ultra-Soft Washable Smart Textiles for All-around Ballistocardiograph, Respiration, and Posture Monitoring During Sleep. *Biosens. Bioelectron.* **2020**, *155*, 112064. [CrossRef]
18. Su, Y.; Chen, C.; Pan, H.; Yang, Y.; Chen, G.; Zhao, X.; Li, W.; Gong, Q.; Xie, G.; Zhou, Y.; et al. Muscle Fibers Inspired High-Performance Piezoelectric Textiles for Wearable Physiological Monitoring. *Adv. Funct. Mater.* **2021**, *31*, 2010962. [CrossRef]
19. Fang, Y.; Li, Y.; Li, Y.; Ding, M.; Xie, J.; Hu, B. Solution-Processed Submicron Free-Standing, Conformal, Transparent, Breathable Epidermal Electrodes. *ACS Appl. Mater. Interfaces* **2020**, *12*, 23689–23696. [CrossRef]
20. Fang, Y.; Li, Y.; Wang, X.; Zhou, Z.; Zhang, K.; Zhou, J.; Hu, B. Cryo-Transferred Ultrathin and Stretchable Epidermal Electrodes. *Small* **2020**, *16*, 2000450. [CrossRef]
21. Heikenfeld, J. Non-Invasive Analyte Access and Sensing through Eccrine Sweat: Challenges and Outlook Circa 2016. *Electroanalysis* **2016**, *28*, 1242–1249. [CrossRef]
22. Su, Y.; Yang, T.; Zhao, X.; Cai, Z.; Chen, G.; Yao, M.; Chen, K.; Bick, M.; Wang, J.; Li, S.; et al. A Wireless Energy Transmission Enabled Wearable Active Acetone Biosensor for Non-Invasive Prediabetes Diagnosis. *Nano Energy* **2020**, *74*, 104941. [CrossRef]
23. Jin, L.; Xiao, X.; Deng, W.; Nashalian, A.; He, D.; Raveendran, V.; Yan, C.; Su, H.; Chu, X.; Yang, T.; et al. Manipulating Relative Permittivity for High-Performance Wearable Triboelectric Nanogenerators. *Nano Lett.* **2020**, *20*, 6404–6411. [CrossRef] [PubMed]
24. Zhou, Z.; Chen, K.; Li, X.; Zhang, S.; Wu, Y.; Zhou, Y.; Meng, K.; Sun, C.; He, Q.; Fan, W.; et al. Sign-to-Speech Translation Using Machine-Learning-Assisted Stretchable Sensor Arrays. *Nat. Electron.* **2020**, *3*, 571–578. [CrossRef]
25. Zou, Y.; Libanori, A.; Xu, J.; Nashalian, A.; Chen, J. Triboelectric Nanogenerator Enabled Smart Shoes for Wearable Electricity Generation. *Research* **2020**, *2020*, 7158953. [CrossRef]
26. Mahmud, M.A.P.; Zolfagharian, A.; Gharaie, S.; Kaynak, A.; Farjana, S.H.; Ellis, A.V.; Chen, J.; Kouzani, A.Z. 3d-Printed Triboelectric Nanogenerators: State of the Art, Applications, and Challenges. *Adv. Energy Sustain. Res.* **2021**, *2*, 2000045. [CrossRef]
27. Meng, K.; Zhao, S.; Zhou, Y.; Wu, Y.; Zhang, S.; He, Q.; Wang, X.; Zhou, Z.; Fan, W.; Tan, X.; et al. A Wireless Textile-Based Sensor System for Self-Powered Personalized Health Care. *Matter* **2020**, *2*, 896–907. [CrossRef]
28. Chen, G.; Li, Y.; Bick, M.; Chen, J. Smart Textiles for Electricity Generation. *Chem. Rev.* **2020**, *120*, 3668–3720. [CrossRef] [PubMed]
29. Liu, Z.; Li, H.; Shi, B.; Fan, Y.; Wang, Z.L.; Li, Z. Wearable and Implantable Triboelectric Nanogenerators. *Adv. Funct. Mater.* **2019**, *29*, 1808820. [CrossRef]

30. Kwak, S.S.; Yoon, H.-J.; Kim, S.-W. Textile-Based Triboelectric Nanogenerators for Self-Powered Wearable Electronics. *Adv. Funct. Mater.* **2019**, *29*, 1804533. [CrossRef]
31. Chen, X.; Song, Y.; Su, Z.; Chen, H.; Cheng, X.; Zhang, J.; Han, M.; Zhang, H. Flexible Fiber-Based Hybrid Nanogenerator for Biomechanical Energy Harvesting and Physiological Monitoring. *Nano Energy* **2017**, *38*, 43–50. [CrossRef]
32. Song, Y.; Wang, H.; Cheng, X.; Li, G.; Chen, X.; Chen, H.; Miao, L.; Zhang, X.; Zhang, H. High-Efficiency Self-Charging Smart Bracelet for Portable Electronics. *Nano Energy* **2019**, *55*, 29–36. [CrossRef]
33. Xiao, X.; Chen, G.; Libanori, A.; Chen, J. Wearable Triboelectric Nanogenerators for Therapeutics. *Trends Chem.* **2021**. [CrossRef]
34. Abiri, P.; Duarte-Vogel, S.; Chou, T.C.; Abiri, A.; Gudapati, V.; Yousefi, A.; Roustaei, M.; Chang, C.C.; Cui, Q.; Hsu, J.J.; et al. In Vivo Intravascular Pacing Using a Wireless Microscale Stimulator. *Ann. Biomed. Eng.* **2021**. [CrossRef] [PubMed]
35. Chen, G.; Fang, Y.; Zhao, X.; Tat, T.; Chen, J. Textiles for Learning Tactile Interactions. *Nat. Electron.* **2021**, *4*, 175–176. [CrossRef]
36. Davoodi, E.; Montazerian, H.; Haghniaz, R.; Rashidi, A.; Ahadian, S.; Sheikhi, A.; Chen, J.; Khademhosseini, A.; Milani, A.S.; Hoorfar, M.; et al. 3d-Printed Ultra-Robust Surface-Doped Porous Silicone Sensors for Wearable Biomonitoring. *ACS Nano* **2020**, *14*, 1520–1532. [CrossRef] [PubMed]
37. Fang, Y.; Chen, G.; Bick, M.; Chen, J. Smart Textiles for Personalized Thermoregulation. *Chem. Soc. Rev.* **2021**. [CrossRef]
38. Zhou, Z.; Weng, L.; Tat, T.; Libanori, A.; Lin, Z.; Ge, L.; Yang, J.; Chen, J. Smart Insole for Robust Wearable Biomechanical Energy Harvesting in Harsh Environments. *ACS Nano* **2020**, *14*, 14126–14133. [CrossRef]
39. Li, X.; Dunn, J.; Salins, D.; Zhou, G.; Zhou, W.; Schussler-Fiorenza Rose, S.M.; Perelman, D.; Colbert, E.; Runge, R.; Rego, S.; et al. Digital Health: Tracking Physiomes and Activity Using Wearable Biosensors Reveals Useful Health-Related Information. *PLoS Biol.* **2017**, *15*, e2001402. [CrossRef]
40. Xiao, X.; Xiao, X.; Nashalian, A.; Libanori, A.; Fang, Y.; Li, X.; Chen, J. Triboelectric Nanogenerators for Self-Powered Wound Healing. *Adv. Healthc. Mater.* **2021**. [CrossRef] [PubMed]
41. Zhang, N.; Huang, F.; Zhao, S.; Lv, X.; Zhou, Y.; Xiang, S.; Xu, S.; Li, Y.; Chen, G.; Tao, C.; et al. Photo-Rechargeable Fabrics as Sustainable and Robust Power Sources for Wearable Bioelectronics. *Matter* **2020**, *2*, 1260–1269. [CrossRef]
42. An, Y.; Ren, Y.; Bick, M.; Dudek, A.; Hong-Wang Waworuntu, E.; Tang, J.; Chen, J.; Chang, B. Highly Fluorescent Copper Nanoclusters for Sensing and Bioimaging. *Biosens. Bioelectron.* **2020**, *154*, 112078. [CrossRef]
43. Bandodkar, A.J.; Wang, J. Non-Invasive Wearable Electrochemical Sensors: A Review. *Trends Biotechnol.* **2014**, *32*, 363–371. [CrossRef]
44. Gao, W.; Emaminejad, S.; Nyein, H.Y.; Challa, S.; Chen, K.; Peck, A.; Fahad, H.M.; Ota, H.; Shiraki, H.; Kiriya, D.; et al. Fully Integrated Wearable Sensor Arrays for Multiplexed in Situ Perspiration Analysis. *Nature* **2016**, *529*, 509–514. [CrossRef] [PubMed]
45. Schazmann, B.; Morris, D.; Slater, C.; Beirne, S.; Fay, C.; Reuveny, R.; Moyna, N.; Diamond, D. A Wearable Electrochemical Sensor for the Real-Time Measurement of Sweat Sodium Concentration. *Anal. Methods* **2010**, *2*, 342. [CrossRef]
46. Koh, A.; Kang, D.; Xue, Y.; Lee, S.; Pielak, R.M.; Kim, J.; Hwang, T.; Min, S.; Banks, A.; Bastien, P.; et al. A Soft, Wearable Microfluidic Device for the Capture, Storage, and Colorimetric Sensing of Sweat. *Sci. Transl. Med.* **2016**, *8*, 366ra165. [CrossRef]
47. Rose, D.P.; Ratterman, M.E.; Griffin, D.K.; Hou, L.; Kelley-Loughnane, N.; Naik, R.R.; Hagen, J.A.; Papautsky, I.; Heikenfeld, J.C. Adhesive Rfid Sensor Patch for Monitoring of Sweat Electrolytes. *IEEE Trans. Biomed. Eng.* **2015**, *62*, 1457–1465. [CrossRef]
48. Sonner, Z.; Wilder, E.; Gaillard, T.; Kasting, G.; Heikenfeld, J. Integrated Sudomotor Axon Reflex Sweat Stimulation for Continuous Sweat Analyte Analysis with Individuals at Rest. *Lab Chip* **2017**, *17*, 2550–2560. [CrossRef] [PubMed]
49. Bandodkar, A.J.; Jia, W.; Yardimci, C.; Wang, X.; Ramirez, J.; Wang, J. Tattoo-Based Noninvasive Glucose Monitoring: A Proof-of-Concept Study. *Anal. Chem.* **2015**, *87*, 394–398. [CrossRef] [PubMed]
50. Jia, W.; Bandodkar, A.J.; Valdes-Ramirez, G.; Windmiller, J.R.; Yang, Z.; Ramirez, J.; Chan, G.; Wang, J. Electrochemical Tattoo Biosensors for Real-Time Noninvasive Lactate Monitoring in Human Perspiration. *Anal. Chem.* **2013**, *85*, 6553–6560. [CrossRef] [PubMed]
51. Kim, J.; Imani, S.; de Araujo, W.R.; Warchall, J.; Valdes-Ramirez, G.; Paixao, T.R.; Mercier, P.P.; Wang, J. Wearable Salivary Uric Acid Mouthguard Biosensor with Integrated Wireless Electronics. *Biosens. Bioelectron.* **2015**, *74*, 1061–1068. [CrossRef]
52. Huang, X.; Liu, Y.; Chen, K.; Shin, W.J.; Lu, C.J.; Kong, G.W.; Patnaik, D.; Lee, S.H.; Cortes, J.F.; Rogers, J.A. Stretchable, Wireless Sensors and Functional Substrates for Epidermal Characterization of Sweat. *Small* **2014**, *10*, 3083–3090. [CrossRef] [PubMed]
53. Lee, H.; Song, C.; Hong, Y.S.; Kim, M.S.; Cho, H.R.; Kang, T.; Shin, K.; Choi, S.H.; Hyeon, T.; Kim, D.H. Wearable/Disposable Sweat-Based Glucose Monitoring Device with Multistage Transdermal Drug Delivery Module. *Sci. Adv.* **2017**, *3*, e1601314. [CrossRef] [PubMed]
54. Kudo, H.; Sawada, T.; Kazawa, E.; Yoshida, H.; Iwasaki, Y.; Mitsubayashi, K. A Flexible and Wearable Glucose Sensor Based on Functional Polymers with Soft-Mems Techniques. *Biosens. Bioelectron.* **2006**, *22*, 558–562. [CrossRef]
55. Yu, Y.; Nassar, J.; Xu, C.; Min, J.; Yang, Y.; Dai, A.; Doshi, R.; Huang, A.; Song, Y.; Gehlhar, R.; et al. Biofuel-Powered Soft Electronic Skin with Multiplexed and Wireless Sensing for Human-Machine Interfaces. *Sci. Robot.* **2020**, *5*, eaaz7946. [CrossRef] [PubMed]
56. De Giovanni, N.; Fucci, N. The Current Status of Sweat Testing for Drugs of Abuse: A Review. *Curr. Med. Chem.* **2013**, *20*, 545–561.
57. Yao, H.; Shum, A.J.; Cowan, M.; Lahdesmaki, I.; Parviz, B.A. A Contact Lens with Embedded Sensor for Monitoring Tear Glucose Level. *Biosens. Bioelectron.* **2011**, *26*, 3290–3296. [CrossRef]
58. Liao, Y.-T.; Yao, H.; Lingley, A.; Parviz, B.; Otis, B.P. A 3-μW CMOS Glucose Sensor for Wireless Contact-Lens Tear Glucose Monitoring. *IEEE J. Solid State Circuits* **2012**, *47*, 335–344. [CrossRef]

59. Xu, M.; Jiang, Y.; Pradhan, S.; Yadavalli, V.K. Use of Silk Proteins to Form Organic, Flexible, Degradable Biosensors for Metabolite Monitoring. *Front. Mater.* **2019**, *6*, 331. [CrossRef]

60. He, W.; Wang, C.; Wang, H.; Jian, M.; Lu, W.; Liang, X.; Zhang, X.; Yang, F.; Zhang, Y. Integrated Textile Sensor Patch for Real-Time and Multiplex Sweat Analysis. *Sci. Adv.* **2019**, *5*, eaax0649. [CrossRef]

61. Brothers, M.C.; DeBrosse, M.; Grigsby, C.C.; Naik, R.R.; Hussain, S.M.; Heikenfeld, J.; Kim, S.S. Achievements and Challenges for Real-Time Sensing of Analytes in Sweat within Wearable Platforms. *Acc. Chem. Res.* **2019**, *52*, 297–306. [CrossRef]

62. Jeyaram, A.; Jay, S.M. Preservation and Storage Stability of Extracellular Vesicles for Therapeutic Applications. *AAPS J.* **2017**, *20*, 1. [CrossRef] [PubMed]

63. Torrente-Rodriguez, R.M.; Tu, J.; Yang, Y.; Min, J.; Wang, M.; Song, Y.; Yu, Y.; Xu, C.; Ye, C.; IsHak, W.W.; et al. Investigation of Cortisol Dynamics in Human Sweat Using a Graphene-Based Wireless Mhealth System. *Matter* **2020**, *2*, 921–937. [CrossRef] [PubMed]

64. Kidwell, D.A.; Holland, J.C.; Athanaselis, S. Testing for Drugs of Abuse in Saliva and Sweat. *J. Chromatogr. B* **1998**, *713*, 111–135. [CrossRef]

65. Choi, J.; Ghaffari, R.; Baker, L.B.; Rogers, J.A. Skin-Interfaced Systems for Sweat Collection and Analytics. *Sci. Adv.* **2018**, *4*, eaar3921. [CrossRef] [PubMed]

66. Desax, M.C.; Ammann, R.A.; Hammer, J.; Schoeni, M.H.; Barben, J.; Swiss Paediatric Respiratory Research Group. Nanoduct Sweat Testing for Rapid Diagnosis in Newborns, Infants and Children with Cystic Fibrosis. *Eur. J. Pediatr.* **2008**, *167*, 299–304. [CrossRef]

67. Bandodkar, A.J.; Jeang, W.J.; Ghaffari, R.; Rogers, J.A. Wearable Sensors for Biochemical Sweat Analysis. *Annu. Rev. Anal. Chem.* **2019**, *12*, 1–22. [CrossRef]

68. Choi, J.; Kang, D.; Han, S.; Kim, S.B.; Rogers, J.A. Thin, Soft, Skin-Mounted Microfluidic Networks with Capillary Bursting Valves for Chrono-Sampling of Sweat. *Adv. Healthc. Mater.* **2017**, *6*, 1601355. [CrossRef]

69. Bariya, M.; Nyein, H.Y.Y.; Javey, A. Wearable Sweat Sensors. *Nat. Electron.* **2018**, *1*, 160–171. [CrossRef]

70. Zhao, J.; Lin, Y.; Wu, J.; Nyein, H.Y.Y.; Bariya, M.; Tai, L.C.; Chao, M.; Ji, W.; Zhang, G.; Fan, Z.; et al. A Fully Integrated and Self-Powered Smartwatch for Continuous Sweat Glucose Monitoring. *ACS Sens.* **2019**, *4*, 1925–1933. [CrossRef]

71. Park, S.; Heo, S.W.; Lee, W.; Inoue, D.; Jiang, Z.; Yu, K.; Jinno, H.; Hashizume, D.; Sekino, M.; Yokota, T.; et al. Self-Powered Ultra-Flexible Electronics Via Nano-Grating-Patterned Organic Photovoltaics. *Nature* **2018**, *561*, 516–521. [CrossRef] [PubMed]

72. Morris, D.; Coyle, S.; Wu, Y.; Lau, K.T.; Wallace, G.; Diamond, D. Bio-Sensing Textile Based Patch with Integrated Optical Detection System for Sweat Monitoring. *Sens. Actuators B* **2009**, *139*, 231–236. [CrossRef]

73. Bandodkar, A.J.; You, J.-M.; Kim, N.-H.; Gu, Y.; Kumar, R.; Mohan, A.M.V.; Kurniawan, J.; Imani, S.; Nakagawa, T.; Parish, B.; et al. Soft, Stretchable, High Power Density Electronic Skin-Based Biofuel Cells for Scavenging Energy from Human Sweat. *Energy Environ. Sci.* **2017**, *10*, 1581–1589. [CrossRef]

74. Yang, Y.; Gao, W. Wearable and Flexible Electronics for Continuous Molecular Monitoring. *Chem. Soc. Rev.* **2018**, *48*, 1465–1491. [CrossRef] [PubMed]

75. Chu, M.X.; Miyajima, K.; Takahashi, D.; Arakawa, T.; Sano, K.; Sawada, S.; Kudo, H.; Iwasaki, Y.; Akiyoshi, K.; Mochizuki, M.; et al. Soft Contact Lens Biosensor for in Situ Monitoring of Tear Glucose as Non-Invasive Blood Sugar Assessment. *Talanta* **2011**, *83*, 960–965. [CrossRef] [PubMed]

76. Heikenfeld, J. Bioanalytical Devices: Technological Leap for Sweat Sensing. *Nature* **2016**, *529*, 475–476. [CrossRef] [PubMed]

77. Corrie, S.R.; Coffey, J.W.; Islam, J.; Markey, K.A.; Kendall, M.A. Blood, Sweat, and Tears: Developing Clinically Relevant Protein Biosensors for Integrated Body Fluid Analysis. *Analyst* **2015**, *140*, 4350–4364. [CrossRef]

78. Kahn, N.; Lavie, O.; Paz, M.; Segev, Y.; Haick, H. Dynamic Nanoparticle-Based Flexible Sensors: Diagnosis of Ovarian Carcinoma from Exhaled Breath. *Nano Lett.* **2015**, *15*, 7023–7028. [CrossRef] [PubMed]

79. Mannoor, M.S.; Tao, H.; Clayton, J.D.; Sengupta, A.; Kaplan, D.L.; Naik, R.R.; Verma, N.; Omenetto, F.G.; McAlpine, M.C. Graphene-Based Wireless Bacteria Detection on Tooth Enamel. *Nat. Commun.* **2012**, *3*, 763. [CrossRef] [PubMed]

80. Lee, H.; Choi, T.K.; Lee, Y.B.; Cho, H.R.; Ghaffari, R.; Wang, L.; Choi, H.J.; Chung, T.D.; Lu, N.; Hyeon, T.; et al. A Graphene-Based Electrochemical Device with Thermoresponsive Microneedles for Diabetes Monitoring and Therapy. *Nat. Nanotechnol.* **2016**, *11*, 566–572. [CrossRef] [PubMed]

81. Sonner, Z.; Wilder, E.; Heikenfeld, J.; Kasting, G.; Beyette, F.; Swaile, D.; Sherman, F.; Joyce, J.; Hagen, J.; Kelley-Loughnane, N.; et al. The Microfluidics of the Eccrine Sweat Gland, Including Biomarker Partitioning, Transport, and Biosensing Implications. *Biomicrofluidics* **2015**, *9*, 031301. [CrossRef] [PubMed]

82. Windmiller, J.R.; Wang, J. Wearable Electrochemical Sensors and Biosensors: A Review. *Electroanalysis* **2013**, *25*, 29–46. [CrossRef]

83. Windmiller, J.R.; Bandodkar, A.J.; Valdes-Ramirez, G.; Parkhomovsky, S.; Martinez, A.G.; Wang, J. Electrochemical Sensing Based on Printable Temporary Transfer Tattoos. *Chem. Commun.* **2012**, *48*, 6794–6796. [CrossRef] [PubMed]

84. Kim, J.; de Araujo, W.R.; Samek, I.A.; Bandodkar, A.J.; Jia, W.; Brunetti, B.; Paixão, T.R.L.C.; Wang, J. Wearable Temporary Tattoo Sensor for Real-Time Trace Metal Monitoring in Human Sweat. *Electrochem. Commun.* **2015**, *51*, 41–45. [CrossRef]

85. Munje, R.D.; Muthukumar, S.; Panneer Selvam, A.; Prasad, S. Flexible Nanoporous Tunable Electrical Double Layer Biosensors for Sweat Diagnostics. *Sci. Rep.* **2015**, *5*, 14586. [CrossRef]

86. Kinnamon, D.; Ghanta, R.; Lin, K.C.; Muthukumar, S.; Prasad, S. Portable Biosensor for Monitoring Cortisol in Low-Volume Perspired Human Sweat. *Sci. Rep.* **2017**, *7*, 13312. [CrossRef]

87. Shields, J.B.; Johnson, B.C.; Hamilton, T.S.; Mitchell, H.H. The Excretion of Ascorbic Acid and Dehydroascorbic Acid in Sweat and Urine under Different Environmental Conditions. *J. Biol. Chem.* **1945**, *161*, 351–356. [CrossRef]
88. Cizza, G.; Marques, A.H.; Eskandari, F.; Christie, I.C.; Torvik, S.; Silverman, M.N.; Phillips, T.M.; Sternberg, E.M.; Group, P.S. Elevated Neuroimmune Biomarkers in Sweat Patches and Plasma of Premenopausal Women with Major Depressive Disorder in Remission: The Power Study. *Biol. Psychiatry* **2008**, *64*, 907–911. [CrossRef]
89. Yoshizumi, J.; Kumamoto, S.; Nakamura, M.; Yamana, K. Target-Induced Strand Release (Tisr) from Aptamer-DNA Duplex: A General Strategy for Electronic Detection of Biomolecules Ranging from a Small Molecule to a Large Protein. *Analyst* **2008**, *133*, 323–325. [CrossRef]
90. Mehrvar, M.; Abdi, M. Recent Developments, Characteristics, and Potential Applications of Electrochemical Biosensors. *Anal. Sci.* **2004**, *20*, 1113–1126. [CrossRef]
91. Seshadri, D.R.; Li, R.T.; Voos, J.E.; Rowbottom, J.R.; Alfes, C.M.; Zorman, C.A.; Drummond, C.K. Wearable Sensors for Monitoring the Physiological and Biochemical Profile of the Athlete. *NPJ Digital Med.* **2019**, *2*, 72. [CrossRef]
92. Brasier, N.; Eckstein, J. Sweat as a Source of Next-Generation Digital Biomarkers. *Digit. Biomark.* **2019**, *3*, 155–165. [CrossRef]
93. Emaminejad, S.; Gao, W.; Wu, E.; Davies, Z.A.; Yin Yin Nyein, H.; Challa, S.; Ryan, S.P.; Fahad, H.M.; Chen, K.; Shahpar, Z.; et al. Autonomous Sweat Extraction and Analysis Applied to Cystic Fibrosis and Glucose Monitoring Using a Fully Integrated Wearable Platform. *Proc. Natl. Acad. Sci. USA* **2017**, *114*, 4625–4630. [CrossRef]
94. Patterson, M.J.; Galloway, S.D.R.; Nimmo, M.A. Variations in Regional Sweat Composition in Normal Human Males. *Exp. Physiol.* **2000**, *85*, 869–875. [CrossRef]
95. Nyein, H.Y.; Gao, W.; Shahpar, Z.; Emaminejad, S.; Challa, S.; Chen, K.; Fahad, H.M.; Tai, L.C.; Ota, H.; Davis, R.W.; et al. A Wearable Electrochemical Platform for Noninvasive Simultaneous Monitoring of Ca^{2+} and Ph. *ACS Nano* **2016**, *10*, 7216–7224. [CrossRef] [PubMed]
96. Anastasova, S.; Crewther, B.; Bembnowicz, P.; Curto, V.; Ip, H.M.; Rosa, B.; Yang, G.Z. A Wearable Multisensing Patch for Continuous Sweat Monitoring. *Biosens. Bioelectron.* **2017**, *93*, 139–145. [CrossRef] [PubMed]
97. Guinovart, T.; Bandodkar, A.J.; Windmiller, J.R.; Andrade, F.J.; Wang, J. A Potentiometric Tattoo Sensor for Monitoring Ammonium in Sweat. *Analyst* **2013**, *138*, 7031–7038. [CrossRef] [PubMed]
98. Gao, W.; Nyein, H.Y.Y.; Shahpar, Z.; Fahad, H.M.; Chen, K.; Emaminejad, S.; Gao, Y.; Tai, L.-C.; Ota, H.; Wu, E.; et al. Wearable Microsensor Array for Multiplexed Heavy Metal Monitoring of Body Fluids. *ACS Sens.* **2016**, *1*, 866–874. [CrossRef]
99. Tai, L.C.; Liaw, T.S.; Lin, Y.; Nyein, H.Y.Y.; Bariya, M.; Ji, W.; Hettick, M.; Zhao, C.; Zhao, J.; Hou, L.; et al. Wearable Sweat Band for Noninvasive Levodopa Monitoring. *Nano Lett.* **2019**, *19*, 6346–6351. [CrossRef]
100. Tai, L.C.; Gao, W.; Chao, M.; Bariya, M.; Ngo, Q.P.; Shahpar, Z.; Nyein, H.Y.Y.; Park, H.; Sun, J.; Jung, Y.; et al. Methylxanthine Drug Monitoring with Wearable Sweat Sensors. *Adv. Mater.* **2018**, *30*, e1707442. [CrossRef]
101. Kim, J.; Jeerapan, I.; Imani, S.; Cho, T.N.; Bandodkar, A.; Cinti, S.; Mercier, P.P.; Wang, J. Noninvasive Alcohol Monitoring Using a Wearable Tattoo-Based Iontophoretic-Biosensing System. *ACS Sens.* **2016**, *1*, 1011–1019. [CrossRef]
102. Abellan-Llobregat, A.; Jeerapan, I.; Bandodkar, A.; Vidal, L.; Canals, A.; Wang, J.; Morallon, E. A Stretchable and Screen-Printed Electrochemical Sensor for Glucose Determination in Human Perspiration. *Biosens. Bioelectron.* **2017**, *91*, 885–891. [CrossRef]
103. Kim, B.-Y.; Lee, H.-B.; Lee, N.-E. A Durable, Stretchable, and Disposable Electrochemical Biosensor on Three-Dimensional Micro-Patterned Stretchable Substrate. *Sens. Actuators B* **2019**, *283*, 312–320. [CrossRef]
104. Nemiroski, A.; Christodouleas, D.C.; Hennek, J.W.; Kumar, A.A.; Maxwell, E.J.; Fernandez-Abedul, M.T.; Whitesides, G.M. Universal Mobile Electrochemical Detector Designed for Use in Resource-Limited Applications. *Proc. Natl. Acad. Sci. USA* **2014**, *111*, 11984–11989. [CrossRef]
105. Mugweru, A.; Clark, B.L.; Pishko, M.V. Electrochemical Sensor Array for Glucose Monitoring Fabricated by Rapid Immobilization of Active Glucose Oxidase within Photochemically Polymerized Hydrogels. *J. Diabetes Sci. Technol.* **2007**, *1*, 366–371. [CrossRef]
106. Chung, M.; Fortunato, G.; Radacsi, N. Wearable Flexible Sweat Sensors for Healthcare Monitoring: A Review. *J. R. Soc. Interface* **2019**, *16*, 20190217. [CrossRef]
107. Song, Y.; Min, J.; Yu, Y.; Wang, H.; Yang, Y.; Zhang, H.; Gao, W. Wireless Battery-Free Wearable Sweat Sensor Powered by Human Motion. *Sci. Adv.* **2020**, *6*, eaay9842. [CrossRef] [PubMed]
108. Peng, Z.; Song, J.; Gao, Y.; Liu, J.; Lee, C.; Chen, G.; Wang, Z.; Chen, J.; Leung, M.K.H. A Fluorinated Polymer Sponge with Superhydrophobicity for High-Performance Biomechanical Energy Harvesting. *Nano Energy* **2021**, *85*, 106021. [CrossRef]
109. Zou, Y.; Raveendran, V.; Chen, J. Wearable Triboelectric Nanogenerators for Biomechanical Energy Harvesting. *Nano Energy* **2020**, *77*, 105303. [CrossRef]
110. Zou, Y.; Xu, J.; Chen, K.; Chen, J. Advances in Nanostructures for High-Performance Triboelectric Nanogenerators. *Adv. Mater. Technol.* **2021**, *6*, 2000916. [CrossRef]
111. Zhou, Y.; Deng, W.; Xu, J.; Chen, J. Engineering Materials at the Nanoscale for Triboelectric Nanogenerators. *Cell Rep. Phys. Sci.* **2020**, *1*, 100142. [CrossRef]
112. Deng, W.; Zhou, Y.; Zhao, X.; Zhang, S.; Zou, Y.; Xu, J.; Yeh, M.H.; Guo, H.; Chen, J. Ternary Electrification Layered Architecture for High-Performance Triboelectric Nanogenerators. *ACS Nano* **2020**, *14*, 9050–9058. [CrossRef] [PubMed]
113. Chu, X.; Zhao, X.; Zhou, Y.; Wang, Y.; Han, X.; Zhou, Y.; Ma, J.; Wang, Z.; Huang, H.; Xu, Z.; et al. An Ultrathin Robust Polymer Membrane for Wearable Solid-State Electrochemical Energy Storage. *Nano Energy* **2020**, *76*, 105179. [CrossRef]

114. Xu, J.; Zou, Y.; Nashalian, A.; Chen, J. Leverage Surface Chemistry for High-Performance Triboelectric Nanogenerators. *Front. Chem.* **2020**, *8*, 577327. [CrossRef]
115. Bakker, E.; Pretsch, E. Potentiometric Sensors for Trace-Level Analysis. *Trends Analyt. Chem.* **2005**, *24*, 199–207. [CrossRef]
116. Terse-Thakoor, T.; Punjiya, M.; Matharu, Z.; Lyu, B.; Ahmad, M.; Giles, G.E.; Owyeung, R.; Alaimo, F.; Shojaei Baghini, M.; Brunyé, T.T.; et al. Thread-Based Multiplexed Sensor Patch for Real-Time Sweat Monitoring. *NPJ Flexible Electron.* **2020**, *4*, 1–10. [CrossRef]
117. Karbelkar, A.A.; Furst, A.L. Electrochemical Diagnostics for Bacterial Infectious Diseases. *ACS Infect. Dis.* **2020**, *6*, 1567–1571. [CrossRef] [PubMed]

 *biosensors*

Review

REASSURED Multiplex Diagnostics: A Critical Review and Forecast

Jonas A. Otoo and Travis S. Schlappi *

Keck Graduate Institute, Claremont, CA 91711, USA; jotoo18@students.kgi.edu
* Correspondence: tschlappi@kgi.edu

Abstract: The diagnosis of infectious diseases is ineffective when the diagnostic test does not meet one or more of the necessary standards of affordability, accessibility, and accuracy. The World Health Organization further clarifies these standards with a set of criteria that has the acronym ASSURED (Affordable, Sensitive, Specific, User-friendly, Rapid and robust, Equipment-free and Deliverable to end-users). The advancement of the digital age has led to a revision of the ASSURED criteria to REASSURED: Real-time connectivity, Ease of specimen collection, Affordable, Sensitive, Specific, User-friendly, Rapid and robust, Equipment-free or simple, and Deliverable to end-users. Many diagnostic tests have been developed that aim to satisfy the REASSURED criteria; however, most of them only detect a single target. With the progression of syndromic infections, coinfections and the current antimicrobial resistance challenges, the need for multiplexed diagnostics is now more important than ever. This review summarizes current diagnostic technologies for multiplexed detection and forecasts which methods have promise for detecting multiple targets and meeting all REASSURED criteria.

Keywords: diagnostics; multiplex; point-of-care diagnostics; REASSURED

1. Introduction

Clinical diagnostics are devices or methods that are used to detect biomarkers in the genome, proteome and metabolome for diagnosis, subclassification, prognosis, susceptibility risk assessment, treatment selection, and response to therapy monitoring [1,2]. Biomarker analytes include nucleic acids, proteins, peptides, lipids, metabolites, and other small molecules [3,4]. Diagnostic tests are generally carried out in central labs, clinics, hospitals, doctors' offices, and point-of-care (POC) settings. Thousands of diagnostic tests have been developed over the years, with varying levels of complexity, turnaround time, cost, and other factors. While diagnostics account for less than 5% of hospital costs and ~1.6% of all Medicare costs, they influence up to 60–70% of healthcare decision making [5]. There are several stakeholders in diagnostics, each with their own priorities: patients, healthcare providers, payers, pharmaceutical companies, diagnostic device manufacturers, local and international health organizations, governments, public health agencies, and regulatory bodies [6,7].

In order to be FDA (Food and Drug Administration) approved, diagnostic tests need to meet certain standards for analytical and clinical validation. Analytical validation assesses the sensitivity, specificity, accuracy, and precision of the test. Clinical validation assesses the ability of the test to achieve its intended aim. Diagnostic tests in hospitals or reference labs are able to meet analytical and clinical standards for accuracy and performance because complexity and cost are not an issue. It is much more difficult for point-of-care diagnostics, however, which must also minimize cost and complexity in their design and manufacturing. The World Health Organization Special Program for Research and Training in Tropical Diseases (WHO/TDR) concluded, in a study in 2003, that POC diagnostics should meet the ASSURED (Affordable, Sensitive, Specific, User-Friendly, Rapid, Equipment-Free,

Citation: Otoo, J.A.; Schlappi, T.S. REASSURED Multiplex Diagnostics: A Critical Review and Forecast. *Biosensors* **2022**, *12*, 124. https://doi.org/10.3390/bios12020124

Received: 11 January 2022
Accepted: 11 February 2022
Published: 16 February 2022

Publisher's Note: MDPI stays neutral with regard to jurisdictional claims in published maps and institutional affiliations.

Delivered) criteria [8]. In 2006, the WHO/TDR further recommended the ASSURED criteria as a benchmark to decide whether diagnostic tests address disease control needs [9]. The ASSURED criteria represent three main attributes that are significant for a diagnostic test. These attributes are accessibility, affordability, and accuracy. While all three attributes are important, it is very challenging for any diagnostic test to adequately possess all three. The different stakeholders in diagnostics may have varying orders of priority among the three attributes. Patients may want diagnostic tests that are first, affordable, second, accessible, and third, accurate. Healthcare providers likely prefer accuracy, accessibility, and then affordability. Governments may prioritize accessibility over affordability and accuracy. Manufacturers of diagnostics that are maximizing profits probably emphasize accessibility, accuracy, and then affordability.

In the face of the SARS-CoV-2 pandemic, the role and importance of diagnostics has become increasingly apparent. Diagnostics help to track, contain, and control the spread of infectious diseases. Several diagnostic tests were developed in the wake of the SARS-CoV-2 pandemic [10–13], which guided the formulation and implementation of measures that were used to protect the public, find new variants, track the disease, and slow its spread. Diagnostics have also played a major role in non-infectious diseases. Early detection of biomarkers of cancer, cardiovascular disease and metabolic diseases, such as diabetes and hypertension, have reduced the mortality rate of humans over the years [14–17].

1.1. Multiplexed Diagnostics

Multiplexing is the process of simultaneously detecting or identifying multiple biomarkers in a single diagnostic test, which can be valuable for several different types of diseases. For example, pharmacogenomic studies in patients with cardiovascular disease have indicated that the presence of polymorphisms affects patients' response to various drugs [18]. Therefore, the multiplex detection of relevant biomarkers will not only provide insight of the pathophysiology of cardiovascular disease, but also provide a guide for the most efficient treatment option. Most cancers have biomarkers in common with other cancers, hence detecting multiple biomarkers is needed for the accurate differentiation of cancer types or location [19,20]. Hermann et al. [21] demonstrated that several biomarkers are significantly elevated in breast cancer patients versus patients with benign breast tumor disease. The multiplexed detection of these biomarkers enables oncologists to accurately diagnose their patients and select the appropriate therapy, thus improving patient outcomes and decreasing healthcare costs. Cytokines are important in the mediation of immune responses, such inflammation and mobilization of immune cells [22]. They are secreted by different cell types and are very diverse [23]. Multiplexed detection of cytokines is key to the better understanding of the immune response. Abdullah et al. [24] demonstrated that multiplexed detection of cytokines was important to understand whether neural stem cell rosette morphologies had an impact on the profile of cytokine signals and therefore had different outcomes in neurodegenerative disease cell therapies.

Infectious disease is another area where multiplexed diagnostics are extremely valuable. Most infectious diseases, such as urinary tract infections and respiratory infections, have multiple causative pathogens, but the resulting symptoms do not indicate the causative pathogen. On the other hand, different types of infections that have shared symptoms could be misdiagnosed or incompletely diagnosed. For example, SARS-CoV-2 and influenza A or B present with many of the same symptoms and clinical features [25,26]. Studies show that there is the prevalence of influenza coinfection among people with SARS-CoV-2 is 0.4% in the United States of America and 4.5% in Asia [27]. In a case study of 1986 patients that presented with Severe Acute Respiratory Infection (SARI), 14.3%, 8.8% and 0.3% had SARS-CoV-2, influenza and SARS-CoV-2/influenza coinfection, respectively [28]. In another study, 40% of a cohort of Kenyans who sought treatment for fever were presumed to have malaria and received malaria medicines even though they actually had HIV [29]. Incomplete diagnosis of infectious disease leads to inefficient treatments by exposing some pathogens to sub-lethal doses or the wrong antibiotics. This contributes to the emergence of antimicrobial resistance and recurrent infections as well as persistent secondary infections [30,31]. The last two classes of antibiotics were discovered in 1987 and 2004 [32], and since then, we are in a period of discovery void while there is rapid emergence of antimicrobial pathogens to the antibiotics that currently exist (Figure 1). According to O'Neil [33], 10 million people will die annually due to antimicrobial resistance (AMR) by 2050. Furthermore, AMR-related costs and the associated loss of productivity amount to about USD 55 billion annually in the U.S. alone [34]. Better diagnostics and treatment for tuberculosis could save 770,000 lives over the course of 2015 to 2025 [33], while a malaria test could save ~2.2 million lives and prevent ~447 million unnecessary treatments per year [35]. The introduction of antibiotics increased the average lifespan of humankind by 23 years since the first introduction of antibiotics, thus showing the drastic consequences if we were to lose the use of antibiotics that we currently have [32]. Another instance where multiplexing is crucial is the diagnosis of blood infections. Sepsis resulting from blood infections can be caused by many pathogens and becomes increasingly fatal over time, with mortality increasing by 7.6% for every hour that passes without receiving the correct antibiotic [36]. Accurately identifying which pathogen(s) is responsible for the blood infection is therefore a race against time to start the antibiotic therapy before sepsis becomes fatal [37]. The diagnosis of infections should therefore be approached by syndromic diagnosis, wherein all the potential pathogens for an infection or symptom are investigated rather than tested for just the most likely pathogen and then conducting other tests if negative [38,39]. Multiplexed diagnostic tests—wherein one sample is simultaneously tested for multiple pathogens in the same device—are essential for blood infections nowadays and important to combat AMR for all types of infections in the future. A query on the PubMed database of the National Center for Biotechnology Information (NCBI) suggests that researchers have become increasingly more interested in multiplex diagnostics (Figure 2).

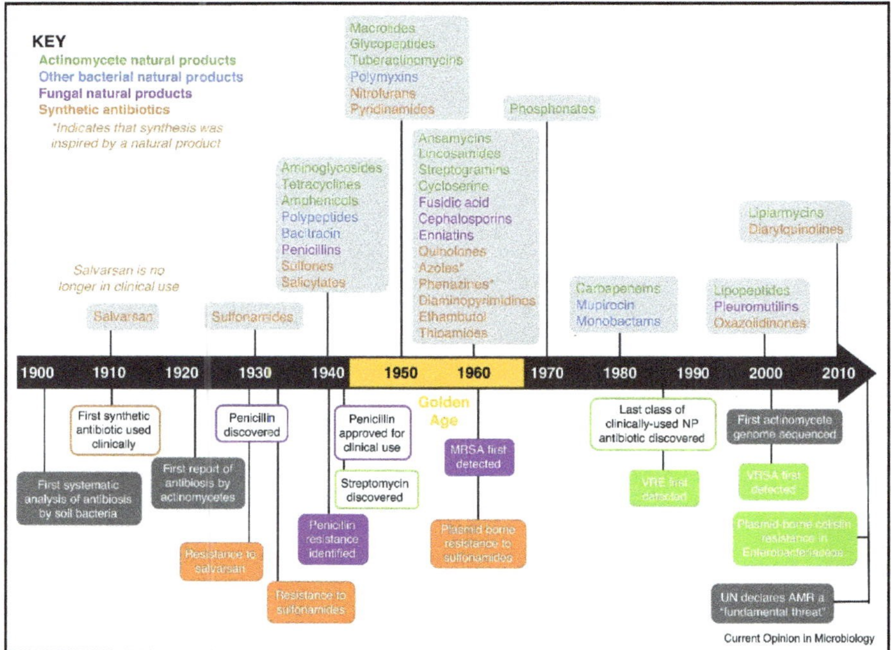

Figure 1. The timeline of antibiotics class discovery/development and onset of antimicrobial resistance reproduced from [29].

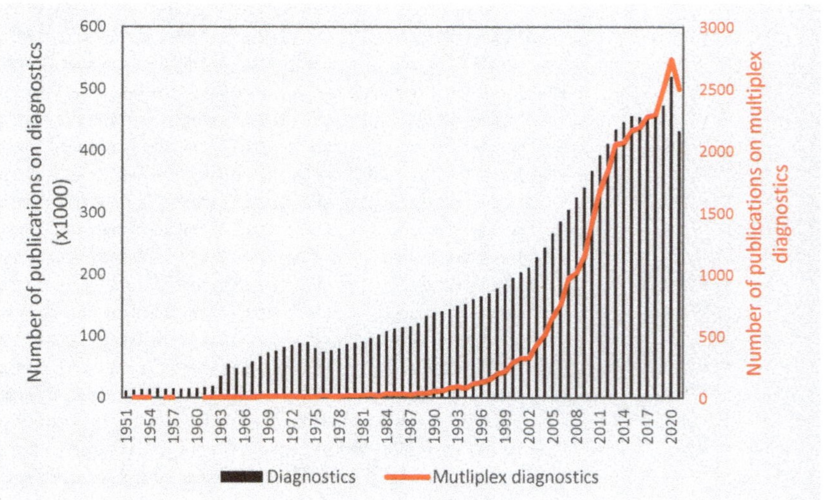

Figure 2. Annual publications related to diagnostics compared to annual publications related to multiplex diagnostics from 1950 to 2021 from the PubMed database (National Center for Biotechnology Information).

1.2. REASSURED Diagnostics

Considering the advances in digital technology and mobile health, a new REAS-SURED (Real-time connectivity, Ease of specimen collection, Affordable, Sensitive, Specific, User-Friendly, Rapid and Robust, Equipment free or simple Environmentally friendly, Deliverable to end-users) framework has been proposed as the benchmark for diagnostic systems [40]. The diagnosis of a disease is just the first step. The information from the diagnosis results needs to be used to inform actionable steps to treat or manage the disease. In a remote setting where a healthcare professional is not readily accessible, real-time connectivity provides the avenue to transmit the results to the healthcare professional for medical advice. Furthermore, having a reader that can provide the results of a diagnostic test is important especially in ambiguous cases where there is uncertainty due to variation in the interpretation of the results. A reader will serve as a standardized way to state the results of the diagnostic test [41–43].

The development of diagnostic tests that meet all the ASSURED criteria, but uses hard-to-obtain samples, such as venous blood, will not be very helpful in the absence of a trained professional to obtain the sample. It is therefore very crucial that, when possible, diagnostic tests should be developed to use easy-to-obtain and non-invasive samples, such as finger pricks, nasal or oral swabs, or urine samples.

While all the elements of the REASSURED criteria are important for POC diagnostics, it is challenging for any diagnostic device to embody all of these elements and trade-offs are often made in one or more elements to achieve other elements. For instance, nucleic acid testing (NAT) is very sensitive and specific, but often requires purification or isolation of the nucleic acid, concentration of the nucleic acid, amplification, and detection of the nucleic acid [44–46]. These processes can be achieved through user steps or by the introduction of equipment components that can execute them. On the other hand, antigen-based diagnostics, such as a lateral flow assay, are not as sensitive and specific as NAT, but are far more user-friendly, affordable, rapid, and deliverable [47]. In these two scenarios, some degree of sensitivity and specificity could be traded for the affordability, user-friendliness, and equipment complexity of the diagnostic test by detecting antigens instead of nucleic acids.

Naseri et al. [48] have summarized POC devices based on lateral flow assays (LFAs) and paper-based analytical devices (PADS) technology that were developed in the last 10 years for common human viral infection diagnostics. Dincer et al. [49] presented a survey of the existing multiplexed POC tests in academia and industry, while Kim et al. [50] summarized current POC tests for multiplex molecular testing of syndromic infections; however, these reviews focused mainly on POC diagnostics rather than summarizing devices that meet REASSURED criteria. In this paper, we present the current state of multiplexed diagnostic technology that meet REASSURED criteria based on an in-house developed scoring scheme. This review summarizes multiplexed diagnostics in three categories: (i) clinically used, (ii) in academia or research only, and (iii) next-generation technology. We then discuss the limitations in developing multiplexed REASSURED diagnostics, present current gaps in technology, and describe the needs for future research and development. For the purpose of this review, clinical diagnostics refer to diagnostics that have been approved by the FDA (including Emergency Use Authorization) or have a CE marking and are available for patient diagnosis.

2. Clinically Available Multiplexed Diagnostics

2.1. Proteins and Peptides

Multiplex detection of select protein or peptide biomarkers in human samples, such as blood, serum, saliva and urine for clinical diagnosis, while very important, presents with a challenging puzzle: human samples typically have a myriad of diverse proteins and peptides [51], only some of which are the protein of interest. Accurately differentiating the select protein biomarkers out of the matrix is challenging due the occurrence of cross-reactivity [52]. The advancement in technology has made it possible for some immunoassays to be adapted to the point-of-care setting for multiplex peptide and protein biomarker detection. LFAs use a variety of detection techniques, such as fluorescent immunoassays (FIA), chemiluminescence immunoassay [53] and colorimetric immune assays [54], for the detection of protein and peptide biomarkers. While LFAs have lower sensitivity compared to molecular diagnostic tests [55], they are rapid and relatively cheaper to fabricate compared to other diagnostics [56]. LFAs were the first tests that meet the WHO ASSURED criteria [47,57]. They are typically equipment free or are accompanied by a simple reader with a digital interface. When immunoassays, such as LFAs, have a colorimetric read-out, the interpretation of the results is subjective to the person who is reading the results. This may be problematic in cases where the biomarkers being detected are present in low concentrations. Utilizing a simple reader in conjunction with these LFAs will promote an objective and a more accurate interpretation of the results. This will also enable the LFAs to satisfy the REASSURED criteria.

Enzyme-Linked Immunosorbent Assays (ELISAs) are a highly sensitive method for the detection of protein and peptide biomarkers. ELISAs are very prone to interferences [58], which pose challenges to developing a multiplex test. This challenge is overcome through the use of spatial multiplexing approaches, such as wells and microarrays [59,60]. To avoid false positive tests as a result of non-specific interactions, there are multiple wash steps in ELISA assays. The automation of ELISAs for adaption to the POC and limited resource settings is therefore challenging because complex equipment components are required for fluid handling to execute wash steps. Furthermore, to avoid false negative tests, there are lengthy incubation periods in ELISA assays. It is therefore very challenging to adapt ELISAs for point-of-care diagnostics that fit the REASSURED criteria.

The BinaxNOW influenza A and B card 2 developed by Abbott is a multiplex immunochromatographic LFA that is able to provide rapid differential diagnosis of influenzas A and B infection [61]. This test is designed to be read by the DIGIVAL reader for result interpretation. The DIGIVAL reader is portable and battery powered, making it suitable for limited resource settings. Becton and Dickinson's (BD) Veritor™ Flu A + B with analyzer distinguishes between influenzas A and B as well. The BD test analyzer is palm sized and battery powered and hence suitable for use at remote and limited resource settings [62].

Acucy influenza A and B test developed by Sekisui diagnostics comes with a portable battery-powered reader as well [63]. Quidel's Sofia 2 Flu + SARS antigen FIA test is a multiplex fluorescent immunoassay for the detection of and differentiating SARS-CoV-2, influenzas A and B [64]. The Sofia 2 reader is portable, but it is not battery powered. It is suitable for point-of-care settings, but it may not be fitting for a remote or limited resource setting. There appears to be a trend of LFA diagnostics being accompanied by readers and real-time connectivity [41–43], hence rapidly adapting and meeting the REASSURED criteria.

2.2. Nucleic Acids

Polymerase chain reaction (PCR) is the gold standard amplification method for molecular diagnostic assays for clinical use. PCR-based diagnostics assays are robust and can use crude samples, such as blood [65]. The key obstacle preventing PCR NATs from meeting all of the ASSURED criteria is that multiple temperatures are required for the amplification of target NAs. Device components that can perform thermal cycling are therefore necessary when developing a PCR-based diagnostic device. It is also challenging to develop multiplex PCR diagnostics. The existence of multiple primers for multiple targets increases the rate of formation of primer dimers, which then leads to non-specific amplification [66]. There is therefore a need for the stringent optimization of reaction conditions and parameters in order to achieve a multiplex PCR [67]. On the other hand, isothermal amplification, such as loop-mediated isothermal amplification (LAMP) and recombinase polymerase amplification (RPA), do not require thermal cycling [68,69]. The sensitivity of LAMP is not affected when the nucleic acid sample is impure and has other crude components, such as proteins and other cellular components [70]. However, a LAMP reaction requires four to six primers for each target, and hence poses a challenge when multiplexing due the occurrence of non-specific amplification [69,71].

The Accula dock developed by Mesa Biotech (now a part of Thermo Fisher Scientific (Waltham, MA, USA.)) is a portable sample-to-answer molecular diagnostic device that uses Mesa Biotech's proprietary PCR technology OSCillating amplification reaction (OSCAR) [72]. The Accula systems operates with a test cassette in which the multiplexed nucleic acid detection occurs. The Accula Flu A and Flu B is CLIA waived the multiplexed test for the detection of influenzas A and B, and the device has a 510K FDA clearance [73]. The disposable test cassette together with the dock are a portable system that checks nearly all the criteria for REASSURED diagnostics.

The Visby Medical Sexual Health (Figure 3A) developed by Visby Medical is a handheld device that is capable of a rapid multiplexed PCR for the detection of *Chlamydia trachomatis*, *Neisseria gonorrhoeae*, and *Trichomonas vaginalis* [74]. The Visby Medical Sexual Health device recently received CLIA waiver and FDA clearance. The device is a disposable sample-to-answer diagnostic, which makes it adaptable for point-of-care testing and in remote settings. Visby medical's diagnostic device can be adaptable to any form of multiplexed molecular diagnostic test, as the Visby Medical COVID-19 test has been granted Emergency Use Authorization (EUA) by the FDA for use by authorized labs [75].

Biomeme's Franklin three9 is a rechargeable battery-operated mobile thermocycler that is capable of conducting a multiplexed detection of nucleic acids and is adaptable to limited resource settings. It is capable of PCR, (Reverse Transcriptase) RT-PCR, (quantitative) qPCR and isothermal amplification. Franklin is not a sample-to-answer platform as it requires upstream steps sample preparation. However, the sample preparation steps can be achieved in about 1–2 min using Biomeme's M1 sample-prep cartridge kits. The Franklin system has Bluetooth and a wireless connection capability and is accompanied by an intuitive companion mobile app that facilitates wireless programing and managing of experiments. The Franklin three9 is capable of simultaneously testing nine samples with three targets each [76].

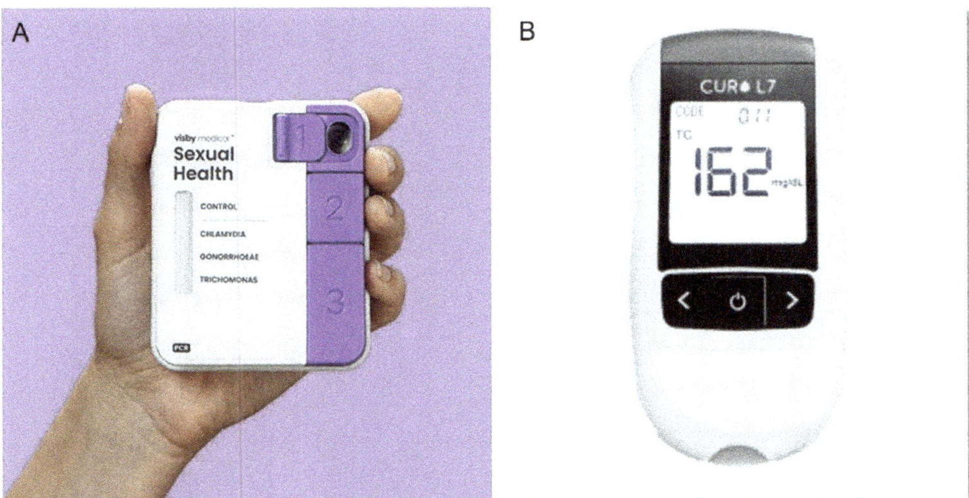

Figure 3. Clinical diagnostics devices scored on the REASSURED scoring scheme: (**A**) Visby Medical Sexual Health reproduced with permission from Visby Medical (https://www.visbymedical.com/resources/press-kit/ (accessed on 10 February 2022)). (**B**). Curo L7 reproduced with permission from Curofit (https://curofit.com/ (accessed on 10 February 2022)).

2.3. Small Molecules, Lipids, and Other Biomarkers

CardioChek PA Analyzer by PTS Diagnostics is a portable handheld diagnostic device that is battery operated. It works in conjunction with panels test strips to measure single and multiplex analytes. The CardioChek PA analyzer and test strips can measure total cholesterol, high density lipoproteins, triglycerides and glucose and provide results in 45 to 90 s. The test strips are stable at room temperature [77].

Curofit's Curo L7 m (Figure 3B) is capable of multiplex runs with up to six simultaneous tests with a cholesterol test strip. The device is handheld and battery-powered and is able to deliver results directly from sample. The Curo L7 m is suitable for point-of-care and low resource settings [78].

3. Multiplexed Diagnostics in Research or Academia

3.1. Proteins and Peptides

There are many multiplex immunoassays (MIAs) under development and only a few have been commercialized [79]. Chen et al. [80] demonstrated the use of a smartphone camera for reading ELISA-on-a-chip assays (Figure 4C). Berg et al. [59] published a cellphone-based hand-held microplate reader (Figure 4A) that used optical fibers to transmit data from ELISA plates to a cell-phone camera for diagnostics at the point of care. Mobile-phone-based ELISA (MELISA) is a portable system published by Zhdanov et al. [60] (Figure 4D). It is a miniature version of ELISA that is capable of executing all ELISA steps as well as providing a phone-based read-out of the results. The MELISA system has multiple reaction wells and has the potential to developed into a multiplexed system. According to the publishers, the total assembly of the MELISA system cost about USD 35. The system does not require any complex instrumentation; however, it uses plasma and hence requires an upstream sample preparation step. Ghosh et al. [81] described a microchannel capillary flow assay that detected malaria by a smartphone-assisted chemiluminescence-based ELISA. Perhaps, mobile phone-based ELISA platforms are the future direction for REASSURED diagnostics for protein and peptide biomarker detection.

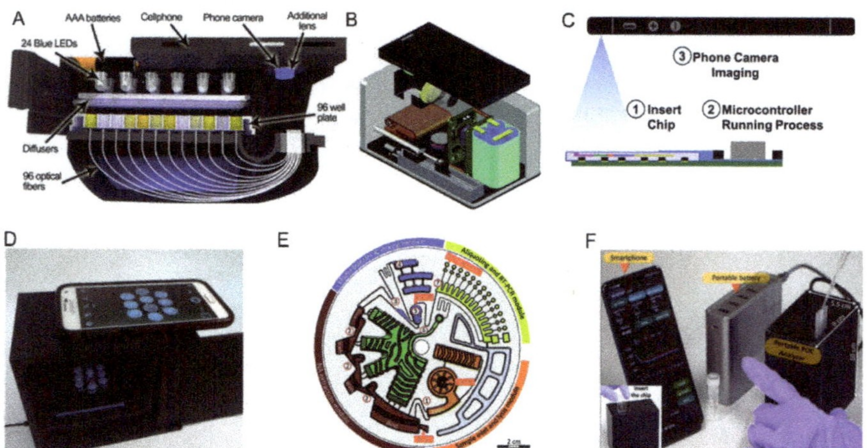

Figure 4. Multiplex diagnostics systems in *research and academia*. (**A**) Smartphone ELISA plate reader system reproduced from [56]. (**B**) FGAS system reproduced from [76] with permission from the Royal Society of Chemistry. (**C**) Smartphone-based ELISA-on-a-chip reproduced from [74], with the permission of AIP Publishing. (**D**) MELISA platform reproduced from [57] with permission from the Biosensors and Bioelectronics. (**E**) RespiDisk system reproduced from [77]. (**F**) IoT-based diagnostic system reproduced [78] with permission from the Biosensors and Bioelectronics.

3.2. Nucleic Acids

Shu et al. [82] proposed rapid multiplexed molecular diagnostic system dubbed flow genetic analysis system (FGAS) that is capable of conducting quantitative detection of nucleic acids (Figure 4B). FGAS is portable and battery powered, making it suitable for low resource settings. It is coupled with a smartphone, which is used for fluorescent imaging. RespiDisk (Figure 4E) is a fully automated multiplex molecular diagnostic device for respiratory tract infections [83]. The platform is based on RT-PCR and capable of automated sample-to-answer analysis, with a turnaround time of 3 h and 20 min. The RespiDisk system operates by centrifugal microfluidics. An Internet of things (IoT)-based diagnostic device is presented by Nguyen et al. [84] (Figure 4F). This platform is accompanied by an integrated microfluidic chip that is capable of running a multiplexed reverse-transcriptase LAMP (RT-LAMP) reaction. In addition, this battery-powered portable device has optical detection capability and was able to accurately detect SARS-CoV-2 from clinical samples in 33 min. The advanced IoT based device can be operated with a smartphone and provides real-time data to the user. It is capable of sample-to-answer analysis and hence there are only few user steps. Carter et al. [85] presented a multiplex lateral flow microarray platform for the detection nucleic acids. This platform combined the desirable qualities of an isothermal nucleic acid test (high sensitivity, high specificity, and no thermal cycling) with the best qualities LFAs (inexpensive, rapid, and equipment-free).

4. Next Generation Multiplex Diagnostics

The development of microfluidics and nanofluidics has inspired the emergence of several miniaturized platforms, such as lab-on-a-chip and lab-on-a-disk. These platforms present the capabilities of molecular-scale sensitivity on low-cost and rapidly fabricated devices [86–88]. However, the adoption of these platforms into clinical diagnostics are yet to be realized. Yeh et al. [89] presented a microfluidic chip called SIMPLE (Self-powered Integrated Microfluidic Point-of-care Low-cost Enabling). The SIMPLE chip is portable and completely integrated, allowing the accurate quantitative detection of nucleic acids from whole blood in 30 min. The emergence of microfluidic technologies propelled the development of digital PCR (dPCR). dPCR offers advantages, such as excellent precision [90], single copy detection, high sensitivity and absolute quantification [91]. Droplet microfluidics [92–94] and microarray [95,96] are some of the techniques used to achieve multiplexing by dPCR. While not able to meet all REASSURED criteria, some dPCR techniques show potential by using a mobile phone for detection and using simple fluid handling methods [97,98]. While very promising, the development and commercialization of microfluidic platforms are hindered by setbacks, such as the high cost and complexity of manufacturing on large scale, and challenges of integration from sample to answer [99,100].

In recent years, a number of studies are migrating towards the application of CRISPR/Cas systems for multiplex molecular diagnostics [101–104]. Gootenberg et al. [102] presents SHERLOCKv2, a multiplex platform for nucleic acid detection with high sensitivity and specificity and is integrated with a lateral flow read out. This presents the potential for SHERLOCKv2 to be developed into a multiplex and portable platform for diagnostics. Recently, Ackerman et al. [105] proposed a high throughput multiplex nucleic acid detection microarray system called CARMEN-Cas13. The high sensitivity and specificity of CARMEN combined with its incredibly high throughput, endows it with the potential of being the ultimate point of care diagnostic device when integrated with upstream sample preparation and concentration steps. Rezaei et al. [106] recently developed a portable device for the screening of SARS-CoV-2 by RT-LAMP and followed by CRISPR/Cas12a reaction and FAM-biotin system to give a fluorescent readout in a LFA. The device is semiautomated and battery operated. It has the potential for multiplexing and is able to produce results in about an hour. Yi et al. presented a similar system termed CRICOLAP for the detection of SARS-CoV-2 and also employs an amplification step by RT-LAMP, which is followed by a CRISPR/Cas12a collateral cleavage system for target recognition [107]. The paper reports a real-time parallel fluorescent readout system.

In the current digital age, next-generation diagnostics are combined with machine learning capabilities for high throughput and highly accurate results. Ballard et al. [108] demonstrated a multiplexed paper-based Vertical Flow Assay (VFA) platform that used a deep learning-based framework for sensing and quantifying high sensitivity C-Reactive Protein. This platform represents a low-cost device that can be adapted for molecular diagnostics at the POC and low resource settings. Machine-learning-assisted dPCR has also improved diagnostic outcomes as demonstrated by Liu [109] and Miglietta [110].

5. Discussion

In the REASSURED scoring scheme (Table 1), LFAs with an in-built or a combined reader had low sensitivity and specificity scores compared to molecular diagnostics, but they had high overall scores. LFAs have been widely adopted for rapid diagnostics for decades and while they are more affordable and simpler to develop and/or use, they do not have good sensitivity and have low multiplex capacity. Most LFAs can only multiplex two or three types of biomarkers. The limitations to multiplexing capability of LFAs are due to technical and operational challenges, such cross-reactivity and selection of appropriate diluents [56,111]. Most proteins or peptides have unique charges and pH and hence, unique isoelectric points in different buffer conditions. There is therefore a challenge of selecting the appropriate buffer for the select protein and peptide biomarkers to multiplexed. In infectious diseases, acquired immune responses do not occur until several days after exposure, and the antibodies linger in the body for days after the pathogen has been cleared [112]. This makes it difficult for LFAs to distinguish between an active and inactive infection.

The reviewed molecular diagnostics demonstrated much higher multiplex capacity compared to the LFAs. Molecular diagnostics are easier to multiplex than LFAs because biomarker recognition is achieved through the highly specific complementary hybridization of primers and/or probes. The quest to bring molecular diagnostic devices to the point-of-care setting has led to an increased focus on the miniaturization of the test systems. A major challenge that is often encountered by the miniaturization of the molecular diagnostic test platforms is the integration of sample preparation steps. Sample preparation include steps for isolation, purification, and concentration of nucleic acids from crude samples, such as blood and saliva. While the execution of these steps increases the sensitivity and specificity of molecular diagnostics, they are a major driver in the cost and complexity of these devices. Molecular diagnostics that have in-built readers or connectivity to smartphones were completely integrated from sample to answer, and handheld and battery-powered devices generally scored the highest points on the multiplexed REASSURED scoring scheme.

Table 1. REASSURED scores of 9 clinically available multiplex diagnostics. The scoring was assigned on a 1 to 3 scale based on developed criteria (Supplementary Materials, Table S1). The total score was obtained by finding the average score across all elements of REASSURED and dividing by 3.

Test (Multiplex Capacity)	R	E	A	S	S	U	R	E	D	Score
Accula dock Flu A/ Flu B Test (2)	3	3	1	2	2	3	3	1	3	78%
Visby Medical Sexual Health (3)	3	3	-	3	3	3	3	3	3	100%
Franklin three9 COVID-19 (27)	3	3	3	3	3	1	3	3	3	93%
Binax Now Influenza A & B with DIGIVAL (2)	3	3	1	1	2	3	3	3	3	81%
BD Veritor™ Flu A + B with analyzer (2)	3	3	2	1	3	3	3	3	3	89%
Sofia® 2 Flu + SARS antigen FIA (3)	3	2	1	1	2	3	3	3	3	81%
Acucy influenza A and B (2)	3	3	2	1	3	3	3	3	3	89%
CardioChek PA Analyzer with CHOL + HDL + GLU Panel (3)	3	3	3	-	-	3	3	3	3	100%
CuroL7 (6)	3	3	3	-	-	3	3	3	3	100%

There is a need for technology that is highly accurate, but also is affordable and accessible, especially in the developing world. Such a technology will not only help to address the need for increased access to diagnostics, but also ensure endemic and pandemic preparedness for the future. More funds need to be allocated to the development of multiplexed REASSURED diagnostics through funding by research and academic institutions and the incentivizing of research and development efforts of industry.

Point-of-care diagnostics development should gravitate towards more syndromic test panels, such as respiratory infection panels, urinary tract infection panels, blood protein panel and STI panels. Multiplexed panel measurements rather than single panel measurements are important because they facilitate the efficient and effective diagnosis of syndromic infections, accurately indicate the correct antibiotic or treatment, and minimize the number of tests that need to be run to diagnose coinfections.

Novel technologies in development that meet the REASSURED criteria should be incentivized by governments and international organizations to bring them to the market. Gene Xpert Omni, unveiled by Cepheid in 2015 and dubbed as the world's most portable molecular diagnostic system, was predicted to decentralize and increase access to TB diagnosis [113,114]. However, the commercialization plans for the Gene Xpert Omni were aborted, and Cepheid has received petitions to reinstate the plan to commercialize the diagnostic system [115,116]. The development of the Cepheid's Gene Xpert systems was supported by the Foundation for Innovative New Diagnostics (FIND) and the National Institutes of Health (NIH), among other investors [117]. According to Gotham et al., FIND is currently evaluating the Gene Xpert Omni, and it is expected to be commercially available in 2022 [117]. Cost is still an issue, however, as the lowest cost of the GeneXpert instrument is USD 11,530 [118] and the per test cost averaged USD 21 [119].

An ideal diagnostic case for SARS-CoV-2/Flu A & B would be a test of $\leq$ USD 1 that can simultaneously detect and differentiate between SARS-CoV-2/Flu A & B RNA in 15 to 60 min with a sensitivity and specificity of >98%. This test would have ≤ 2 user steps, all reagents prepackaged within, be equipment-free (or operated by a simple, portable, and handheld device $\leq$USD 10), be made of environmentally friendly material, and disposable. Moreover, the device, test and its reagents would be stable at room temperature with a shelf-life of about a year. Finally, if a device is necessary beyond the disposable test itself, it would be battery or solar powered, and able to transmit results remotely or by USB connection to a mobile phone. While this ideal use case is for differentiating SARS-CoV-2 from Influenza A/B, a similar multiplexed and inexpensive test would help greatly with other infections, such as UTIs, blood infections, and diarrheal disease [120,121]. Cancer resistance genes identification, cardiovascular disease prognosis, cytokines profiling, and epigenetic modification profiling are other areas where multiplex detection of biomarkers will be invaluable [18,24,122–124].

Lateral flow assays meet the standards for affordability and accessibility, so improving their accuracy could be the answer. Molecular tests already have high accuracy, so a different approach would be adapting molecular tests into a REASSURED format and decreasing their cost/complexity. While there is currently no such diagnostic device, the rapid emergence of new technology, such as machine-learning-assisted diagnostics, CRISPR-based diagnostics and nanofluidic technology, places such ideals within reach with further research and innovation.

Supplementary Materials: The following supporting information can be downloaded at: https://www.mdpi.com/article/10.3390/bios12020124/s1, Table S1: Scoring scheme for assessing diagnostics on the REASSURED criteria. The scoring ranges from 3 to 1, 3 being the highest score and 1 being the lowest score; Table S2: Scoring scheme of clinical diagnostics on the REASSURED criteria [125]. Averages were calculated from the scores of the individual elements of the REASSURED criteria. The Overall score was calculated by expressing the average score as a percentage of 3, the highest achievable average score (Reference [125] is cited in the supplementary information under Table S2).

Author Contributions: Conceptualization, J.A.O.; methodology, J.A.O.; investigation, J.A.O.; resources, T.S.S.; data curation, J.A.O.; writing—original draft preparation, J.A.O.; writing—review and editing. J.A.O. and T.S.S.; visualization, J.A.O.; supervision, T.S.S.; funding acquisition, T.S.S. All authors have read and agreed to the published version of the manuscript.

Funding: This research and the APC was funded by the National Institute of Biomedical Imaging and Bioengineering of the National Institutes of Health under Award Number K01EB027718. The content is solely the responsibility of the authors and does not necessarily represent the official views of the National Institutes of Health.

Institutional Review Board Statement: Not applicable.

Informed Consent Statement: Not applicable.

Data Availability Statement: Not applicable.

Conflicts of Interest: The authors declare no conflict of interest. The funders had no role in the design of the study; in the collection, analyses, or interpretation of data; in the writing of the manuscript, or in the decision to publish the results.

References

1.   Patrinos, G.P.; Ansorge, W.; Danielson, P.B. (Eds.) *Molecular Diagnostics*, 3rd ed.; Elsevier/Academic Press: Amsterdam, The Netherlands; Boston, MA, USA, 2017.

2.   Mayeux, R. Biomarkers: Potential uses and limitations. *Neurotherapeutics* **2004**, *1*, 182–188. [CrossRef]

3.   Nimse, S.B.; Sonawane, M.D.; Song, K.-S.; Kim, T. Biomarker detection technologies and future directions. *Analyst* **2016**, *141*, 740–755. [CrossRef]

4. Chandler, P.D.; Song, Y.; Lin, J.; Zhang, S.; Sesso, H.D.; Mora, S.; Giovannucci, E.L.; Rexrode, K.E.; Moorthy, M.V.; Li, C.; et al. Lipid biomarkers and long-term risk of cancer in the Women's Health Study. *Am. J. Clin. Nutr.* **2016**, *103*, 1397–1407. [CrossRef]

5. The Lewin Group, Inc. The Value of Diagnostics Innovation, Adoption and Diffusion into Health Care. July 2005. Available online: https://www.lewin.com/content/dam/Lewin/Resources/Site_Sections/Publications/ValueofDiagnostics.pdf (accessed on 10 February 2022).

6. Mugambi, M.L.; Peter, T.; Martins, S.F.; Giachetti, C. How to implement new diagnostic products in low-resource settings: An end-to-end framework. *BMJ Glob. Health* **2018**, *3*, e000914. [CrossRef]

7. Roundtable on Translating Genomic-Based Research for Health and Board on Health Sciences Policy; Institute of Medicine. Genome-Based Diagnostics: Demonstrating Clinical Utility in Oncology: Workshop Summary. Washington (DC). National Academies Press (US). 27 December 2013. Available online: https://www.ncbi.nlm.nih.gov/books/NBK195902/ (accessed on 10 February 2022).

8. Kettler, H.; White, K.; Hawkes, S. Mapping the Landscape of Diagnostics for Sexually Transmitted Infections: Key Findings and Recommendations. World Health Organization. 2004. Available online: https://apps.who.int/iris/handle/10665/68990 (accessed on 10 February 2022).

9. Peeling, R.W.; Holmes, K.K.; Mabey, D.; Ronald, A. Rapid tests for sexually transmitted infections (STIs): The way forward. *Sex. Transm. Infect.* **2006**, *82* (Suppl. S5), v1–v6. [CrossRef]

10. Afzal, A.; Iqbal, N.; Feroz, S.; Ali, A.; Ehsan, M.A.; Khan, S.A.; Rehman, A. Rapid antibody diagnostics for SARS-CoV-2 adaptive immune response. *Anal. Methods* **2021**, *13*, 4019–4037. [CrossRef]

11. Zhou, Y.; Wu, Y.; Ding, L.; Huang, X.; Xiong, Y. Point-of-care COVID-19 diagnostics powered by lateral flow assay. *Trends Anal. Chem.* **2021**, *145*, 116452. [CrossRef]

12. Hogan, C.A.; Garamani, N.; Lee, A.S.; Tung, J.K.; Sahoo, M.K.; Huang, C.; Stevens, B.; Zehnder, J.; Pinsky, B.A. Comparison of the Accula SARS-CoV-2 Test with a Laboratory-Developed Assay for Detection of SARS-CoV-2 RNA in Clinical Nasopharyngeal Specimens. *J. Clin. Microbiol.* **2020**, *58*, e01072-20. [CrossRef]

13. Rhoads, D.D.; Cherian, S.S.; Roman, K.; Stempak, L.M.; Schmotzer, C.L.; Sadri, N. Comparison of Abbott ID Now, DiaSorin Simplexa, and CDC FDA Emergency Use Authorization Methods for the Detection of SARS-CoV-2 from Nasopharyngeal and Nasal Swabs from Individuals Diagnosed with COVID-19. *J. Clin. Microbiol.* **2020**, *58*, e00760-20. [CrossRef]

14. Trinidad, C.V.; Tetlow, A.L.; Bantis, L.E.; Godwin, A.K. Reducing ovarian cancer mortality through early detection: Approaches using circulating biomarkers. *Cancer Prev. Res.* **2020**, *13*, 241–252. [CrossRef]

15. McCormack, V.; Aggarwal, A. Early cancer diagnosis: Reaching targets across whole populations amidst setbacks. *Br. J. Cancer* **2021**, *124*, 1181–1182. [CrossRef] [PubMed]

16. Herman, W.H.; Ye, W.; Griffin, S.J.; Simmons, R.K.; Davies, M.J.; Khunti, K.; Rutten, G.E.; Sandbaek, A.; Lauritzen, T.; Borch-Johnsen, K.; et al. Early detection and treatment of type 2 diabetes reduce cardiovascular morbidity and mortality: A simulation of the results of the anglo-danish-dutch study of intensive treatment in people with screen-detected diabetes in primary care (ADDITION-Europe). *Diabetes Care* **2015**, *38*, 1449–1455. [CrossRef] [PubMed]

17. Capotosto, L.; Massoni, F.; de Sio, S.; Ricci, S.; Vitarelli, A. Early Diagnosis of Cardiovascular Diseases in Workers: Role of Standard and Advanced Echocardiography. *BioMed Res. Int.* **2018**, *2018*, e7354691. [CrossRef] [PubMed]

18. Adamcova, M.; Šimko, F. Multiplex biomarker approach to cardiovascular diseases. *Acta Pharm. Sin.* **2018**, *39*, 1068–1072. [CrossRef] [PubMed]

19. Witte, M.H.; Dellinger, M.T.; Papendieck, C.M.; Boccardo, F. Overlapping biomarkers, pathways, processes and syndromes in lymphatic development, growth and neoplasia. *Clin. Exp. Metastasis* **2012**, *29*, 707–727. [CrossRef]

20. Kumar, S.; Mohan, A.; Guleria, R. Biomarkers in cancer screening, research and detection: Present and future: A review. *Biomarkers* **2006**, *11*, 385–405. [CrossRef]

21. Hermann, N.; Dressen, K.; Schroeder, L.; Debald, M.; Schildberg, F.A.; Walgenbach-Bruenagel, G.; Hettwer, K.; Uhlig, S.; Kuhn, W.; Hartmann, G.; et al. Diagnostic relevance of a novel multiplex immunoassay panel in breast cancer. *Tumour Biol.* **2017**, *39*, 1010428317711381. [CrossRef]

22. Furman, D.; Davis, M.M. New approaches to understanding the immune response to vaccination and infection. *Vaccine* **2015**, *33*, 5271–5281. [CrossRef]

23. Young, H.A. Cytokine Multiplex Analysis. *Methods Mol. Biol.* **2009**, *511*, 85–105. [CrossRef]

24. Abdullah, M.A.A.; Amini, N.; Yang, L.; Paluh, J.L.; Wang, J. Multiplexed analysis of neural cytokine signaling by a novel neural cell-cell interaction microchip. *Lab Chip* **2020**, *20*, 3980–3995. [CrossRef]

25. Tang, J.W.; Tambyah, P.A.; Hui, D.S.C. Emergence of a novel coronavirus causing respiratory illness from Wuhan, China. *J. Infect.* **2020**, *80*, 350–371. [CrossRef] [PubMed]

26. Khorramdelazad, H.; Kazemi, M.H.; Najafi, A.; Keykhaee, M.; Emameh, R.Z.; Falak, R. Immunopathological similarities between COVID-19 and influenza: Investigating the consequences of Co-infection. *Microb. Pathog.* **2021**, *152*, 104554. [CrossRef]

27. Dadashi, M.; Khaleghnejad, S.; Abedi Elkhichi, P.; Goudarzi, M.; Goudarzi, H.; Taghavi, A.; Vaezjalali, M.; Hajikhani, B. COVID-19 and Influenza Co-infection: A Systematic Review and Meta-Analysis. *Front. Med.* **2021**, *8*, 681469. [CrossRef] [PubMed]

28. Akhtar, Z.; Islam, M.A.; Aleem, M.A.; Mah-E-Muneer, S.; Ahmmed, M.K.; Ghosh, P.K.; Rahman, M.; Rahman, M.Z.; Sumiya, M.K.; Rahman, M.M.; et al. SARS-CoV-2 and influenza virus coinfection among patients with severe acute respiratory infection during the first wave of COVID-19 pandemic in Bangladesh: A hospital-based descriptive study. *BMJ Open* **2021**, *11*, e053768. [CrossRef] [PubMed]

29. Sanders, E.J.; Wahome, E.; Mwangome, M.; Thiong'o, A.N.; Okuku, H.S.; Price, M.A.; Wamuyu, L.; Macharia, M.; McClelland, R.S.; Graham, S.M. Most adults seek urgent healthcare when acquiring HIV-1 and are frequently treated for malaria in coastal Kenya. *AIDS* **2011**, *25*, 1219–1224. [CrossRef] [PubMed]

30. Multani, A.; Allard, L.S.; Wangjam, T.; Sica, R.A.; Epstein, D.J.; Rezvani, A.R.; Ho, D.Y. Missed diagnosis and misdiagnosis of infectious diseases in hematopoietic cell transplant recipients: An autopsy study. *Blood Adv.* **2019**, *3*, 3602–3612. [CrossRef]

31. Levin-Reisman, I.; Brauner, A.; Ronin, I.; Balaban, N.Q. Epistasis between antibiotic tolerance, persistence, and resistance mutations. *PNAS* **2019**, *116*, 14734–14739. [CrossRef]

32. Hutchings, M.I.; Truman, A.W.; Wilkinson, B. Antibiotics: Past, present and future. *Curr. Opin. Microbiol.* **2019**, *51*, 72–80. [CrossRef]

33. O'Neill, J. Tackling Drug-Resistant Infections Globally: Final Report and Recommendations. May 2016. Available online: https://amr-review.org/sites/default/files/160525_Final%20paper_with%20cover.pdf (accessed on 10 February 2022).

34. Nanayakkara, A.K.; Boucher, H.W.; Fowler, V.G., Jr.; Jezek, A.; Outterson, K.; Greenberg, D.E. Antibiotic resistance in the patient with cancer: Escalating challenges and paths forward. *A Cancer J. Clin.* **2021**, *71*, 488–504. [CrossRef]

35. Urdea, M.; Penny, L.A.; Olmsted, S.S.; Giovanni, M.Y.; Kaspar, P.; Shepherd, A.; Wilson, P.; Dahl, C.A.; Buchsbaum, S.; Moeller, G.; et al. Requirements for high impact diagnostics in the developing world. *Nature* **2006**, *444*, 73–79. [CrossRef]

36. Kumar, A.; Roberts, D.; Wood, K.E.; Light, B.; Parrillo, J.E.; Sharma, S.; Suppes, R.; Feinstein, D.; Zanotti, S.; Taiberg, L.; et al. Duration of hypotension before initiation of effective antimicrobial therapy is the critical determinant of survival in human septic shock. *Crit. Care Med.* **2006**, *34*, 1589–1596. [CrossRef] [PubMed]

37. Westphal, G.A.; Koenig, Á.; Caldeira Filho, M.; Feijó, J.; de Oliveira, L.T.; Nunes, F.; Fujiwara, K.; Martins, S.F.; Gonçalves, A.R.R. Reduced mortality after the implementation of a protocol for the early detection of severe sepsis. *J. Crit. Care* **2011**, *26*, 76–81. [CrossRef] [PubMed]

38. Wiwanitkit, V.; Err, H. "Syndromic approach" to diagnosis and treatment of critical tropical infections. *Indian J. Crit. Care Med.* **2014**, *18*, 479. [CrossRef]

39. Djomand, G.; Gao, H.; Singa, B.; Hornston, S.; Bennett, E.; Odek, J.; McClelland, R.S.; John-Stewart, G.; Bock, N. Genital infections and syndromic diagnosis among HIV-infected women in HIV care programmes in Kenya. *Int. J. STD AIDS* **2016**, *27*, 19–24. [CrossRef] [PubMed]

40. Land, K.J.; Boeras, D.I.; Chen, X.-S.; Ramsay, A.R.; Peeling, R.W. REASSURED diagnostics to inform disease control strategies, strengthen health systems and improve patient outcomes. *Nat. Microbiol.* **2019**, *4*, 46–54. [CrossRef] [PubMed]

41. Mahmoudi, T.; Pourhassan-Moghaddam, M.; Shirdel, B.; Baradaran, B.; Morales-Narváez, E.; Golmohammadi, H. On-Site Detection of Carcinoembryonic Antigen in Human Serum. *Biosensors* **2021**, *11*, 392. [CrossRef] [PubMed]

42. Ruppert, C.; Phogat, N.; Laufer, S.; Kohl, M.; Deigner, H.-P. A smartphone readout system for gold nanoparticle-based lateral flow assays: Application to monitoring of digoxigenin. *Microchim. Acta* **2019**, *186*, 119. [CrossRef]

43. Saisin, L.; Amarit, R.; Somboonkaew, A.; Gajanandana, O.; Himananto, O.; Sutapun, B. Significant Sensitivity Improvement for Camera-Based Lateral Flow Immunoassay Readers. *Sensors* **2018**, *18*, 4026. [CrossRef]

44. Scott, L.J. Verigene®gram-positive blood culture nucleic acid test: An in vitro diagnostic assay for identification of gram-positive bacteria associated with bloodstream infections and bacterial resistance markers. *Mol. Diagn. Ther.* **2013**, *17*, 117–122. [CrossRef]

45. Xie, C.; Jiang, L.; Huang, G.; Pu, H.; Gong, B.; Lin, H.; Ma, S.; Chen, X.; Long, B.; Si, G.; et al. Comparison of different samples for 2019 novel coronavirus detection by nucleic acid amplification tests. *Int. J. Infect. Dis.* **2020**, *93*, 264–267. [CrossRef]

46. Nguyen, H.V.; Nguyen, V.D.; Nguyen, H.Q.; Chau, T.H.T.; Lee, E.Y.; Seo, T.S. Nucleic acid diagnostics on the total integrated lab-on-a-disc for point-of-care testing. *Biosens. Bioelectron.* **2019**, *141*, 111466. [CrossRef]

47. di Nardo, F.; Chiarello, M.; Cavalera, S.; Baggiani, C.; Anfossi, L. Ten years of lateral flow immunoassay technique applications: Trends, challenges and future perspectives. *Sensors* **2021**, *21*, 5185. [CrossRef]

48. Naseri, M.; Ziora, Z.M.; Simon, G.P.; Batchelor, W. ASSURED-compliant point-of-care diagnostics for the detection of human viral infections. *Rev. Med. Virol.* **2021**. [CrossRef]

49. Dincer, C.; Bruch, R.; Kling, A.; Dittrich, P.S.; Urban, G.A. Multiplexed Point-of-Care Testing—xPOCT. *Trends Biotechnol.* **2017**, *35*, 728–742. [CrossRef]

50. Kim, H.; Huh, H.J.; Park, E.; Chung, D.-R.; Kang, M. Multiplex Molecular Point-of-Care Test for Syndromic Infectious Diseases. *BioChip J.* **2021**, *15*, 14–22. [CrossRef]

51. Rusling, J.F. Multiplexed Electrochemical Protein Detection and Translation to Personalized Cancer Diagnostics. *Anal. Chem.* **2013**, *85*, 5304–5310. [CrossRef]

52. Felix, A.C.; Souza, N.C.S.; Figueiredo, W.M.; Costa, A.A.; Inenami, M.; da Silva, R.M.; Levi, J.E.; Pannuti, C.S.; Romano, C.M. Cross reactivity of commercial anti-dengue immunoassays in patients with acute Zika virus infection. *J. Med. Virol.* **2017**, *89*, 1477–1479. [CrossRef]

53. Deng, J.; Yang, M.; Wu, J.; Zhang, W.; Jiang, X. A self-contained chemiluminescent lateral flow assay for point-of-care testing. *Anal. Chem.* **2018**, *90*, 9132–9137. [CrossRef]

54. Park, J. An optimized colorimetric readout method for lateral flow immunoassays. *Sensors* **2018**, *18*, 4084. [CrossRef]
55. Montesinos, I.; Gruson, D.; Kabamba, B.; Dahma, H.; Van den Wijngaert, S.; Reza, S.; Carbone, V.; Vandenberg, O.; Gulbis, B.; Wolff, F.; et al. Evaluation of two automated and three rapid lateral flow immunoassays for the detection of anti-SARS-CoV-2 antibodies. *J. Clin. Virol.* **2020**, *128*, 104413. [CrossRef]
56. Koczula, K.M.; Gallotta, A. Lateral flow assays. *Essays Biochem.* **2016**, *60*, 111–120. [CrossRef] [PubMed]
57. Hanafiah, K.M.; Arifin, N.; Bustami, Y.; Noordin, R.; Garcia, M.; Anderson, D. Development of multiplexed infectious disease lateral flow assays: Challenges and opportunities. *Diagnostics* **2017**, *7*, 51. [CrossRef]
58. Souf, S. Recent advances in diagnostic testing for viral infections. *Biosci. Horiz.* **2016**, *9*, hzw010. [CrossRef]
59. Berg, B.; Cortazar, B.; Tseng, D.; Ozkan, H.; Feng, S.; Wei, Q.; Chan, R.Y.L.; Burbano, J.; Farooqui, Q.; Lewinski, M.; et al. Cellphone-Based Hand-Held Microplate Reader for Point-of-Care Testing of Enzyme-Linked Immunosorbent Assays. *ACS Nano* **2015**, *9*, 7857–7866. [CrossRef] [PubMed]
60. Zhdanov, A.; Keefe, J.; Franco-Waite, L.; Konnaiyan, K.R.; Pyayt, A. Mobile phone based ELISA (MELISA). *Biosens. Bioelectron.* **2018**, *103*, 138–142. [CrossRef]
61. BinaxNowTM Influenza A&B Card 2. Available online: https://www.globalpointofcare.abbott/en/product-details/binaxnow-influenza-a-and-b-2.html (accessed on 5 January 2022).
62. BD VeritorTM Plus System for Flu A + B. Available online: https://bdveritor.bd.com/en-us/main/rapid-antigen-testing/flu-a-b (accessed on 5 January 2022).
63. Acucy®Influenza A&B. Available online: https://sekisuidiagnostics.com/products-all/acucy-influenza-ab/ (accessed on 5 January 2022).
64. Sofia SARS Antigen FIA. Available online: https://www.quidel.com/immunoassays/rapid-sars-tests/sofia-sars-antigen-fia (accessed on 5 January 2022).
65. Kermekchiev, M.B.; Kirilova, L.I.; Vail, E.E.; Barnes, W.M. Mutants of Taq DNA polymerase resistant to PCR inhibitors allow DNA amplification from whole blood and crude soil samples. *Nucleic Acids Res.* **2009**, *37*, e40. [CrossRef] [PubMed]
66. Elnifro, E.M.; Ashshi, A.M.; Cooper, R.J.; Klapper, P.E. Multiplex PCR: Optimization and application in diagnostic virology. *Clin. Microbiol. Rev.* **2000**, *13*, 559–570. [CrossRef]
67. Henegariu, O.; Heerema, N.A.; Dlouhy, S.R.; Vance, G.H.; Vogt, P.H. Multiplex PCR: Critical parameters and step-by-step protocol. *Biotechniques* **1997**, *23*, 504–511. [CrossRef]
68. Daher, R.K.; Stewart, G.; Boissinot, M.; Bergeron, M.G. Recombinase polymerase amplification for diagnostic applications. *Clin. Chem.* **2016**, *62*, 947–958. [CrossRef]
69. Notomi, T. Loop-mediated isothermal amplification of DNA. *Nucleic Acids Res.* **2000**, *28*, e63. [CrossRef]
70. Francois, P.; Tangomo, M.; Hibbs, J.; Bonetti, E.J.; Boehme, C.C.; Notomi, T.; Perkins, M.D.; Schrenzel, J. Robustness of a loop-mediated isothermal amplification reaction for diagnostic applications. *FEMS Immunol. Med. Microbiol.* **2011**, *62*, 41–48. [CrossRef]
71. Moonga, L.C.; Hayashida, K.; Kawai, N.; Nakao, R.; Sugimoto, C.; Namangala, B.; Yamagishi, J. Development of a Multiplex Loop-Mediated Isothermal Amplification (LAMP) Method for Simultaneous Detection of Spotted Fever Group Rickettsiae and Malaria Parasites by Dipstick DNA Chromatography. *Diagnostics* **2020**, *10*, 897. [CrossRef] [PubMed]
72. Actionable. Accessible. Affordable. SARS-CoV-2 (COVID-19) Rapid PCR Testing. Available online: https://www.mesabiotech.com/ (accessed on 5 January 2022).
73. 510(k) Substantial Equivalence Determinatin Decision Summary: Accula Flu A/Flu B Test. U.S. Food and Drug Administration. Available online: https://www.accessdata.fda.gov/cdrh_docs/reviews/K171641.pdf (accessed on 5 January 2022).
74. Visby Medical Sexual Health Test. Available online: https://www.visbymedical.com/sexual-health-test/ (accessed on 5 January 2022).
75. Visby Medical COVID-19 Test Fast, Accurate, and Actionable PC. Available online: https://www.visbymedical.com/covid-19-test/ (accessed on 5 January 2022).
76. Mobile qPCR Thermocyclers. Available online: https://info.biomeme.com/mobile-qpcr-thermocyclers (accessed on 5 January 2022).
77. CardioChek PA Analyzer. Available online: https://ptsdiagnostics.com/cardiochek-pa-analyzer/ (accessed on 5 January 2022).
78. Curo L7. Available online: https://curofit.com/s/lipidocare-l7-cholesterol-test-kit/ (accessed on 5 January 2022).
79. Vashist, S.K.; Luppa, P.B.; Yeo, L.Y.; Ozcan, A.; Luong, J.H.T. Emerging technologies for next-generation point-of-care testing. *Trends Biotechnol.* **2015**, *33*, 692–705. [CrossRef] [PubMed]
80. Chen, A.; Wang, R.; Bever, C.R.S.; Xing, S.; Hammock, B.D.; Pan, T. Smartphone-interfaced lab-on-a-chip devices for field-deployable enzyme-linked immunosorbent assay. *Biomicrofluidics* **2014**, *8*, 064101. [CrossRef] [PubMed]
81. Ghosh, S.; Aggarwal, K.; Vinitha, T.U.; Nguyen, T.; Han, J.; Ahn, C.H. A new microchannel capillary flow assay (MCFA) platform with lyophilized chemiluminescence reagents for a smartphone-based POCT detecting malaria. *Microsyst. Nanoeng.* **2020**, *6*, 5. [CrossRef]
82. Shu, B.; Zhang, C.; Xing, D. A handheld flow genetic analysis system (FGAS): Towards rapid, sensitive, quantitative and multiplex molecular diagnosis at the point-of-care level. *Lab Chip* **2015**, *15*, 2597–2605. [CrossRef] [PubMed]
83. Rombach, M.; Hin, S.; Specht, M.; Johannsen, B.; Lüddecke, J.; Paust, N.; Zengerle, R.; Roux, L.; Sutcliffe, T.; Peham, J.R.; et al. RespiDisk: A point-of-care platform for fully automated detection of respiratory tract infection pathogens in clinical samples. *Analyst* **2020**, *145*, 7040–7047. [CrossRef] [PubMed]

84. Nguyen, H.Q.; Bui, H.K.; Phan, V.M.; Seo, T.S. An internet of things-based point-of-care device for direct reverse-transcription-loop mediated isothermal amplification to identify SARS-CoV-2. *Biosens. Bioelectron.* **2022**, *195*, 113655. [CrossRef]

85. Carter, D.J.; Cary, R.B. Lateral flow microarrays: A novel platform for rapid nucleic acid detection based on miniaturized lateral flow chromatography. *Nucleic Acids Res.* **2007**, *35*, e74. [CrossRef]

86. Lin, Y.-L.; Huang, Y.-J.; Teerapanich, P.; Leïchlé, T.; Chou, C.-F. Multiplexed immunosensing and kinetics monitoring in nanofluidic devices with highly enhanced target capture efficiency. *Biomicrofluidics* **2016**, *10*, 034114. [CrossRef]

87. Ye, X.; Li, Y.; Wang, L.; Fang, X.; Kong, J. All-in-one microfluidic nucleic acid diagnosis system for multiplex detection of sexually transmitted pathogens directly from genitourinary secretions. *Talanta* **2021**, *221*, 121462. [CrossRef] [PubMed]

88. Mortelmans, T.; Kazazis, D.; Padeste, C.; Berger, P.; Li, X.; Ekinci, Y. A nanofluidic device for rapid and multiplexed SARS-CoV-2 serological antibody detection. *Res. Square* **2021**. [CrossRef]

89. Yeh, E.-C.; Fu, C.-C.; Hu, L.; Thakur, R.; Feng, J.; Lee, L.P. Self-powered integrated microfluidic point-of-care low-cost enabling (SIMPLE) chip. *Sci. Adv.* **2017**, *3*, e1501645. [CrossRef] [PubMed]

90. Kuypers, J.; Jerome, K.R. Applications of Digital PCR for Clinical Microbiology. *J. Clin. Microbiol.* **2017**, *55*, 1621–1628. [CrossRef]

91. Mao, X.; Liu, C.; Tong, H.; Chen, Y.; Liu, K. Principles of digital PCR and its applications in current obstetrical and gynecological diseases. *Am. J. Transl. Res.* **2019**, *11*, 7209–7222.

92. Badran, S.; Chen, M.; Coia, J.E. Multiplex Droplet Digital Polymerase Chain Reaction Assay for Rapid Molecular Detection of Pathogens in Patients With Sepsis: Protocol for an Assay Development Study. *JMIR Res. Protoc.* **2021**, *10*, e33746. [CrossRef]

93. Wouters, Y.; Dalloyaux, D.; Christenhusz, A.; Roelofs, H.M.; Wertheim, H.F.; Bleeker-Rovers, C.P.; Te Morsche, R.H.; Wanten, G.J. Droplet digital polymerase chain reaction for rapid broad-spectrum detection of bloodstream infections. *Microb. Biotechnol.* **2020**, *13*, 657–668. [CrossRef]

94. Chen, B.; Jiang, Y.; Cao, X.; Liu, C.; Zhang, N.; Shi, D. Droplet digital PCR as an emerging tool in detecting pathogens nucleic acids in infectious diseases. *Clin. Chim. Acta* **2021**, *517*, 156–161. [CrossRef]

95. Gheit, T.; Landi, S.; Gemignani, F.; Snijders, P.J.; Vaccarella, S.; Franceschi, S.; Canzian, F.; Tommasino, M. Development of a sensitive and specific assay combining multiplex PCR and DNA microarray primer extension to detect high-risk mucosal human papillomavirus types. *J. Clin. Microbiol* **2006**, *44*, 2025–2031. [CrossRef]

96. Tian, M.; Tian, Y.; Li, Y.; Lu, H.; Li, X.; Li, C.; Xue, F.; Jin, N. Microarray Multiplex Assay for the Simultaneous Detection and Discrimination of Influenza A and Influenza B Viruses. *Indian J. Microbiol.* **2014**, *54*, 211–217. [CrossRef]

97. Selck, D.A.; Karymov, M.A.; Sun, B.; Ismagilov, R.F. Increased Robustness of Single-Molecule Counting with Microfluidics, Digital Isothermal Amplification, and a Mobile Phone versus Real-Time Kinetic Measurements. *Anal. Chem.* **2013**, *85*, 11129–11136. [CrossRef] [PubMed]

98. Shen, F.; Sun, B.; Kreutz, J.E.; Davydova, E.K.; Du, W.; Reddy, P.L.; Joseph, L.J.; Ismagilov, R.F. Multiplexed Quantification of Nucleic Acids with Large Dynamic Range Using Multivolume Digital RT-PCR on a Rotational SlipChip Tested with HIV and Hepatitis C Viral Load. *J. Am. Chem. Soc.* **2011**, *133*, 17705–17712. [CrossRef] [PubMed]

99. Ghosh, R.; Gopalakrishnan, S.; Savitha, R.; Renganathan, T.; Pushpavanam, S. Fabrication of laser printed microfluidic paper-based analytical devices (LP-µPADs) for point-of-care applications. *Sci. Rep.* **2019**, *9*, 7896. [CrossRef] [PubMed]

100. Abafogi, A.T.; Kim, J.; Lee, J.; Mohammed, M.O.; van Noort, D.; Park, S. 3D-Printed Modular Microfluidic Device Enabling Preconcentrating Bacteria and Purifying Bacterial DNA in Blood for Improving the Sensitivity of Molecular Diagnostics. *Sensors* **2020**, *20*, 1202. [CrossRef]

101. Agrawal, S.; Fanton, A.; Chandrasekaran, S.S.; Charrez, B.; Escajeda, A.M.; Son, S.; Mcintosh, R.; Bhuiya, A.; de León Derby, M.D.; Switz, N.A.; et al. Rapid, Point-of-Care Molecular Diagnostics with Cas13. 4 April 2021. Available online: https://www.medrxiv.org/content/10.1101/2020.12.14.20247874v2 (accessed on 16 December 2021).

102. Gootenberg, J.S.; Abudayyeh, O.O.; Kellner, M.J.; Joung, J.; Collins, J.J.; Zhang, F. Multiplexed and portable nucleic acid detection platform with Cas13, Cas12a, and Csm6. *Science* **2018**, *360*, 439–444. [CrossRef] [PubMed]

103. Quan, J.; Langelier, C.; Kuchta, A.; Batson, J.; Teyssier, N.; Lyden, A.; Caldera, S.; McGeever, A.; Dimitrov, B.; King, R.; et al. FLASH: A next-generation CRISPR diagnostic for multiplexed detection of antimicrobial resistance sequences. *Nucleic Acids Res.* **2019**, *47*, e83. [CrossRef]

104. Dhar, B.C.; Steimberg, N.; Mazzoleni, G. Point-of-Care Pathogen Detection with CRISPR-based Programmable Nucleic Acid Binding Proteins. *ChemMedChem* **2021**, *16*, 1566–1575. [CrossRef]

105. Ackerman, C.M.; Myhrvold, C.; Thakku, S.G.; Freije, C.A.; Metsky, H.C.; Yang, D.K.; Ye, S.H.; Boehm, C.K.; Kosoko-Thoroddsen, T.S.F.; Kehe, J.; et al. Massively multiplexed nucleic acid detection with Cas13. *Nature* **2020**, *582*, 277–282. [CrossRef]

106. Rezaei, M.; Razavi Bazaz, S.; Morshedi Rad, D.; Shimoni, O.; Jin, D.; Rawlinson, W.; Ebrahimi Warkiani, M. A Portable RT-LAMP/CRISPR Machine for Rapid COVID-19 Screening. *Biosensors* **2021**, *11*, 369. [CrossRef]

107. Yi, Z.; de Dieu Habimana, J.; Mukama, O.; Li, Z.; Odiwuor, N.; Jing, H.; Nie, C.; Hu, M.; Lin, Z.; Wei, H.; et al. Rational Programming of Cas12a for Early-Stage Detection of COVID-19 by Lateral Flow Assay and Portable Real-Time Fluorescence Readout Facilities. *Biosensors* **2021**, *12*, 11. [CrossRef]

108. Ballard, Z.S.; Joung, H.A.; Goncharov, A.; Liang, J.; Nugroho, K.; Di Carlo, D.; Garner, O.B.; Ozcan, A. Deep learning-enabled point-of-care sensing using multiplexed paper-based sensors. *NPJ Digit. Med.* **2020**, *3*, 66. [CrossRef]

109. Liu, C.; Li, B.; Lin, H.; Yang, C.; Guo, J.; Cui, B.; Pan, W.; Feng, J.; Luo, T.; Chu, F.; et al. Multiplexed analysis of small extracellular vesicle-derived mRNAs by droplet digital PCR and machine learning improves breast cancer diagnosis. *Biosens. Bioelectron.* **2021**, *194*, 113615. [CrossRef] [PubMed]
110. Miglietta, L.; Moniri, A.; Pennisi, I.; Malpartida-Cardenas, K.; Abbas, H.; Hill-Cawthorne, K.; Bolt, F.; Jauneikaite, E.; Davies, F.; Holmes, A.; et al. Coupling Machine Learning and High Throughput Multiplex Digital PCR Enables Accurate Detection of Carbapenem-Resistant Genes in Clinical Isolates. *Front. Mol. Biosci.* **2021**, *8*, 775299. [CrossRef] [PubMed]
111. Ellington, A.A.; Kullo, I.J.; Bailey, K.R.; Klee, G.G. Antibody-Based Protein Multiplex Platforms: Technical and Operational Challenges. *Clin. Chem.* **2010**, *56*, 186–193. [CrossRef]
112. Institute of Medicine (US) Forum on Microbial Threats and Global Infectious Disease Surveillance and Detection. In *Assessing the Challenges—Finding Solutions, Workshop Summary*; National Academies Press: Washington, DC, USA, 2007. Available online: https://www.ncbi.nlm.nih.gov/books/NBK52875/ (accessed on 20 December 2021).
113. World's Most Portable Molecular Diagnostics System Unveiled at AACC. 28 July 2015. Available online: https://www.tbonline.info/posts/2015/7/28/worlds-most-portable-molecular-diagnostics-system-/ (accessed on 6 January 2022).
114. Cepheid Targets Development of a Point of Care HIV Viral Load Test From a Few Drops of Blood. 8 September 2016. Available online: https://www.prnewswire.com/news-releases/cepheid-targets-development-of-a-point-of-care-hiv-viral-load-test-from-a-few-drops-of-blood-300324397.html (accessed on 6 January 2022).
115. Advocates Urge Cepheid to Reinstate Plans to Commercialize GeneXpert Omni. 14 October 2021. Available online: https://www.tbonline.info/posts/2021/10/14/advocates-urge-cepheid-reinstate-plans-commerciali/ (accessed on 6 January 2022).
116. Branigan, D, Time for $5 Coalition Urges Cepheid to Reinstate Plans to Launch GeneXpert Omni. 14 October 2021. Available online: https://www.tbonline.info/media/uploads/documents/time_for_$5_coalition_open_letter_to_cepheid_14oct2021.pdf (accessed on 6 January 2022).
117. Gotham, D.; McKenna, L.; Deborggraeve, S.; Madoori, S.; Branigan, D. Public investments in the development of GeneXpert molecular diagnostic technology. *PLoS ONE* **2021**, *16*, e0256883. [CrossRef] [PubMed]
118. Buss, B.A.; Baures, T.J.; Yoo, M.; Hanson, K.E.; Alexander, D.P.; Benefield, R.J.; Spivak, E.S. Impact of a multiplex PCR assay for bloodstream infections with and without antimicrobial stewardship intervention at a cancer hospital. *Open Forum Infect. Dis.* **2018**, *5*, ofy258. [CrossRef] [PubMed]
119. Huang, S.H.; Lin, Y.F.; Tsai, M.H.; Yang, S.; Liao, M.L.; Chao, S.W.; Hwang, C.C. Detection of common diarrhea-causing pathogens in Northern Taiwan by multiplex polymerase chain reaction. *Medicine* **2018**, *97*, e11006. [CrossRef]
120. Lam, D.; Luu, P.L.; Song, J.Z.; Qu, W.; Risbridger, G.P.; Lawrence, M.G.; Lu, J.; Trau, M.; Korbie, D.; Clark, S.J.; et al. Comprehensive evaluation of targeted multiplex bisulphite PCR sequencing for validation of DNA methylation biomarker panels. *Clin. Epigenetics* **2020**, *12*, 90. [CrossRef]
121. Olkhov-Mitsel, E.; Zdravic, D.; Kron, K.; van der Kwast, T.; Fleshner, N.; Bapat, B. Novel Multiplex MethyLight Protocol for Detection of DNA Methylation in Patient Tissues and Bodily Fluids. *Sci. Rep.* **2014**, *4*, 4432. [CrossRef]
122. Lin, S.Y.; Huang, S.K.; Huynh, K.T.; Salomon, M.P.; Chang, S.C.; Marzese, D.M.; Lanman, R.B.; Talasaz, A.; Hoon, D.S. Multiplex gene profiling of cell-free DNA in patients with metastatic melanoma for monitoring disease. *JCO Precis. Oncol.* **2018**, *2*, 1–30. [CrossRef]
123. "GeneXpert". Available online: https://www.finddx.org/pricing/genexpert/ (accessed on 8 January 2022).
124. Hsiang, E.; Little, K.M.; Haguma, P.; Hanrahan, C.F.; Katamba, A.; Cattamanchi, A.; Davis, J.L.; Vassall, A.; Dowdy, D. Higher cost of implementing Xpert($^{®}$) MTB/RIF in Ugandan peripheral settings: Implications for cost-effectiveness. *Int. J. Tuberc. Lung Dis.* **2016**, *20*, 1212–1218. [CrossRef] [PubMed]
125. COVID-19 Testing. Available online: https://info.biomeme.com/covid-19 (accessed on 6 January 2022).

Review

Novel Biorecognition Elements against Pathogens in the Design of State-of-the-Art Diagnostics

Maria G. Sande [1], Joana L. Rodrigues [1], Débora Ferreira [1], Carla J. Silva [2,3] and Ligia R. Rodrigues [1,*]

[1] CEB—Centre of Biological Engineering, Campus de Gualtar, Universidade do Minho, 4710-057 Braga, Portugal; msande@centi.pt (M.G.S.); joanarodrigues@ceb.uminho.pt (J.L.R.); deboraferreira@ceb.uminho.pt (D.F.)

[2] CENTI—Center for Nanotechnology and Smart Materials, Rua Fernando Mesquita 2785, 4760-034 Vila Nova de Famalicão, Portugal; cjsilva@citeve.pt

[3] CITEVE—Technological Center for the Textile and Clothing Industries of Portugal, Rua Fernando Mesquita 2785, 4760-034 Vila Nova de Famalicão, Portugal

*   Correspondence: lrmr@deb.uminho.pt; Tel.: +351-253601978

Abstract: Infectious agents, especially bacteria and viruses, account for a vast number of hospitalisations and mortality worldwide. Providing effective and timely diagnostics for the multiplicity of infectious diseases is challenging. Conventional diagnostic solutions, although technologically advanced, are highly complex and often inaccessible in resource-limited settings. An alternative strategy involves convenient rapid diagnostics which can be easily administered at the point-of-care (POC) and at low cost without sacrificing reliability. Biosensors and other rapid POC diagnostic tools which require biorecognition elements to precisely identify the causative pathogen are being developed. The effectiveness of these devices is highly dependent on their biorecognition capabilities. Naturally occurring biorecognition elements include antibodies, bacteriophages and enzymes. Recently, modified molecules such as DNAzymes, peptide nucleic acids and molecules which suffer a selective screening like aptamers and peptides are gaining interest for their biorecognition capabilities and other advantages over purely natural ones, such as robustness and lower production costs. Antimicrobials with a broad-spectrum activity against pathogens, such as antibiotics, are also used in dual diagnostic and therapeutic strategies. Other successful pathogen identification strategies use chemical ligands, molecularly imprinted polymers and Clustered Regularly Interspaced Short Palindromic Repeats-associated nuclease. Herein, the latest developments regarding biorecognition elements and strategies to use them in the design of new biosensors for pathogens detection are reviewed.

Keywords: biorecognition; diagnosis; biosensor; pathogens; aptamers; antibodies; peptides; enzymes; DNAzymes; peptide nucleic acids

Citation: Sande, M.G.; Rodrigues, J.L.; Ferreira, D.; Silva, C.J.; Rodrigues, L.R. Novel Biorecognition Elements against Pathogens in the Design of State-of-the-Art Diagnostics. *Biosensors* **2021**, *11*, 418. https://doi.org/10.3390/bios11110418

Received: 17 September 2021
Accepted: 22 October 2021
Published: 26 October 2021

Publisher's Note: MDPI stays neutral with regard to jurisdictional claims in published maps and institutional affiliations.

1. Introduction

Infectious diseases remain a significant global health concern and cause of mortality. Lower-respiratory infections, diarrhoeal diseases and tuberculosis are currently among the top ten global causes of mortality [1]. A worrying development that will greatly hamper treatment and control of infectious diseases is the emergence of drug-resistant pathogens in recent decades. Currently, at least 700,000 people around the world die each year due to infections caused by drug-resistant organisms [2].

For centuries, clinical manifestations were the most common means to establish a diagnosis for many infections. Remarkably, this is still often the case today because diagnosis of an infection can take several days before a result is delivered. The classical methods for microbial identification are based on culture, which is a slow process providing low sensitivity and has limited use in viral identification. Other traditional microbial diagnostics use microscopy and staining, and serological methods, such as hemagglutination assays. Although inexpensive and rapidly performed, these methods suffer from the disadvantages associated with culturing [3,4]. More recently, standard molecular methods have been adopted for microbial diagnosis, such as Polymerase chain reaction (PCR) and Matrix-assisted laser desorption/ionisation time-of-flight mass spectrometry (MALDI-TOF). These methods, although highly sensitive and relatively fast to perform, are complex and require expensive sophisticated equipment with low portability, often making them inaccessible [3].

To successfully manage infectious diseases in a clinical setting, a rapid diagnosis and administration of targeted antimicrobial or antiviral therapy is crucial [5]. Because current solutions can fail to quickly identify serious infections, they can prove fatal due to a delay in treatment. Examples of such conditions include sepsis, a potentially life-threatening condition [6], and *Clostridium botulinum* infection, which produces lethal neurotoxins that can lead to rapid deterioration of a patient but is sometimes misdiagnosed as stroke or other conditions. To improve diagnostic outcomes, the World Health Organization (WHO) defined that the ideal diagnostic tests should be affordable, sensitive, specific, rapid, equipment-free and delivered to those who need them [7]. Most conventional methods do not meet all or most of these criteria. With the aim of meeting these criteria, point-of-care (POC) diagnostics such as biosensors for mass application are under development because they are inherently flexible, easy to use, minimally instrumented and can be produced at low costs [8].

Biorecognition elements, also referred to as bioreceptors, have the essential function of providing analyte specificity to a biosensor. An ideal biorecognition element possesses a selective and potent affinity towards the bioanalyte, thus endowing a biosensor with good specificity [9]. This review provides an update on the different types of biorecognition elements being studied and the strategies employed for the identification of pathogens that can be used for POC diagnostics. Although the use of antibodies for biorecognition dominates the landscape due to their inherent abilities for antigen detection, there are a few challenges associated with their use in biosensors, such as instability and high cost. Hence, recent research has been more focused on nucleic acid derivatives such as aptamers and peptide nucleic acids (PNAs), which can be modified to improve their biorecognition capabilities while being arguably more stable than antibodies. Numerous studies have been conducted on aptamers and the most recent ones are herein summarised. In addition, theranostic approaches which combine therapy and diagnosis [10] make use of the dual properties of antibiotics and certain types of peptides to selectively detect and simultaneously inactivate pathogens, making them extremely useful, and several examples are discussed. Enzymes, similarly to antibodies, have intrinsic properties for pathogen identification and have also been adapted to some extent, although they also suffer from instability. Other recent strategies for pathogen identification are also explored, such as Clustered regularly interspaced short palindromic repeats/associated nuclease (CRISPR-Cas) and molecularly imprinted polymers (MIPs).

This plethora of promising biorecognition elements are versatile enough to be used by themselves, routinely coated onto magnetic nanoparticles (MNPs), microparticles and microfluidic devices or included in detection probes where they are integrated into biosensors for pathogen identification. All these strategies and methods are thoroughly discussed in this review with a focus on sensing platforms used especially in POC diagnostics. Although most of the reviews related to biosensors are more focused on the detection strategies and principles [11,12], our review addresses the important aspect of how exactly the pathogenic targets are identified or recognised prior to detection. In addition, a number of examples of commercially available biosensors are examined. This aspect is often overlooked in other reviews, which usually only consider results from research outputs.

2. Recent Progress in Platforms for Pathogen Detection

In the past few decades, biosensors have emerged as an alternative to conventional diagnostic methods. Biosensors are highly advantageous due to their high sensitivity, low-cost, ease-of-use and rapid response time for analyte detection [12]. A biosensor comprises a biorecognition element that binds to a target analyte (e.g. bacteria, virus, protein, polysaccharide) and transducer elements that translate the binding occurrence into a detectable signal with a detector providing readout (Figure 1) [13]. The translation of prototype biosensors to commercial development has hitherto been a slow process. However, in response to the emerging need for POC diagnostics and the manifold benefits of sensor-based testing, numerous new commercial biosensors are being introduced to worldwide markets.

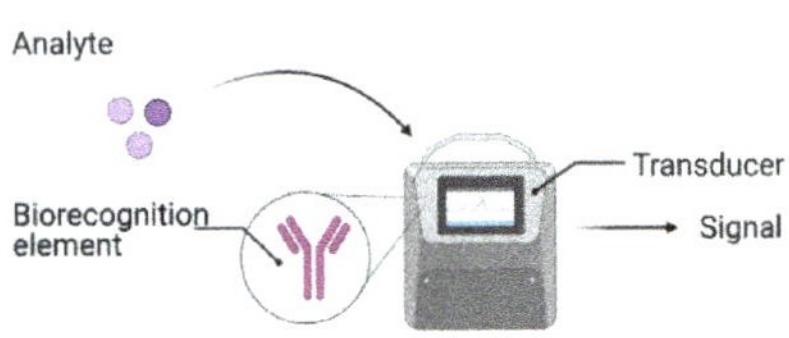

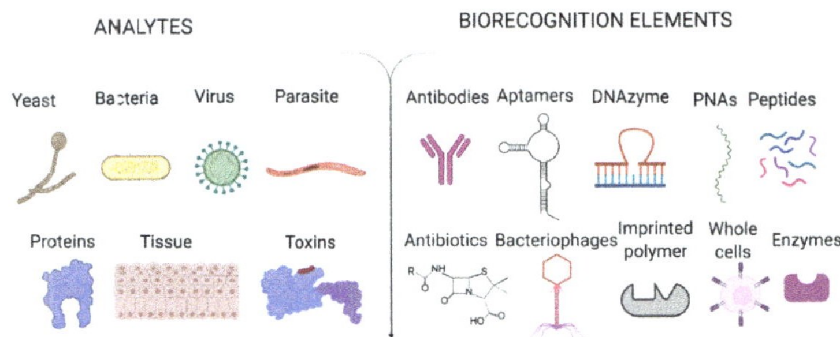

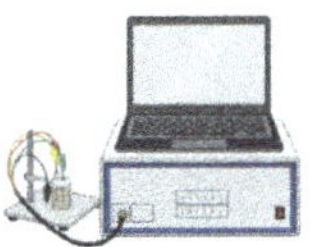

Figure 1. Main components of a biosensor: analyte, biorecognition element and transducer which produces a detectable signal. The major types of biorecognition elements and analytes in the context of clinical pathogen detection are included. PNAs: peptide nucleic acids. Created with BioRender.com.

Table 1 highlights some sensor-based products for the identification of pathogens that are currently available on the market. Most of these products are based on immunological principles using antibodies to recognise the analyte. These commercial sensor platforms are usually in the form of lateral flow immunoassays (LFA), latex agglutination assays and rapid antigen tests. Other classes of novel biorecognition elements that have shown some promise in pathogen identification are still not commonly used in commercial sensors. Regarding the detection systems, biosensors based on electrochemical, piezoelectric and optical principles are still not widely commercialised.

Table 1. Commercially available clinical biosensors for identification of human pathogens.

Pathogen Class	Biorecognition Element	Company	Product	Type of Test	Refs.
Clostridium difficile	Antibodies specific to *C. difficile* antigen glutamate dehydrogenase	Corisbio	*Clostridium* K-SeT	Immunochromatographic assay	[14]
	N/A	Thermo Scientific	Oxoid™ *Clostridium difficile* Test Kit	Rapid latex agglutination	[15]
Escherichia coli serogroup O157	N/A	Meridian Bioscience, USA	ImmunoCard STAT! *E. coli* O157 Plus	Lateral flow immunoassay	[16]
	Antibodies specific to O157 serogroup antigen	Thermo Scientific	*Escherichia coli* O157 Latex Test	Colorimetric latex agglutination	[17]
Haemophilus influenzae b	Antibodies specific to *H. influenzae* type b antigen	Thermo Scientific	Wellcogen™ *Haemophilus influenzae b*	Rapid Latex Agglutination	[18]
Helicobacter pylori	Antibodies specific to *H. pylori*	Quidel, USA	QuickVue *H. pylori* Test	Solid phase immunoassay	[19]
	Anti-*H. pylori* IgG antibody	Beckman Coulter, USA	Icon HP	Immunoassay	[20]
	Antibodies specific to *H. pylori* antigens	Corisbio	Pylori-Strip/Pylori K-SeT	Immunochromatographic assay	[21,22]
Moraxella catarrhalis	Indoxyl butyrate detects the enzyme butyrate esterase of *M. catarrhalis*	Thermo Scientific	Remel™ *Catarrhalis* Test Disc	Colorimetric chemical identification	[23]
Staphylococcus aureus	A protein detects clumping factor and protein A of *S. aureus*	Thermo Scientific	BactiStaph™ Latex Agglutination Test Kit	Agglutination of protein-coated latex particles	[24]
Streptococcus pneumoniae	Specific rabbit antibody	Thermo Scientific	DrySpot™ Pneumo Latex Agglutination Test	Agglutination of antibody-coated latex particles	[25]
	N/A	Abbott	BinaxNOW™ *Streptococcus pneumoniae* Antigen Card	Lateral flow immunoassay	[26]
Epstein-Barr virus	Heterophile antibody detects bovine red cell mononucleosis antigen	Thermo Scientific	Infectious Mononucleosis Test using Latex Agglutination	Latex Agglutination	[27]
Influenza A and B virus	Antibodies specific to Influenza type A and type B antigens	Quidel, USA	QuickVue Influenza A + B test	Immunoassay	[28]
	Antibody detects nucleoprotein antigens	Abbott	Alere BinaxNOW® Influenza A & B Card	Immunochromatographic assay	[29]
Severe acute respiratory syndrome coronavirus 2 (SARS-CoV-2)	N/A	Lucira Health	COVID-19 All-In-One Test Kit (FDA approved)	Amplification of viral genetic material	[30]
	Antibody	Abbott	BinaxNOW COVID-19 Ag Card Home Test (FDA Emergency Use Authorization)	Rapid antigen	[31]
	Antibody	Ellume Health	Ellume COVID-19 Home Test (FDA Emergency Use Authorization)	Rapid antigen	[32]
Plasmodium falciparum	Antibodies detect histidine-rich protein II antigen specific to *P. falciparum* and a pan-malarial antigen	Abbott	BinaxNOW® Malaria	Immunochromatographic assay	[33]

N/A—Information Not Available.

Other advancements to improve diagnostics for infectious diseases make use of computational methods and tools. More recently, computational modelling has been used to design biorecognition elements with an effectiveness unlikely to be matched by conventional wet laboratory techniques. Molecular docking is an example of a computational modelling technique that has been used to design high-affinity biorecognition elements towards the detection of pathogenic targets [34]. For ligand discovery, in silico simulations have been used to design "superselective" multivalent probes by optimising the multiplicity (binding to the target DNA at multiple sites), instead of the strength of the probe-target bond, wherein the target can be pathogens, to enhance detection sensitivity and specificity [35]. Technologies such as Next-Generation Sequencing (NGS) or massively parallel sequencing, which are high throughput sequencing methods, have the potential to establish hypothesis-free diagnostic approaches to detect virtually any pathogen leading to a paradigm shift in how infections can be diagnosed [3]. Metagenomic NGS (mNGS) has proven capabilities of detecting a wide range of pathogens present in a patient sample [36]. Additionally, the enrichment of target nucleic acids prior to mNGS enhances the sensitivity of detection [37]. To illustrate, Wylie and collaborators [38] developed a capture probe called ViroCap, constituted by a panel of 2 million sequences able to enrich viral nucleic acids of a large set of eukaryotic viruses preceding mNGS to increase the sensitivity of virus detection. All the expected viruses from 26 clinical samples and 30 additional viruses were detected, whereas when only mNGS was used some of the viruses were not detected. Although mNGS can be successfully implemented, it is time- and resource-intensive and requires bioinformatics expertise. Therefore, its potential in diagnostics is based on specific cases where unsuspected or unculturable organisms are present and may be not detected (false negative results) with standard assays.

3. Biorecognition Elements

The essential requirement of an effective biorecognition element is to provide specificity for the (bio)analyte (also referred to as the target) [9]. The most important characteristics of a biorecognition element for pathogen identification are sensitivity (few false negatives) and selectivity (few false positives) [39]. Widely used as biorecognition elements are the antibodies' immunoglobulins (Ig) due to their exceptionally high affinity and specificity towards their target analytes. Moreover, aptamers, peptides, bacteriophages and enzymes, among others, have also been established as being effective for biorecognition purposes. The targets of biorecognition elements, which enable pathogen identification, are usually specific surface molecules (e.g., biomarkers), such as proteins and epitopes or their by-products (toxins and metabolites) or nucleic acids. Exploring the various biorecognition elements and their inherent characteristics is essential to improve and develop novel POC diagnostics.

3.1. Antibodies

Igs, generally referred to as antibodies, are large glycoproteins produced by white blood cells with strong affinity and specificity towards their target analytes. These qualities make them a natural and popular choice as biorecognition elements and consequently, have been adapted for use in pathogen identification. In addition to whole monoclonal antibodies (mAbs), which are laboratory made, antibody-derived single-chain variable fragments (scFv) and fragment antigen-binding (Fab') units are commonly used for biorecognition [40]. They are more cost-effective than mAbs while providing similar specificity values as conventional antibody approaches. Antibody-based probes are most commonly used for the detection of specific proteins and whole cells [41]. Because numerous methods of pathogen identification using antibodies as biorecognition elements have been reported, this review provides only a brief overview of a few recent reported strategies.

Antibodies are identified within the relevant biochemical pathways and then produced in animals and purified for various applications, including for use as capture probes in sensors platforms [9]. Most examples of antibody-based biosensors for detection of pathogens use commercially produced antibodies, as in the case of Park et al. [42] who developed a 3D printed microfluidic device with antibodies conjugated to MNPs (Ab-MNPs) as capture probes against the common pathogen enterohaemorrhagic *Escherichia coli* O157:H7. The device, termed a magnetic pre-concentrator (3DµFMP) was tested using human blood mixed with 10^2–10^7 colony forming units (CFU) of *E. coli* O157:H7 and Ab-MNPs (10^{13} particles/mL). The mixture was treated with adenosine triphosphate (ATP) eliminating reagent before being injected into the 3DµFMP device. Captured cells were then magnetically separated and quantified using an ATP luminescence assay achieving an excellent limit of detection (LOD) of 10 CFU/mL in blood. The 3DµFMP device was efficient, selectively enriching *E. coli* O157:H7 700-fold in a volume of 100 mL in one hour.

Antibodies were also used in multiplex detection of pathogenic bacteria by Jang et al. [43]. Figure 2 illustrates the synthesis of modified capture magnetic beads (MBs) and dual nanoprobes for detection, which are based on both surface-enhanced Raman spectroscopy (SERS) and fluorescence. Two different mAbs that recognise two pathogens, *E. coli* J5 and *Francisella tularensis*, were conjugated to respective MB-clusters for selective magnetic capture (Figure 2A). The dual nanoprobes used for detection (Figure 2B,C), comprised silver nanoparticles (AgNPs) conjugated with SERS reporters and fluorescence dyes and, subsequently, the AgNP clusters were encapsulated. They were conjugated to another pair of mAbs that recognise the respective bacterial targets (Figure 2A). In the presence of *F. tularensis* and *E. coli* J5, the MBs-clusters selectively bind to the respective bacterial target (Figure 2D). Following magnetic enrichment, the dual nanoprobes were added, which formed sandwich-type immunocomplexes containing the bacteria, MB-clusters and the dual nanoprobes (Figure 2E). Detection was performed by SERS (Figure 2F) and fluorescence microscopy. A linear correlation was verified between the Raman intensity and the pathogen concentration from 10^2 to 10^6 cells/mL, and the LOD was found to be less than 10^2 cells/mL.

Another similar example adapted the use of antibody conjugated NP enrichment with SERS detection in an easy-to-use LFA format. Primary antibody-conjugated magnetic gold nanoparticles (AuNPs) were used for enrichment of *E. coli*. The enzyme rennet was used to prevent aggregation of the AuNPs, to facilitate free movement of the bacteria along the paper-based LFA strip followed by SERS detection. The sensitivity of this system was comparable with a plate-counting method and could be completed in 1 h [44].

A POC biosensor for the identification of *Mycobacterium tuberculosis* was reported using monoclonal antibodies with high specificity towards the well-established *M. tuberculosis* heat shock protein X (HspX) [45]. The antibodies were directly immobilised on a plasmonic sensor surface for detection, achieving a LOD of 0.63 ng/mL in pretreated sputum samples. Using a similar strategy, Zika virus was detected by immobilising a monoclonal antibody specific to an envelope protein of Zika virus onto an impedimetric immunosensor [46]. The sensor achieved a low LOD of 10 picomolar.

Although antibodies have been the gold standard as affinity molecules for decades, their limitations include poor stability and reproducibility, which complicate their use in sensing platforms that require a long shelf-life. Other problems relate to lengthy production time and the need for ethical approval, which increase costs [47]. Therefore, research in recent years has also focused on finding alternate biorecognition elements with improved specifications.

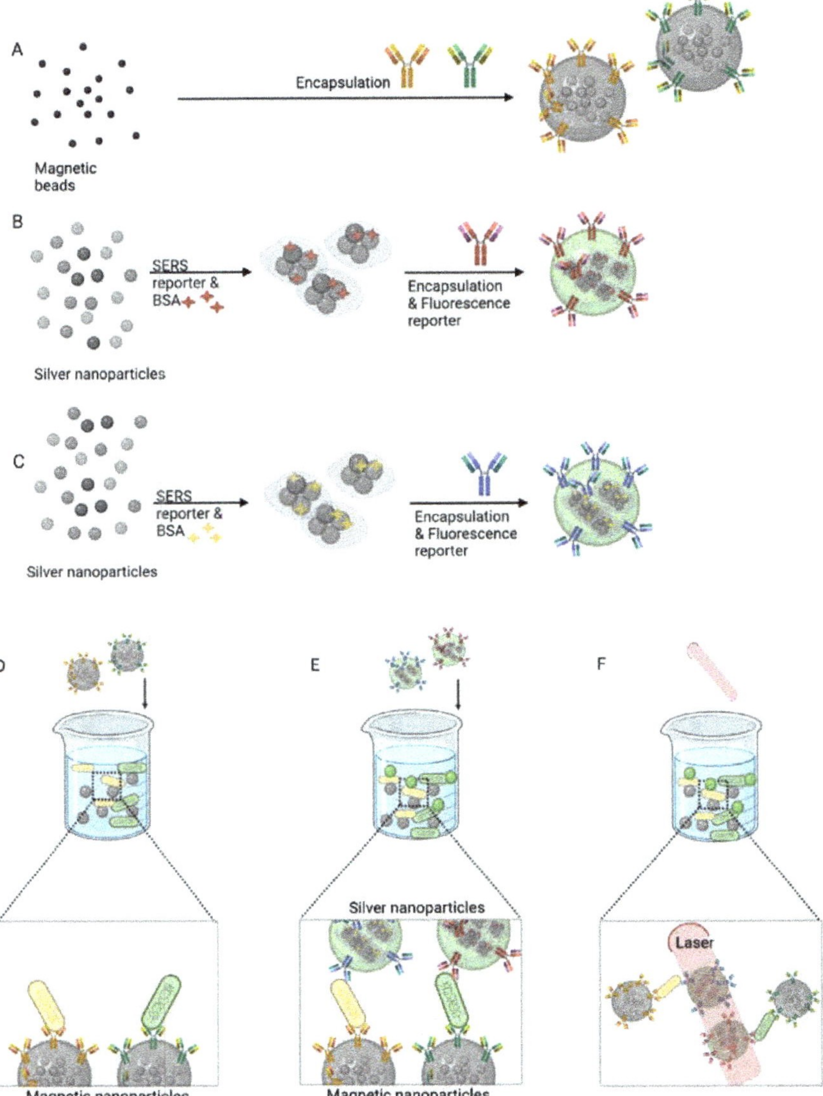

Figure 2. Three-step synthesis of monoclonal antibodies (mABs) conjugated with magnetic beads and SERS (surface-enhanced Raman spectroscopy) and fluorescence-based dual nanoprobes for the multiplex detection of *Escherichia coli* J5 and *Francisella tularensis*: (**A**) Magnetic bead clusters were encapsulated and conjugated to mABs to selectively capture either *E. coli* or *F. tularensis*. (**B,C**) For the subsequent detection step after bacterial capture, AgNP clusters were encoded with SERS reporters (red and yellow stars), stabilised by bovine serum albumin (BSA), conjugated further with fluorescent dyes and encapsulated in a polymer. (**D**) *E. coli* J5 and *F. tularensis* bind to the respective mABs and were magnetically separated. (**E,F**) Multiplex detection of sandwich immunocomplexes composed of bacteria, magnetic bead clusters and the SERS and fluorescence-based dual nanoprobes was achieved. Detection method develop by Jang et al. [43]. Created with BioRender.com.

3.2. Enzymes

Enzymes are natural, biologically derived molecules that have evolved an innate ability to achieve analyte specificity [9]. Enzymes attain specificity to a bioanalyte through binding cavities deep in their structure. Binding can occur by various means such as hydrogen-bonding, electrostatic binding and other non-covalent interactions [48]. However, enzymes, such as antibodies, are sensitive to degradation, which affects their reproducibility and, by extension, their applicability in biosensors. Liu et al. [49] developed a microwave-assisted method for the synthesis of red fluorescence gold nanoclusters (AuNCs) functionalised with lysozyme and demonstrated their potential use in a visual detection nanosensor of *E. coli*. Lysozyme is advantageous as it reduces Au^+ ions and thus stabilises the AuNCs. Additionally, lysozymes are antimicrobial enzymes that can recognise and kill several types of bacteria [50]. Due to lysozyme's high specificity for *E. coli* and the fluorescence enhancement of the lysozyme functionalised AuNCs, the nanosensor detected *E. coli* with good sensitivity (LOD = 2×10^4 CFU/mL). Due to the nature of the fluorescence evolution, the nanosensor can be suitably applied to real-time and fast detection of bacteria.

Label-free and real-time genus-specific detection of bacterial pathogens using lytic enzymes was proposed by Couniot et al. [51]. The lytic enzyme lysostaphin selectively digests staphylococci. Therefore, a compatible microelectrode sensor with *Staphylococcus epidermidis* anchored to its surface was incubated with lysostaphin. A real-time shift in impedance was observed when target bacteria were lysed by the enzyme. The LOD was calculated as 10^8 CFU/mL within a minute of bacterial incubation. Specificity of the lysostaphin-based sensor to *S. epidermidis* was demonstrated in synthetic urine also containing *Enterococcus faecium*.

More recently, Clemente et al. [52] proposed a simple visualised method for rapid diagnosis of the respiratory pathogen *Pseudomonas aeruginosa* due to enzymatic liquefaction of infected sputum samples. The catalytic conversion of hydrogen peroxide, when added to an infected sample, by the *P. aeruginosa* enzyme catalase results in disruption of biofilms and generates an effervescence of oxygen bubbles which can be clearly seen. Catalase is produced only by *P. aeruginosa* and so other bacteria do not produce a reaction on addition of hydrogen peroxidase. An LOD of 10^5 cells/mL was reported, which is considered the clinical threshold for respiratory infections detection. An important advantage of using enzymes instead of affinity molecules is that non-specific binding on sensors is eliminated, which could greatly facilitate its use for real samples. Moreover, lytic enzymes can be potentially applied in the detection of all Gram-positive bacteria.

3.3. Peptides

Peptides typically consist of short chains of 10–40 residues [53]. Antimicrobial peptides (AMPs) are part of the innate immune system in many eukaryotes. They usually bind to the negatively charged cell membranes of bacteria prompting cell lysis. AMPs have been successfully applied as biorecognition elements, in part due to their broad-spectrum activity against a variety of pathogenic bacteria. Cell-penetrating peptides (CPPs) are another class of short peptide sequences that find application as biomolecular delivery vehicles due to their ability to breach cellular membranes [54]. Both AMPs and CPPs have potential as biosensors and therapeutics due to their ability to inactivate pathogens [55]. Synthetic peptides can be selected randomly by phage display from phage libraries for application in biosensors and often perform better than antibodies [53,54]. Their relatively easy synthesis and intrinsic stability render them as suitable candidates to increase the shelf-life of sensor diagnostic platforms [56]. However, some key challenges are related to detection of bacteria in real samples, and their relatively low sensitivity and selectivity for diagnostic applications remain [53,57].

Certain pathogens can be identified indirectly based on their distinct by-products. For example, *E. coli* and some other bacterial species produce high levels of alkaline phosphatase (ALP). Zhang et al. [58] constructed a fluorescent probe that senses bacterial ALP activity. The novel probe comprised a controlled aggregation-induced emission luminogen (AIEgen) conjugated with a self-assembling peptide. The AIEgen-peptide probe designated as TPEPy-$^DF^DF_PY^DEG^DK$ (TPEPy-pY) was designed containing a phosphorylated tyrosine, which, in the presence of bacterial ALP, is dephosphorylated. The dephosphorylation of TPEPy-pY reduces its hydrophilicity, causing the probe to assemble on the bacterial surface, triggering the AIEgen to fluoresce. The probe demonstrated good sensitivity and selectivity for ALP activity, with an LOD of 3.38×10^6 CFU/mL.

Numerous examples of biorecognition methods make use of AMPs. In a recent study, Yuan et al. [59], reported MNPs modified with an AMP bacitracin A were able to capture bacteria. Their work revealed that the interactions between bacitracin A and bacteria are due to a pyrophosphate group present in the lipid target on bacteria and other indirect interactions mediated by sodium and zinc ions. After magnetic separation, SERS tags bound to the captured bacteria were used for their detection. The SERS spectra allowed to distinguish between *E. coli, Staphylococcus aureus* and *P. aeruginosa*.

Seminal work related to AMPs employed Magainin I to semi-selectively identify pathogenic bacteria [56]. Magainin I (GIGKFLHSAGKFGKAFVGEIMKS), which is naturally present on the skin of African clawed frogs, exhibits antibiotic activity against numerous species of bacteria. A micro-capacitive electrode sensor was functionalised with Magainin I. The sensor demonstrated adequate selectivity to distinguish strains of specific pathogenic Gram-negative bacteria, while Magainin I retained broadband detection abilities. A sensitivity to *E. coli* of 1 bacterium/μL was achieved. Another enrichment and detection platform comprised of a chemically modified microfluidic platform immobilised with Magainin I to capture bacteria achieved an LOD of 5 CFU/mL of *Salmonella* and 10 CFU/mL of *Brucella* from urine samples within 60 min [60].

Class IIa bacteriocin AMPs such as leucocin A are noted for their anti-*Listeria* activity. A study by Azmi et al. [61] highlights these attributes of leucocin A. A peptide array was used to screen short peptides from a synthetic peptide library with selectivity for *Listeria monocytogenes* for use as biorecognition elements in a sensor. By employing this screening method, the Leucocin A fragment Leu10 (GEAFSAGVHRLANG) exhibited the highest affinity to target bacteria relative to the other peptide fragments. Similarly, leucocin A was used to selectively capture *L. monocytogenes* from among other Gram-positive strains followed by impedimetric detection, as illustrated in Figure 3. An LOD of 10^3 CFU/mL was achieved by this method [62].

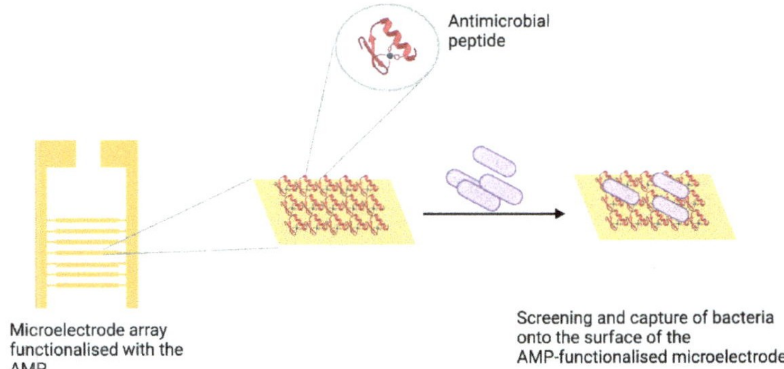

Figure 3. Graphical representation of an antimicrobial peptide (AMP)-based biosensor used for impedimetric detection of bacteria. The AMP is immobilised on a microelectrode array. The functionalised sensor selectively captures the target cells due to the immobilised AMP. Created with BioRender.com.

In some cases AMPs perform better than antibodies, as illustrated by Arcidiacono et al. [63], who evaluated the potential of fluorescently labelled AMPs as an alternative to labelled antibodies in the detection of *E. coli* O157:H7. AMPs cecropin P1, SMAP29 and PGQ were labelled with a fluorescent dye Cy5 and screened using a cell binding assay. It was revealed that Cy5-cecropin P1 improved the detection of target bacteria 10-fold when compared to a Cy5 labelled with an anti-*E. coli* O157:H7 antibody.

Other miscellaneous AMPs that have been researched for their biorecognition capabilities include Clavanin A and Ubiquicidin. Clavanin A is isolated from the marine tunicate *Styela clava*. An electrochemical biosensor was constructed using AuNPs chemically modified with cysteine and functionalised with Clavanin A for the detection of *Salmonella typhimurium* and *E. coli*. The sensor displayed moderate sensitivity and was able to differentiate signals for bacterial concentrations between 10^1 and 10^4 CFU/mL [64]. In a different study, a chemically modified version of Ubiquicidin enhanced with a fluorophore for detection by optical endomicroscopy demonstrated selectivity for pathogenic bacteria, and for the pathogenic fungus *Aspergillus fumigatus* (that causes pulmonary infections) over human cells in an ex vivo human lung model [65].

Multiplex systems based on arrays of various AMPs can be used for biorecognition of various analytes such as viruses. Fluorescently labelled viruses can be identified based on their characteristic response pattern upon interaction with the AMP panel [66]. Similarly, the goal of a study conducted by Kulagina et al. [67] was to establish such a multiplex detection system for pathogenic bacteria and viruses relevant to biodefence. The authors developed an array of different types of AMPs immobilised on a sensing substrate. The pathogens were fluorescently labelled and their identification was based on an evaluation of their binding pattern to the immobilised AMPs. The pathogens tested were Venezuelan equine encephalitis virus (VEE), vaccinia virus, *Brucella melitensis* and *Coxiella burnetii*. The AMPs array comprised polymyxins B and E, cecropins A, B, and P, melittin, parasin, bactenecin and Magainin I, and antibodies were used as controls. After the binding assays, it was observed that most of the immobilised AMPs bound labelled vaccinia virus, VEE and *C. burnetii* proportional to their concentration, and *B. melitensis* bound to bactenecin, polymyxin B and E. By establishing the characteristic binding patterns of various microbial pathogens, AMP arrays can be used to distinguish between them.

In a recent example, Fu et al. [68], developed a sensor array comprised of fluorogenic peptide probes for the differential sensing of the Ebola virus (EBV). The probes were constructed based on self-assembly between graphene oxide, which is a strong fluorescence quencher, and three fluorescently labelled peptide fragments T-RS5, T-QY7 and T-ED17, which were derived from antibodies. In the presence of pseudo-viruses (not able to replicate), the probes displayed an increase in fluorescence proportional to virus concentrations. This suggests that the peptide probes are removed from the graphene oxide surface to form peptide-virus complexes, resulting in fluorescence recovery. Based on the analysis of the fluorescence signals, differential sensing is evident in spite of the similarity of the viral capsid glycoproteins of EBV, Marburg virus and vesicular stomatitis virus.

A peptide microarray approach has also been developed to improve the accuracy of COVID-19 diagnosis. For example, Li et al. [69] prepared a microarray of spike protein (S1)-derived peptides from SARS-CoV-2 with full S1 coverage and analysed the immunological response from 2434 serum samples of COVID-19 patients including asymptotic patients. Based on the results, several 12-mer peptides were identified as suitable antigens to detect antibodies against SARS-CoV-2. While monitoring the IgG response, one of the peptides exhibited a sensitivity of 95.5% and specificity of 96.7%, which is comparable to the performance of the S1 itself for detection of symptomatic and asymptomatic COVID-19 cases. Additionally, a panel of four selected peptides was constructed with capabilities to prevent potential cross-reactivity with serum containing other coronaviruses. Additionally, Cai et al. [70] also evaluated a serological method for COVID-19 diagnosis and developed a peptide-based magnetic chemiluminescence enzyme immunoassay (MCLIA). Twenty candidate peptide epitopes of antigens of SARS-CoV-2 were predicted in silico and synthesised. The peptides were then linked to MBs and tested in an MCLIA with serum from COVID-19 patients to detect their binding to IgG and IgM antibodies. A peptide derived from the S1 protein of the virus showed the best performance and was further evaluated in an MCLIA with serum from 276 patients with COVID-19. The IgG and IgM positive detection rates were found to be 71.4% and 57.2%, respectively. By comparison, Pomplun et al. [71] screened peptides with high affinity towards the receptor-binding domain (RBD) of SARS-CoV-2 S1 towards the development of an efficient SARS-CoV-2 diagnostics. A method based on affinity selection using mass spectrometry (MS) was used to rapidly screen a library composed of 800 million synthetic peptides. Three sequences were identified with dissociation constants in the range 80–970 nM for the RBD. It was also shown that the RBD was selectively enriched by the selected peptides from a complex matrix comprising human serum proteins.

Because numerous peptides have been successfully tested for their biorecognition capabilities in sensing platforms and in arrays, they are promising candidates for use in biosensors, especially when taken together with their various other stated benefits.

3.4. Nucleic Acid Derivatives

The development of in vitro selection methodologies, such as the so-called "Systematic evolution of ligands by EXponential enrichment" (SELEX), to screen for single stranded DNA or RNA (ssDNA or ssRNA) from random-sequence nucleic acid libraries with high-affinity binding properties gave rise to intensive investigation on synthetic nucleic acids with special properties [72]. Since then, a variety of functional synthetic nucleic acids and their analogues have been developed for various diagnostics and therapeutic applications, including aptamers, DNAzymes and PNAs [73,74]. These synthetic molecules have distinct advantages over traditional antibodies as they are arguably more stable, versatile and cheaper to produce, making them the current preferred choice across sensing platforms [47].

3.4.1. Aptamers

Aptamers are short ssDNA or ssRNA molecules, having a length of 25–100 bases, that fold into stable three-dimensional conformations. Due to their structure they recognise and bind to targets via hydrogen bonding, van der Waals forces and/or electrostatic interactions [54,75,76]. This behaviour makes them ideal biorecognition elements for targeting pathogens with high affinity and specificity. Compared to antibodies, aptamers exhibit significant advantages, including lower molecular weight, easier and cheaper production methods and good chemical stability [77–79].

In addition, another advantage of aptamers is that they can be raised against an extensive variety of targets from small molecules to big proteins and also live cells, such as whole bacteria, for example [80]. The general steps of SELEX are illustrated in Figure 4. They can be briefly described as: (1) the target molecules (or cells) are incubated for a defined period with a ssDNA library pool; (2) non-binding sequences are then rinsed off; (3) next, bound sequences are recovered and PCR-amplified—for example, by using fluorescein isothiocyanate-labelled sense primers and biotin-labelled antisense primers; (4) finally, the antisense strands are removed to generate ssDNA pools for subsequent cycles of selection. As the cycles progress, various parameters, such as incrementing the number rinses, can be performed to increase stringency, allowing retention of aptamers with the strongest affinity from the pool. Flow cytometry is often used to monitor the enrichment of the selected pools by binding assays with the target, whereby selected pools with increased fluorescence are compared to the DNA library. To increase specificity to the target, the recovered pools obtained after some of the rounds are incubated with control molecules/cells to filter out sequences that bind to common sites present on the target, in addition to on the control [81,82].

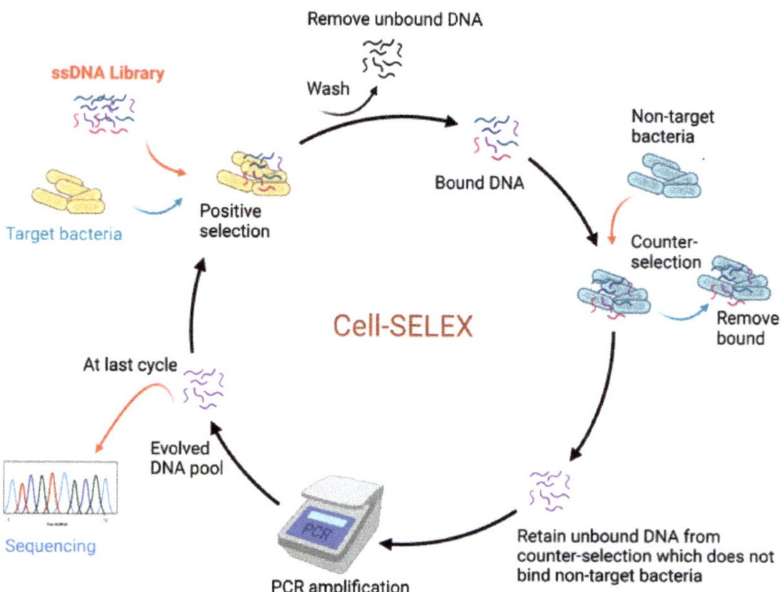

Figure 4. The general steps involved in the selection of candidate aptamers towards a target bacterium by cell SELEX (systematic evolution of ligands by exponential enrichment). These include incubation of the target bacteria with a single stranded DNA (ssDNA) library, washing steps, counter-selection and PCR (polymerase chain reaction) amplification before the cycle is repeated. After all cycles are completed, candidate aptamers are sequenced. Created with BioRender.com.

The selected aptamers are being used as biorecognition elements in sensing platforms to detect a wide variety of pathogens in patient samples. For example, an innovative approach was developed by Wang et al. [75] to produce an aptamer-based device for application in bacterial infection diagnostics. The authors screened aptamers using cell-SELEX and identified highly specific aptamers capable of recognising three nosocomial and antibiotic-resistant bacteria, namely *E. coli*, *Acinetobacter baumannii* and the multidrug-resistant *S. aureus*. These aptamers were further integrated into a microfluidic system to form a paper-based dual-aptamer microfluidic chip. Compared with traditional laboratory techniques, this microfluidic system exhibited many advantages, including faster detection times, smaller size, higher specificity and multiplex capabilities.

In another study, Savory and collaborators [83] used bacterial cell-SELEX to screen for aptamers against *Proteus mirabilis*, which causes catheter-associated urinary tract infections, in combination with in silico maturation (ISM) to improve aptamer specificity. ISM uses a genetic algorithm to predict aptamer sequences with stronger target affinity than the promising parent sequences raised by cell-SELEX. This is achieved through successive rounds of sequence scrambling and random mutation in silico, followed by functional screening in vitro and selection of improved aptamers. After two cycles of ISM, one aptamer displayed a 36% higher specificity value than the original sequence selected by cell-SELEX [83].

Aptamers also find use in the detection of tuberculosis [84]. During the early stages of infection by virulent *M. tuberculosis*, culture filtrate protein 10 (CFP10) and early secreted antigen target-6 (ESAT6) antigens are secreted and the detection of such proteins can be used for the early and specific diagnosis of tuberculosis [85]. Therefore, Tang et al. [84] raised aptamers against CFP10 and ESAT6 antigenic targets using SELEX. The selected screened aptamers (CE24 and CE15) were used in an enzyme-linked oligonucleotide assay (ELONA) to detect the proteins CFP10 and ESAT6 in serum samples of patients with active pulmonary tuberculosis, extra-pulmonary tuberculosis and healthy donors. The results demonstrated a specificity and sensitivity of 94.1% and 100% (using CE24-based ELONA) and, 94.1% and 89.6% (using CE15-based ELONA), respectively [84].

Recently, aptamers have also been conjugated with MNPs to improve diagnosis. For example, Wang et al. [86] developed POC diagnostic systems to diagnose sepsis and perform blood disinfection using aptamers. The commercial aptamers able to recognise bacterial species were conjugated with iron oxide MNPs functionalised with chlorin e6 (Fe_3O_4-Ce$_6$-Apt). This nano-system allowed successful diagnosis of sepsis in mouse models caused only by *S. aureus* or by multiple bacterial species, namely *S. aureus* and *E. coli*, with a detection sensitivity comparable with serological techniques and a shorter turnaround time. Moreover, a total extracorporeal disinfection of blood was achieved due to the strong photodynamic effect of the Fe_3O_4-Ce$_6$-Apt system. Another example of Fe_3O_4 MNPs functionalised with aptamers was reported by Hao et al. [87] for the specific capture of enteropathogenic *S. typhimurium*. Commercially available specialised aptamer complexes containing a sequence specific to capture *S. typhimurium* and a primer region used to assist in detection were used. These primer sequences play a role in rolling circle amplification (RCA) to produce long ssDNA with hundreds of tandem-repeat sequences [88]. Therefore, while the bacteria are captured, simultaneously several signal probes are assembled on the RCA products for enhanced amplification of the recognition event. This unique method was reported to have a high selectivity for *S. typhimurium*, with an excellent LOD equal to 10 CFU/mL.

In addition to the abovementioned studies, there are currently many other examples of aptamers used to identify various pathogens and their biomarkers [89–91]. Tables 2 and 3 summarise other examples of aptamers used to identify human bacterial pathogens and viral human pathogens, respectively, for the diagnosis of infections.

Table 2. Summary of recent aptamer-based methods for the identification of bacterial pathogens.

Bacteria	Target (Analyte)	Detection System	Refs.
Escherichia coli O157:H7 and *Salmonella typhimurium*	Whole bacteria	Aptamer-modified fluorescent magnetic multifunctional nanoprobes	[92]
Staphylococcus aureus and *Escherichia coli*	Whole bacteria	Multiplex aptamer-based hydrogel barcodes	[93,94]
Escherichia coli O157:H7	Whole bacteria	Electrochemical biosensor with amino-functionalised metal–organic frame integrated with aptamers	[95]
Staphylococcus aureus	Whole bacteria	Electrochemical detection by dual-aptamer-based sandwich with silver nanoparticles	[96]
Staphylococcus aureus	Enterotoxin A protein	Graphene oxide-based fluorescent bioassay	[97]
Staphylococcus aureus	Whole bacteria	Surface-enhanced Raman spectroscopy (SERS) biosensor	[98]
Campylobacter jenuni	Whole bacteria	N/A	[77]
Streptococcus pyogenes	M11 M-type serotype whole bacteria	N/A	[99]
Mycobacterium tuberculosis	MPT64 secreted protein	Enzyme linked oligonucleotide assay	[100]
Methicillin-resistant *Staphylococcus aureus* (MRSA)	Penicillin binding protein 2a (PBP2a)	Fluorometric assay	[101]
Pseudomonas aeruginosa	Whole bacteria	Fluorometric assay	[102]

N/A—Not Applicable.

Table 3. Summary of recent aptamer-based methods for the identification of viral pathogens.

Virus	Target (Analyte)	Detection System	Refs.
Avian influenza strain H5Nx	Whole virus	Sandwich-type surface plasmon resonance (SPR)-based biosensor assay	[103]
Influenza A strain H3N2	Globular region of hemagglutinin	Aptamer-functionalised magnetic microparticle-based colorimetric method	[104,105]
Zika	Zika NS1 Protein	Aptamer-Based enzyme-linked immunosorbent assay (ELISA)	[106]
Norovirus	Murine norovirus and capsids of a human norovirus strain GII.3	Electrochemical sensor	[107]
H1N1	Inactivated H1N1 virus particles	Electrochemical impedance sensor	[108]
Human immuno-deficiency virus Type 1 (HIV-1)	Glycoprotein-120 (gp-120)	Liquid crystal optical sensor	[109,110]
Human papillomavirus (HPV)	L1-major capsid protein of HPV	Electrochemical impedance sensor	[111,112]
SARS-CoV-2	Receptor-binding domain (RBD) of the spike glycoprotein	N/A	[113]
SARS-CoV-2	Nucleocapsid protein	ELISA and a gold nanoparticle immunochromatographic strip	[114]
SARS-CoV-2	Nucleocapsid protein	N/A	[115]
SARS-CoV-2	Spike glycoprotein	N/A	[116]
Ebola	Viral RNA	Antiresonant reflecting optical waveguide	[117]

N/A—Not Applicable.

Despite aptamers' numerous advantages and recognised potential, the translation of aptamer-based products to the clinics or other markets has been slow [118,119]. Reasons for this include the high financial investment made in the research and production of antibodies, and the general unfamiliarity regarding aptamers and their interesting performance [120]. In addition, although aptamers may be cheaper to produce, in general, the affinity properties of antibodies in comparison to aptamers remain superior. Furthermore, SELEX can be a very time-consuming process. However, their advantages together with an increasing awareness about them could lead, in the near future, to a wider use of aptamers in the increasingly relevant POC diagnostics and therapeutics field.

3.4.2. DNAzymes

DNAzymes, also called deoxyribozymes, are synthetic ssDNA oligonucleotides that display catalytic activities [121]. Inspired by the existence of naturally occurring ribozymes (RNAzymes), Breaker and Joyce identified, in 1994, the first DNA-like enzymes by SELEX [122]. DNAzymes were initially produced for the detection of lead contaminations [123]. More recently, they have been generated to identify cancer cells [124], pathogenic bacteria [125] and other biomarkers. For example, Zheng et al. [125] created a sensing platform using MNPs functionalised with a fluorescently-responsive DNAzyme for the detection of pathogenic *E. coli*. The *E. coli*-specific DNAzyme was synthesised by template-mediated ligation and further modified with MNPs and acetylcholinesterase (AChE) to form a complex. In the presence of bacterial lysate, the DNAzyme domain binds to target molecules from the bacterial content, triggering a cleavage event which releases AChE. The free AChE subsequently plays a role in enhancing the fluorescence signal of the detection system. DNAzymes were found to bind to the target with high specificity and sensitivity, exhibiting a LOD of 60 CFU/mL with a linear range from 10^2 to 10^7 CFU/mL.

A highly innovative system for the detection of bacteria in blood was also developed by Kang et al. [126]. SELEX was used to identify DNAzymes with specificity to the *E. coli* lysates. The construction of the DNAzyme is shown in Figure 5. The DNAzyme domain was enzymatically ligated with a DNA–RNA chimeric substrate. This substrate contained a ribonucleotide cleavage site flanked by a fluorophore and a quencher. In the presence of the target *E. coli* lysate, the DNAzyme binds the target molecule, changes its conformation and cleaves the fluorophore from its quencher, generating a high fluorescence detectable signal. During detection, DNAzyme sensors were mixed with blood constituents and then encapsulated in hundreds of millions of picolitre droplets. These were analysed by a 3D high-throughput particle counter to detect fluorescent particles. Using this method, *E. coli* was detected in a range of very low concentrations from a single cell up to 10^4 cells per mL within a span of 1.5 to 4 h. Afterwards, this biosensor was adapted for detection of *Klebsiella pneumoniae* in a fluorescent paper sensor [127].

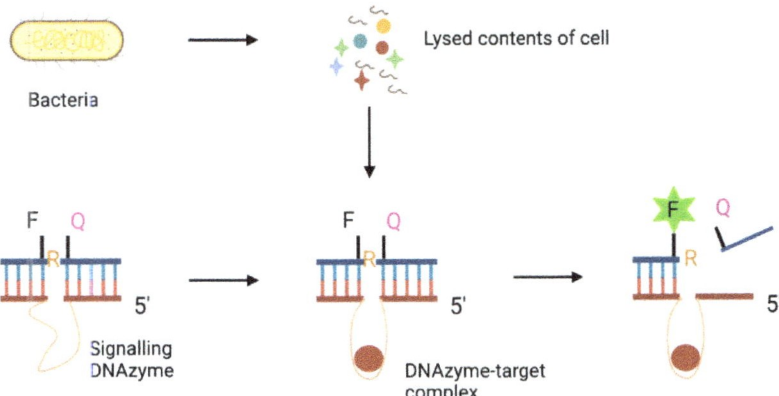

Figure 5. DNAzyme biosensor proposed by Kang et al. [126]. The target(s) from lysed bacteria bind(s) to the DNAzyme sequence (orange), which undergoes a change in conformational triggering the activation of the DNAzyme. The activated DNAzyme cleaves the fluorogenic substrate at the ribonucleotide connection (R), this releases the fluorophore (F) and quencher (Q) to produce a high-fluorescence signal. Created with BioRender.com.

DNAzymes were also successfully employed in a colorimetric paper sensor for the sensitive detection of another human pathogen, *Helicobacter pylori*, from human stool samples achieving a LOD of 10^4 CFU/mL [128]. The RNA-cleaving properties of the DNAzyme were activated by a protein biomarker of *H. pylori* and the DNAzyme was identified by an in vitro selection process similar to SELEX. The method required minimal sample processing and was completed in a few minutes.

In relation to virus detection, Kim et al. [129] devised a fast and simple colorimetric assay to detect the human immunodeficiency virus (HIV) from human serum using a previously established functional DNAzyme motif generated by conventional PCR. The key aspect is the design of the primers that target the HIV-1 *gag* gene and, which during PCR amplification, insert a functional DNAzyme sequence in the PCR product. After amplification, hemin is added to assist in the formation of a G-quadruplex structure within the DNAzyme sensor to catalyse the oxidation of 2,2′-azino-bis(3-ethylbenzothiazoline-6-sulfonate) (ABTS). The oxidation of ABTS yields a change from colourless to blueish green which can be visualised and quantified. Recently, Anantharaj et al. [130]. designed a biosensor for the detection of SARS-CoV-2 RNA using the same working principle. This DNAzyme sensor is highly sensitive because it selectively targets the N gene of SARS-CoV-2, which is not present in the genomes of other viruses. The LOD of the sensor was 10^3 copies of viral RNA. Other recent advances of G-quadruplex DNAzyme based biosensors were revised by Xi et al. [131].

Similar to aptamers, DNAzymes exhibit remarkable selectivity to their targets and therefore continue to find use as recognition molecules for a wide range of applications.

3.4.3. Peptide Nucleic Acids (PNAs)

PNAs are artificial molecules composed of a polypeptide backbone with nucleic acid bases attached as side chains. They are notable nucleic acid analogues due to their unique physicochemical and biochemical attributes, stability and striking hybridisation attributes [74]. Due to these characteristics, PNAs have wide application in molecular diagnosis [132].

A paper-based colorimetric multiplex sensor using a Pyrrolidinyl PNA (acpcPNA) probe for the detection of Middle East Respiratory Syndrome coronavirus (MERS-CoV), human papillomavirus (HPV) and *M. tuberculosis* was successfully developed by Teengam et al. [133]. The probe was based on a previously described motif [134], wherein a PNA conjugated with modified AgNPs induces aggregation of the AgNPs in the absence of complementary target DNA. When the target DNA is present, a DNA–acpcPNA duplex is formed, originating the dispersion of the AgNPs and a concomitant detectable colour change.

PNA Fluorescence In Situ Hybridisation (PNA-FISH) has become useful for the specific, reliable and rapid detection of human pathogens. Machado et al. [135] developed a novel PNA-FISH method with specificity for *Lactobacillus* and *Gardnerella vaginalis*. Specificity and sensitivity of the PNA probes were 98% and 100% for *Lactobacillus* and *G. vaginalis*, respectively. Furthermore, the probes were evaluated in samples mimicking the vaginal microflora of patients with a bacterial vaginosis infection, to demonstrate their applicability for diagnosis. PNA-FISH was also recently used by Rocha et al. [136] in the detection of *L. monocytogenes* using a previously developed PNA probe (LmPNA1253) selected by Almeida et al. [137] coupled with a new blocker probe. The method was able to detect *L. monocytogenes* with an LOD of 0.5 CFU/mL in certain food samples.

PNA probes have also been used in the detection of clinical viruses, such as the hepatitis C virus (HCV), which causes chronic liver disease. Ahour et al. [138] developed an electrochemical sensor for HCV detection based on a 20-mer PNA probe that targets a highly conserved consensus sequence present in core/E1 domain from HCV genome. This sequence was cloned in a recombinant plasmid that was used to test the electrode. The PNA was able to hybridise to this sequence without being necessary to denature the plasmid. This represents an advantage because the DNA in nature is in a double-stranded form. Therefore, this method can be used for identification of all HCV genotypes through direct detection. For detection, a gold (Au) electrode modified with cysteine and conjugated with the PNA probe was used. The hybridisation detection was performed by monitoring the difference between the voltametric response of methylene blue (which serves as an electroactive indicator) accumulated on the PNA–modified Au electrode before and after the hybridisation event.

It can be concluded that PNAs hybridise more efficiently with complementary DNA and RNA because they are neutral and their interactions lack charge repulsion which is present in other nucleic acid-based probes. Therefore, they are highly useful when used in conjunction with technologies such as FISH to create simple, rapid and highly accurate microbial detection sensors [139]. In addition, sensors employing PNAs have also been suggested to have great potential in the detection and diagnosis of COVID-19, and can thus aid in curtailing its spread [140].

3.5. CRISPR-Cas

CRISPR are a family of DNA sequences originating in bacteriophages that have previously infected prokaryotes and subsequently become incorporated into their genomes as a defence mechanism to recognise foreign nucleic acid sequences and eliminate them by using the endonuclease activity associated with an enzyme called Cas. Cas evolved in prokaryotes for defence against invading viruses by cleaving their nucleic acid like a pair of scissors [141,142]. CRISPR RNA (crRNA) guides Cas to recognise and cleave target nucleic acids; thus, crRNA can be programmed towards any specific DNA or RNA of interest such as pathogenic genetic material, for instance, by hybridising to a complementary sequence [143]. In this manner, CRISPR-Cas has been repurposed as a gene editing tool, in disease treatment and diagnosis [144]. For diagnostic applications, CRISPR-Cas systems have been used to sense nucleic acid-based pathogenic biomarkers with single-base resolution.

An example of such a system is CRISPR-Cas9, which leverages its sequence-specific nuclease activity to distinguish between viral lineages. To illustrate, a recent Zika virus outbreak prompted Pardee et al. [145] to develop a widely regarded workflow for a portable, low-cost colorimetric sensor which couples isothermal RNA amplification and toehold switches on a paper-based platform. The sensor was further coupled with a CRISPR-Cas9 module that has the ability to distinguish between strains of a virus with single-base resolution. Briefly, in the presence of the target RNA, a Cas9-mediated cleavage is triggered, resulting in a truncated RNA product that is unable to activate the sensor toehold switch and does not produce a colour change in the test paper. While in the presence of non-target viral RNA, the full-length RNA product comprising the sequence to trigger the sensor is generated, thus activating the sensor and producing a colour change on test paper. Detection of Zika virus from monkey plasma infected with the virus was achieved in the low femtomolar range. Additionally, Ai et al. [146] also developed a rapid assay using CRISPR for detection of *M. tuberculosis*. The method combined an amplification step of the target bacterial sequence by recombinase polymerase amplification (RPA) followed by a Cas12a detection step. After the amplification step, the presence of target RNA activates the Cas12a cleavage system, which in turn triggers a colour change in an ssDNA reporter that is present. The detection system is extremely sensitive with almost single-copy sensitivity. More recently, Kellner et al. [147] established a platform that also combines RPA with CRISPR-Cas to detect target RNA or DNA sequences. This platform was named SHERLOCK, which stands for specific high-sensitivity enzymatic reporter unlocking. In addition to being portable and extremely sensitive in the detection of DNA or RNA from real clinical samples, this platform is able to detect multiple targets through the use of a multiplex fluorescence-based detection system.

For the detection of SARS-CoV-2, Hou et al. [148] proposed an alternative to the standard reverse transcription quantitative polymerase chain reaction (RT-qPCR) detection by means of a rapid assay based on polymerase-mediated amplification and CRISPR/Cas13a. Figure 6 illustrates the steps of the assay. This isothermal method is highly advantageous because it does not require expensive and bulky thermocycler equipment and only takes 40 min. To test the novel assay, 52 RNA samples from patients with COVID-19 were subjected to mNGS and three potential target sequences were identified. Subsequently, CRISPR gRNAs and RPA primers were designed and screened. A primer set that targeted open reading frame 1ab (*orf1ab*) displayed the best specificity and sensitivity and was used to develop the CRISPR assay, which was based on T7 transcription and a Cas13 detection step. To evaluate the specificity of the CRISPR assay, target viral DNA was substituted with human DNA and a panel of bacterial and viral pathogens. None of these test samples caused a false positive reaction. Further, the CRISPR assay demonstrated 100% sensitivity because it was able to detect all 52 cases of COVID-19. In the future, the role of CRISPR-associated nucleases can be expanded for direct diagnostic testing of nucleic acids due to their exceptional single molecule sensitivity.

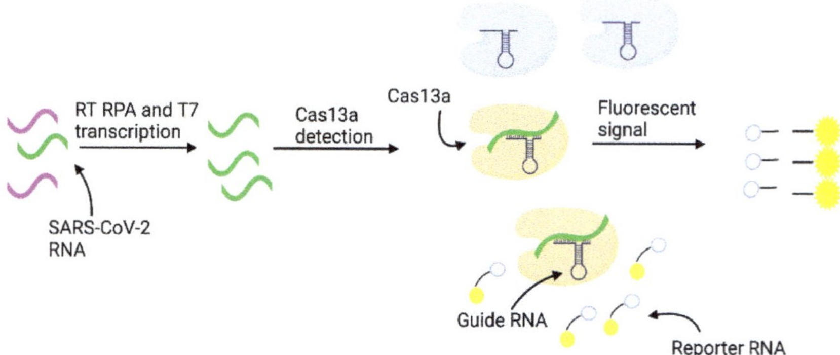

Figure 6. Schematic illustration of the Clustered Regularly Interspaced Short Palindromic Repeats/Cas13a (CRISPR/Cas13a) system for detection of SARS-CoV-2 ribonucleic acid (RNA) proposed by Hou et al. [148]. Reverse-Transcription Recombinase Polymerase Amplification (RT-RPA) followed by a T7 transcription is used to amplify the SARS-CoV-2 RNA. In the next step, the nuclease activity of Cas13a is activated when the guide RNA binds specifically to the open reading frame 1ab (*orf1ab*) gene and triggers the cleavage of the RNA reporter. The cleaved RNA reporter produces a fluorescent signal for detection of SARS-CoV-2. Created with BioRender.com.

3.6. Bacteriophages

Bacteriophages (also termed phages) are a type of viruses that infect and replicate within their target bacteria. Due to their high specificity, conferred by receptor binding proteins (RBPs) on the bacteriophage surfaces with which they target bacteria, they have potential for application in diagnostic tools and treatments against bacterial infection [149–152]. For instance, Liana et al. [153] functionalised MNPs with a high density of T4 bacteriophages and subsequently used them to capture *E. coli*. T4 bacteriophages infect *E. coli* and the one used in the reported study had specificity to *E. coli* type B (ATCC 11303) by means of their tail fibres. In addition to the density of bacteriophage loaded on the MNPs, the authors reported the important effects of tryptone presence in the medium and the incubation temperature used to grow *E. coli* in the capture capabilities of the MNP probes. The T4 bacteriophages were found to bind irreversibly to *E. coli* at 37 °C in tryptone-containing media (rich in tryptophane) during an incubation time of just 10 min.

In a different study, tosyl-activated MBs were functionalised with a bacteriophage PAP1, which is highly specific to *P. aeruginosa*, to establish a bacteriophage-affinity strategy for its detection [151]. The bacteriophage tail fibres and baseplate identified and captured *P. aeruginosa* onto the MBs. Subsequently, the bacteriophage replication cycle proceeded for about 100 min after which the progenies lysed the bacteria causing the release of intracellular ATP. A firefly luciferase-ATP bioluminescence system was used to quantify the captured bacteria. The LOD was determined to be 2×10^2 CFU/mL and the process of capture and detection was completed within 2 h. To estimate the suitability of this strategy for POC diagnosis, glucose, human urine and rat plasma samples were spiked with *P. aeruginosa* at various concentrations and bacteria recovery tests with the functionalised MBs were performed. Recovery rates ranged from 77.4% to 96.9%, demonstrating good reliability for the detection of bacteria in complex samples [151]. In an example of multiplex detection [150], two phage RBPs, *gp*18 and *gp*109, with potential specificity to the genera *Enterococcus* or *Staphylococcus*, respectively, were identified in silico. The RBPs were fused with fluorescent proteins and used in spectrofluorometric assays with the target bacteria for their multiplex detection by fluorescence. Additionally, *gp18* also showed high sensitivity to *E. faecium* and *E. faecalis* by binding to 80% and 100% of tested strains, respectively.

The exceptional specificity displayed by bacteriophages for their target render them prime candidates for diagnostic platforms. However, more bacteriophages that target other dangerous pathogens need to be further discovered or engineered to build promising diagnostic systems that can also function as therapeutics.

3.7. Molecularly Imprinted Polymers (MIPs)

Artificial material-based biorecognition elements rely on the specific morphology or shape of the target for selective capture [11,154]. MIPs include cell imprinted polymers (CIPs), which are the most common examples of MIPs in the context of biorecognition. Some processes used to produce MIPs include micro-contact stamping, bacteria-mediated lithography and colloid imprints. For example, Khan et al. [155] fabricated a MIP by imprinting the bacterial flagella of *P. mirabilis* onto electropolymerised phenol. The flagellar protein imprint sites have rebinding ability in the presence of a sample containing *P. mirabilis*. The MIP was evaluated for biorecognition in an electrochemical biosensor to detect *P. mirabilis* during a rebinding event. The sensor was found to be highly sensitive with an LOD of 0.7 ng/mL.

There are many more examples on the use of CIP in diagnosis. For example, Golabi et al. [156] reported an electrochemical biosensor employing CIPs that specifically recognise *S. epidermidis*. The CIPs' fabrication is based on the polymerisation of 3-aminophenylboronic acid. The features of the imprinted surface include complementary cavities at the polymer surface, presenting structural specificity in terms of shape and size, but also chemical specificity via diol molecules which are present on the cell walls of *S. epidermidis*. The sensor response was reported as being proportional to log 10^3–10^7 CFU/mL of *S. epidermidis* and was highly specific for the target strain when compared to non-target species such as *E. coli, Deinococcus proteolyticus* and *S. pneumoniae*. Similarly, a polydopamine-based CIP was imprinted with template *E. coli* O157:H7 for capturing bacteria together with a polyclonal antibody [157]. By electro-chemiluminescent detection, a very low LOD of 8 CFU/mL was measured. There are other successful demonstrations of bacterial biorecognition by employing CIPs reporting high specificity and sensitivity [158,159].

CIPs are also widely applied in the detection of virus in diagnostics [160]. Furthermore, Cai et al. [161] developed an MIP-based sensor that exhibited exceptional specificity for E7 protein derived from the HPV [162]. The MIP was built by imprinting the tips of nanotube arrays with a polyphenol nanocoating. The coating was non-conducting and detection was performed by electrochemical impedance spectroscopy. The high specificity of the sensor was confirmed when another similar protein of HPV called E6 was not recognised by the E7 MIPs. The E7 protein was recognised with sub picogram per litre sensitivity, surpassing that achieved by conventional MIPs and comparable with nanosensors based on biomolecular recognition with ligands. A similarly constructed MIP system was developed by Ma et al. [163] for specific recognition of the HIV-p24 capsid protein of the HIV virus.

Detection by MIPs is label-free and can be even more affordable than ligand-based biorecognition platforms. However, during a binding event between template and molecules, conversion of the resulting signal by detection systems needs to be improved, in addition to elimination of noise in the signal.

3.8. Antibiotics

Although most antibiotics lack high selectivity due to their broad-spectrum activity, in combination with certain upgrades that provide or improve selectivity, they have proven to be effective for biorecognition of pathogenic targets, as illustrated by the examples presented in this section [164,165]. Antibiotics are widely available, highly stable and exhibit strong binding capabilities to bacteria, making them excellent for combinatorial strategies in biorecognition or in pre-enrichment steps in sensing platforms [164]. However, this approach can only be considered for bacterial species that have not developed resistance to the antibiotic under consideration.

3.8.1. Vancomycin

Vancomycin (Van) is a well-known broad-spectrum lycopeptide antibiotic that interacts with a range of Gram-positive bacteria. Lycopeptide antibiotics are known to bind to the D-alanyl-D-alanine (D-Ala-D-Ala) dipeptide groups present in the cell wall of bacterial strains through hydrogen bonds [166]. Therefore, Van is popularly used in the functionalisation of MBs as bacterial capture probes and in combination with secondary recognition molecules to improve specificity [164]. For example, Yang et al. [167] designed multivalent "brush-like magnetic nanoprobes" and demonstrated their potential for the efficient enrichment of pathogens. To construct the brush-like magnetic nanoprobes, commercial amino-MBs were modified with poly-L-lysine (PLL), followed by the connection of polyethylene glycol (PEG) to the amine sites of PLL. Van was previously linked to the carboxyl molecule of PEG. By using these nanoprobes, an enrichment efficiency greater than 94% was achieved, and also an excellent recovery of *L. monocytogenes* was obtained within 20 min, at a bacterial concentration of 10^2 CFU/mL. Moreover, Meng et al. [168] synthesised similar PEGylated MNPs functionalised with Van (Van-PMs), also to capture and enrich the virulent foodborne pathogen *L. monocytogenes* from spiked lettuce samples prior to detection by PCR. The Van-PMs displayed a high capture efficiency of around 83% and 90% with LOD of 30 CFU/g and 30 CFU/mL in lettuce samples and in PBS, respectively. All steps, including enrichment and PCR, were accomplished in 4 h.

On another study, Yang et al. [164] synthesised MBs with a size of 100 nm functionalised with Van to capture *S. aureus* prior to chemiluminescent detection. To improve the specificity of these Van-MBs to *S. aureus*, rabbit Ig G (IgG) tagged with ALP was used as a second recognition molecule, because the Fc region of rabbit IgG binds to protein A in *S. aureus* surface. The resulting sandwich complex of Van-MBs-*S. aureus*-IgG significantly improved the specificity due to the recognition of *S. aureus* at two distinct sites. In addition, it facilitated ultrasensitive chemiluminescent detection of *S. aureus* in a linear range of $1.2–12 \times 10^6$ CFU/mL and with a very low LOD (3.3 CFU/mL). The entire process was completed in 75 min.

More recently, Wang et al. [169], developed an efficient SERS biosensor that combined Van-modified Fe_3O_4 AgNPs and gold and silver (Au-Ag) NPs for enrichment and sensitive and specific detection of bacteria. The high-performance Van-Fe_3O_4@Ag MNPs served as effective capture probes for Gram-positive bacteria such as *E. coli*, *S. aureus* and methicillin-resistant *Staphylococcus aureus* (MRSA), achieving a LOD of 5×10^2 cells/mL. Then, the plasmonic Au-Ag NPs were used as secondary NPs to increase the detection sensitivity.

3.8.2. Amoxicillin

Some bacteria possess so-called penicillin binding proteins (PBPs) which generate β-lactamase enzymes to resist β-lactam producing antibiotics such as amoxicillin and penicillin by breaking the lactam amide bond. Due to this interaction, the β-lactam in amoxicillin has selective affinity to PBPs present on the bacterial cell membrane [170]. Based on this chemical interaction, Hasan et al. [171] synthesised an effective enrichment probe against *S. aureus* and *E. coli* by modifying MNPs with 3-aminopropyltriethoxysilane and amoxicillin to target PBPs. The modified MNPs successfully captured bacteria forming aggregates that are then separated using an external magnet. MALDI-TOF was used to confirm the interaction between the amoxicillin and bacterial PBPs. The LOD for *S. aureus* and *E. coli* was in the range of $10^3–10^4$ CFU/mL using MALDI-TOF after 5 min incubation with amoxicillin functionalised MNPs.

3.8.3. Ampicillin

Ampicillin has all the stated advantages of antibiotics for use in biorecognition. It is widely available, bringing down costs, and has broad-spectrum high-affinity binding properties for bacteria. These strong affinity properties of ampicillin and the specificity of antibodies were combined to design a LFA for detection of *Salmonella enteritidis* [165]. Ampicillin coated on MNPs facilitated the initial enrichment of *S. enteritidis* from spiked food samples by capture and magnetic separation. Subsequently, the enriched sample was applied to the LFA. The anti-*S. enteritidis* mAB with high specificity towards *S. enteritidis* dispensed on the LFA functioned as a specific enrichment agent, forming a sandwich complex with the ampicillin-MNPs. The colour change was detected by the naked eye at an LOD of 10^2–10^3 CFU/mL.

3.8.4. Neomycin

Neomycin is an aminoglycoside broad spectrum oligosaccharide antibiotic that targets the 30S ribosomal subunit and interferes with decoding and translocation, which, consequently, inhibits protein synthesis in bacteria [172]. Zhang et al. [173] developed lipid conjugated neomycin-based probes that selectively identify and label antibiotic-resistant bacteria simultaneously without disrupting host immune cells. In addition, this probe has theranostic abilities, such as the presence of neomycin, which inactivates the bacteria in addition to detection. This is useful because the probe can have therapeutic and diagnostic applications [10]. The lipidated probes exhibited highly specific and strong fluorescent signals against MRSA and an increased inhibition effect compared to probes with only neomycin. This increased specificity and decreased bacterial resistance is related to lipid chains being able to increase the membrane permeability to the antibiotic. Therefore, this strategy has the potential to detect bacterial infections with high selectivity and sensitivity, and enhance the antibacterial effect of antibiotics without harm to host cells.

Clearly, a number of antibiotics have been tested for their potential as capture probes and, in combination with selective ligands, can be very effective for bacterial identification.

3.9. Chemical Compounds

In addition to the classes of biorecognition elements exhaustively discussed above, various molecules not covered by these classes also exhibit affinity and selectivity for pathogens and can be exploited for use in sensing platforms. In one such example, Pang et al. [174] designed a platform partly composed of maltohexaose-decorated cholesterol that is able to target bacteria both in vitro and in vivo through a bacteria-specific maltodextrin transporter pathway. They investigated the theranostic capabilities of this smart nanoliposome-based platform (MLP18) for targeting of multi-drug resistant strains of MRSA and β-lactamase *E. coli* in different mouse models and subsequent delivery of purpurin 18 (P18), a sonosensitiser for sonodynamic therapy. After injection in a mouse-model, the MLP18 selectively directs P18 to the bacterial infection site. Fluorescence imaging showed that the MLP18 produced a specific fluorescence signal at the site of infection that remained for 24 h post-injection. At the site of sterile inflammation only a negligible fluorescence signal was detected, demonstrating the bacterial-targeting abilities of MLP18.

Moreover, Fu et al. [175] investigated the considerable potential of chemical ligand discovery from natural sources by the coupling of affinity MS and metabolomic approaches. They identified two ligands, 18β-glycyrrhetinic acid and licochalcone A derived from liquorice root used in traditional Chinese medicine which were able to bind and disrupt the nucleoprotein of EBV and Marburg virus. Ligand binding assays and various biophysical analyses were used to evaluate the interactions of the ligands with the bacterial nucleoproteins.

During infection, pathogens interact with host tissue and some of these interactions are specific, such as when pathogens bind to specific host cell membrane proteins. For example, human extracellular matrix proteins, such as collagen and fibronectin, are known receptors for outer membrane protein adhesins present on various microbial pathogens during the initial stages of infection [176]. These interactions can be exploited by coating collagen or fibronectin on MBs, for example, and using them to capture and detect pathogens using sensing platforms.

4. Future Perspectives

A paradigm shift is taking place in the way infectious diseases, predominantly caused by pathogenic bacteria and viruses, are being diagnosed. However, even though highly technological methods that can highly sensitively identify pathogens from samples emerged a few decades ago, this has not been translated into a faster, more efficient and accurate diagnosis for most people suffering with various infections due to their inaccessibility. As discussed, it is evident that diagnostics in the form of simple POC tests, biosensors and kits are the future and can greatly improve outcomes for patients due to a timely diagnosis. Development of workflows to detect the myriad pathogens is underway and is dominated by various sensing platforms. In this review, the latest biorecognition elements in workflows and detection systems are categorised and a number of recent application examples for each are elucidated. The biorecognition elements that hold the greatest potential for incorporation into POC diagnostics are the nucleic acid derivatives, if they can equal antibodies in performance parameters. The incorporation of such elements can drive down the cost of biosensors. Antibody-based sensors will continue to be popular but their market share may shrink. Alternatively, CRISPR technology has an emerging potential for diagnostics among all the biorecognition elements herein discussed due to their impressive sensitivity. However, extensive research is needed for each infectious disease to choose the right combination of Cas enzyme and to design the best system.

Nevertheless, there are some bottlenecks in the translation of sensing platforms comprising these biorecognition elements from bench-top to clinical. A vast array of development in the field is available, although not well organised. There is no standard method to determine which platforms have the potential to achieve commercial development and need more funding, for instance. Furthermore, biorecognition elements need to be subject to some kind of standardisation procedures, in addition to parameters related to their targets, such as LOD values. Collaboration between academia, clinical practitioners and industries that can invest in the further development of POC sensing platforms is critical, followed by consideration of regulatory clearance and introduction to commercial markets.

5. Conclusions

In conclusion, based on the evidence presented in this review, antibodies currently remain the biorecognition element of choice, but alternatives, especially in the form of nucleic acid derivatives, are being sought.

Antibodies are still dominantly used in commercial immunoassays and immunosensors, and are at various stages of research. This is due to their ability to provide a combination of strong target affinity and exceptional selectivity. Furthermore, due to their extensive use over many decades, the field of antibody selection is well developed with a vast number of molecules available and well characterised for selection based on the envisaged application. However, alternatives are increasingly being sought due to their disadvantages, including a cumbersome development and selection process that is expensive, in addition to their inherent instability, which can make them difficult to incorporate in sensor kits. The advantages and disadvantages of using enzymes as biorecognition elements are similar to those of antibodies. Due to their natural origin, they display superior affinity properties to targets, but they also suffer from instability issues, and are difficult to isolate and process in a laboratory environment. Possibly due to this issue, the implementation of enzymes as biorecognition elements for diagnostic applications is still limited and needs further development.

As mentioned, the main alternatives to antibodies being studied are derivatives of nucleic acids such as aptamers, PNAs, DNAzymes and antibody-derived fragments. The main advantage of those molecules is that they can be developed much more cheaply than antibodies. In some instances, they have demonstrated target affinity comparable with their antibody counterparts while exhibiting excellent stability and reproducibility, which is a vital requirement for POC diagnostics. They are also versatile enough to be combined with most POC sensor detection platforms, including electrochemical, optical, colorimetric and LFA. Their main disadvantage is that they require a selection process that is time-consuming and sometimes difficult to perform. However, as illustrated in this review, there are numerous excellent examples of nucleic acid derivative-based biorecognition elements in sensing platforms, which will certainly translate to their extensive application in POC sensors in the near future.

Classes of compounds and molecules such as antibiotics and peptides, which were traditionally used for anti-infective and therapeutic purposes, have seen their broad-spectrum activity against pathogens re-purposed for pathogen identification in diagnostics. Due to their wide range, they are usually used for pre-enrichment or combined with other biorecognition elements to enhance the sensitivity of the sensor. A huge number of AMPs are under scrutiny; however, their emerging disadvantage is the lack of selectivity.

Bacteriophages present excellent selectivity due to their innate targeting function, which is similar to that of antibodies. Development of bacteriophage-based sensing platforms is in a nascent stage compared to the other biorecognition elements, but their high target specificity makes them promising candidates for biorecognition.

MIPs have been included in this review due to their importance and demonstrated capabilities in pathogen-sensing platforms, although ligand molecules are not usually used for biorecognition in these cases. Their advantages include lower production cost compared to antibodies, high batch-to-batch reproducibility, and chemical and mechanical robustness, and do not involve animals' sacrifice. MIPs can be extremely versatile because almost any target can be imprinted for their detection and have great potential in sensing platforms.

CRISPR-Cas technology, although still emerging in the diagnostic landscape, has already shown immense potential with platforms such as SHERLOCK-enabled multiplexed and ultra-sensitive detection of DNA or RNA from clinical samples. The main advantage of CRISPR is that its single-base resolution selectivity is unmatched by any other biorecognition element. The nature of the technology enables it to be leveraged in POC diagnostic sensing platforms. Finally, chemical compounds such as metabolites, polysaccharides and several other chemicals are useful for targeting specific pathogenic targets for which they have a known affinity interaction. In conclusion, depending on the application, one or a combination of more suitable biorecognition element(s) investigated in this review can provide high performance biorecognition capabilities in POC diagnostics.

Author Contributions: Literature review, M.G.S.; Writing—original draft preparation, M.G.S.; Writing—Review & Editing, J.L.R., D.F., C.J.S. and L.R.R.; Conceptualization, Supervision and Funding Acquisition, L.R.R. All authors have read and agreed to the published version of the manuscript.

Funding: This research is affiliated with the VibrANT project that received funding from the EU Horizon 2020 Research and Innovation Programme under the Marie Sklowdowska-Curie Grant, agreement no 765042. In addition, the authors acknowledge the financial support from Fundação para a Ciência e Tecnologia (FCT) under the scope of the strategic funding of UID/BIO/04469/2020 unit. Débora Ferreira (DF) is the recipient of a fellowship supported by a doctoral advanced training (call NORTE-69-2015-15) funded by the European Social Fund under the scope of Norte2020.

References

1. World Health Organization. The Top 10 Causes of Death. Available online: http://www.who.int/en/news-room/fact-sheets/detail/the-top-10-causes-of-death (accessed on 16 March 2021).
2. World Health Organisation. New Report Calls for Urgent Action to Avert Antimicrobial Resistance Crisis. Available online: https://www.who.int/news/item/29-04-2019-new-report-calls-for-urgent-action-to-avert-antimicrobial-resistance-crisis (accessed on 16 March 2021).
3. Gu, W.; Miller, S.; Chiu, C.Y. Clinical Metagenomic Next-Generation Sequencing for Pathogen Detection. *Annu. Rev. Pathol. Mech. Dis.* **2019**, *14*, 319–338. [CrossRef] [PubMed]
4. Mahony, J.B.; Petrich, A.; Smieja, M. Molecular diagnosis of respiratory virus infections. *Crit. Rev. Clin. Lab. Sci.* **2011**, *48*, 217–249. [CrossRef]
5. Sin, M.L.; Mach, K.E.; Wong, P.K.; Liao, J.C. Advances and challenges in biosensor-based diagnosis of infectious diseases. *Expert Rev. Mol. Diagn.* **2014**, *14*, 225–244. [CrossRef] [PubMed]

6. Rudd, K.E.; Johnson, S.C.; Agesa, K.M.; Shackelford, K.A.; Tsoi, D.; Kievlan, D.R.; Colombara, D.V.; Ikuta, K.S.; Kissoon, N.; Finfer, S.; et al. Global, regional, and national sepsis incidence and mortality, 1990–2017: Analysis for the Global Burden of Disease Study. *Lancet* **2020**, *395*, 200–211. [CrossRef]

7. Urdea, M.; Penny, L.A.; Olmsted, S.S.; Giovanni, M.Y.; Kaspar, P.; Shepherd, A.; Wilson, P.; Dahl, C.A.; Buchsbaum, S.; Moeller, G.; et al. Requirements for high impact diagnostics in the developing world. *Nature* **2006**, *444*, 73–79. [CrossRef]

8. Dhiman, A.; Kalra, P.; Bansal, V.; Bruno, J.G.; Sharma, T.K. Aptamer-based point-of-care diagnostic platforms. *Sens. Actuators B Chem.* **2017**, *246*, 535–553. [CrossRef]

9. Morales, M.A.; Halpern, J.M. Guide to Selecting a Biorecognition Element for Biosensors. *Bioconjug. Chem.* **2018**, *29*, 3231–3239. [CrossRef]

10. Lim, E.K.; Kim, T.; Paik, S.; Haam, S.; Huh, Y.M.; Lee, K. Nanomaterials for theranostics: Recent advances and future challenges. *Chem. Rev.* **2015**, *115*, 327–394. [CrossRef] [PubMed]

11. Cesewski, E.; Johnson, B.N. Electrochemical biosensors for pathogen detection. *Biosens. Bioelectron.* **2020**, *159*, 112214. [CrossRef]

12. Bahadir, E.B.; Sezgintürk, M.K. Applications of commercial biosensors in clinical, food, environmental, and biothreat/biowarfare analyses. *Anal. Biochem.* **2015**, *478*, 107–120. [CrossRef]

13. Sharma, A.; Sharma, N.; Kumari, A.; Lee, H.J.; Kim, T.Y.; Tripathi, K.M. Nano-carbon based sensors for bacterial detection and discrimination in clinical diagnosis: A junction between material science and biology. *Appl. Mater. Today* **2020**, *18*, 100467. [CrossRef]

14. Coris Bioconcept Clostridium K-SeT. Available online: https://www.corisbio.com/products/clostridium-k-set (accessed on 1 June 2021).

15. Thermofisher Scientific Oxoid™ Clostridium difficile Test Kit. Available online: https://www.thermofisher.com/order/catalog/product/DR1107A#/DR1107A (accessed on 1 June 2021).

16. Meridian Bioscience ImmunoCard STAT! E. coli O157 Plus. Available online: https://www.meridianbioscience.com/human-condition/gastrointestinal/e-coli/immunocard-stat-e-coli-o157-plus/ (accessed on 1 June 2021).

17. Thermofisher Scientific Escherichia coli O157 Latex Test. Available online: https://www.thermofisher.com/order/catalog/product/DR0620M#/DR0620M (accessed on 1 June 2021).

18. Thermofisher Scientific Wellcogen™ *Haemophilus influenzae* b Rapid Latex Agglutination. Test. Available online: https://www.thermofisher.com/order/catalog/product/R30858801#/R30858801 (accessed on 1 June 2021).

19. Quidel QuickVue, H. pylori Test. Available online: https://www.quidel.com/immunoassays/rapid-h-pylori-tests (accessed on 1 June 2021).

20. Wilburn Medical USA Beckman Coulter 395160A Icon HP (H.Pylori) Test Kit. Available online: https://wilburnmedicalusa.com/beckman-coulter-395160a-icon-hp-h-pylori-test-kit/ (accessed on 1 June 2021).

21. Coris Bioconcept Pylori-Strip. Available online: https://www.corisbio.com/products/pylori-strip (accessed on 1 June 2021).

22. Coris Bioconcept Pylori K-SeT. Available online: https://www.corisbio.com/products/pylori-k-set (accessed on 1 June 2021).

23. Thermofisher Scientific Remel™ Catarrhalis Test Disc. Available online: https://www.thermofisher.com/order/catalog/product/R21121#/R21121 (accessed on 1 June 2021).

24. Thermofisher Scientific BactiStaph™ Latex Agglutination Test Kit. Available online: https://www.thermofisher.com/order/catalog/product/R21143#/R21143 (accessed on 1 June 2021).

25. Thermofisher Scientific DrySpot™ Pneumo Latex Agglutination Test. Available online: https://www.thermofisher.com/order/catalog/product/DR0420M#/DR0420M (accessed on 1 June 2021).

26. Abbott BinaxNOW™ Streptococcus Pneumoniae Antigen Card. Available online: https://www.globalpointofcare.abbott/en/product-details/binaxnow-streptococcus-pneumoniae-ww.html (accessed on 1 June 2021).

27. Thermofisher Scientific Infectious Mononucleosis Test Using Latex. Agglutination. Available online: https://www.thermofisher.com/order/catalog/product/DR0780M#/DR0780M (accessed on 1 June 2021).

28. Quidel QuickVue Influenza A + B Test. Available online: https://www.quidel.com/immunoassays/rapid-influenza-tests/quickvue-influenza-test (accessed on 1 June 2021).

29. Abbott Alere BinaxNOW® Influenza A & B Card. Available online: https://www.globalpointofcare.abbott/en/product-details/binaxnow-influenza-a-and-b.html (accessed on 1 June 2021).

30. Lucira Health COVID-19 All-In-One Test Kit. Available online: https://www.lucirahealth.com/ (accessed on 1 June 2021).

31. Abbott BinaxNOW COVID-19 Ag Card Home Test. Available online: https://www.globalpointofcare.abbott/en/product-details/binaxnow-covid-19-home-test-us.html (accessed on 1 June 2021).

32. Ellume Health Ellume COVID-19 Home Test. Available online: https://www.ellumehealth.com/products/consumer-products/covid-home-test (accessed on 1 June 2021).

33. Abbott BinaxNOW® Malaria. Available online: https://www.globalpointofcare.abbott/en/product-details/binaxnow-malaria.html (accessed on 1 June 2021).

34. Prada, Y.A.; Soler, M.; Guzmán, F.; Castillo, J.J.; Lechuga, L.M.; Mejía-Ospino, E. Design and characterization of high-affinity synthetic peptides as bioreceptors for diagnosis of cutaneous leishmaniasis. *Anal. Bioanal. Chem.* **2021**, *413*, 4545–4555. [CrossRef]

35. Curk, T.; Brackley, C.A.; Farrell, J.D.; Xing, Z.; Joshi, D.; Direito, S.; Bren, U.; Angioletti-Uberti, S.; Dobnikar, J.; Eiser, E.; et al. Computational design of probes to detect bacterial genomes by multivalent binding. *Proc. Natl. Acad. Sci. USA* **2020**, *117*, 8719–8726. [CrossRef] [PubMed]

36. Chiang, A.D.; Dekker, J.P. From the pipeline to the bedside: Advances and challenges in clinical metagenomics. *J. Infect. Dis.* **2020**, *221*, S331–S340. [CrossRef] [PubMed]

37. O'Flaherty, B.M.; Li, Y.; Tao, Y.; Paden, C.R.; Queen, K.; Zhang, J.; Dinwiddie, D.L.; Gross, S.M.; Schroth, G.P.; Tong, S. Comprehensive viral enrichment enables sensitive respiratory virus genomic identification and analysis by next generation sequencing. *Genome Res.* **2018**, *28*, 869–877. [CrossRef]

38. Wylie, K.M.; Wylie, T.N.; Buller, R.; Herter, B.; Cannella, M.T.; Storch, G.A. Detection of viruses in clinical samples by use of metagenomic sequencing and targeted sequence capture. *J. Clin. Microbiol.* **2018**, *56*. [CrossRef] [PubMed]

39. Singh, A.; Arutyunov, D.; Szymanski, C.M.; Evoy, S. Bacteriophage based probes for pathogen detection. *Analyst* **2012**, *137*, 3405–3421. [CrossRef]

40. Crivianu-Gaita, V.; Thompson, M. Aptamers, antibody scFv, and antibody Fab' fragments: An overview and comparison of three of the most versatile biosensor biorecognition elements. *Biosens. Bioelectron.* **2016**, *85*, 32–45. [CrossRef] [PubMed]

41. Foudeh, A.M.; Fatanat Didar, T.; Veres, T.; Tabrizian, M. Microfluidic designs and techniques using lab-on-a-chip devices for pathogen detection for point-of-care diagnostics. *Lab Chip* **2012**, *12*, 3249–3266. [CrossRef] [PubMed]

42. Park, C.; Lee, J.; Kim, Y.; Kim, J.; Lee, J.; Park, S. 3D-printed microfluidic magnetic preconcentrator for the detection of bacterial pathogen using an ATP luminometer and antibody-conjugated magnetic nanoparticles. *J. Microbiol. Methods* **2017**, *132*, 128–133. [CrossRef]

43. Jang, H.; Hwang, E.Y.; Kim, Y.; Choo, J.; Jeong, J.; Lim, D.W. Surface-enhanced raman scattering and fluorescence-based dual nanoprobes for multiplexed detection of bacterial pathogens. *J. Biomed. Nanotechnol.* **2016**, *12*, 1938–1951. [CrossRef] [PubMed]

44. Ilhan, H.; Guven, B.; Dogan, U.; Torul, H.; Evran, S.; Çetin, D.; Suludere, Z.; Saglam, N.; Boyaci, İ.H.; Tamer, U. The coupling of immunomagnetic enrichment of bacteria with paper-based platform. *Talanta* **2019**, *201*, 245–252. [CrossRef]

45. Peláez, E.C.; Estevez, M.C.; Mongui, A.; Menéndez, M.C.; Toro, C.; Herrera-Sandoval, O.L.; Robledo, J.; García, M.J.; Del Portillo, P.; Lechuga, L.M. Detection and Quantification of HspX Antigen in Sputum Samples Using Plasmonic Biosensing: Toward a Real Point-of-Care (POC) for Tuberculosis Diagnosis. *ACS Infect. Dis.* **2020**, *6*, 1110–1120. [CrossRef]

46. Kaushik, A.; Yndart, A.; Kumar, S.; Jayant, R.D.; Vashist, A.; Brown, A.N.; Li, C.Z.; Nair, M. A sensitive electrochemical immunosensor for label-free detection of Zika-virus protein. *Sci. Rep.* **2018**, *8*, 3–7. [CrossRef]

47. Reid, R.; Chatterjee, B.; Das, S.J.; Ghosh, S.; Sharma, T.K. Application of aptamers as molecular recognition elements in lateral flow assays. *Anal. Biochem.* **2020**, *593*, 113574. [CrossRef] [PubMed]

48. Oue, S.; Okamoto, A.; Yano, T.; Kagamiyama, H. Redesigning the substrate specificity of an enzyme by cumulative effects of the mutations of non-active site residues. *J. Biol. Chem.* **1999**, *274*, 2344–2349. [CrossRef]

49. Liu, J.; Lu, L.; Xu, S.; Wang, L. One-pot synthesis of gold nanoclusters with bright red fluorescence and good biorecognition Abilities for visualization fluorescence enhancement detection of *E. coli*. *Talanta* **2015**, *134*, 54–59. [CrossRef] [PubMed]

50. Vocadlo, D.J.; Davies, G.J.; Laine, R.; Withers, S.G. Catalysis by hen egg-white lysozyme proceeds via a covalent intermediate. *Nature* **2001**, *412*, 835–838. [CrossRef] [PubMed]

51. Couniot, N.; Vanzieleghem, T.; Rasson, J.; Van Overstraeten-Schlögel, N.; Poncelet, O.; Mahillon, J.; Francis, L.A.; Flandre, D. Lytic enzymes as selectivity means for label-free, microfluidic and impedimetric detection of whole-cell bacteria using ALD-Al$_2$O$_3$ passivated microelectrodes. *Biosens. Bioelectron.* **2015**, *67*, 154–161. [CrossRef]

52. Clemente, A.; Alba-Patiño, A.; Rojo-Molinero, E.; Russell, S.M.; Borges, M.; Oliver, A.; De La Rica, R. Rapid Detection of Pseudomonas aeruginosa Biofilms via Enzymatic Liquefaction of Respiratory Samples. *ACS Sens.* **2020**, *5*, 3956–3963. [CrossRef]

53. Pavan, S.; Berti, F. Short peptides as biosensor transducers. *Anal. Bioanal. Chem.* **2012**, *402*, 3055–3070. [CrossRef]

54. Rodrigues, J.L.; Ferreira, D.; Rodrigues. L.R. Synthetic biology strategies towards the development of new bioinspired technologies for medical applications. In *Bioinspired Materials for Medical Applications*; Rodrigues, L., Mota, M., Eds.; Woodhead Publishing: Cambridge, UK, 2017; pp. 451–497. ISBN 9780081007464.

55. Xu, C.; Akakuru, O.U.; Zheng, J.; Wu, A. Applications of iron oxide-based magnetic nanoparticles in the diagnosis and treatment of bacterial infections. *Front. Bioeng. Biotechnol.* **2019**, *7*, 141. [CrossRef] [PubMed]

56. Mannoor, M.S.; Zhang, S.; Link, A.J.; McAlpine, M.C. Electrical detection of pathogenic bacteria via immobilized antimicrobial peptides. *Proc. Natl. Acad. Sci. USA* **2010**, *107*, 19207–19212. [CrossRef] [PubMed]

57. Hoyos-Nogués, M.; Gil, F.J.; Mas-Moruno, C. Antimicrobial peptides: Powerful biorecognition elements to detect bacteria in biosensing technologies. *Molecules* **2018**, *23*, 1683. [CrossRef] [PubMed]

58. Zhang, X.; Ren, C.; Hu, F.; Gao, Y.; Wang, Z.; Li, H.; Liu, J.; Liu, B.; Yang, C. Detection of Bacterial Alkaline Phosphatase Activity by Enzymatic in Situ Self-Assembly of the AIEgen-Peptide Conjugate. *Anal. Chem.* **2020**, *92*, 5185–5190. [CrossRef] [PubMed]

59. Yuan, K.; Mei, Q.; Guo, X.; Xu, Y.; Yang, D.; Sánchez, B.J.; Sheng, B.; Liu, C.; Hu, Z.; Yu, G.; et al. Antimicrobial peptide based magnetic recognition elements and Au@Ag-GO SERS tags with stable internal standards: A three in one biosensor for isolation, discrimination and killing of multiple bacteria in whole blood. *Chem. Sci.* **2018**, *9*, 8781–8795. [CrossRef]

60. Dao, T.N.T.; Lee, E.Y.; Koo, B.; Jin, C.E.; Lee, T.Y.; Shin, Y. A microfluidic enrichment platform with a recombinase polymerase amplification sensor for pathogen diagnosis. *Anal. Biochem.* **2018**, *544*, 87–92. [CrossRef]

61. Azmi, S.; Jiang, K.; Stiles, M.; Thundat, T.; Kaur, K. Detection of *Listeria monocytogenes* with short peptide fragments from class IIa bacteriocins as recognition elements. *ACS Comb. Sci.* **2015**, *17*, 156–163. [CrossRef]

62. Etayash, H.; Jiang, K.; Thundat, T.; Kaur, K. Impedimetric detection of pathogenic gram-positive bacteria using an antimicrobial peptide from class IIa bacteriocins. *Anal. Chem.* **2014**, *86*, 1693–1700. [CrossRef]

63. Arcidiacono, S.; Pivarnik, P.; Mello, C.M.; Senecal, A. Cy5 labeled antimicrobial peptides for enhanced detection of *Escherichia coli* O157:H7. *Biosens. Bioelectron.* **2008**, *23*, 1721–1727. [CrossRef]

64. De Miranda, J.L.; Oliveira, M.D.L.; Oliveira, I.S.; Frias, I.A.M.; Franco, O.L.; Andrade, C.A.S. A simple nanostructured biosensor based on clavanin A antimicrobial peptide for gram-negative bacteria detection. *Biochem. Eng. J.* **2017**, *124*, 108–114. [CrossRef]

65. Akram, A.R.; Avlonitis, N.; Scholefield, E.; Vendrell, M.; McDonald, N.; Aslam, T.; Craven, T.H.; Gray, C.; Collie, D.S.; Fisher, A.J.; et al. Enhanced avidity from a multivalent fluorescent antimicrobial peptide enables pathogen detection in a human lung model. *Sci. Rep.* **2019**, *9*, 8422. [CrossRef]

66. Wright, A.T.; Anslyn, E.V. Differential receptor arrays and assays for solution-based molecular recognition. *Chem. Soc. Rev.* **2006**, *35*, 14–28. [CrossRef]

67. Kulagina, N.V.; Shaffer, K.M.; Ligler, F.S.; Taitt, C.R. Antimicrobial peptides as new recognition molecules for screening challenging species. *Sens. Actuators B Chem.* **2007**, *121*, 150–157. [CrossRef] [PubMed]

68. Fu, M.Q.; Wang, X.C.; Dou, W.T.; Chen, G.R.; James, T.D.; Zhou, D.M.; He, X.P. Supramolecular fluorogenic peptide sensor array based on graphene oxide for the differential sensing of ebola virus. *Chem. Commun.* **2020**, *56*, 5735–5738. [CrossRef]

69. Li, Y.; Lai, D.; Lei, Q.; Xu, Z.; Wang, F.; Hou, H.; Chen, L.; Wu, J.; Ren, Y.; Liang, M.M.; et al. Systematic evaluation of IgG responses to SARS-CoV-2 spike protein-derived peptides for monitoring COVID-19 patients. *Cell. Mol. Immunol.* **2021**, *18*, 621–631. [CrossRef] [PubMed]

70. Cai, X.F.; Chen, J.; Hu, J.l.; Long, Q.X.; Deng, H.J.; Liu, P.; Fan, K.; Liao, P.; Liu, B.Z.; Wu, G.C.; et al. A peptide-based magnetic chemiluminescence enzyme immunoassay for serological diagnosis of coronavirus disease 2019. *J. Infect. Dis.* **2020**, *222*, 189–195. [CrossRef]

71. Pomplun, S.; Jbara, M.; Quartararo, A.J.; Zhang, G.; Brown, J.S.; Lee, Y.C.; Ye, X.; Hanna, S.; Pentelute, B.L. De Novo Discovery of High-Affinity Peptide Binders for the SARS-CoV-2 Spike Protein. *ACS Cent. Sci.* **2021**, *7*, 156–163. [CrossRef]

72. Tuerk, C.; Gold, L. Systematic evolution of ligands by exponential enrichment: RNA ligands to bacteriophage T4 DNA polymerase. *Science* **1990**, *249*, 505–510. [CrossRef]

73. Mcconnell, E.M.; Cozma, I.; Morrison, D.; Li, Y. Biosensors made of synthetic functional nucleic acids toward better human health. *Anal. Chem.* **2020**, *92*, 327–344. [CrossRef] [PubMed]

74. Saadati, A.; Hassanpour, S.; de la Guardia, M.; Mosafer, J.; Hashemzaei, M.; Mokhtarzadeh, A.; Baradaran, B. Recent advances on application of peptide nucleic acids as a bioreceptor in biosensors development. *TrAC Trends Anal. Chem.* **2019**, *114*, 56–68. [CrossRef]

75. Wang, C.H.; Wu, J.J.; Lee, G. Bin Screening of highly-specific aptamers and their applications in paper-based microfluidic chips for rapid diagnosis of multiple bacteria. *Sens. Actuators B Chem.* **2019**, *284*, 395–402. [CrossRef]

76. Sousa, D.; Ferreira, D.; Rodrigues, J.L.; Rodrigues, L.R. Nanotechnology in Targeted Drug Delivery and Therapeutics. In *Applications of Targeted Nano Drugs and Delivery Systems*; Mohapatra, S.S., Ranjan, S., Dasgupta, N., Mishra, R.K., Thomas, S., Eds.; Elsevier: Amsterdam, The Netherlands, 2019; pp. 357–409.

77. Dwivedi, H.P.; Smiley, R.D.; Jaykus, L.A. Selection and characterization of DNA aptamers with binding selectivity to Campylobacter jejuni using whole-cell SELEX. *Appl. Microbiol. Biotechnol.* **2010**, *87*, 2323–2334. [CrossRef]

78. Meirinho, S.G.; Dias, L.G.; Peres, A.M.; Rodrigues, L.R. Voltammetric aptasensors for protein disease biomarkers detection: A review. *Biotechnol. Adv.* **2016**, *34*, 941–953. [CrossRef]

79. Dias, L.G.; Meirinho, S.G.; Veloso, A.C.A.; Rodrigues, L.R.; Peres, A.M. Electronic tongues and aptasensors. In *Bioinspired Materials for Medical Applications*; Rodrigues, L., Mota, M., Eds.; Woodhead Publishing: Cambridge, UK, 2017; pp. 371–402. ISBN 9780081007464.

80. Meirinho, S.G.; Dias, L.G.; Peres, A.M.; Rodrigues, L.R. Electrochemical aptasensor for human osteopontin detection using a DNA aptamer selected by SELEX. *Anal. Chim. Acta* **2017**, *987*, 25–37. [CrossRef] [PubMed]

81. Sefah, K.; Shangguan, D.; Xiong, X.; O'Donoghue, M.B.; Tan, W. Development of DNA aptamers using cell-selex. *Nat. Protoc.* **2010**, *5*, 1169–1185. [CrossRef]

82. Meirinho, S.G.; Dias, L.G.; Peres, A.M.; Rodrigues, L.R. Development of an electrochemical RNA-aptasensor to detect human osteopontin. *Biosens. Bioelectron.* **2015**, *71*, 332–341. [CrossRef]

83. Savory, N.; Lednor, D.; Tsukakoshi, K.; Abe, K.; Yoshida, W.; Ferri, S.; Jones, B.V.; Ikebukuro, K. In silico maturation of binding-specificity of DNA aptamers against Proteus mirabilis. *Biotechnol. Bioeng.* **2013**, *110*, 2573–2580. [CrossRef] [PubMed]

84. Tang, X.L.; Zhou, Y.X.; Wu, S.M.; Pan, Q.; Xia, B.; Zhang, X.L. CFP10 and ESAT6 aptamers as effective Mycobacterial antigen diagnostic reagents. *J. Infect.* **2014**, *69*, 569–580. [CrossRef]

85. Weldingh, K.; Andersen, P. ESAT-6/CFP10 skin test predicts disease in *M. tuberculosis*-infected Guinea pigs. *PLoS ONE* **2008**, *3*, 1978. [CrossRef] [PubMed]

86. Wang, J.; Wu, H.; Yang, Y.; Yan, R.; Zhao, Y.; Wang, Y.; Chen, A.; Shao, S.; Jiang, P.; Li, Y.Q. Bacterial species-identifiable magnetic nanosystems for early sepsis diagnosis and extracorporeal photodynamic blood disinfection. *Nanoscale* **2018**, *10*, 132–141. [CrossRef] [PubMed]

87. Hao, L.; Gu, H.; Duan, N.; Wu, S.; Ma, X.; Xia, Y.; Wang, H.; Wang, Z. A chemiluminescent aptasensor based on rolling circle amplification and Co^{2+}/N-(aminobutyl)-N-(ethylisoluminol) functional flowerlike gold nanoparticles for *Salmonella typhimurium* detection. *Talanta* **2017**, *164*, 275–282. [CrossRef]

88. Murakami, T.; Sumaoka, J.; Komiyama. M. Sensitive isothermal detection of nucleic-acid sequence by primer generation-rolling circle amplification. *Nucleic Acids Res.* **2009**, *37*, e19. [CrossRef]

89. Trunzo, N.E.; Hong, K.L. Recent progress in the identification of aptamers against bacterial origins and their diagnostic applications. *Int. J. Mol. Sci.* **2020**, *21*, 5074. [CrossRef]

90. Park, K.S. Nucleic acid aptamer-based methods for diagnosis of infections. *Biosens. Bioelectron.* **2018**, *102*, 179–188. [CrossRef] [PubMed]

91. Zou, X.; Wu, J.; Gu, J.; Shen, L.; Mao, L. Application of aptamers in virus detection and antiviral therapy. *Front. Microbiol.* **2019**, *10*, 1462. [CrossRef]

92. Li, L.; Li, Q.; Liao, Z.; Sun, Y.; Cheng, Q.; Song, Y.; Song, E.; Tan, W. Magnetism-Resolved Separation and Fluorescence Quantification for Near-Simultaneous Detection of Multiple Pathogens. *Anal. Chem.* **2018**, *90*, 9621–9628. [CrossRef]

93. Xu, Y.; Wang, H.; Luan, C.; Liu, Y.; Chen, B.; Zhao, Y. Aptamer-based hydrogel barcodes for the capture and detection of multiple types of pathogenic bacteria. *Biosens. Bioelectron.* **2018**, *100*, 404–410. [CrossRef]

94. Sande, M.G.; Çaykara, T.; Silva, C.J.; Rodrigues, L.R. New solutions to capture and enrich bacteria from complex samples. *Med. Microbiol. Immunol.* **2020**, *209*, 335–341. [CrossRef]

95. Shahrokhian, S.; Ranjbar, S. Aptamer immobilization on amino-functionalized metal-organic frameworks: An ultrasensitive platform for the electrochemical diagnostic of: *Escherichia coli* O157:H7. *Analyst* **2018**, *143*, 3191–3201. [CrossRef]

96. Abbaspour, A.; Norouz-Sarvestani, F.; Noori, A.; Soltani, N. Aptamer-conjugated silver nanoparticles for electrochemical dual-aptamer-based sandwich detection of *Staphylococcus aureus*. *Biosens. Bioelectron.* **2015**, *68*, 149–155. [CrossRef] [PubMed]

97. Huang, Y.; Chen, X.; Xia, Y.; Wu, S.; Duan, N.; Ma, X.; Wang, Z. Selection, identification and application of a DNA aptamer against Staphylococcus aureus enterotoxin A. *Anal. Methods* **2014**, *6*, 690–697. [CrossRef]

98. Pang, Y.; Wan, N.; Shi, L.; Wang, C.; Sun, Z.; Xiao, R.; Wang, S. Dual-recognition surface-enhanced Raman scattering(SERS)biosensor for pathogenic bacteria detection by using vancomycin-SERS tags and aptamer-Fe3O4@Au. *Anal. Chim. Acta* **2019**, *1077*, 288–296. [CrossRef]

99. Hamula, C.L.A.; Peng, H.; Wang, Z.; Tyrrell, G.J.; Li, X.F.; Le, X.C. An improved SELEX technique for selection of DNA aptamers binding to M-type 11 of Streptococcus pyogenes. *Methods* **2016**, *97*, 51–57. [CrossRef] [PubMed]

100. Sypabekova, M.; Bekmurzayeva, A.; Wang, R.; Li, Y.; Nogues, C.; Kanayeva, D. Selection, characterization, and application of DNA aptamers for detection of *Mycobacterium tuberculosis* secreted protein MPT64. *Tuberculosis* **2017**, *104*, 70–78. [CrossRef] [PubMed]

101. Qiao, J.; Meng, X.; Sun, Y.; Li, Q.; Zhao, R.; Zhang, Y.; Wang, J.; Yi, Z. Aptamer-based fluorometric assay for direct identification of methicillin-resistant Staphylococcus aureus from clinical samples. *J. Microbiol. Methods* **2018**, *153*, 92–98. [CrossRef] [PubMed]

102. Zhong, Z.; Gao, X.; Gao, R.; Jia, L. Selective capture and sensitive fluorometric determination of Pseudomonas aeruginosa by using aptamer modified magnetic nanoparticles. *Microchim. Acta* **2018**, *185*. [CrossRef] [PubMed]

103. Nguyen, V.T.; Seo Bin, H.; Kim, B.C.; Kim, S.K.; Song, C.S.; Gu, M.B. Highly sensitive sandwich-type SPR based detection of whole H5Nx viruses using a pair of aptamers. *Biosens. Bioelectron.* **2016**, *86*, 293–300. [CrossRef]

104. Chen, C.; Zou, Z.; Chen, L.; Ji, X.; He, Z. Functionalized magnetic microparticle-based colorimetric platform for influenza A virus detection. *Nanotechnology* **2016**, *27*, 435102. [CrossRef]

105. Sung, H.J.; Kayhan, B.; Ben-Yedidia, T.; Arnon, R. A DNA aptamer prevents influenza infection by blocking the receptor binding region of the viral hemagglutinin. *J. Biol. Chem.* **2004**, *279*, 48410–48419. [CrossRef]

106. Lee, K.H.; Zeng, H. Aptamer-Based ELISA Assay for Highly Specific and Sensitive Detection of Zika NS1 Protein. *Anal. Chem.* **2017**, *89*, 12743–12748. [CrossRef]

107. Giamberardino, A.; Labib, M.; Hassan, E.M.; Tetro, J.A.; Springthorpe, S.; Sattar, S.A.; Berezovski, M.V.; DeRosa, M.C. Ultrasensitive norovirus detection using DNA aptasensor technology. *PLoS ONE* **2013**, *8*, 79087. [CrossRef]

108. Bai, C.; Lu, Z.; Jiang, H.; Yang, Z.; Liu, X.; Ding, H.; Li, H.; Dong, J.; Huang, A.; Fang, T.; et al. Aptamer selection and application in multivalent binding-based electrical impedance detection of inactivated H1N1 virus. *Biosens. Bioelectron.* **2018**, *110*, 162–167. [CrossRef] [PubMed]

109. Dey, A.K.; Griffiths, C.; Lea, S.M.; James, W. Structural characterization of an anti-gp120 RNA aptamer that neutralizes R5 strains of HIV-1. *RNA* **2005**, *11*, 873–884. [CrossRef] [PubMed]

110. Abbasi, A.D.; Hussain, Z.; Yang, K.L. Aptamer laden liquid crystals biosensing platform for the detection of HIV-1 glycoprotein-120. *Molecules* **2021**, *26*, 2893. [CrossRef]

111. Chekin, F.; Bagga, K.; Subramanian, P.; Jijie, R.; Singh, S.K.; Kurungot, S.; Boukherroub, R.; Szunerits, S. Nucleic aptamer modified porous reduced graphene oxide/MoS2 based electrodes for viral detection: Application to human papillomavirus (HPV). *Sens. Actuators B Chem.* **2018**, *262*, 991–1000. [CrossRef]

112. Leija-Montoya, A.G.; Benítez-Hess, M.L.; Toscano-Garibay, J.D.; Alvarez-Salas, L.M. Characterization of an RNA aptamer against HPV-16 L1 virus-like particles. *Nucleic Acid Ther.* **2014**, *24*, 344–355. [CrossRef]

113. Song, Y.; Song, J.; Wei, X.; Huang, M.; Sun, M.; Zhu, L.; Lin, B.; Shen, H.; Zhu, Z.; Yang, C. Discovery of Aptamers Targeting the Receptor-Binding Domain of the SARS-CoV-2 Spike Glycoprotein. *Anal. Chem.* **2020**, *92*, 9895–9900. [CrossRef] [PubMed]

114. Zhang, L.; Fang, X.; Liu, X.; Ou, H.; Zhang, H.; Wang, J.; Li, Q.; Cheng, H.; Zhang, W.; Luo, Z. Discovery of sandwich type COVID-19 nucleocapsid protein DNA aptamers. *Chem. Commun.* **2020**, *56*, 10235–10238. [CrossRef]

115. Chen, Z.; Wu, Q.; Chen, J.; Ni, X.; Dai, J. A DNA Aptamer Based Method for Detection of SARS-CoV-2 Nucleocapsid Protein. *Virol. Sin.* **2020**, *35*, 351–354. [CrossRef] [PubMed]
116. Schmitz, A.; Weber, A.; Bayin, M.; Breuers, S.; Fieberg, V.; Famulok, M.; Mayer, G. A SARS-CoV-2 Spike Binding DNA Aptamer that Inhibits Pseudovirus Infection by an RBD-Independent Mechanism. *Angew. Chem. Int. Ed.* **2021**, *60*, 10279–10285. [CrossRef] [PubMed]
117. Du, K.; Cai, H.; Park, M.; Wall, T.A.; Stott, M.A.; Alfson, K.J.; Griffiths, A.; Carrion, R.; Patterson, J.L.; Hawkins, A.R.; et al. Multiplexed efficient on-chip sample preparation and sensitive amplification-free detection of Ebola virus. *Biosens. Bioelectron.* **2017**, *91*, 489–496. [CrossRef] [PubMed]
118. Bruno, J.G. Predicting the uncertain future of aptamer-based diagnostics and therapeutics. *Molecules* **2015**, *20*, 6866–6887. [CrossRef] [PubMed]
119. Lakhin, A.V.; Tarantul, V.Z.; Gening, L.V. Aptamers: Problems, solutions and prospects. *Acta Nat.* **2013**, *5*, 34–43. [CrossRef]
120. Bauer, M.; Strom, M.; Hammond, D.S.; Shigdar, S. Anything you can do, I can do better: Can aptamers replace antibodies in clinical diagnostic applications? *Molecules* **2019**, *24*, 4377. [CrossRef]
121. Silverman, S.K. DNA as a versatile chemical component for catalysis, encoding, and stereocontrol. *Angew. Chem. Int. Ed.* **2010**, *49*, 7180–7201. [CrossRef]
122. Breaker, R.R.; Joyce, G.F. A DNA enzyme that cleaves RNA. *Chem. Biol.* **1994**, *1*, 223–229. [CrossRef]
123. Swearingen, C.B.; Wernette, D.P.; Cropek, D.M.; Lu, Y.; Sweedler, J.V.; Bohn, P.W. Immobilization of a catalytic DNA molecular beacon on Au for Pb(II) detection. *Anal. Chem.* **2005**, *77*, 442–448. [CrossRef]
124. Geng, X.; Zhang, M.; Wang, X.; Sun, J.; Zhao, X.; Zhang, L.; Wang, X.; Shen, Z. Selective and sensitive detection of chronic myeloid leukemia using fluorogenic DNAzyme probes. *Anal. Chim. Acta* **2020**, *1123*, 28–35. [CrossRef]
125. Zheng, L.; Qi, P.; Zhang, D. DNA-templated fluorescent silver nanoclusters for sensitive detection of pathogenic bacteria based on MNP-DNAzyme-AChE complex. *Sens. Actuators B Chem.* **2018**, *276*, 42–47. [CrossRef]
126. Kang, D.K.; Ali, M.M.; Zhang, K.; Huang, S.S.; Peterson, E.; Digman, M.A.; Gratton, E.; Zhao, W. Rapid detection of single bacteria in unprocessed blood using Integrated Comprehensive Droplet Digital Detection. *Nat. Commun.* **2014**, *5*, 5427. [CrossRef] [PubMed]
127. Ali, M.M.; Slepenkin, A.; Peterson, E.; Zhao, W. A Simple DNAzyme-Based Fluorescent Assay for Klebsiella pneumoniae. *ChemBioChem* **2019**, *20*, 906–910. [CrossRef] [PubMed]
128. Ali, M.M.; Wolfe, M.; Tram, K.; Gu, J.; Filipe, C.D.M.; Li, Y.; Brennan, J.D. A DNAzyme-Based Colorimetric Paper Sensor for *Helicobacter pylori*. *Angew. Chem.* **2019**, *131*, 10012–10016. [CrossRef]
129. Kim, S.U.; Batule, B.S.; Mun, H.; Byun, J.Y.; Shim, W.B.; Kim, M.G. Colorimetric molecular diagnosis of the HIV gag gene using DNAzyme and a complementary DNA-extended primer. *Analyst* **2018**, *143*, 695–699. [CrossRef]
130. Anantharaj, A.; Das, S.J.; Sharanabasava, P.; Lodha, R.; Kabra, S.K.; Sharma, T.K.; Medigeshi, G.R. Visual Detection of SARS-CoV-2 RNA by Conventional PCR-Induced Generation of DNAzyme Sensor. *Front. Mol. Biosci.* **2020**, *7*, 444. [CrossRef] [PubMed]
131. Xi, H.; Juhas, M.; Zhang, Y. G-quadruplex based biosensor: A potential tool for SARS-CoV-2 detection. *Biosens. Bioelectron.* **2020**, *167*, 112494. [CrossRef]
132. Lee, H.T.; Kim, S.K.; Yoon, J.W. Antisense peptide nucleic acids as a potential anti-infective agent. *J. Microbiol.* **2019**, *57*, 423–430. [CrossRef]
133. Teengam, P.; Siangproh, W.; Tuantranont, A.; Vilaivan, T.; Chailapakul, O.; Henry, C.S. Multiplex Paper-Based Colorimetric DNA Sensor Using Pyrrolidinyl Peptide Nucleic Acid-Induced AgNPs Aggregation for Detecting MERS-CoV, MTB, and HPV Oligonucleotides. *Anal. Chem.* **2017**, *89*, 5428–5435. [CrossRef]
134. Lee, J.S.; Lytton-Jean, A.K.R.; Hurst, S.J.; Mirkin, C.A. Silver nanoparticle—Oligonucleotide conjugates based on DNA with triple cyclic disulfide moieties. *Nano Lett.* **2007**, *7*, 2112–2115. [CrossRef]
135. Machado, A.; Almeida, C.; Salgueiro, D.; Henriques, A.; Vaneechoutte, M.; Haesebrouck, F.; Vieira, M.J.; Rodrigues, L.; Azevedo, N.F.; Cerca, N. Fluorescence in situ Hybridization method using Peptide Nucleic Acid probes for rapid detection of *Lactobacillus* and *Gardnerella* spp. *BMC Microbiol.* **2013**, *13*, 82. [CrossRef]
136. Rocha, R.; Sousa, J.M.; Cerqueira, L.; Vieira, M.J.; Almeida, C.; Azevedo, N.F. Development and application of Peptide Nucleic Acid Fluorescence in situ Hybridization for the specific detection of *Listeria monocytogenes*. *Food Microbiol.* **2019**, *80*, 1–8. [CrossRef]
137. Almeida, C.; Azevedo, N.F.; Santos, S.; Keevil, C.W.; Vieira, M.J. Discriminating multi-species populations in biofilms with peptide nucleic acid fluorescence in situ hybridization (PNA FISH). *PLoS ONE* **2011**, *6*, 14786. [CrossRef] [PubMed]
138. Ahour, F.; Pournaghi-Azar, M.H.; Alipour, E.; Hejazi, M.S. Detection and discrimination of recombinant plasmid encoding hepatitis C virus core/E1 gene based on PNA and double-stranded DNA hybridization. *Biosens. Bioelectron.* **2013**, *45*, 287–291. [CrossRef]
139. Oliveira, R.; Almeida, C.; Azevedo, N.F. Detection of Microorganisms by Fluorescence In Situ Hybridization Using Peptide Nucleic Acid. In *Methods in Molecular Biology*; Nielsen, P., Ed.; Humana Press Inc.: New York, NY, USA, 2020; Volume 2105, pp. 217–230.
140. Talebian, S.; Wallace, G.G.; Schroeder, A.; Stellacci, F.; Conde, J. Nanotechnology-based disinfectants and sensors for SARS-CoV-2. *Nat. Nanotechnol.* **2020**, *15*, 618–621. [CrossRef] [PubMed]
141. Wang, X.; Shang, X.; Huang, X. Next-generation pathogen diagnosis with CRISPR/Cas-based detection methods. *Emerg. Microbes Infect.* **2020**, *9*, 1682–1691. [CrossRef]

142. Cong, L.; Ran, F.A.; Cox, D.; Lin, S.; Barretto, R.; Habib, N.; Hsu, P.D.; Wu, X.; Jiang, W.; Marraffini, L.A.; et al. Multiplex genome engineering using CRISPR/Cas systems. *Science* **2013**, *339*, 819–823. [CrossRef]
143. Kaminski, M.M.; Abudayyeh, O.O.; Gootenberg, J.S.; Zhang, F.; Collins, J.J. CRISPR-based diagnostics. *Nat. Biomed. Eng.* **2021**, *5*, 643–656. [CrossRef]
144. Khambhati, K.; Bhattacharjee, G.; Singh, V. Current progress in CRISPR-based diagnostic platforms. *J. Cell. Biochem.* **2019**, *120*, 2721–2725. [CrossRef] [PubMed]
145. Pardee, K.; Green, A.A.; Takahashi, M.K.; Braff, D.; Lambert, G.; Lee, J.W.; Ferrante, T.; Ma, D.; Donghia, N.; Fan, M.; et al. Rapid, Low-Cost Detection of Zika Virus Using Programmable Biomolecular Components. *Cell* **2016**, *165*, 1255–1266. [CrossRef]
146. Ai, J.W.; Zhou, X.; Xu, T.; Yang, M.; Chen, Y.; He, G.Q.; Pan, N.; Cai, Y.; Li, Y.; Wang, X.; et al. CRISPR-based rapid and ultra-sensitive diagnostic test for *Mycobacterium tuberculosis*. *Emerg. Microbes Infect.* **2019**, *8*, 1361–1369. [CrossRef]
147. Kellner, M.J.; Koob, J.G.; Gootenberg, J.S.; Abudayyeh, O.O.; Zhang, F. SHERLOCK: Nucleic acid detection with CRISPR nucleases. *Nat. Protoc.* **2019**, *14*, 2986–3012. [CrossRef]
148. Hou, T.; Zeng, W.; Yang, M.; Chen, W.; Ren, L.; Ai, J.; Wu, J.; Liao, Y.; Gou, X.; Li, Y.; et al. Development and evaluation of a rapid CRISPR-based diagnostic for COVID-19. *PLoS Pathog.* **2020**, *16*, e1008705. [CrossRef]
149. Eisenstein, M. My enemy's enemy is my friend. *Nat. Methods* **2006**, *3*, 338. [CrossRef]
150. Santos, S.B.; Cunha, A.P.; Macedo, M.; Nogueira, C.L.; Brandão, A.; Costa, S.P.; Melo, L.D.R.; Azeredo, J.; Carvalho, C.M. Bacteriophage-receptor binding proteins for multiplex detection of Staphylococcus and Enterococcus in blood. *Biotechnol. Bioeng.* **2020**, *117*, 3286–3298. [CrossRef] [PubMed]
151. He, Y.; Wang, M.; Fan, E.; Ouyang, H.; Yue, H.; Su, X.; Liao, G.; Wang, L.; Lu, S.; Fu, Z. Highly Specific Bacteriophage-Affinity Strategy for Rapid Separation and Sensitive Detection of Viable *Pseudomonas aeruginosa*. *Anal. Chem.* **2017**, *89*, 1916–1921. [CrossRef]
152. Ackermann, H.W. Tailed bacteriophages: The order caudovirales. *Adv. Virus Res.* **1998**, *51*, 135–201. [CrossRef] [PubMed]
153. Liana, A.E.; Marquis, C.P.; Gunawan, C.; Gooding, J.J.; Amal, R. T4 bacteriophage conjugated magnetic particles for *E. coli* capturing: Influence of bacteriophage loading, temperature and tryptone. *Colloids Surf. B Biointerfaces* **2017**, *151*, 47–57. [CrossRef] [PubMed]
154. Pan, J.; Chen, W.; Ma, Y.; Pan, G. Molecularly imprinted polymers as receptor mimics for selective cell recognition. *Chem. Soc. Rev.* **2018**, *47*, 5574–5587. [CrossRef]
155. Khan, M.A.R.; Aires Cardoso, A.R.; Sales, M.G.F.; Merino, S.; Tomás, J.M.; Rius, F.X.; Riu, J. Artificial receptors for the electrochemical detection of bacterial flagellar filaments from *Proteus mirabilis*. *Sens. Actuators B Chem.* **2017**, *244*, 732–741. [CrossRef]
156. Golabi, M.; Kuralay, F.; Jager, E.W.H.; Beni, V.; Turner, A.P.F. Electrochemical bacterial detection using poly(3-aminophenylboronic acid)-based imprinted polymer. *Biosens. Bioelectron.* **2017**, *93*, 87–93. [CrossRef]
157. Chen, S.; Chen, X.; Zhang, L.; Gao, J.; Ma, Q. Electrochemiluminescence Detection of *Escherichia coli* O157:H7 Based on a Novel Polydopamine Surface Imprinted Polymer Biosensor. *ACS Appl. Mater. Interfaces* **2017**, *9*, 5430–5436. [CrossRef]
158. Idil, N.; Hedström, M.; Denizli, A.; Mattiasson, B. Whole cell based microcontact imprinted capacitive biosensor for the detection of *Escherichia coli*. *Biosens. Bioelectron.* **2017**, *87*, 807–815. [CrossRef]
159. Shan, X.; Yamauchi, T.; Yamamoto, Y.; Shiigi, H.; Nagaoka, T. A rapid and specific bacterial detection method based on cell-imprinted microplates. *Analyst* **2018**, *143*, 1568–1574. [CrossRef] [PubMed]
160. Piletsky, S.; Canfarotta, F.; Poma, A.; Bossi, A.M.; Piletsky, S. Molecularly Imprinted Polymers for Cell Recognition. *Trends Biotechnol.* **2020**, *38*, 368–387. [CrossRef] [PubMed]
161. Cai, D.; Ren, L.; Zhao, H.; Xu, C.; Zhang, L.; Yu, Y.; Wang, H.; Lan, Y.; Roberts, M.F.; Chuang, J.H.; et al. A molecular-imprint nanosensor for ultrasensitive detection of proteins. *Nat. Nanotechnol.* **2010**, *5*, 597–601. [CrossRef]
162. Cui, F.; Zhou, Z.; Zhou, H.S. Molecularly imprinted polymers and surface imprinted polymers based electrochemical biosensor for infectious diseases. *Sensors* **2020**, *20*, 996. [CrossRef] [PubMed]
163. Ma, Y.; Shen, X.L.; Zeng, Q.; Wang, H.S.; Wang, L.S. A multi-walled carbon nanotubes based molecularly imprinted polymers electrochemical sensor for the sensitive determination of HIV-p24. *Talanta* **2017**, *164*, 121–127. [CrossRef]
164. Yang, S.; Ouyang, H.; Su, X.; Gao, H.; Kong, W.; Wang, M.; Shu, Q.; Fu, Z. Dual-recognition detection of Staphylococcus aureus using vancomycin-functionalized magnetic beads as concentration carriers. *Biosens. Bioelectron.* **2016**, *78*, 174–180. [CrossRef]
165. Bu, T.; Yao, X.; Huang, L.; Dou, L.; Zhao, B.; Yang, B.; Li, T.; Wang, J.; Zhang, D. Dual recognition strategy and magnetic enrichment based lateral flow assay toward Salmonella enteritidis detection. *Talanta* **2020**, *206*, 120204. [CrossRef]
166. Kell, A.J.; Stewart, G.; Ryan, S.; Peytavi, R.; Boissinot, M.; Huletsky, A.; Bergeron, M.G.; Simard, B. Vancomycin-modified nanoparticles for efficient targeting and preconcentration of gram-positive and gram-negative bacteria. *ACS Nano* **2008**, *2*, 1777–1788. [CrossRef] [PubMed]
167. Yang, X.; Zhou, X.; Zhu, M.; Xing, D. Sensitive detection of *Listeria monocytogenes* based on highly efficient enrichment with vancomycin-conjugated brush-like magnetic nano-platforms. *Biosens. Bioelectron.* **2017**, *91*, 238–245. [CrossRef] [PubMed]
168. Meng, X.; Li, F.; Li, F.; Xiong, Y.; Xu, H. Vancomycin modified PEGylated-magnetic nanoparticles combined with PCR for efficient enrichment and detection of *Listeria monocytogenes*. *Sens. Actuators B Chem.* **2017**, *247*, 546–555. [CrossRef]
169. Wang, C.W.; Gu, B.; Liu, Q.Q.; Pang, Y.F.; Xiao, R.; Wang, S.Q. Combined use of vancomycin-modified Ag-coated magnetic nanoparticles and secondary enhanced nanoparticles for rapid surface-enhanced Raman scattering detection of bacteria. *Int. J. Nanomed.* **2018**, *13*, 1159–1178. [CrossRef]

170. Kong, K.F.; Schneper, L.; Mathee, K. Beta-lactam antibiotics: From antibiosis to resistance and bacteriology. *Apmis* **2010**, *118*, 1–36. [CrossRef]
171. Hasan, N.; Guo, Z.; Wu, H.F. Large protein analysis of *Staphylococcus aureus* and *Escherichia coli* by MALDI TOF mass spectrometry using amoxicillin functionalized magnetic nanoparticles. *Anal. Bioanal. Chem.* **2016**, *408*, 6269–6281. [CrossRef]
172. Carter, A.P.; Clemons, W.M.; Brodersen, D.E.; Morgan-Warren, R.J.; Wimberly, B.T.; Ramakrishnan, V. Functional insights from the structure of the 30S ribosomal subunit and its interactions with antibiotics. *Nature* **2000**, *407*, 340–348. [CrossRef]
173. Zhang, Q.; Wang, Q.; Xu, S.; Zuo, L.; You, X.; Hu, H.Y. Aminoglycoside-based novel probes for bacterial diagnostic and therapeutic applications. *Chem. Commun.* **2017**, *53*, 1366–1369. [CrossRef]
174. Pang, X.; Xiao, Q.; Cheng, Y.; Ren, E.; Lian, L.; Zhang, Y.; Gao, H.; Wang, X.; Leung, W.; Chen, X.; et al. Bacteria-responsive nanoliposomes as smart sonotheranostics for multidrug resistant bacterial infections. *ACS Nano* **2019**, *13*, 2427–2438. [CrossRef]
175. Fu, X.; Wang, Z.; Li, L.; Dong, S.; Li, Z.; Jiang, Z.; Wang, Y.; Shui, W. Novel Chemical Ligands to Ebola Virus and Marburg Virus Nucleoproteins Identified by Combining Affinity Mass Spectrometry and Metabolomics Approaches. *Sci. Rep.* **2016**, *6*, 29680. [CrossRef]
176. Vaca, D.J.; Thibau, A.; Schütz, M.; Kraiczy, P.; Happonen, L.; Malmström, J.; Kempf, V.A.J. Interaction with the host: The role of fibronectin and extracellular matrix proteins in the adhesion of Gram-negative bacteria. *Med. Microbiol. Immunol.* **2020**, *209*, 277–299. [CrossRef] [PubMed]

biosensors

Review

Fiber Optic Sensors for Vital Signs Monitoring. A Review of Its Practicality in the Health Field

Christian Perezcampos Mayoral [1,*], Jaime Gutiérrez Gutiérrez [2,*], José Luis Cano Pérez [1], Marciano Vargas Treviño [2], Itandehui Belem Gallegos Velasco [3], Pedro António Hernández Cruz [3], Rafael Torres Rosas [4], Lorenzo Tepech Carrillo [2], Judith Arnaud Ríos [5], Edmundo López Apreza [2] and Roberto Rojas Laguna [6]

[1] Doctorado en Biociencias, Facultad de Medicina y Cirugía, Universidad Autónoma "Benito Juárez" de Oaxaca, Ex Hacienda de Aguilera S/N, Calz. San Felipe del Agua, 68050 Oaxaca de Juárez, Mexico; jcano@uabjo.mx

[2] Escuela de Sistemas Biológicos e Innovación Tecnológica, Universidad Autónoma "Benito Juárez" de Oaxaca (SBIT-UABJO), Av. Universidad S/N, Ex-Hacienda 5 Señores, 68120 Oaxaca de Juárez, Mexico; mvargas.cat@uabjo.mx (M.V.T.); ltepech@uabjo.mx (L.T.C.); elopez.cat@uabjo.mx (E.L.A.)

[3] Centro de Investigación Facultad de Medicina UNAM-UABJO, Facultad de Medicina y Cirugía, Universidad Autónoma "Benito Juárez" de Oaxaca, Ex Hacienda de Aguilera S/N, Calz. San Felipe del Agua, 68050 Oaxaca de Juárez, Mexico; itanbel@hotmail.com (I.B.G.V.); phernandez.cat@uabjo.mx (P.A.H.C.)

[4] Facultad de Odontología, Universidad Autónoma "Benito Juárez" de Oaxaca, Av. Universidad S/N, Ex-Hacienda 5 Señores, 68120 Oaxaca de Juárez, Mexico; rtorres.cat@uabjo.mx

[5] Doctorado en Ciencias en Desarrollo Regional y Tecnológico, Tecnológico Nacional de México Campus Oaxaca, Avenida Ing. Víctor Bravo Ahuja No. 125 Esquina Calzada Tecnológico, 68030 Oaxaca de Juárez, Mexico; judith.ar23@gmail.com

[6] Departamento de Electrónica, División de Ingeniería, Universidad de Guanajuato, Carretera Salamanca-Valle de Santiago km 3.5 + 1.8, Comunidad de Palo Blanco, 36885 Salamanca, Mexico; rlaguna@ugto.mx

* Correspondence: cpecmay@uabjo.mx (C.P.M.); jgutierrez.cat@uabjo.mx (J.G.G.)

Citation: Perezcampos Mayoral, C.; Gutiérrez Gutiérrez, J.; Cano Pérez, J.L.; Vargas Treviño, M.; Gallegos Velasco, I.B.; Hernández Cruz, P.A.; Torres Rosas, R.; Tepech Carrillo, L.; Arnaud Ríos, J.; Apreza, E.L.; et al. Fiber Optic Sensors for Vital Signs Monitoring. A Review of Its Practicality in the Health Field. *Biosensors* **2021**, 11, 58. https://doi.org/10.3390/bios11020058

Received: 30 December 2020
Accepted: 19 February 2021
Published: 23 February 2021

Publisher's Note: MDPI stays neutral with regard to jurisdictional claims in published maps and institutional affiliations.

Abstract: Vital signs not only reflect essential functions of the human body but also symptoms of a more serious problem within the anatomy; they are well used for physical monitoring, caloric expenditure, and performance before a possible symptom of a massive failure—a great variety of possibilities that together form a first line of basic diagnosis and follow-up on the health and general condition of a person. This review includes a brief theory about fiber optic sensors' operation and summarizes many research works carried out with them in which their operation and effectiveness are promoted to register some vital sign(s) as a possibility for their use in the medical, health care, and life support fields. The review presents methods and techniques to improve sensitivity in monitoring vital signs, such as the use of doping agents or coatings for optical fiber (OF) that provide stability and resistance to the external factors from which they must be protected in in vivo situations. It has been observed that most of these sensors work with single-mode optical fibers (SMF) in a spectral range of 1550 nm, while only some work in the visible spectrum (Vis); the vast majority, operate through fiber Bragg gratings (FBG), long-period fiber gratings (LPFG), and interferometers. These sensors have brought great advances to the measurement of vital signs, especially with regard to respiratory rate; however, many express the possibility of monitoring other vital signs through mathematical calculations, algorithms, or auxiliary devices. Their advantages due to miniaturization, immunity to electromagnetic interference, and the absence of a power source makes them truly desirable for everyday use at all times.

Keywords: fiber optic sensor; vital signs; biosensor; human body; body temperature; heart rate; respiratory rate; blood pressure

1. Introduction

The penetration of fiber optic sensors in the medical or health care market is very low or almost null in home devices due to their high cost or too many regulations, which hinder their entry; however, there have been small advances as these sensors are used in high precision surgery or in magnetic resonance imaging; their use is increasing in various technologies or ancillary equipment, so it has advanced one step at a time [1].

Since the 1980s [2], optical fiber sensors have been used for real-time pressure measurement of tendons [3] and thereafter in a wide variety of possible medical applications as pressure pads in contact with the skin [4], the cardiovascular system, or even invasive sensors during urodynamic analysis [5]. Many researchers have focused their studies on monitoring and detecting vital signs through fiber optic sensors that are located in some medium that is in contact with the skin [6] and with the capacity to be composed of two or more sensors on the same fiber [7]. Such applications may be placed in textile vestments [8,9] that are very novel in their shape or that have an ingenious way of installing them, while at the same time providing protection to the optical fiber [10]. Because vital signs are the first way to test a person's health and stability, heart pulse monitoring, being a well-studied field [11], is one area that lends itself to continuing ways of innovating and developing improvements in measurement devices and equipment [12,13].

Fiber optic sensors have been used in many applications for the measurement of chemical parameters, liquid flow and levels, and gas detection. Nonetheless, they have been mainly used in the electrical and mechanical fields due to the great advantages that have been attributed to them, such as immunity to electromagnetic interference, apart from the small size of the fiber that makes them perfect for the development of lightweight and mechanically robust sensors [14]. However, in the medical field they have not yet gained wide acceptance since conventional sensors have also made great advances, giving them good user characteristics such as reliability, maintenance and support, and technological integration [15]. There is a great need for a simple system for measuring vital signs for home health monitoring.

2. Sensor Principles

In the next analysis, we considered various studies based on the premise of dividing sensor measurement of four vital signs that are: body temperature, respiration or breathing rate (BR), pulse or heart rate (HR), and blood pressure. Here, we illustrate the main characteristics of each sensor, how measurements can be obtained from each of them, and the peculiarities of these sensors. Furthermore, we examine the common characteristics among other methods of vital signs monitoring. In addition, the basic technologies used by each sensor are briefly presented, including common characteristics, features, and qualitative techniques for their characterization. The tables present the most complete articles in terms of the parameters that we took into consideration for this review, such as the optical fiber (OF) type and wavelength at which the sensor operates, type of technology used, operating ranges, sensitivity, and medium of contact with the test subject.

2.1. Fiber Bragg Gratings (FBGs)

FBGs have been widely used in various sectors of industry to measure deformations [16]. They are periodic diffraction modifications printed on the core of an OF, behaving as selective filters that reflect only the components of the spectrum of a packet propagating in the core according to the Bragg relation

$$\lambda = 2n\Lambda_L \tag{1}$$

where the wavelength is represented as λ, n is the effective refractive index of the core mode, and Λ_L the spatial period of the refractive index modulation. If the FBG is strained along the fiber axis, it changes as does the Bragg wavelength as it moves, resulting in a measurement of strain or temperature [17]; this allows for high resolution, accuracy, and dynamic range suitable for applications such as biomedicine and can be calculated according to the equation:

$$\Delta\lambda_\beta = 2\left(\Lambda\frac{\partial n_{eff}}{\partial\varepsilon} + n_{eff}\frac{\partial\Lambda}{\partial\varepsilon}\right)\Delta\varepsilon_z + 2\left(\Lambda\frac{\partial n_{eff}}{\partial T} + n_{eff}\frac{\partial\Lambda}{\partial T}\right)\Delta T \tag{2}$$

Here, the first term represents the dependence of the pressure on the Bragg wavelength and the second term represents the effect of temperature on the same parameter. It is shown that an FBG can be used as a sensor by observing the light reflected through the FBG for the longitudinal mechanical strain ε_z, and the temperature T. For dynamic loading, the heat input can be neglected and the sensitivity of the FBG for ε_z, is expressed according to the equation:

$$\frac{\Delta\lambda_\beta}{\lambda_\beta} = \left\{\left(1 - \left(\frac{n^2}{2}\right)\right)[p_{12} - v(p_{11} + p_{12})]\right\}\varepsilon_z \tag{3}$$

where v and p_{ij} are mechanical properties of the coating; therefore, Equation (3) can be simplified as:

$$\frac{\Delta\lambda_\beta}{\lambda_\beta} = (1 - p_e)\varepsilon_z \tag{4}$$

When p_e is the optical pressure coefficient of the OF, the Bragg wavelength is directly proportional to the length of the grating because the deformation of the Bragg wavelength can be observed, and the induced deformation can be controlled [18]. Figure 1 shows how an FBG operates, where the core of the fiber is modified through periodic inscriptions that will act as a filter for a certain wavelength of light. Of the transmitted light, the grating will only return the reflected light.

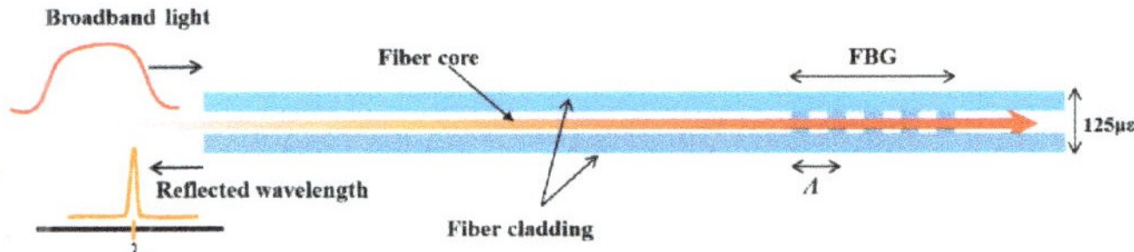

Figure 1. Graphical mode of operation of a fiber Bragg grating (FBG) sensor [19]. In this type of sensor, the grating is inserted into the core of the optical fiber (OF), as shown in a signal where the wavelength filtering is done by such a sensor.

2.1.1. Long-Period Fiber Gratings (LPFGs)

Another common technique to change the fiber nanostructure is through a mechanically induced LPFG (MLPFG). Such a sensor has been shown to perform well as a mode-coupled device, pressure sensor, and filter [20]. Its working principle turns out to be quite as simple as pressing an OF between a ribbed plate and a flat plate, where the ribbed plate has to carry inscribed periods ranging in size from 100 μm to 1000 μm. Figure 2 shows how gratings with these periods inscribed in the core only let through certain light at some wavelength; its feature is that it promotes coupling between LP_{01} core mode and jointly propagating antisymmetric LP_{1m} ($m > 0$) and OF cladding modes, resulting in transmission spectra containing a series of attenuation notches for discrete wavelengths, which satisfy the phase matching condition

$$\lambda_{res}^n = \left(n_{effcore} - n_{effclad(n)}\right)\Lambda \tag{5}$$

where $n_{effcore}$ and $n_{effclad(n)}$ are the effective refractive indices of the core and the nth mode of the cladding at the resonant wavelength λ_{res}^n, and Λ is the period of the grating [21].

However all fiber optic-based devices are considered immune to electromagnetic interference [22]. Fiber gratings can be affected in their wavelength by deformation and temperature. This same cross sensitivity to pressure and temperature has proven to be a major obstacle to the development of practical applications, but, in recent years, interferometer-based sensors to measure pressure have become a major object of analysis and study, becoming quite popular for different laboratory tests [23]. Among those that stand out are interferometers that use the interference of light waves to be able to accurately measure wavelengths of light themselves, very small distances, and certain optical phenomena through two light beams that travel different optical paths, determined by a system of mirrors or plates that in the end converge to form an interference pattern. The Sagnac, Mach–Zehnder interferometer (MZI), Fabry–Pérot interferometer (FPI), and Michelson interferometer stand out as OF sensors based on multimode interference, offering advantages for their low production cost, high sensitivity, and robustness compared to other fiber sensors [24].

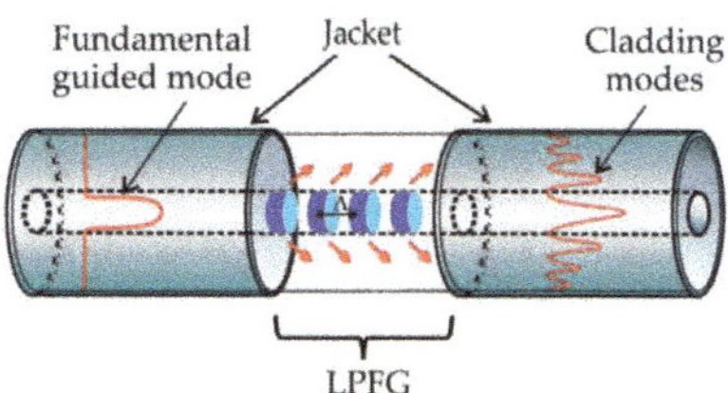

Figure 2. Schematic of a long-period fiber grating (LPFG) and the mode of propagation of light through the core, where gratings separated by defined periods perform the filtering function [25].

These gratings are a periodic modulation induced by mechanical action or through some chemical or laser to generate such periods by modifying the refractive index of the fiber core. The effect of these LPFGs is the induction of a series of attenuation bands in the transmission shape of the fiber that produce wavelengths at which the LPFGs induce a phase coupling between the guided and cladding modes. The study of LPFG attenuation bands has led to many potential applications, including the medical and sensing area, since the wavelengths of the attenuation bands are sensitive to the strain, temperature, and refractive index of the surrounding medium [26].

2.1.2. Mechanical Induction and Fiber Optic Sensors

Mechanical induction through microbend is one of the earliest forms of sensor characterization; the losses caused by these microbends have been a major problem for fiber cable designers, but it has been these same losses that have helped many sensor designers to adapt to the effects caused by the bends to measure many physical parameters and variables, such as temperature and pressure. With some outstanding performance characteristics, they have been successful for some commercial applications. Additionally, there have been many advances in understanding microbend sensors and investigating how to increase the dynamic range and improve sensitivity to the measurement parameter of interest while reducing sensitivity to unwanted variables. Among the great advantages of microbend sensors are mechanical and optical efficiency that allow for low part count and low cost and easy mechanical assembly that does not require fiber optic splices with other components and therefore avoids the problems of thermal expansion difference. The Figure 3, shows an example of how the fiber (yellow) is crushed as force is exerted through the plates when they are brought together, an attempt to inscribe the periods of the plate ribs to the OF. These are fail-safe sensors, as they produce a calibrated output signal in its correct function or when fail completely, immediately going to a no-light output state [27].

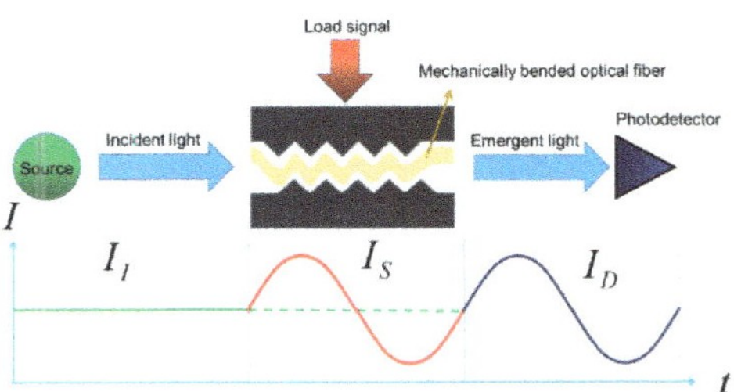

Figure 3. Schematic of a microbend fiber sensor bay mechanical induction [28]. The figure shows the mechanism of how the fiber undergoes microbending to create a sensor and microbending losses caused by disturbances in the optical fiber.

2.2. Optical Fiber Interferometers

Continuing with the operation principles, we can analyze the utility of the interferometers as devices whose operation is in effect caused by the interference of two beams of light, which propagate by different optical paths through one or more optical fibers [29]. It is the form of operation in which a light beam is separated or is grouped that is required for these modules to work with optical fibers; we can basically emphasize four types of interferometers whose configurations are those of Fabry–Pérot, Mach–Zehnder, Michelson, and Sagnac—all of which have been widely used and demonstrated [30]. The main characteristics and differences among them in their application as optical fiber sensors will be described.

Used for extremely precise measurements in very small distances, its design also is known as "etalon". Most conventional optical interferometers have their equivalent in OF. In these interferometers the light is guided by the fibers and the splitter plates are replaced by directional couplers for OF. The mirrors in some cases can be replaced by the fiber ends themselves properly terminated. Figure 4, shows the shape of an optical arrangement of an FPI, where the light beam from the source is reflected multiple times (causing interference) through the parallel surfaces, and the reflected light is sent to the detector by an optical lens.

In optical fibers, one way that it is possible to be created is through a ring (gyroscope) that works as a guide of a wave; this is possible if the optical fiber conserves a defined front of affluent wave and if the effectiveness in the connection of the light in the fiber is not very small [31].

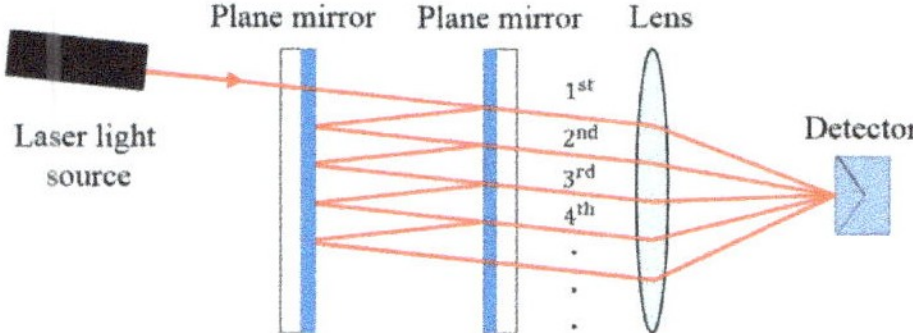

Figure 4. Sample of the operation of the Fabry–Pérot interferometer (FPI) [32].

2.2.1. Optical Fiber Mach–Zehnder Interferometer

Its structure consists of two couplers that divide the optical power of entrance in two equal parts; thus, the light travels through two different ways—one way operating as a reference while the other is used for sensing, it's desirable that OF have the same length so that the interference pattern is constructive, and the phases are equal. In Figure 5a it is possible to observe that since the beam of incident light is divided in two arms by means of a coupler later to recombine itself by means of a second coupler, later to recombine itself by means of a second coupler, the recombined beam of light has the interference component depending on the path optical between the arms. When this happens, for the detection, the reference arm stays isolated from any external variation, and another arm (the sensing region) is exposed to the changes. The changes in the arm (the sensing region) are due the mode propagation of the beam, that is sensitive to the refractive index changes of the materials of the sensor. This refractive index changes, are what originate the difference between paths optical, with which it is possible to later analyze the variation in the interference signal with photodetector [33].

2.2.2. Optical Fiber Michelson Interferometer

In the optical fiber sensor that is presented in Figure 5b, the coupler splits the beam of light by two different optical methods, where the light reflected by the mirrors recombines them by means of the coupler, giving origin to an interference pattern that arrives for analysis at the photodetector. It can be said that this interferometer's method of operation is very similar to the MZI; in fact, a Michelson is like half of an MZI as far as its configuration, since it uses a single coupler and a photodetector. Basically, the interference takes place between the beams of both arms, but each one of the light beams is reflected at the end of each arm in a Michelson interferometer. The main difference between these two interferometers are their mirrors —but their manufacture and principles of operation are almost equal. In this device, Faraday rotator mirrors are used to maintain the polarization of the separated beams in the fiber [34].

2.2.3. Optical Fiber Sagnac Interferometer

In this type of interferometer, which is seen in Figure 5c, the light beam enters the coupler by one of the entrance fibers, as it is possible to observe in the figure, giving rise to the light that divides in two beams with the same intensity, but with each one of them moving in opposing directions through the optical fiber, which in the end is going to be on the side of the configuration of the Sagnac—the one that goes toward the photodetector. This phenomenon is known as detection of rotation, which is obtained by placing the device on a spin table, which is when these turns happen and the lines of the interference pattern are moved [35].

2.2.4. Optical Fiber Fabry–Pérot Interferometer

This interferometer is considered the simplest since its single configuration needs only a circulator and a photodetector. For this interferometer, the interference plays a fundamental role since it is originated by means of a cavity in some of the ends of the optical fiber, as is in Figure 5d, where a light beam that is originated at the source travels through the fiber toward the circulator. When the light bean arriving at the cavity, the interference due to the light beam is reflected into cavity, to returns toward the photodetector. The cavity of a Fabry–Pérot interferometer is made up of two separated parallel reflecting surfaces; the interference happens due to the multiple superpositions of the reflected and transmitted beams of light in these parallel surfaces [36].

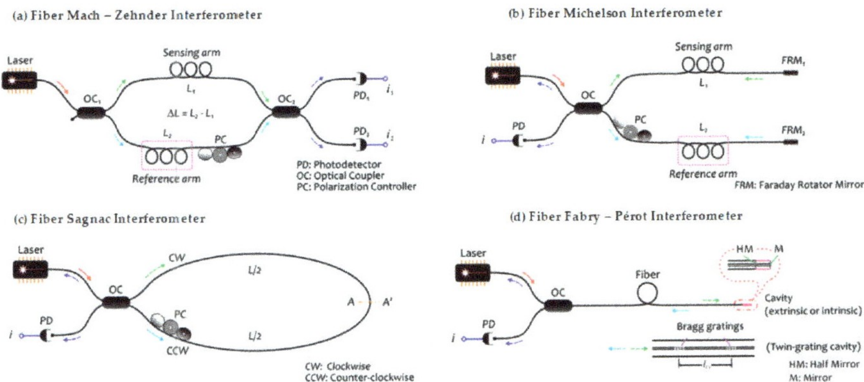

Figure 5. Configuration of different optical fiber interferometers [37]. (**a**) The figure exemplifies the basic operation of a fiber Mach–Zehnder interferometer (MZI) type, from the exit of the beam of the light source to its analysis through the photodetectors. (**b**) The figure shows a sensor using a fiber Michelson interferometer, which if analyzed is very similar in configuration to MZI, save for the fewer rotatory devices and mirrors. (**c**) The figure shows the configuration of the fiber Sagnac interferometer, where the opposite ways that the beam takes from light are observed with clarity. (**d**) The figure shows a fiber FPI type and the way in which the light beam travels from the source to its return to photodetector.

In Figure 6 are different configurations of as much an extrinsic as an intrinsic FPI sensors type, which is important since it shows us with greater detail and clarity the form in which the optical fiber can be characterized so that it works like a sensor. Here, it is possible to indicate that each interferometer has its different cavity configurations like: SMF with internal and external mirror, FBG cavity, SMF-Photonic Crystal Fiber(PCF)-SMF there are designed form measurement different signals or applications.

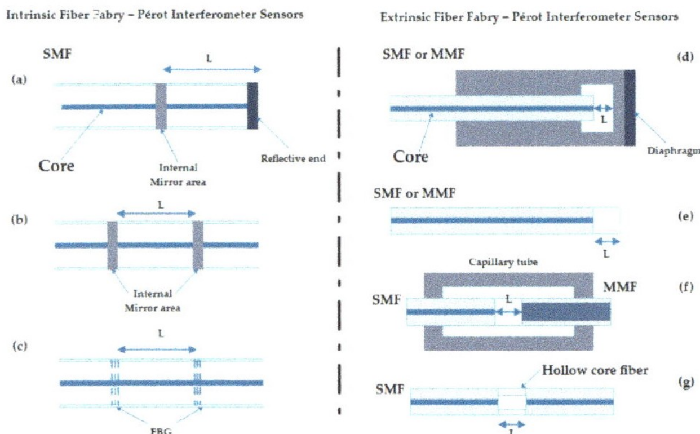

Figure 6. Configurations of FPI sensors [38]. The (adapted) figure shows the intrinsic and extrinsic configurations of FPI fiber sensors, where (**a**) represents a cavity formed by an internal mirror at the end of a fiber; (**b**) a cavity formed by two internal mirrors; and (**c**) a cavity formed by two FBGs. For each case, L represents the length of the optical cavity. Configuration (**d**) represents a cavity formed by a diaphragm at the end of a fiber; (**e**) a cavity formed by the surfaces of a cover in the end of the fiber; (**f**) a cavity formed in the end of a single-mode optical fiber (SMF) and an aligned multi-mode fiber (MMF) through a capillary; (**g**) a cavity formed by a SMF that the end is joined with a hollow core fiber. Additionally, the equal one in each case L represents the length of the optical cavity, which is an air bubble in the configurations (**d**), (**f**), and (**g**).

In Figure 7, showed some most common configurations on the fiber MZI sensors type, the changes in the microstructure of the fiber are by taper, joints with other fibers, and displacements of the cores between one or more fibers or with multicore fibers, where the fundamental issue is to make a difference of optical methods through the interference.

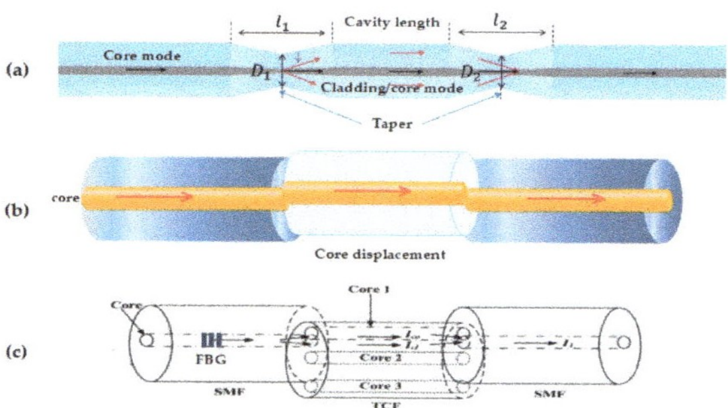

Figure 7. Different configurations from MZI fiber sensor (adapted) [39,40]. Here are different fiber MZI where we point out: (**a**) shows an interferometer created through a taper in the optical fiber, the same that is obtained by applying force in opposite directions when an unloading by electrical arc is made to modify the nano structure of optical fiber [41]; (**b**) shows joints of optical fiber but with a space displacement in the cores, which originates interference [42]; (**c**) shows joints of several types of optical fiber, where in the first, a fiber Bragg grating type created on a single mode fiber that is joined with an optical fiber with three cores that also hold a concealed knife with another single mode fiber segment [43]. The above are different techniques that are used to generate the sensors and, like the configurations of Fabry–Pérot, these are made to obtain greater sensitivity, specificity, stability, or exactitude in the sensor that one wants to create.

3. Health Field Sensing

In this section we describe and analyze according to their characteristics the research on OF sensors that seems to us the most important, which we highlight in each of the tables on each vital sign and where we compare the characteristics, operating ranges, sensitivity, and technology applied to the sensor. All this will help us to give a more accurate perspective of how most of them are good sensors, but they are still far from being able to operate as personal device or gadget in the home.

Cardiovascular diseases are a public health problem and the number one cause of death in the world [44], and for this very reason the constant monitoring of vital signs is important since many of the keys to this problem such as heart rate and respiratory rate, blood pressure, volume measurement, and oxygenation have specific applications and methods for monitoring. An example is the measurement of systolic pumping capacity, which is an important parameter for patients with congestive heart failure. Another example is when the heart stroke volume is multiplied by the heart rate, since it is possible to obtain the cardiac output, which is defined as the total volume of blood pumped per minute. Due to a reduced pumping capacity of the heart, patients with heart failure have a decreased cardiac output [45,46]. Like these examples, there is a wide variety of conditions that the human body presents due to cardiovascular problems, which, if detected in time and with constant monitoring, can be overcome or even eliminated.

The normal range of body temperature is between 36.5 °C and 37.3 °C. Temperature can be measured by tympanic route, orally, axillary, or rectally. Elevated temperature can be a sign of infection, arthritis, heat stroke, etc. Some useful temperature terms are hypothermia, which is a reduced body temperature of 35 °C or less. Pyrexia which is an elevated temperature, the three types of pyrexia are low (normal temperature up to 38 °C), moderate/high (38 to 40 °C) and hyperpyrexia (40 °C and above) [47].

Normal pulse or heart rate (HR) is between 60 and 100 beats per minute (BPM) in persons older than 10 years. In children and infants, the pulse is faster. Figure 8, shows the most common locations for easy acquisition of cardiac pulse and blood pressure. Several factors can affect pulse rate, such as fever, heart problems, infections, pyrexia, hypovolemia, hypovolemia, physical condition, anxiety, and medications. Some useful terms are tachycardia, a resting pulse greater than 100 BPM in adults, and bradycardia, a pulse less than 60 BPM [47].

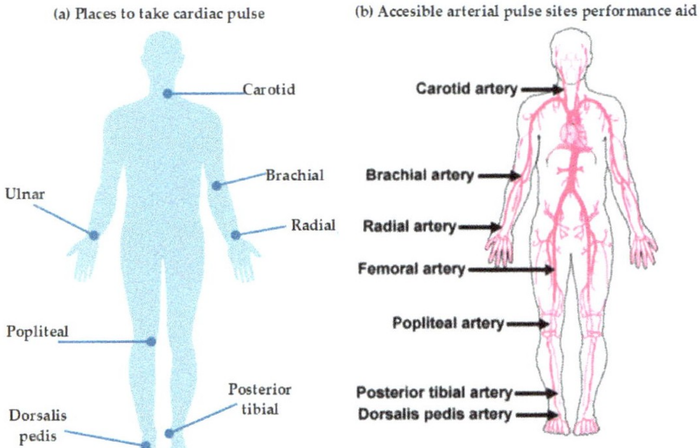

Figure 8. (adapted) [48] Human body areas for heart rate and blood pressure measurement. In (**a**) are the main areas where the cardiac pulse can be taken in the human body, since it is there where the main arteries are located and therefore the measurements are easier and more accurate; (**b**) shows the most accessible places for sampling blood pressure; it is observed that almost the same arteries are used for blood pressure and cardiac pulse, so that some sensors are able to perform both samples through the same signal [49].

For respiration, the normal respiratory rate is between 12 and 18 breaths per minute (bpm). Some of the factors that affect respiratory rate are anemia, pneumonia, asthma, chronic obstructive pulmonary disease (COPD), severe bleeding, stress/anxiety, medications, etc. Useful terminology includes bradypnea, or breathing slower than normal; tachypnea, or breathing faster than normal (greater than 20 bpm); dyspnea, or difficult breathing in conscious patients; orthopnea, dyspnea that occurs when the patient is lying down; and apnea, the temporary suspension of breathing [47].

For blood pressure, normal data for an adult human at rest is a blood pressure of 90/60 (systolic/diastolic) mmHg up to 120/80 mmHg; therefore, hypertension is blood pressure above 140/90mmHg. Figure 9 shows a scale of blood pressure values for adults, according to what is considered normal in systolic and diastolic pressure as well as the minimum and maximum limits (the ideal pressure represented in the green colored positions). The main causes of hypertension are obesity, chronic kidney disease, high alcohol consumption, smoking, and adrenal/thyroid disorders. Hypotension is low blood pressure (systolic less than 90 mmHg). Causes include pregnancy, dehydration, underactive thyroid, heart failure, blood loss, and anaphylaxis. [47].

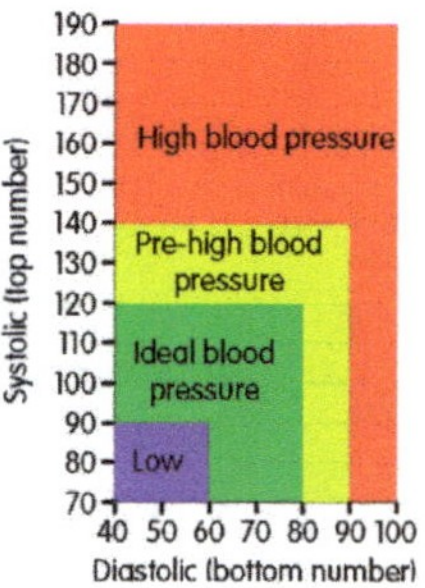

Figure 9. Main ranges for blood pressure [50]. The figure shows the ranges of blood pressure in adults, ranging from low, ideal, acceptable high, and severe high, both systolic and diastolic pressure.

3.1. Body Temperature

The body temperature of a healthy adult at rest is between 36.5°C and 37.3°C [47], with an average of 37 °C. In most of the articles that were analyzed, a clear tendency to use SMFs for the characterization of OF sensors to measure temperature was observed. Additionally, analyses of experiments within the near infrared spectrum (NIR), working mostly between 1500 nm and 1600 nm, facilitated the use of FBG to work with those periods and wavelengths. The sensitivity of results changes according to the other elements and techniques used for the creation of the respective sensor [51]. The research proposed a temperature sensor in biological tissues using an SMF, with the photo-thermal deflection of a laser that travels through an OF and its temperature gradient. Measurements have been taken in chicken organs (heart, liver, and gizzard) and where they have managed to obtain a heat pulse with an output power, all working at 1550 nm. Continuing with a project where researchers worked with an SMF at 1561 nm [14], they tested a sensor for human breathing detection that also used an FBG for temperature during the respiration monitoring process and measurements performed on the nose for obtaining sensitivity results of 11.4 pm/°C between 10 °C and 44.8 °C. A good research study was [52], in which they exposed a fiber sensor with possible applications in biomedicine, immunology, and biophysics, based on FBG working within the visible spectrum at 673nm with an SMF and using several types of lasers such as argon-ion, argon fluoride, and titanium–sapphire as a laser amplifier, finding a very wide temperature range between 242.45 °C and −211.55 °C. In another work [53], we found that researchers used the spectrum ranging from 1500 nm to 1600 nm; their sensor was an FBG to measure temperature during human mechanical ventilation, and the sensor was placed in an invasive mechanical ventilation tube coated with agar and polyamide acrylate: they obtained a total error of 3%, where the percentage represented the level of agar or agarose, obtaining sensitivity values of 114.7 pm/%, 0.12 nm/%, and 0.14 nm/%. Other work in which an SMF was first used to characterize the sensor obtained very large temperature ranges [54]: they presented a temperature sensor based on a mechanically induced fiber grating through a photoelastic effect with a heat-shrinkable tube, and its range of effectiveness within the spectrum was in the 1445 nm and 1570 nm, obtaining sensitivity results from de 3 nm/100 °C to 10 nm/100 °C, with a range between −20 °C and 50 °C. Continuing, [55], with a sensor where they showed its multiple possible applications but tested it with temperature, was based on an MZI with a pair of LPFG's created with a CO_2 laser, but measurements were performed in ethyl alcohol, with a wavelength of 1050 nm and 1239.4 nm, sensitivity of 11.7 pm/°C, and temperatures ranging from 30 °C to 110 °C.

In Table 1, the most complete works are exhibited only on body temperature. The table's columns identify the fiber optic type and operating wavelength in nanometers, where the sensor was tested, sensitivity in nanometers/Celsius degrees, characterization of the fiber optic nanostructure or utilization of doping agents, detection ranges on Celsius degrees, and the reference from where the information was taken. Additionally, the table shows that similarities can be clearly observed in the operating ranges of the first two articles [15,56], and the two [57,58] in the table using in all cases SMFs and the NIR spectra for their work. Based on the technology, two concatenated LPGs acting as an MZI for the design of a temperature refractometer [15] are presented to start with. In addition, a fiber sensor was found with possible applications for strain, temperature, and curvature. With characterization of the same with an LPFG through electrical discharges, a photonic OF is spliced with SMF, and the measurements are performed in the environment (Figure 10) [56]. In Figure 11, we can see a similar work, also used to join different types of fiber, the union of the fibers can be done through electric arcs or by means of some chemical compound. The entries continue with a humidity and temperature sensor characterized with only an LPFG created by an argon laser source [57]. Furthermore, another entry presents a very complex arrangement on an optical sensor based on the splicing of two pieces of SMF with a coreless fiber between them and where there are LPFGs to measure the temperature [58]. Next, a very recent study is presented on a sensor with an MMF to measure temperature through the surface plasmon resonance (SPR) technique, with integration for its utility with a smartphone; here, the device presented a flash along with a camera for signal processing through an application for the smartphone [59]—a great work with many possibilities. Finally we must point out a different sensor [14], as it is able to measure both body temperature and respiratory rate; in relation to the temperature yields, it has quite favorable results, although they varies greatly over time as it is affected by external factors.

Figure 10. (Adapted) Schematic design of the sensor proposed by Agostino Iadicicco. A view of the hollow core fiber is displayed with an LPFG and in turn spliced with conventional SMF and their respective fiber optic / physical contact (FC/PC) connectors [56]. The use of multiple splices together with different fiber types is common to obtain higher sensitivity or sensors capable of high-precision filtering; however, the right materials and tools are needed to achieve so much precision in the OF nanostructure.

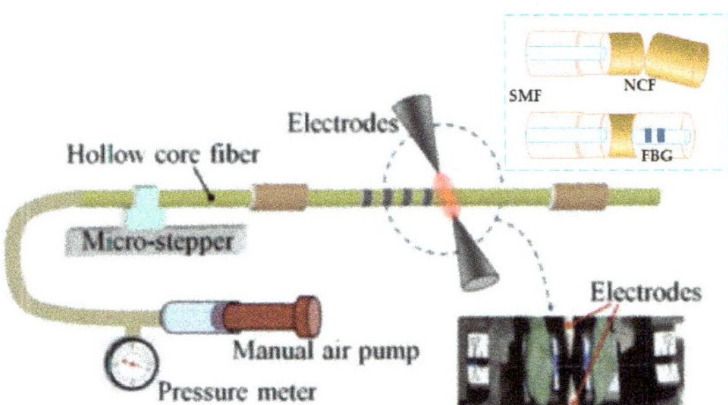

Figure 11. **(Adaptation)** Example of sensor non-core fiber – long period grating (SNS-LPG) experimental arrangement proposed by Agostino Iadicicco. This is the experimental arrangement they propose for splicing two SMF parts with a non-core fiber (NCF) in the middle and there create a LPBG, resulting in the SNS-LPG arrangement [56]. This is another work that hints at the accuracy of the equipment to characterize OF nanostructure; no doubt these are innovative ideas, but outside of a laboratory for our purposes they are not feasible—they are rather sensors for research.

Table 1. Body temperature, research with optical fiber sensors.

Fiber Type and Sensor Operating Wavelength (nm)	Sensor Tested on	Sensitivity (nm/°C)	Sensor Technology	Detection Ranges (°C)	Ref.
SMF 1550	Thorax, chest wall	0.31	FBG Encapsulated in PDMS	33 to 37	[6]
SMF 1561.07 to 1561.467	Nose	0.0114	FBG	10 to 44.8	[14]
SMF 1557 and 1627	Heating tube filled with water	0.95 to 1.03	LPFG H and KrF excimer laser	20 to 100	[15]
SMF and Photonic hollow core 1495.4 and 1520.3	Temperature chamber	0.0119 to 0.0138	LPFG Electric arc discharges	30 to 80	[56]
SMF 1515 and 1547	Climatic chamber	0.41066 and 0.40509	LPFG $NaOH$, KOH, H_2SO_4, H_2O, $C_3H_4O_2$, $NaCl$, C_3H_8ClN. argon-ion laser (244 nm—UV)	25 to 85	[57]
SMF and Non-core 1588	Thermostatic furnace	-0.00643	LPFG	34 to 154	[58]
MMF 630	distilled water	-9.52519E-05	SPR sensor Au film	30 to 70	[59]
POF 1552.45–1552.65	Chest	−0.055	He–Cd laser for inscription, taped fiber, DPDS dopant and UV irradiation	20 to 50	[60]
SMF 1550	Thorax and neck	-	FBG	35.5 to 39.5	[61]
SMF 1554.1207 1512.5 to 1587.5	Chest	0.010378 to 0.03844	FBG, encapsulated on PDMS, Sylgard 184	35.5 to 37	[62]
1550.218 and 1550.208	Thorax	10.3 to 11.3	FBG, acrylic and fiberglass encapsulated	20 to 90	[63]

SMF: single-mode fiber; MMF: multi-mode fiber; POF: plastic optical fiber; FBG: fiber Bragg grating; PDMS: polydimethylsiloxane; LPFG: long-period fiber grating; H: hydrogen; KrF: krypton fluoride; NaOH: sodium hydroxide; KOH: potassium hydroxide; H_2SO_4: sulfuric acid; H_2O: water; $C_3H_4O_2$: methylglyoxal; NaCl: sodium chloride; C_3H_8ClN: 3-chloropropylamine; SPR: surface plasmon resonance; Au: gold; DPDS: diphenyl disulfide; He-Cd: helium cadmium; UV: ultraviolet.

3.2. Respiration or Breathing Rate

This is the vital sign that most researchers study with fiber optic sensors; temperature also has many research works but not as a vital sign [64]. Several things can be highlighted, such as the large number of creative works when performing experiments, the characterization of the OF, and that the vast majority use SMF; however, there are a wide variety of doping agents that are used to increase sensitivity or techniques to apply torque, force, or expose the same core of the fiber. Splicing and discharging are also found as the most used techniques to modify the nanostructure of the OF [65]. However, almost all the works use lasers and the NIR spectrum, thus decreasing the complexity or variables that could affect and intervene in the experiments. Besides, the calculations are more accurate with reference to reality [66]. Large and complex studies were observed that leave the door open for possible diversification and mass production of these sensors and use within the home market in the near future, since their characteristics and designs would allow them to compete against the electronic devices for monitoring vital signs that are currently available or would serve together to strengthen the quality of the same. [67].

Taking into consideration that a healthy resting adult human has a respiratory rate of 12 to 18 breaths per minute [47] (premature infants have the highest respiratory rates at 40–60 breaths per minute). Here we can observe an OF sensor interferometer that measures nasal curvature during the respiratory process, through inhalation and exhalation, with an overall performance accuracy of 5.62% (with variations of ±2.26% in inhalation and ±3.19% in exhalation) and a sensitivity of 94.4% for inhalation and 93.93% for exhalation. [68]. They proposed a noninvasive OF sensor to monitor respiration, blood pressure, and movement in the human body using MMF and SMF. Another [69] presented a model OF sensor [70] with an interferometer for detecting heart pulse and respiration by putting it on a mattress, where the measurements ranged from 1561.07 nm to 1561.467 nm. Continuing, a work was shown for the monitoring of respiration through a FBG sensor [71] with a plastic optical fiber (POF) implanted in a flexible textile that measures abdominal curvature. The properties of polymer composite fibers make them unique and very good for their ease of handling, as they can withstand rough treatment compared to traditional fibers; they use a polymethylmethacrylate in the core of the plastic fiber and a fluoropolymer in the coating with wavelengths ranging from 650 nm to 655.9 nm, with a maximum error of 1 % with a very short fiber of 6.6 cm in length. Here, we continue analyzing [72], an FBG sensor to measure humidity, temperature, and respiration during the process of invasive mechanical ventilation through endotracheal intubation, obtaining BR readings from 11.5 breaths per minute (bpm) to 24 bpm. It can be observed (Figure 12) that its OF is given a layer of hygroscopic material, agar, and an acrylate cover for protection since the characterized fiber goes inside a hypodermic needle, achieving a confidence level during monitoring of 95%. Another project that draws attention was [73], where they presented an OF sensor generated with microcurves forming a mesh for the measurement of the perioperative pulse and respiratory rate in infants [74] from 01 to 12 months of age. Monitoring of the infant was performed through a mat where he remained lying down to obtain readings of 23.5 breaths per minute (bpm) to 23.5 bpm all with an MMF working in the 1310 nm of the NIR spectrum.

In Table 2, the most complete articles with respect to our parameters are highlighted but only those where the OFs used did not receive any modification in their nanostructure by any dopant or coupling agent. Heading the table [75] we find the proposal of a series of parallel optical sensors to measure respiration by means of thoracic and abdominal curvature during the respiratory process, using gratings with periods of 532 nm. The following article is a continuation of the same Allsop study from 2004, where they made measurements in the human rib cage and abdomen. Another work that presented an LPFG is one that monitored curvature during chest movements on breathing; it also measured the amount of air during stimulation and exhalation to make the results more accurate measurements of the chest and abdomen [76]. Additionally seen was a comparison of three sensors to measure continuous breathing during anesthesia application in a magnetic resonance imaging (MRI)) [77]; this allowed continuous sampling of the motion caused by breathing in the chest and abdomen of the human body [78]. There are even sensors with FBG to measure the volume of air during the breathing process by curvature of the chest [79], and there are also those but with fibers attached to tissues [80] that monitor almost the same movements during magnetic resonance imaging (MRI) examinations [81]. As a complex with 12 FBGs for monitoring respiratory parameters and heart rate (that is, the volumes and movements involved in the chest and abdomen), the minimal movement associated with human breathing is monitored throughout the system [82]. Some have been built to measure a particular element or substance but have been successful in detecting human respiration through testing in other areas. Here they use a temperature FBG in the breathing monitoring process. [83]. Figure 13, shows a representation of smart textiles, where the aim is to weave OF into clothing and present design alternatives so that they can be in contact with the skin and do their monitoring work. This is one of the great challenges of smart textiles with OF sensors.

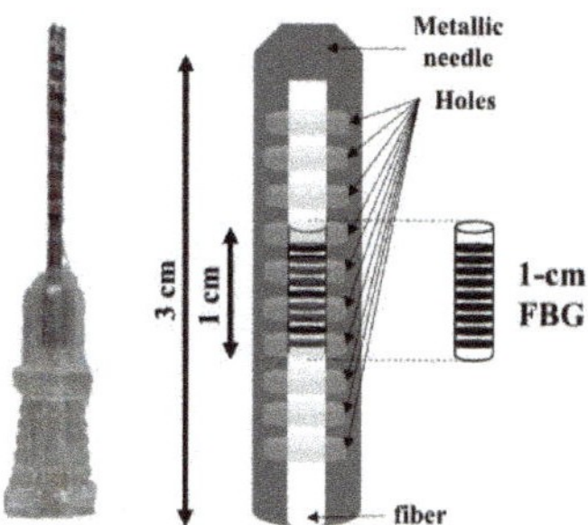

Figure 12. C. Massaroni guide and sensor protection scheme. This is an example of the hypodermic needle. In the ventilation process of mechanical invasion, the hypodermic needle is used to measure respiration, temperature, and humidity. In addition, with a really simple and innovative design, the hypodermic needle is used to protect the OF [72].

Table 2. Respiration or breathing rate, research with optical fiber sensors.

Fiber Type and Sensor Operating Wavelength (nm)	Sensor Test Position	Respiratory Rate—Breaths per Minute BR—(bpm)	Sensor Technology	Error or Variation (%)	Ref.
SMF 1550	Thorax, chest wall	13.4 to 19.5 men 14.5 to 19 women	FBG Encapsulated in PDMS	±1.96	[6]
SMF	Tricuspid area and carotid artery	18	MZI, FBG	-	[7]
SMF 1550	Baby upper abdominal area	10 to 100	Curves on the fiber	0.25	[55]
SMF 1510	Thorax, torso, and abdomen	10	LPFG	6 to 8	[56]
SMF 1547.77	Thorax and abdomen (precordium area)	9 to 11	FBG Bending effects OTDR	10	[58]
SMF, hetero-core 1310	Smart textile for the upper abdominal area	16	Macro-bending on the fiber	1	[60]
SMF 1550	Thorax and neck	12 to 24	FBG	2	[61]
SMF 1533 and 1557	Thoraco-abdominal and chest wall surface	6 to 11 on standing, 5 to 11 on supine	FBG	0.38	[62]
SMF 1299 and 1548	Thoraco-abdominal on resuscitation manikin	-	LPFG, FBG	0.4 and 0.8	[76]
SMF 1470.73	Human chest wall on supine	5 to 11 on natural, 6 to 12 on shallow	LPFG	4.4 to 8.7 and 5.8 to 10.1	[78]
SMF 1532 and 1541	Upper thorax	14	FBG	8.3	[81]
MMF 1310 nm	Body back	6 to 14 Average 12.31	Microbendings on the fiber	9.8	[84]

SFM: single-mode fiber; MMF: multi-mode fiber; FBG: fiber Bragg grating; PDMS: polydimethylsiloxane; MZI: Mach–Zehnder interferometer; LPFG: long-period fiber grating; OTDR: optical time-domain reflectometry; BR: breathing rate.

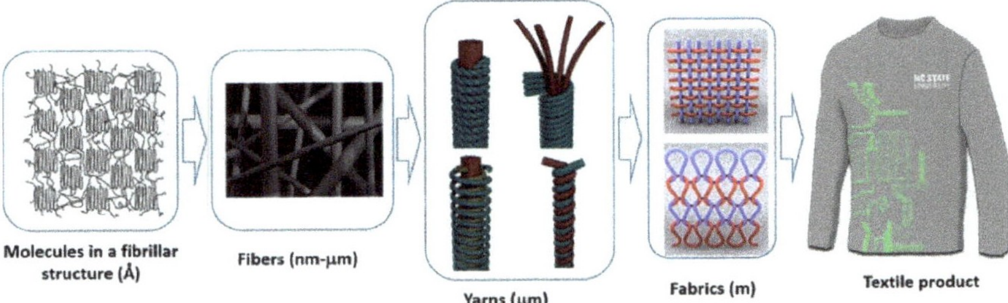

Figure 13. OF textile material representation by Kony Chatterjee. Here you can see the great work of twisting the optical fiber and weaving it to a textile, all for transport and protection, as it provides an ingenious way to move it and place it in a textile garment to be in constant contact with the sample patient. This is a clear example of how you could make different textile garments depending on what you want to monitor and the part of the human body from which you want to sample data [85].

Continuing with a study where [86] showed an optical balistocardiogram technique to measure cardiac and respiratory activity noninvasively, they used an optical FBG device with a germanium-doped SMF placed on the thorax, obtaining sensitivity of 1.20 pm $(\mu\varepsilon)^{-1}$, corresponding to 17 Min^{-1}. They presented an SMF sensor based on BFG, where several methods were proposed for obtaining vital signs [87], as well as modifications for the characterization of the same fiber with doping agents such as germanium, indium arsenide, and gallium; measurements were performed at the waist, neck, elbow, and ankle of the human body as they sought to determine where it was best to place their sensor. They obtained detection ranges of 1550 ± 0.5 and 1525–1570 nm ± 0.1, or the wavelength at which it operated. In another work, we can observe in Figure 14 how through the characterization of an SMF they created a twin core that operated from 1551.85 nm to 1552.15 nm, where they used a dopant agent of pulverized gold [88]. The result was a noninvasive sensor to measure respiration and HR with readings in the human chest, obtaining a sensitivity of 18 nm/m^{-1}.

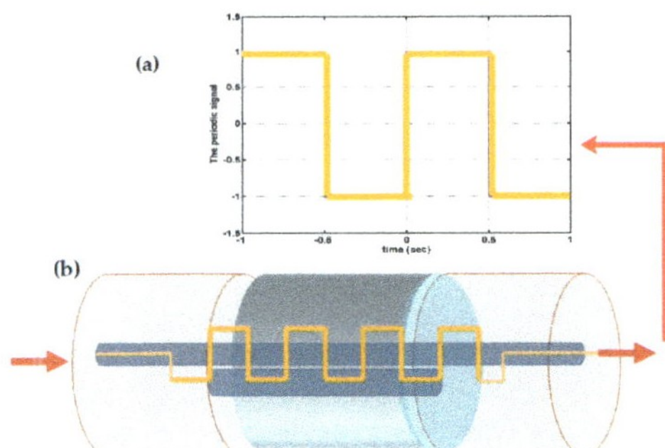

Figure 14. (Adapted) Design of the twin core fiber and its operation. In (**a**) you can also observe the operating wavelength of the SMF once it is characterized as a twin core, where it is also doped with sprayed gold to increase sensitivity and in (**b**) the schematic design and representation of the characterized fiber—a great job with a lot of creativity to create single mode fiber—two core fiber—single mode fiber (SMF-TCF-SMF) array [88].

In Table 3, we present data from work done with respiration but which involved a modification in the nanostructure of the OF by using chemical doping agents, which is why we differentiate them from the studies in Table 2—apart from the fact that it can be observed they are the most investigated fiber optic sensors, with the highest number of studies that are talking about vital signs.

Table 3. Respiration or breathing rate, research with optical fiber sensors with characterization of the nanostructure through doping agents. Characterization with chemicals in the nanostructure.

Fiber Type and Sensor Operating Wavelength (nm)	Sensor Test Position	Respiratory Rate—Breaths per Minute BR—(bpm)	Sensor Technology	Error or Variation (%)	Ref.
SMF 1561.07 to 1561.467	Nose	12 to 18 Average 15	FBG created with H	-	[14]
POF 1552.45–1552.65	Chest	18	He–Cd laser for inscription, taped fiber, DPDS dopant and UV irradiation	-	[60]
SMF 1554.1203 1513.444–1585.787	Thorax	16.22	FBG, PDMS, Sylgard 184	3.9	[62]
SMF 1550.218 and 1550.208	Thorax	16	FBG, acrylic and fiberglass for protection	4.64	[63]
SMF, MMF, TMF	–	18	MZI		[69]
SMF 1440 to 1550, best on 1519.95	Chest of respiratory manikin	–	LPFG, FBG UV and argon-ion laser induced Er doped and core of GeO_2/SiO_2, inner cladding of SiO_2, outer cladding of $SiO_2/F/P_2O_5$	-	[79]
SMF 1533 to 1553	Chest wall	11 to 12	FBG polymeric glue and POF	0.3	[83]
SMF	Chest	Average of 17	FBG, Ge doped fiber		[86]
SMF 1554.1204 nm	Chest	15.4925 standing, 15.5119 supine, 15.7638 sitting	FBG PDMS for encapsulation, Sylgard 184, CH_3Cl as dopant	4.4	[89]
MMF, ECF 1550	Back of the body seated	16 to 20	FBG Microbending, He-Ne laser, mechanical induction	15	[90]
SMF 1550	Back of the body on supine	14 to 19	FBG, PDMS	4.41	[91]
SMF 1536 and 1548	Nose bridge	9.64 to 10.76	FBG Ge doped, strain variations	1.3	[92]
SMF 1554.1207 nm	Chest	21.6676 standing 20.6386 seated 14.8741 back	FBG, encapsulated on PDMS, Sylgard 184	-	[93]

SFM: single-mode fiber; MMF: multi-mode fiber; TMF: two-mode fiber; POF: plastic optical fiber; FBG: fiber Bragg grating; MZI: Mach–Zehnder interferometer; Ge: germanium; H: hydrogen; Er: erbium; PDMS: polydimethylsiloxane; LPFG: long-period fiber grating; ECF: eccentric core fiber; DPDS: diphenyl disulfide; UV: ultraviolet. GeO_2: germanium dioxide; SiO_2: silicon dioxide; F: fluorine; P_2O_5: phosphorus pentoxide. CH_3Cl: chloromethane; He–Ne: helium neon; He–Cd: helium cadmium.

An LPFG sensor with three SMF fibers that analyzes the curvature and heat generated during breath monitoring was observed, even though the measurements were performed on a resuscitation training dummy—on the chest [79]. Using FBG, there was a sensor encapsulated in a compound with excellent thermal and elastic properties to monitor respiration and measure data from the sensor placed on the human chest [89]. Continuing with a study of a fiber optic sensor with bends for breathing monitoring and a sensor on the back and chest [90], so far it can be seen that almost all studies on breathing with OF were achieved through the measurements they obtained on the chest and the chest during the same process of human breathing—something logical and relatively simple since it is just a count of the breaths that occur in a given time. Another work is that of an FBG sensor to measure respiration and cardiac pulse during MRI examinations; as the sensor is immune to electromagnetic interference, the measurements were performed on the body through a mattress for a patient in supine position [91]. Likewise, a really striking and creative sensor is presented as is a smart textile, composed of 12 FBG sensors to measure all trunk movements, including breathing; being a whole garment, its measurements were obtained in chest, rib cage, and human abdomen [83] (Figure 15, shows various representations of smart textiles with OF sensors). Finally there is a noninvasive device (Figure 16) for respiratory monitoring through a fiber with two FBG, which measures nasal airflow [92].

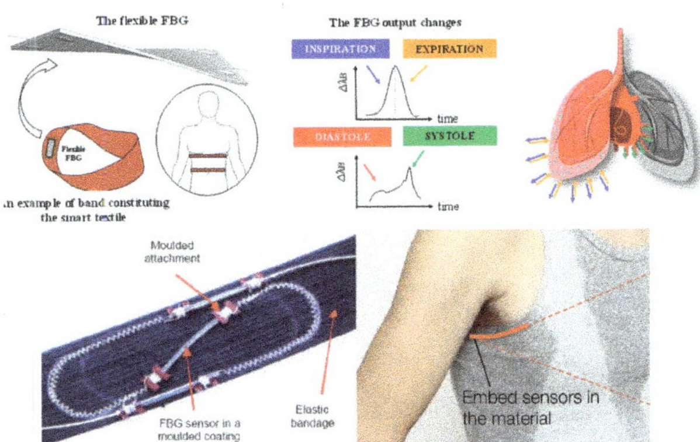

Figure 15. (Adapted) An example of smart textile sensors [94–96]. A representation of some very novel sensor designs is shown, based on a study to determine the location and configuration of the FBGs in the textile in order to measure the multiple details of breathing and its involved movements. It is intended to show the research that there is for the development of this type of garments that help constant monitoring of vital signs or physical activity.

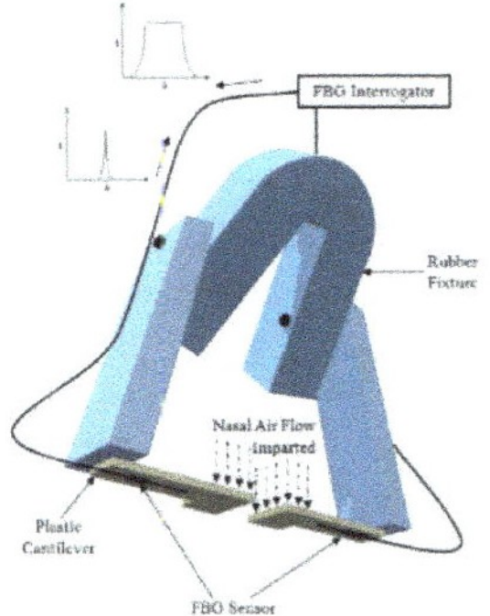

Figure 16. An example of the Pant Shweta prototype, a breathing measurement device. Another design to highlight where, through a very simple prototype, it is possible to measure breathing through a sensor with a FBG for its task. The prototype is most ingenious as it is a nasal bridge that measures the respiratory flow and can measure the respiratory rate from the volume without problems—a totally different and novel proposal. [92].

3.3. Pulse or Heart Rate

With the following vital sign, it is clear to us from the studies presented and the literature that pulse or heart rate is almost always accompanied by some other vital sign, either body temperature or respiratory rate, since the similarity for obtaining it through pressure or temperature (more common) is the same and only the parameters for its interpretation or mathematical calculations based on the heartbeat change. However, the usefulness of the sensor or characterization changes completely when we analyze the techniques in depth; although they are similar, they all have small details that identify them and give a sensor its unique character. Here we also begin to glimpse the problem of replication and mass production that would be desirable due to the size and characteristics of the sensors. A healthy adult at rest has on average a HR between 60 and 100 beats per minute (BPM) [47]. We can also say that a good way to know the physical health of an individual is through the monitoring and measurement of the heart rate, since this way it is possible to know the intensity and physical work of an individual, strength, expenditure, and calorie intake [92], among other things—in addition to how essential it is for the diagnosis of diseases, constant monitoring and treatment of people efficiently [97]. FBG-based sensors have been studied in processes. Since it is possible to establish the pressure needed during the perforation of the cardiac wall, in a study of cardiac arrhythmias, in vivo tests were performed in the hearts of two sheep. Working with an SMF in the range of 1561 nm to 1564 nm, an invasive procedure was used where the sensor was protected with a steel cylinder and urethane with epoxy glue [98]. Another way to find the cardiac pulse is through an optical sensor based on the reflectivity of light, where its change can be measured through the movement of the chest. This was done with a POF (plastic optical fiber) based on the principles of light intensity change every time the heart emits some vibration. With these studies it was possible to detect the peaks of a harmonic signal at a frequency of 15 Hz, 7.5 Hz, 10.5 Hz, and 22.5 Hz in a healthy adult with a mean value of 78 BPM; these results were so promising that the researchers made a working prototype [99].

Starting with the analysis of the most complete work, the best elements for the equipment are those presented in Table 4. First, a physical sensor that measured the heart pulse through a pillow placed on the back of the head will be analyzed; to improve the sensitivity, microbends were made in the fiber. It was found that the accuracy of the sensor decreased as the weight of the head increased. To obtain the results, the same sensor was compared with a commercial sensor [100]. The following is a paper in which they present an FBG sensor for the detection of HR through vibrations: the sensor was capable of being used during MRI procedures due to its low signal-to-noise ratio and being immune to electromagnetic interference [101]. Another study is one in which they showed the development of a FBG sensor for monitoring vital signs; this was performed superficially in different parts of the body, where different results were obtained compared to other sensors using the techniques of Bland Altman. Pulse and blood pressure were finally obtained by performing calculations and measurements in the temple, finger, ankle, and dorsum of the foot. This turned out to be a very complete work where different parts of the body were taken into account and where it was possible to measure the HR correctly [102]. We can also observe another fiber optic sensor with FBG for HR monitoring, where the use of an elastomeric material with measurements on the chest was tested. This type of material is very useful in the field of health and medicine because its properties are very friendly to human skin [12]. Another team showed a fiber sensor to measure HR, where patients were placed in a massage chair to monitor the data that were delivered by a pad on their resting heads—an effective method and also very useful because it helped to relax and lower the HR [103]. A very complete work is that of an OF sensor based on FBG where several methods for obtaining vital signs were proposed, as well as modifications for the characterization of the fiber with measurements in the waist, neck, elbow, and ankle of the human body; here, a measurement of temperature obtained during the exhalation process was performed [87]. This work is a clear example of the possibility and feasibility of measuring another vital sign during the process of obtaining the one that really interests us—it could well be by the hand of some other electronic or auxiliary device. Finally, an HR sensor made with a POF with three different configurations of the same fiber—which were straight, sinusoidal, and spiral-shaped, and were tested on the human hand, chest, and neck under fast, normal, and slow HR conditions—highlighted the photo elastic properties of the fiber [104]. In Figure 17 we can see an example of the work of Bonefacino, J. [60], where through an FBG sensor they managed to obtain measurements of the human pulse and respiration, all through an ultra-fast etching process of the strips. They showed sensitivity data of 150 bpm in HR and 8 BPM in BR.

Table 4. Pulse or heart rate, research with optical fiber sensors.

Fiber Type and Sensor Operating Wavelength (nm)	Sensor Test Position	Heart Rate—Beats per Minute HR—(BPM)	Sensor Technology	Error or Variation (%)	Ref.
SMF 1550	Thorax, chest wall	64 to 81 men 67 to 98 women	FBG Encapsulated in PDMS	±1.96	[6]
SMF	Tricuspid area and carotid artery	61	MZI, FBG	-	[7]
SMF 1550	Chest	57.5	FBG Encapsulated in PDMS	1.96	[12]
POF 1552.45 - 1552.65	Chest	150	He–Cd laser for inscription, taped fiber, DPDS dopant and UV irradiation	±2	[60]
SMF 1550	Thorax and neck	60 to 120	FBG	2	[61]
SMF 1554.1203 1513.444–1585.787	Thorax	78.54	FBG, encapsulated on PDMS, Sylgard 184	1.96	[62]
SMF 1550.218 and 1550.208	Thorax	74.3	FBG, acrylic and fiberglass encapsulated	4.87	[63]
SMF, MMF, TMF	-	66	MZI	-	[69]
MMF 1310 nm	Body back	77 to 83 Average 66.55	Microbendings on the fiber	0.6	[84]
SMF	Chest	Average of 107	FBG, Ge doped	-	[86]
SMF 1545 to 1555	Radial artery at the wrist	51	MZI, FBG, InGaAs	-	[87]
SMF 1554.1207 nm	Chest	62,8363 standing 61,9159 seated 76,8499 back	FBG, PDMS, Sylgard 184	-	[93]
MMF	Back of the head	58 to 74	Microbendings on fiber	2.7 to 3.44	[100]
SMF 1538.4 to 1538.6	Back of the body	76.8	FBG	< 7.4	[101]
SMF 1549.5 to 1550.5	Temple (best), finger, ankle (worst) and dorsum pedis	Average of 66.33, 60.33, 60 and 57.66	FBG, MZI	1.47 (best) 28.33 (worst)	[102]
MMF	Back of the body	84	Mechanical induction	7.31	[103]
POF 950	Neck and chest	68 (best) and 52 (worst)	PDMS and plastic polymer macro-bending and strain	-	[104]
SMF 1550 nm	Wrist	66	Er 12μm thick Al diaphragm	5	[105]

SFM: single-mode fiber; MMF: multi-mode fiber; TMF: two-mode fiber; POF: plastic optical fiber; FBG: fiber Bragg grating; MZI: Mach–Zehnder interferometer; Ge: germanium; Er: erbium; PDMS: polydimethylsiloxane; ECF: eccentric core fiber; DPDS: diphenyl disulfide; UV: ultraviolet; He–Cd: helium cadmium; InGaAs: indium gallium arsenide; Al: aluminum.

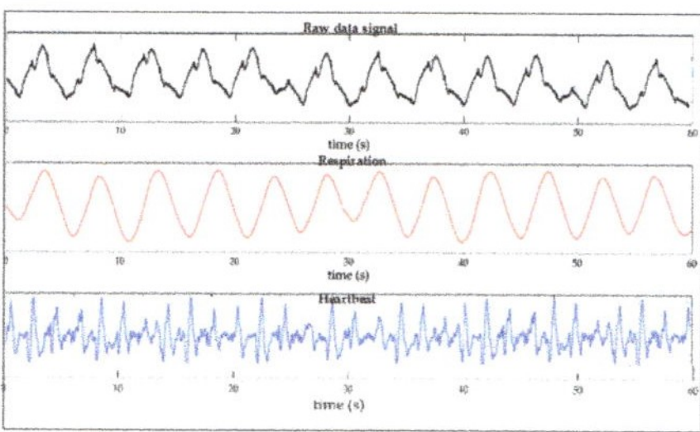

Figure 17. Based on the signals representing the waveforms of the filtered hard data as an example of Bonefacio Julien. It can be seen how the output signal and its respective wavelength appear in black with noise during the analysis performed by the FBG sensor, while the signals in red and blue, BR and HR, respectively, having already been filtered, can be seen in a clearer way due to the elimination of noise—a simple sign that from the same signal can be obtained different data out of phase in time and the feasibility of being able to obtain not only a vital signal to monitor. Here, the complexity lies more in monitoring both in real time [106].

Figure 18 also shows an example of the work done by Chuanglu, C. [107], where you can see that through the fitting of a signal obtained by a FBG sensor and smoothing this same signal, you can get the desired results. Here is shown the behavior of the same blood pressure (systole/diastolic) with a device that can also get more data, such as respiration and pulse.

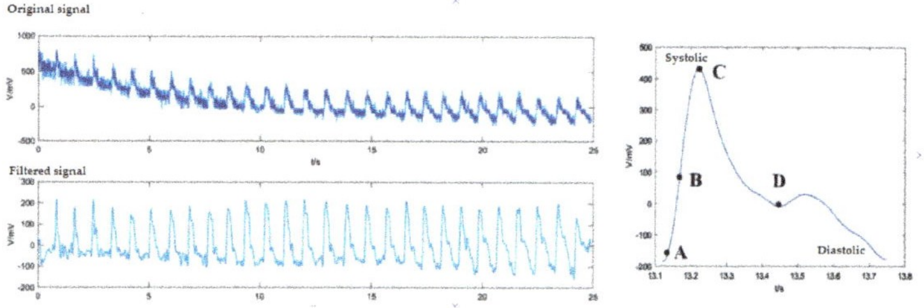

Figure 18. Example of the cardiac pulse flow during monitoring. It shows how the cardiac pulse signal is obtained. At first, background noise is observed in the output signal, as well as thermal and electromagnetic interference, so they proceed to filter the signal and take samples of the pulses to smooth the signal and finally take more samples to process the signal [107].

3.4. Arterial or Blood Pressure

Finally, we have the vital sign of blood pressure, where an adult human at rest has a blood pressure of 90/60 mmHg to 120/80 mmHg [47]. Blood pressure turns out to be the most complicated vital sign to obtain through OF sensors since in a traditional way it is very easy to apply pressure on the arm through a sphygmomanometer or tensiometer and thus stop the flow, which will cause an increase in pressure that can be measured. This same principle applied to fiber optic sensors is complicated since by increasing the sensitivity and characterization of the OF, these data can be confused with BR or HR but, through calculations or an algorithm is easier to obtain, resulting in the best way to obtain these data. Blood pressure also serves as a way to know or estimate multiple causes of cardiovascular conditions or problems. To start, we have research on the feasibility of using a system based on OF for invasive measurement of blood pressure [108] in which they present gain and offset errors of <3.4% and <0.25%, respectively, delay < 1 msec, and maximum rate of change > 30,000 mm Hg/sec, and where the effectiveness of using an OF to withstand electromagnetic interference was obtained. Another example is the study of a prototype sensor to measure Doppler blood flow velocity through a SMF [109] working in the NIR at 790 nm and 25 °C, where the Doppler shift of the radiation of a laser was recorded with an OF, thus beginning to visualize the intervention of possible tools for medical aid or assistance with the characterization of optical sensors.

Starting with the research work presented in Table 5, we have firstly an OF sensor of FPI for medical use (blood pressure) that works with white light; experiments and measurements were performed in a goat in which the internal pressure changes of the heart and the aortic vein were monitored [110]. Next, a work where they present another OF sensor to measure blood pressure in vivo and invasively, using as a sample a pig in which a catheter was introduced into its coronary artery from which blood pressure measurements were obtained in the aortic arch, right coronary artery [111]. We continue with an FBG sensor that measures waves in the deformations of the body surface that turned out to be pulse waves that were generated with the contractions and expansions of the arteries, achieving reliable values on the systolic pressure, where measurements were obtained from the neck and ankles of the human body [112]. Then, we have an OF sensor that looks like an advance from a previous sensor [111], where a complete design and packaging of the sensor was created for in vivo use in a swine coronary artery to measure the pressure of the arch of the aorta and the right coronary artery; it was demonstrated that the method of encapsulating the sensor provided effective protection and flexibility in the handling of the OF sensor in vivo for blood pressure measurements. Basically, it was a feasibility study of the operation of the sensor and how they provided protection and stability to work outside the laboratory—undoubtedly an excellent work as it provides evidence for the future possibility of further mass producing or developing OF sensors for the market [113]. We continue with the development of a FBG sensor for monitoring vital signs; this was performed superficially in different parts of the body, where different results were obtained compared with other sensors using the Bland–Altman technique. Pulse and blood pressure were finally obtained by performing calculations; the measurements were performed at different parts of the human body: temple, finger, ankle, and dorsum of the foot [102]. We also observed a portable sensor that used an FBG to measure tensions with high accuracy as the pulse wave signal; that is, together with mathematical calculations as the blood pressure values were obtained, the measurements were performed on the arm. This study validated that when the measurements are at different heights, the results change until the body posture changes. The measurement turned out to be good as long as the height was not changed—at least it was the parameter that affected the blood pressure sensor they had [114]. An FBG sensor in a smart textile made of silk fabric was presented for constant monitoring of vital signs, which can be calculated by analyzing the generated pulse wave, with special emphasis on blood pressure [115]. Another very good work (Figure 19) is from a study in which they showed an FBG sensor to measure HR and blood pressure through an aluminum diaphragm [105] that was placed on the wrist of the arm, which served as an acoustic amplifier when emitting such pulses and could be detected through the characterized fiber. This is a very good idea implemented for obtaining data through the generated sound waves and their analysis to generate results on the activity of pressure, pulse, and blood viscosity. Another group presented an FBG sensor with a POF [116]; in this case they proposed the use of plastic fiber since the common fiber is very delicate and could fracture during the operation of the sensor; therefore, with the POF that curl is omitted, the sensor gets to detect a pulse wave for blood pressure up to 8 times higher than with a silica OF. The results show that the blood pressure had very little error but the acceleration correlation of the plethysmograph was not the desired one. Finally, a sensor based on FBG is presented where several methods for obtaining vital signs were proposed, as well as modifications for fiber characterization. Measurement on the waist, neck, elbow, and ankle of the human body, the temperature measured was the one emanating from the exhalation process and the blood pressure was measured through the variations in the diameter of the artery caused by the same pressure. It was also concluded that the sensor could not effectively measure the pulse and pressure, one or the other should be separately eligible [87].

Table 5. Blood pressure, research with optical fiber sensors.

Fiber Type and Sensor Operating Wavelength (nm)	Sensor Test Position	Blood Pressure (mmHg)	Sensor Technology	Error or Variation (mmHg)	Ref.
SMF 1549.5 to 1550.5	Temple (best), finger, ankle(worst) and dorsum pedis	131/73.5 average	MZI, FBG	± 3	[102]
SMF 1550	Wrist	116.5/71.75 average	Er, Al diaphragm	-	[105]
SMF 550 to 700	On a goat left ventricle, left atrium, right atrium, and aorta	−100 to 400	FPI, PI, $C_8H_{20}O_4Si$	± 4	[110]
SMF, MMF 1547.5	In-vivo coronary artery of a swine	54 to 88 in aortic arch 60 to 100 in right coronary artery	FPI cavity with HF, SiO_2 diaphragm	-	[111]
SMF 1549.5 to 1550.5 1559.5 to 1560.5	Neck and ankle	106 to 119 and 109 to 122	MZI, FBG	7 and 5 on systolic	[112]
SMF, MMF	Tortuous vessels of a swine model in-vivo	54 to 88	FPI, stretched core MMF, SiO_2 diaphragm	-	[113]
SMF 1549.5 to 1550.5	Right wrist	110.7 supine 107.3 sitting 103.8 standing	MZI, FBG, SiO_2, InGaAs	3 supine 2.8 sitting 3.8 standing	[114]
SMF 1525 to 1575	Wrists	123.6 average with no cover 119.2 average with cover	FBG	2 no covered 6 covered	[115]
POF 15543 and 1553	Left arm	106.5/65 average	FBG, glue NORLAND 78	5/3	[116]

SFM: single-mode fiber; MMF: multi-mode fiber; POF: plastic optical fiber; FBG: fiber Bragg grating; MZI: Mach–Zehnder interferometer; FPI: Fabry–Pérot interferometer; PI: polyimide, $C_8H_{20}O_4Si$: tetraethoxysilane; HF: hydrofluoric acid; SiO_2: silicone dioxide; Er: erbium; InGaAs: indium gallium arsenide; Al: aluminum.

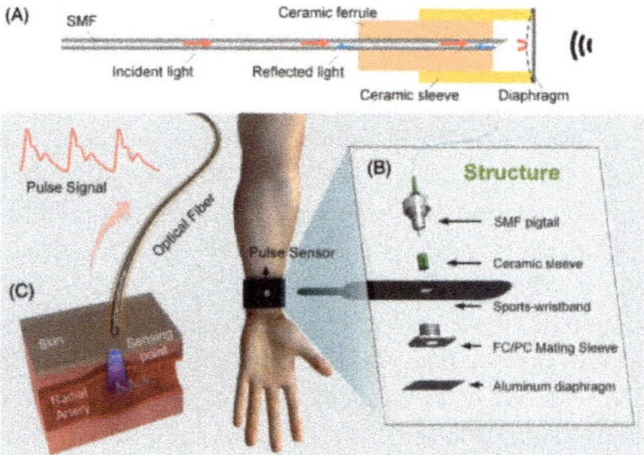

Figure 19. Wang's figure of an example of a fiber sensor showing its structure, composition, and operation. The model of an OF sensor is shown from its components and how each of them are structured. The whole design for its operation is presented and how the characterized fiber works—together with all its components and operation functions as a wrist blood pressure sensor. A bold design of how vibrations originated by a membrane can be used to obtain readings in the wave [105]. (**A**), Schematicdiagram of optical fiber pulsesensor; (**B**), Structural design ofwearable pulse sensor; (**C**), Thetransmission of pulse waves signals.

3.5. Multiparametric

These represent investigative works that are outstanding due to their complexity and their ability to monitor multiple vital signs or some complementary characteristic that serves to obtain the same signs [117]. Although most base their technique on obtaining pressure or temperature along with mathematical calculations to obtain the other vital signs, they do so by comparing their results with sensors that are already present in the market and trading through Bland–Altman techniques.

In Table 6 we start by reviewing a work on an interferometer-type fiber sensor where different configurations of the same for monitoring psychophysical activity were compared to sample HR, motion, and blood pressure, performing measurements on the human body [118]. That study is followed by a proposed noninvasive OF sensor to monitor respiration, blood pressure, and movement with drugs on the human body through a chair with a backrest and a pillow where it carried the sensor [119]. Continuing with another FBG-based sensor to measure respiration, pulse, and movement, but also to obtain data on temperature variations, measurements were taken on the human neck of a sample subject [61]. Next, a work was analyzed where a sensor that handled an FBG with microcurves is presented showing its practicality in measuring heart rate and respiratory rate. This sensor with multiple uses and capabilities through smart objects was tested using mattresses, pillows, chairs, and swings where a sample subject was placed [84]. The review continues with a proposal on a ballistocardiogram technique using optics to measure cardiac and respiratory activity noninvasively—using an optical FBG device with measurements on the human chest [86]. Next is a design with its verified functionality of a noninvasive sensor to measure and monitor pulse and respiratory rate through an FBG measuring rib cage movements and also performing measurements on the human chest [93]. Work continues on a noninvasive multichannel OF sensor for HR, BR, and temperature monitoring through FBG characterization, encapsulated in a polydimethylsiloxane polymer where measurements were performed on the human rib cage [62]. Then an FBG sensor integrating three types of sensing for pulse, HR, and body temperature is analyzed. The study consisted measurements on the human body using a fairly complete vital signs sensor based of OF and was developed by a team of researchers [69]. Our review continues with a sensor for the detection of human respiration that used an FBG for temperature during the process of monitoring respiration, making measurements in the nose [14]. Another FBG sensor for simultaneous measurement of respiration used pulse wave and heart sound with measurements at the tricuspid region of the chest and the carotid artery of the neck [7]. Other techniques created FBGs in an SMF and used them in the biomedical field as HR, BR, and temperature sensors, performing measurements in the human chest. Additionally, techniques are shown to increase the sensitivity of the fibers using different doping agents for the fiber and also exposure to ultraviolet rays [60]. There is also a proposed model on a fiber optic sensor with interferometer for the detection of heart pulse and respiration through its tuning on a mat to measure different parts of the body [69]. The review continues with consideration of a sensor for monitoring heart rate and respiratory rate through a design of a ballistocardiogram composed of FBG—all to be applied during the MRI process where it was shown that the sensor was not affected by any disturbance caused by the resonance analysis, this through measurements on the human chest [63]. Finally, we have a work that showed a sensor with a FBG to measure the heart pulse and blood pressure through an aluminum diaphragm that was placed on the wrist of the body, which served as an acoustic amplifier at the time of emitting such pulses and was able to detect through the characterized fiber [105].

Table 6. Multiparametric—vital signs, research on optical fiber sensors.

Number of Sensors	Body Temperature	Respiration or Breathing Rate	Pulse or Heart Rate	Blood Pressure	Ref.
3	×	×	×		[6]
2		×	×		[7]
1	×	×			[14]
2	×	×	×		[60]
2	×	×	×		[61]
2	×	×	×		[62]
1	×	×	×		[63]
2		×	×		[69]
1		×	×		[84]
1		×	×		[86]
1		×	×		[93]
1			×	×	[105]

Table 6 only shows those works on fiber optic sensors that were capable of obtaining measurements of two or more vital signs; however, for ease of comparison, the data of most interest to us were included in the corresponding tables.

4. Discussion

The preceding provides us with an important starting point and a research base to continue testing more features in sensors that have a competitive potential in the market or can be used to generate knowledge. There is a growing demand for medical devices for constantly monitoring vital signs, health, and stability as a means to prevent deterioration of the patient's condition or the illness, or simply to warn of any elevation that may indicate a more serious disease. This leads to the need for medical sensor application technology in real-time, at low cost, and that is commercially viable, that compete with devices commercial electronics.

A large number of the sensors in the literature reviewed work with FBG-characterized SMFs operating at a wavelength within the NIR (1550 nm), where they are more sensitive to the change in the refractive index of the OF and are for monitoring respiratory rate, as it is the most commonly used, inexpensive, and therefore most accessible fiber type. Apart from the fact that the characterization with FBG at that wavelength is a widely tested that has given multiple results using lasers [120], the complexity is to adjust the sensor as a filter fort to measurement of some vital sign [72,114,115]. All the sensors analyzed, are performed for measurements of physical parameters in the medical field (temperature, position, force, torque, stretch) but is necessary its calibration and categorization, as they are able to perform different measurements depending of the configuration and design. Therefore, they are applied for obtained manage and to obtain a response of vital signs [87,121].

Unlike the systems of traditional detection and monitoring of vital signs, optical fiber sensors have very attractive characteristics, such as excellent sensitivity, great rank of operation, and a high trustworthiness. The advantage of optical fibers use is that they are elaborated with dielectric materials that are chemically inert (that is, compatible and immune to the electromagnetic interference) and many of them can be covered by designed materials to support great temperatures. These properties make them excellent so that in their characterization as sensors, they can be adapted in hostile and aggressive surroundings, such as the interior of some block of construction, materials, or structures, or in a system of generation and transmission of electricity, where conventional sensors can have an unstable operation and at the same time have increases in the possibility of faults [37]. According to the data that appear in the different tables, one can deduce that the measurement or monitoring of vital signs varies depending on the atmosphere where the registries are made. With almost all reported results, were tested in a laboratory. The tests would be much better in real environments since many sensors are very sensitive to pressure, temperature and vibration changes. It would be ideal if the designs that were investigated could be developed as prototypes. In the review of the reported research works, we found the lack of classification of those sensors that reached an operational sensitivity for applications in health care. Because, each research team established its criteria for the characterization of the sensor, such as: units of measure and limit of operation. Since there is no standard characterization criterion, it is difficult to establish with certainty the feasibility of operation of all sensors of each research team. Knowing only the use and operation of the sensor that was considered under certain very specific conditions.

The results obtained in some works show an enormous variation solely with changing the area of positioning of the sensor. For example, in the case of Chino, S. et al. [102], data varied depending on where the samples were obtained from the four parts of the body (temple, finger, ankle or dorsum pedis) for the obtaining cardiac pulse. This is evidence that, if they wanted to obtain good results for locating some of the vital signs, it is necessary to study the medical literature on the different vital signs, such as physiology, which firstly allows to establish parameters for the correct taking of samples and tests.

The sensors in Table 6 are the most robust and complete proposals for measuring vital signs; their sensitivity and reliability make them great candidates for operation in in vivo situations; however, for us the best are still those that do not require human intervention, but those that are placed on some kind of guide, mattress, or pad [69,74,84,91,103,119], since the error factor due to direct human intervention and the practicality of handling the sensors themselves is eliminated. It can be clearly observed in this table that cardiac pulse and respiration are almost always obtained together. Another important characteristic is to see that only in one of them the blood pressure is obtained together with the cardiac pulse. Finally, another characteristic is that although some are large sensor arrays, not having more sensors means that their capacities grow; the number of elements or sensors involved in each research study does not determine their accuracy, reliability, or greater capacities.

The complexity of the sensors lies not only in the characterization of each one of them, but in that there are some [82,92] in which the monitoring is performed through several OFs, different types of OFs, or with several sensors in the same array. These represent complex models if we take into account their nanostructure and fragility to be handled without suffering any damage, since portability is one of their major disadvantages, without considering their light source, since in most of the researcy they turn out to be lasers.

There are also studies that try to analyze [12,62,72,89,105,113,115] the means of transport or protection, since the tests have brought these means to a more practical position, which is desirable if you want to develop and install such sensors in the market, since the tests ground them to a real situation—which is convenient if what is sought is the development and implementation of this type of sensors. It should be noted that OF sensors already exist commercially, but their use is still limited.

We believe that the best works were those in which they could remain as a prototype proposal [99] and were implemented through comparisons by Bland–Altman techniques [102] against others for commercial use [100] or they were presented and put into operation suddenly.

The use of fiber optic sensors in conjunction with new technologies provides a solid structure for the design of new sensors. The use of characterized FO is of vital importance, since almost all research papers document improved sensitivity, managing to overcome Bland–Altman tests against established devices or gold standards to determine sensitivity and operation during laboratory tests. This not only for vital signs [63,91,105,115] but also for a wide variety of optical sensors for use in the medical field, such as bacteria sensors [122], glucose monitors [123], cardiovascular monitors [107] (although this research work fails to present conclusive results due to the small number of samples, but they prove to have better sensitivity values), diabetic foot treatment [124], and also in the field of biochemistry [125] as diagnostic sensors, environmental monitoring, and organic and chemical compounds.

These sensors demonstrate their great effectiveness, presenting innovative and difficult to overcome designs since the characterization of the optical fibers can deliver extreme sensitivity and accurate data. In terms of cost, they have not been able to be introduced as sensors for domestic use since they cannot compete in cost relative to the established electronic vital sign sensors. The trend observed is the integration of these sensors with different devices, such as programmable boards for hardware development (Arduino and Raspberry Pi) that combine the best of each technology, such as miniaturization, free access, high sensitivity, and immunity to electromagnetic interference, among others—which combined have proven to be sensors with exceptional capabilities.

In addition, in some cases they show similar levels of sensitivity to fibers used in some high cost articles; that is to say, an alternative of lower cost is appearing, but even so it is not sufficiently attractive when competing with the traditional vital signs sensing devices [126–128] that often seem to be more precise and reliable. On the other hand, some otherwise very good work has achieved unexpected results [90], since the percentage of error or variation was very high, due to factors such as temperature differences or pressure. Nevertheless, being good studies, they continue to indicate methodologies to follow or to correct for future projects.

With respect to the sensors for human temperature, from all the work presented in Table 1, we can conclude that the best in terms of operating ranges [6,60–62] are those that only measure the ranges proposed for human temperature (36.5 °C and 37.3 °C), since many are able to detect very high temperatures and are really surplus. In terms of sensitivity, all of them have very small ranges (surface/temperature), from which one could well infer their accuracy. Finally, none were placed in the most commonly used parts of the human body to obtain the temperature, such as the armpit, forehead, ear, or rectum.; we infer that the lack of testing in these sites is because they require presenting a sensor, rather than proposing manipulation or live operation.

With respect to the sensors for monitoring respiration presented in Tables 2 and 3, we conclude the following: the error range present could only be acceptable if there were a standard metric to compare all the sensors, since in many cases it is only a simple counting of breaths, while in others the signal is generated by each movement detected by the sensor and analyzed. However, the best are those that have very little variation or error range [58,60,63,65,66,71]. Here we can conclude that, in effect, it is something very difficult to calculate due to the physical activity of the sample subject and, as in the temperature, only sensors and their operation are presented, not an arrangement that allows to take them into real situations for practice or operation—places working against the environment and its variables.

With respect to the sensors for cardiac pulse monitoring presented in Table 4, we conclude the following: the best ones are those whose data conform to the standards (60 BPM to 100 BPM) and whose degree of error or variation is the least [8,93,129]. Taking into account these data, we can assume that some of them do not present maximum and minimum values of operation, or their respective values of error or standard deviation.

For the sensors for blood pressure monitoring presented in Table 5, we conclude: the best results [88–91] are for those that show greater test data due to the different zones of the body where the sensor was placed. This is because a change in blood pressure can be seen in these areas, where it can be said that the best area is the wrist, due to the ease of placement of the sensor and the results shown. The only desirable thing would be to test and prove the feasibility of these sensors in extreme conditions of physical activity (walking or some exercise) and to observe any variation.

5. Conclusions

Fiber optic sensors have great advantages, such as being immune to radiation and electromagnetic interference, chemical corrosion. Their lightness and small size are attributes that make them perfect for monitoring health or for use in the medical field as a great tool capable of providing information in real time—even more so with the monitoring of vital signs, where these sensors can be used as a simple monitor or even life support for the diagnosis or prevention of disease.

The projection of Fiber Optic Sensors [130], has grown very slowly into the markets directly competing with conventional sensor technology from 1980 to 2000, due largely to the high cost and a limited number of sensors. Considering the evolution of fiber optic sensors and a projection for the year 2020, it is expected that the cost of components for their manufacture will be more economical and attractive. This would result in a greater opening and distribution of the market, more economical and affordable sensors for companies, and better quality components.

The trend of our time is the preparation of more accurate and sensitive, faster, and highly specific fiber sensors, not only in the field of medicine and health. On the other hand, these sensors have a tendency to be more complex and therefore financially demanding, making them more expensive to mass produce. The number of publications on sensors for medical and health care has increased in recent years and without a doubt this will grow enormously in 2021 due to the problems that humanity faces with COVID-19. We are at a decisive moment where the constant monitoring of our vital signs must be taken into account and we must be attentive to any changes; in fact, the, respiratory rate monitors and oximeters have seen their commercial demand increase enormously since the vast majority of people who have the possibility of acquiring any of them have no choice but to monitor their vital signs in isolation due to the possibility of contagion of the disease. It has become common to measure temperature in all meeting establishments or places of human recreation. All this favors that these sensors can ultimately see a commercial outlet and not just remain as laboratory projects. Although there are some that as prototypes seem to be very good, one can perhaps find a way to make them more resilient or to improve their portability or to use visible light—since one of the technolgy's main problems is its fragility and light source.

In this article, a review of fiber optic sensors for monitoring vital signs was conducted, where characteristics such as the technologies used for manufacture, compounds or elements that modify the nanostructure of the optical fiber, operating ranges, and sensitivity are presented and discussed. We analyzed which of them are the best proposals for their possible live operation, since it is one of the concerns to be uncovered with this work. The main feature of all of them is the monitoring of vital signs in real time. Some of them even measure oxygen saturation and stress. Some very ingenious or complex designs are illustrated, both as their insertion in textiles and arrangements with various types of fiber—each of them with different modifications in their cores. In conclusion, fiber optic sensors for monitoring vital signs allow the technology enclosed in them to design prototypes for use in various healthcare facilities. These sensors greatly improve the comfort, ease of operation, and accuracy of health monitoring devices.

Funding: This research received no external funding.

Institutional Review Board Statement: Not applicable.

Informed Consent Statement: Not applicable.

Acknowledgments: To the Consejo Nacional de Ciencia y Tecnología (CONACYT) de México for their support during the elaboration of this article, where it benefits through scholarship number 745773.

For Accessibility to the Use of the Figures of the Scientific Journal Publishers: Reproduced from Ref. [105] Journal of BioPhotonics of John Wiley and Sons. Reproduced from Ref. [37] Optical Interferometry of the IntechOpen Publications. Reproduced from Refs. [19,25,28,32,39,40,56,85,94–96,106,107] Sensors Journal, Materials Journal, Applied Sciences Journal, Fibers Journal and Journal of Functional Biomaterials of the MDPI. Reproduced from Refs. [72,92] IEEE Sensors Journal and IEEE MeMeA of the IEEE. Reproduced from Ref. [88] Biomedical Optics Express of the OSA Publishing.

Conflicts of Interest: The authors declare no conflict of interest.

References

1. Pinet, É. Medical Applications: Saving Lives. *Nat. Photonics* **2008**, *2*, 150–153. [CrossRef]
2. Güemes, J.A.; Sierra-Pérez, J. Fiber Optics Sensors. In *New Trends in Structural Health Monitoring*; Ostachowicz, W., Güemes, J.A., Eds.; CISM International Centre for Mechanical Sciences; Springer: Vienna, Austria, 2013; Volume 542, pp. 265–316, ISBN 978-3-7091-1390-5.
3. Roriz, P.; Carvalho, L.; Frazão, O.; Santos, J.L.; Simões, J.A. From Conventional Sensors to Fibre Optic Sensors for Strain and Force Measurements in Biomechanics Applications: A Review. *J. Biomech.* **2014**, *47*, 1251–1261. [CrossRef]
4. Kanellos, G.T.; Papaioannou, G.; Tsiokos, D.; Mitrogiannis, C.; Nianios, G.; Pleros, N. Two Dimensional Polymer-Embedded Quasi-Distributed FBG Pressure Sensor for Biomedical Applications. *Opt. Express* **2010**, *18*, 179–186. [CrossRef]
5. Poeggel, S.; Tosi, D.; Duraibabu, D.; Leen, G.; McGrath, D.; Lewis, E. Optical Fibre Pressure Sensors in Medical Applications. *Sensors* **2015**, *15*, 17115–17148. [CrossRef]
6. Nedoma, J.; Fajkus, M.; Martinek, R.; Vasinek, V. Non-Invasive Fiber-Optic Biomedical Sensor for Basic Vital Sign Monitoring. *Adv. Electr. Electron. Eng.* **2017**, *15*, 336–342. [CrossRef]
7. Ogawa, K.; Koyama, S.; Ishizawa, H.; Fujiwara, S.; Fujimoto, K. Simultaneous Measurement of Heart Sound, Pulse Wave and Respiration with Single Fiber Bragg Grating Sensor. In Proceedings of the 2018 IEEE International Symposium on Medical Measurements and Applications (MeMeA), Rome, Italy, 11–13 June 2018. [CrossRef]
8. Krebber, K.; Liehr, S.; Witt, J. Smart Technical Textiles Based on Fibre Optic Sensors. In Proceedings of the OFS2012 22nd International Conference on Optical Fiber Sensors, Beijing, China, 17 October 2012; Volume 8421, p. 84212A.
9. Du, W.; Tao, X.M.; Tam, H.Y.; Choy, C.L. Fundamentals and Applications of Optical Fiber Bragg Grating Sensors to Textile Structural Composites. *Compos. Struct.* **1998**, *42*, 217–229. [CrossRef]
10. Measures, R.M. Smart Composite Structures with Embedded Sensors. *Compos. Eng.* **1992**, *2*, 597–618. [CrossRef]
11. Lindberg, L.-G.; Ugnell, H.; Öberg, P.Å. Monitoring of Respiratory and Heart Rates Using a Fibre-Optic Sensor. *Med. Biol. Eng. Comput.* **1992**, *30*, 533–537. [CrossRef]
12. Fajkus, M.; Nedoma, J.; Martinek, R.; Novak, M.; Jargus, J.; Vasinek, V. Fiber Optic Sensor Encapsulated in Polydimethylsiloxane for Heart Rate Monitoring. In Proceedings of the Fiber Optic Sensors and Applications XIV, Anaheim, CA, USA, 27 April 2017; Volume 10208, p. 102080W.
13. Fajkus, M.; Nedoma, J.; Martinek, R. Alternative Fiber Optic Sensor Based on Bragg Grating for Heart Rate Monitoring. *Int. J. Biosens. Bioelectron.* **2018**, *4*. [CrossRef]
14. Manujło, A.; Osuch, T. Temperature Fiber Bragg Grating Based Sensor for Respiration Monitoring. In Proceedings of the Photonics Applications in Astronomy, Communications, Industry, and High Energy Physics Experiments 2017, Wilga, Poland, 7 August 2017; Volume 10445, p. 104451A.
15. Tripathi, S.M.; Bock, W.J.; Kumar, A.; Mikulic, P. Temperature Insensitive High-Precision Refractive-Index Sensor Using Two Concatenated Dual-Resonance Long-Period Gratings. *Opt. Lett.* **2013**, *38*, 1666–1668. [CrossRef]
16. Wang, G.; Pran, K.; Sagvolden, G.; Havsgård, G.B.; Jensen, A.E.; Johnson, G.A.; Vohra, S.T. Ship Hull Structure Monitoring Using Fibre Optic Sensors. *Smart Mater. Struct.* **2001**, *10*, 472–478. [CrossRef]
17. Fresvig, T.; Ludvigsen, P.; Steen, H.; Reikerås, O. Fibre Optic Bragg Grating Sensors: An Alternative Method to Strain Gauges for Measuring Deformation in Bone. *Med. Eng. Phys.* **2008**, *30*, 104–108. [CrossRef]
18. Othonos, A. Fiber Bragg Gratings. *Rev. Sci. Instrum.* **1997**, *68*, 4309–4341. [CrossRef]
19. Wu, L.; Maheshwari, M.; Yang, Y.; Xiao, W. Selection and Characterization of Packaged FBG Sensors for Offshore Applications. *Sensors* **2018**, *18*, 3963. [CrossRef] [PubMed]
20. Cho, J.Y.; Lee, K.S. A Birefringence Compensation Method for Mechanically Induced Long-Period Fiber Gratings. *Opt. Commun.* **2002**, *213*, 281–284. [CrossRef]
21. Kumar, S.; Nair, B.; Pillai, V.P.M.; Nayar, U. Amplified Spontaneous Emission–Induced Thermal Nonlinearities in a Mechanically Induced Long-Period Fiber Grating. *Opt. Eng.* **2008**, *47*, 1–7. [CrossRef]
22. Yan, L.S.; Yi, A.; Pan, W.; Luo, B. A Simple Demodulation Method for FBG Temperature Sensors Using a Narrow Band Wavelength Tunable DFB Laser. *IEEE Photonics Technol. Lett.* **2010**, *22*, 1391–1393. [CrossRef]
23. Dong, X.; Tam, H.Y.; Shum, P. Temperature-Insensitive Strain Sensor with Polarization-Maintaining Photonic Crystal Fiber Based Sagnac Interferometer. *Appl. Phys. Lett.* **2007**, *90*, 151113. [CrossRef]
24. Tian, K.; Farrell, G.; Wang, X.; Yang, W.; Xin, Y.; Liang, H.; Lewis, E.; Wang, P. Strain Sensor Based on Gourd-Shaped Single-Mode-Multimode-Single-Mode Hybrid Optical Fibre Structure. *Opt. Express* **2017**, *25*, 18885–18896. [CrossRef]
25. Wright, R.F.; Lu, P.; Devkota, J.; Lu, F.; Ziomek-Moroz, M.; Ohodnicki, P.R. Corrosion Sensors for Structural Health Monitoring of Oil and Natural Gas Infrastructure: A Review. *Sensors* **2019**, *19*, 3964. [CrossRef]
26. Allsop, T.; Reeves, R.; Webb, D.J.; Bennion, I.; Neal, R. A High Sensitivity Refractometer Based upon a Long Period Grating Mach–Zehnder Interferometer. *Rev. Sci. Instrum.* **2002**, *73*, 1702–1705. [CrossRef]
27. Berthold, J.W. Historical Review of Microbend Fiber-Optic Sensors. *J. Light. Technol.* **1995**, *13*, 1193–1199. [CrossRef]
28. Gong, Z.; Xiang, Z.; OuYang, X.; Zhang, J.; Lau, N.; Zhou, J.; Chan, C.C. Wearable Fiber Optic Technology Based on Smart Textile: A Review. *Materials* **2019**, *12*, 3311. [CrossRef]
29. Lee, B.H.; Kim, Y.H.; Park, K.S.; Eom, J.B.; Kim, M.J.; Rho, B.S.; Choi, H.Y. Interferometric Fiber Optic Sensors. *Sensors* **2012**, *12*, 2467–2486. [CrossRef] [PubMed]

30. Zhu, T.; Wu, D.; Liu, M.; Duan, D.-W. In-Line Fiber Optic Interferometric Sensors in Single-Mode Fibers. *Sensors* **2012**, *12*, 10430–10449. [CrossRef]

31. Vali, V.; Shorthill, R.W. Fiber Ring Interferometer. *Appl. Opt.* **1976**, *15*, 1099. [CrossRef] [PubMed]

32. Shih, Y.-C.; Tung, P.-C.; Jywe, W.-Y.; Chang, C.-P.; Shyu, L.-H.; Hsieh, T.-H. Investigation on the Differential Quadrature Fabry–Pérot Interferometer with Variable Measurement Mirrors. *Appl. Sci.* **2020**, *10*, 6191. [CrossRef]

33. Heideman, R.G.; Lambeck, P.V. Remote Opto-Chemical Sensing with Extreme Sensitivity: Design, Fabrication and Performance of a Pigtailed Integrated Optical Phase-Modulated Mach–Zehnder Interferometer System. *Sens. Actuators B Chem.* **1999**, *61*, 100–127. [CrossRef]

34. Kashyap, R.; Nayar, B. An All Single-Mode Fiber Michelson Interferometer Sensor. *J. Light. Technol.* **1983**, *1*, 619–624. [CrossRef]

35. Starodumov, A.N.; Zenteno, L.A.; Monzon, D.; De La Rosa, E. Fiber Sagnac Interferometer Temperature Sensor. *Appl. Phys. Lett.* **1997**, *70*, 19–21. [CrossRef]

36. Flores, C.E.D. *Fabricación y Caracterización de Interferómetros Fabry—Perot de Fibra Óptica Extrínsecos y su Aplicación en Sistemas de Sensado de Variables Físicas*; Centro de Investigaciones en Óptica, A.C.: León, Guanajuato, Mexico, 2018.

37. Wang, L.; Fang, N. Applications of Fiber-Optic Interferometry Technology in Sensor Fields. In *Optical Interferometry*; Banishev, A.A., Bhowmick, M., Wang, J., Eds.; InTech: Shanghai, China, 2008; ISBN 978-953-51-2955-4.

38. Yin, S.; Ruffin, P.B.; Yu, F.T.S. *Fiber Optic Sensors*, 2nd ed.; CRC Press: Boca Raton, FL, USA, 2008; ISBN 978-1-4200-5366-1.

39. Liao, Y.-C.; Liu, B.; Liu, J.; Wan, S.-P.; He, X.-D.; Yuan, J.; Fan, X.; Wu, Q. High Temperature (Up to 950 °C) Sensor Based on Micro Taper In-Line Fiber Mach–Zehnder Interferometer. *Appl. Sci.* **2019**, *9*, 2394. [CrossRef]

40. Ding, L.; Li, Y.; Zhou, C.; Hu, M.; Xiong, Y.; Zeng, Z. In-Fiber Mach-Zehnder Interferometer Based on Three-Core Fiber for Measurement of Directional Bending. *Sensors* **2019**, *19*, 205. [CrossRef]

41. Aref, S.H. Physical Measurement with In-Line Fiber Mach-Zehnder Interferometer Using Differential Phase White Light Interferometry. *Opt. Fiber Technol.* **2017**, *38*, 98–103. [CrossRef]

42. Tian, Z.; Yam, S.S.-H.; Loock, H.-P. Single-Mode Fiber Refractive Index Sensor Based on Core-Offset Attenuators. *IEEE Photonics Technol. Lett.* **2008**, *20*, 1387–1389. [CrossRef]

43. Yu, J.; Xu, S.; Jiang, Y.; Chen, H.; Feng, W. Multi-Parameter Sensor Based on the Fiber Bragg Grating Combined with Triangular-Lattice Four-Core Fiber. *Optik* **2020**, *208*, 164094. [CrossRef]

44. Cardiovascular Diseases (CVDs). Available online: https://www.who.int/news-room/fact-sheets/detail/cardiovascular-diseases-(cvds) (accessed on 15 February 2021).

45. Atrial Fibrillation. Available online: https://www.ahajournals.org/doi/epub/10.1161/01.CIR.0000022730.66617.D9 (accessed on 15 February 2021).

46. Weissler, A.M.; Harris, W.S.; Schoenfeld, C.D. Systolic Time Intervals in Heart Failure in Man. *Circulation* **1968**, *37*, 149–159. [CrossRef]

47. Ball, J.; Dains, J.; Solomon, B.; Stewart, R. *Seidel's Guide to Physical Examination*, 9th ed.; Available online: https://www.elsevier.com/books/seidels-guide-to-physical-examination/ball/978-0-323-48195-3 (accessed on 28 December 2020).

48. The Arterial Pulse. Available online: http://healthcaresciencesocw.wayne.edu/vs/4_2.htm (accessed on 21 February 2021).

49. Introduction to Vital Signs Examination and Evaluation Study Guide. Available online: http://healthcaresciencesocw.wayne.edu/vs/start.htm (accessed on 15 February 2021).

50. Blood Pressure UK. Available online: http://www.bloodpressureuk.org/your-blood-pressure/understanding-your-blood-pressure/what-do-the-numbers-mean/ (accessed on 15 February 2021).

51. Gutierrez-Arroyo, A.; Sanchez-Perez, C.; Aleman-Garcia, N. Optical Sensor for Heat Conduction Measurement in Biological Tissue. *J. Phys. Conf. Ser.* **2013**, *450*, 012027. [CrossRef]

52. Mastrapa, G.C.; Guimarães, G.G.; Inácio, P.L.; de Oliveira, V.; Kalinowski, H.J. Fibre Bragg Gratings in the Visible: Towards Low-Cost Detection. In Proceedings of the Optical Sensing and Detection V, Strasbourg, France, 9 May 2018; Volume 10680, pp. 1–4.

53. Massaroni, C.; Caponero, M.A.; D'Amato, R.; Lo Presti, D.; Schena, E. Fiber Bragg Grating Measuring System for Simultaneous Monitoring of Temperature and Humidity in Mechanical Ventilation. *Sensors* **2017**, *17*, 749. [CrossRef]

54. Tsutsumi, Y.; Ohashi, M.; Miyoshi, Y. Temperature-Sensitive Mechanical LPFG Using Contractive Force of Heat-Shrinkable Tube. *Opt. Fiber Technol.* **2013**, *19*, 55–59. [CrossRef]

55. Tan, Y.; Sun, L.-P.; Jin, L.; Li, J.; Guan, B.-O. Microfiber Mach-Zehnder Interferometer Based on Long Period Grating for Sensing Applications. *Opt. Express* **2013**, *21*, 154–164. [CrossRef] [PubMed]

56. Iadicicco, A.; Campopiano, S. Sensing Features of Long Period Gratings in Hollow Core Fibers. *Sensors* **2015**, *15*, 8009–8019. [CrossRef]

57. Urrutia Azcona, A.; Goicoechea Fernández, J.; Ricchiuti, A.L.; Barrera, D.; Arregui San Martín, F.J. Simultaneous Measurement of Humidity and Temperature Based on a Partially Coated Optical Fiber Long Period Grating. *Sens. Actuators B Chem.* **2016**, *227*, 135–141. [CrossRef]

58. Geng, T.; Zhang, S.; Peng, F.; Yang, W.; Sun, C.; Chen, X.; Zhou, Y.; Hu, Q.; Yuan, L. A Temperature-Insensitive Refractive Index Sensor Based on No-Core Fiber Embedded Long Period Grating. *J. Light. Technol.* **2017**, *35*, 5391–5396. [CrossRef]

59. Lu, L.; Lu, L.; Jiang, Z.; Hu, Y.; Zhou, H.; Liu, G.; Liu, G.; Chen, Y.; Chen, Y.; Luo, Y.; et al. A Portable Optical Fiber SPR Temperature Sensor Based on a Smart-Phone. *Opt. Express* **2019**, *27*, 25420–25427. [CrossRef] [PubMed]

60. Bonefacino, J.; Tam, H.-Y.; Glen, T.S.; Cheng, X.; Pun, C.-F.J.; Wang, J.; Lee, P.-H.; Tse, M.-L.V.; Boles, S.T. Ultra-Fast Polymer Optical Fibre Bragg Grating Inscription for Medical Devices. *Light Sci. Appl.* **2018**, *7*, 17161. [CrossRef] [PubMed]

61. Elsarnagawy, T.; Haueisen, J.; Farrag, M.; Ansari, S.G.; Fouad, H. Embedded Fiber Bragg Grating Based Strain Sensor as Smart Costume for Vital Signal Sensing. *Sens. Lett.* **2014**, *12*, 1669–1674. [CrossRef]

62. Fajkus, M.; Nedoma, J.; Martinek, R.; Vasinek, V.; Nazeran, H.; Siska, P. A Non-Invasive Multichannel Hybrid Fiber-Optic Sensor System for Vital Sign Monitoring. *Sensors* **2017**, *17*, 111. [CrossRef] [PubMed]

63. Nedoma, J.; Fajkus, M.; Martinek, R.; Nazeran, H. Vital Sign Monitoring and Cardiac Triggering at 1.5 Tesla: A Practical Solution by an MR-Ballistocardiography Fiber-Optic Sensor. *Sensors* **2019**, *19*, 470. [CrossRef]

64. Lee, B. Review of the Present Status of Optical Fiber Sensors. *Opt. Fiber Technol.* **2003**, *9*, 57–79. [CrossRef]

65. Fïdanboylu, K.; Efendïoğlu, H. Review of Turn around Point Long Period Fiber Gratings. *J. Sens. Technol.* **2009**, *5*, 81.

66. Sharma, A.K.; Gupta, J.; Basu, R. Simulation and Performance Evaluation of Fiber Optic Sensor for Detection of Hepatic Malignancies in Human Liver Tissues. *Opt. Laser Technol.* **2018**, *98*, 291–297. [CrossRef]

67. Hunsperger, R.G.; Mentzer, M.A. *Photonic Devices and Systems*; Routledge: Boca Raton, FL, USA, 2017; ISBN 978-0-203-74351-5.

68. Dziuda, L.; Skibniewski, F.W.; Krej, M.; Lewandowski, J. Monitoring Respiration and Cardiac Activity Using Fiber Bragg Grating-Based Sensor. *IEEE Trans. Biomed. Eng.* **2012**, *59*, 1934–1942. [CrossRef]

69. Yu, C.; Xu, W.; Shen, Y.; Bian, S.; Yu, C.; You, S. Non-Invasive Vital Signs Monitoring System Based on Smart Sensor Mat Embedded with Optical Fiber Interferometer. In Proceedings of the 2018 27th Wireless and Optical Communication Conference (WOCC), Hualien, Taiwan, 30 April–1 May 2018; pp. 1–3.

70. Zazula, D.; Šprager, S. Monitoring Respiration by Using Fiber-Optic Interferomtery and Maximum A Posteriori Estimation. In Proceedings of the 2013 Africon, Pointe aux Piments, Mauritius, 9–12 September 2013.

71. Krebber, K.; Lenke, P.; Liehr, S.; Witt, J.; Schukar, M. Smart Technical Textiles with Integrated POF Sensors. In Proceedings of the Smart Sensor Phenomena, Technology, Networks, and Systems 2008, San Diego, CA, USA, 7 April 2008; Volume 6933, p. 69330V.

72. Massaroni, C.; Presti, D.L.; Saccomandi, P.; Caponero, M.A.; D'Amato, R.; Schena, E. Fiber Bragg Grating Probe for Relative Humidity and Respiratory Frequency Estimation: Assessment During Mechanical Ventilation. *IEEE Sens. J.* **2018**, *18*, 2125–2130. [CrossRef]

73. Zhang, Y.; Chen, Z.; Hee, H.I. Noninvasive Measurement of Heart Rate and Respiratory Rate for Perioperative Infants. *J. Light. Technol.* **2019**, *37*, 2807–2814. [CrossRef]

74. Hariyanti Devara, K.; Aisyah, F.N.; Nadia, K.V.A.W.; Purnamaningsih, R.W. Design of a Wearable Fiber Optic Respiration Sensor for Application in NICU Incubators. *AIP Conf. Proc.* **2019**, *2092*, 020002. [CrossRef]

75. Allsop, T.; Carroll, K.; Lloyd, G.; Webb, D.J.; Miller, M.; Bennion, I. Application of Long-Period-Grating Sensors to Respiratory Plethysmography. *J. Biomed. Opt.* **2007**, *12*, 064003. [CrossRef]

76. Allsop, T.D.P.; Earthrowl, T.; Revees, R.; Webb, D.J.; Miller, M.; Jones, B.W.; Bennion, I. Application of Long-Period Grating Sensors to Respiratory Function Monitoring. In Proceedings of the Smart Medical and Biomedical Sensor Technology II, Philadelphia, PA, USA, 7 December 2004; Volume 5588, pp. 148–156.

77. Witt, J.; Narbonneau, F.; Schukar, M.; Krebber, K.; Jonckheere, J.D.; Jeanne, M.; Kinet, D.; Paquet, B.; Depre, A.; D'Angelo, L.T.; et al. Medical Textiles With Embedded Fiber Optic Sensors for Monitoring of Respiratory Movement. *IEEE Sens. J.* **2012**, *12*, 246–254. [CrossRef]

78. Petrović, M.D.; Petrovic, J.; Daničić, A.; Vukčević, M.; Bojović, B.; Hadžievski, L.; Allsop, T.; Lloyd, G.; Webb, D.J. Non-Invasive Respiratory Monitoring Using Long-Period Fiber Grating Sensors. *Biomed. Opt. Express* **2014**, *5*, 1136–1144. [CrossRef]

79. Allsop, T.D.P.; Earthrowl-Gould, T.; Webb, D.J.; Bennion, I. Embedded Progressive-Three-Layered Fiber Long-Period Gratings for Respiratory Monitoring. *J. Biomed. Opt.* **2003**, *8*, 552–558. [CrossRef] [PubMed]

80. Koyama, Y.; Nishiyama, M.; Watanabe, K. Smart Textile Using Hetero-Core Optical Fiber for Heartbeat and Respiration Monitoring. *IEEE Sens. J.* **2018**, *18*, 6175–6180. [CrossRef]

81. Ciocchetti, M.; Massaroni, C.; Saccomandi, P.; Caponero, M.A.; Polimadei, A.; Formica, D.; Schena, E. Smart Textile Based on Fiber Bragg Grating Sensors for Respiratory Monitoring: Design and Preliminary Trials. *Biosensors* **2015**, *5*, 602–615. [CrossRef]

82. Presti, D.L.; Massaroni, C.; Formica, D.; Saccomandi, P.; Giurazza, F.; Caponero, M.A.; Schena, E. Smart Textile Based on 12 Fiber Bragg Gratings Array for Vital Signs Monitoring. *IEEE Sens. J.* **2017**, *17*, 6037–6043. [CrossRef]

83. Lo Presti, D.; Massaroni, C.; Saccomandi, P.; Caponero, M.A.; Formica, D.; Schena, E. A Wearable Textile for Respiratory Monitoring: Feasibility Assessment and Analysis of Sensors Position on System Response. *Annu. Int. Conf.* **2017**, *2017*, 4423–4426. [CrossRef]

84. Chen, Z.; Lau, D.; Teo, J.T.; Ng, S.H.; Yang, X.; Kei, P.L. Simultaneous Measurement of Breathing Rate and Heart Rate Using a Microbend Multimode Fiber Optic Sensor. *J. Biomed. Opt.* **2014**, *19*, 057001. [CrossRef] [PubMed]

85. Chatterjee, K.; Tabor, J.; Ghosh, T.K. Electrically Conductive Coatings for Fiber-Based E-Textiles. *Fibers* **2019**, *7*, 51. [CrossRef]

86. Chethana, K.; Prasad, A.S.G.; Omkar, S.N.; Asokan, S. Fiber Bragg Grating Sensor Based Device for Simultaneous Measurement of Respiratory and Cardiac Activities. *J. Biophotonics* **2017**, *10*, 278–285. [CrossRef] [PubMed]

87. Koyama, S.; Ishizawa, H. Vital Sign Measurement Using FBG Sensor for New Wearable Sensor Development. *Fiber Opt. Sens. Princ. Meas. Appl.* **2019**, 1–16. [CrossRef]

88. Tan, F.; Chen, S.; Lyu, W.; Liu, Z.; Liu, Z.; Yu, C.; Yu, C.; Lu, C.; Tam, H.-Y. Non-Invasive Human Vital Signs Monitoring Based on Twin-Core Optical Fiber Sensors. *Biomed. Opt. Express* **2019**, *10*, 5940–5952. [CrossRef]

89. Fajkus, M.; Nedoma, J.; Siska, P.; Vasinek, V. FBG Sensor of Breathing Encapsulated into Polydimethylsiloxane. In Proceedings of the Optical Materials and Biomaterials in Security and Defence Systems Technology XIII, Edinburgh, UK, 24 October 2016; Volume 9994, p. 99940N.

90. Hu, H.; Sun, S.; Lv, R.; Zhao, Y. Design and Experiment of an Optical Fiber Micro Bend Sensor for Respiration Monitoring. *Sens. Actuators Phys.* **2016**, *251*, 126–133. [CrossRef]

91. Nedoma, J.; Fajkus, M.; Novak, M.; Strbikova, N.; Vasinek, V.; Nazeran, H.; Vanus, J.; Perecar, F.; Martinek, R. Validation of a Novel Fiber-Optic Sensor System for Monitoring Cardiorespiratory Activities During MRI Examinations. *Adv. Electr. Electron. Eng.* **2017**, *15*, 536–543. [CrossRef]

92. Pant, S.; Umesh, S.; Asokan, S. Fiber Bragg Grating Respiratory Measurement Device. In Proceedings of the 2018 IEEE International Symposium on Medical Measurements and Applications (MeMeA), Rome, Italy, 11–13 June 2018; pp. 1–5.

93. Nedoma, J.; Fajkus, M.; Siska, P.; Martinek, R.; Vasinek, V. Non-Invasive Fiber Optic Probe Encapsulated Into PolyDiMethylSiloxane for Measuring Respiratory and Heart Rate of the Human Body. *Adv. Electr. Electron. Eng.* **2017**, *15*. [CrossRef]

94. Lo Presti, D.; Romano, C.; Massaroni, C.; D'Abbraccio, J.; Massari, L.; Caponero, M.A.; Oddo, C.M.; Formica, D.; Schena, E. Cardio-Respiratory Monitoring in Archery Using a Smart Textile Based on Flexible Fiber Bragg Grating Sensors. *Sensors* **2019**, *19*, 3581. [CrossRef]

95. Jia, J.; Xu, C.; Pan, S.; Xia, S.; Wei, P.; Noh, H.Y.; Zhang, P.; Jiang, X. Conductive Thread-Based Textile Sensor for Continuous Perspiration Level Monitoring. *Sensors* **2018**, *18*, 3775. [CrossRef]

96. Massaroni, C.; Saccomandi, P.; Schena, E. Medical Smart Textiles Based on Fiber Optic Technology: An Overview. *J. Funct. Biomater.* **2015**, *6*, 204–221. [CrossRef] [PubMed]

97. Anwar Zawawi, M.; O'Keffe, S.; Lewis, E. Intensity-modulated Fiber Optic Sensor for Health Monitoring Applications: A Comparative Review. *Sens. Rev.* **2013**, *33*, 57–67. [CrossRef]

98. Ho, S.C.M.; Razavi, M.; Nazeri, A.; Song, G. FBG Sensor for Contact Level Monitoring and Prediction of Perforation in Cardiac Ablation. *Sensors* **2012**, *12*, 1002–1013. [CrossRef] [PubMed]

99. Yunianto, M.; Marzuki, A.; Riyanto, R.; Lestari, D. Fiber Optic Sensor Based on Reflectivity Configurations to Detect Heart Rate. *J. Phys. Conf. Ser.* **2016**, *776*, 012110. [CrossRef]

100. Chen, Z.; Teo, J.T.; Ng, S.H.; Yim, H. Smart Pillow for Heart-Rate Monitoring Using a Fiber Optic Sensor. In Proceedings of the Optical Fibers, Sensors, and Devices for Biomedical Diagnostics and Treatment XI, Wuhan, China, 16 February 2011; Volume 7894, p. 789402.

101. Krej, M.; Dziuda, L.; Skibniewski, F.W. A Method of Detecting Heartbeat Locations in the Ballistocardiographic Signal from the Fiber-Optic Vital Signs Sensor. *IEEE J. Biomed. Health Inform.* **2015**, *19*, 1443–1450. [CrossRef]

102. Chino, S.; Ishizawa, H.; Hosoya, S.; Koyama, S.; Fujimoto, K.; Kawamura, T. Research for Wearable Multiple Vital Sign Sensor Using Fiber Bragg Grating -Verification of Several Pulsate Points in Human Body Surface. In Proceedings of the 2017 IEEE International Instrumentation and Measurement Technology Conference (I2MTC), Turin, Italy, 22–25 May 2017.

103. Sadek, I.; Biswas, J.; Abdulrazak, B.; Haihong, Z.; Mokhtari, M. Continuous and Unconstrained Vital Signs Monitoring with Ballistocardiogram Sensors in Headrest Position. In Proceedings of the 2017 IEEE EMBS International Conference on Biomedical Health Informatics (BHI), Orlando, FL, USA, 16–19 February 2017; pp. 289–292.

104. Arifin, A.; Lebang, A.K.; Yunus, M.; Dewang, S.; Idris, I.; Tahir, D. Measurement Heart Rate Based on Plastic Optical Fiber Sensor. *J. Phys. Conf. Ser.* **2019**, *1170*, 012074. [CrossRef]

105. Wang, J.; Liu, K.; Sun, Q.; Ni, X.; Ai, F.; Wang, S.; Yan, Z.; Liu, D. Diaphragm-Based Optical Fiber Sensor for Pulse Wave Monitoring and Cardiovascular Diseases Diagnosis. *J. Biophotonics* **2019**, *12*, e201900084. [CrossRef]

106. Sensors/Free Full-Text/Continuous Vital Monitoring During Sleep and Light Activity Using Carbon-Black Elastomer Sensors/HTML. Available online: https://www.mdpi.com/1424-8220/20/6/1583/htm (accessed on 12 February 2021).

107. Chen, C.; Li, Z.; Zhang, Y.; Zhang, S.; Hou, J.; Zhang, H. A 3D Wrist Pulse Signal Acquisition System for Width Information of Pulse Wave. *Sensors* **2019**, *20*, 11. [CrossRef] [PubMed]

108. Reesink, K.D.; van der Nagel, T.; Bovelander, J.; Jansen, J.R.C.; van der Veen, F.H.; Schreuder, J.J. Feasibility Study of a Fiber-Optic System for Invasive Blood Pressure Measurements. *Catheter. Cardiovasc. Interv.* **2002**, *57*, 272–276. [CrossRef] [PubMed]

109. Krasovskii, V.I.; Feofanov, I.N.; Ivashkin, P.I.; Kazaryan, M.A. A Fiber-Optic Doppler Blood Flow-Velocity Sensor. *St Petersburg Polytech. Univ. J. Phys. Math.* **2017**, *3*, 35–38. [CrossRef]

110. Totsu, K.; Haga, Y.; Esashi, M. Ultra-Miniature Fiber-Optic Pressure Sensor Using White Light Interferometry. *J. Micromech. Microeng.* **2004**, *15*, 71–75. [CrossRef]

111. Wu, N.; Tian, Y.; Zou, X.; Zhai, Y.; Barringhaus, K.; Wang, X. A Miniature Fiber Optic Blood Pressure Sensor and Its Application in in Vivo Blood Pressure Measurements of a Swine Model. *Sens. Actuators B Chem.* **2013**, *181*, 172–178. [CrossRef]

112. Miyauchi, Y.; Koyama, S.; Ishizawa, H. Basic Experiment of Blood-Pressure Measurement Which Uses FBG Sensors. In Proceedings of the 2013 IEEE International Instrumentation and Measurement Technology Conference (I2MTC), Minneapolis, MN, USA, 6–9 May 2013. [CrossRef]

113. Tian, Y.; Wu, N.; Zou, X.; Zhang, Y.; Barringhaus, K.; Wang, X. A Study on Packaging of Miniature Fiber Optic Sensors for In-Vivo Blood Pressure Measurements in a Swine Model. *IEEE Sens. J.* **2014**, *14*, 629–635. [CrossRef]

114. Koyama, S.; Ishizawa, H.; Sakaguchi. A.; Hosoya, S.; Kawamura, T. Influence on Calculated Blood Pressure of Measurement Posture for the Development of Wearable Vital Sign Sensors. *J. Sens.* **2017**, *2017*, 1–10. [CrossRef]

115. Koyama, S.; Sakaguchi, A.; Ishizawa, H.; Yasue, K.; Oshiro, H.; Kimura, H. Vital Sign Measurement Using Covered FBG Sensor Embedded into Knitted Fabric for Smart Textile. *J. Fiber Sci. Technol.* **2017**, *73*, 300–308. [CrossRef]
116. Haseda, Y.; Bonefacino, J.; Tam, H.-Y.; Chino, S.; Koyama, S.; Ishizawa, H. Measurement of Pulse Wave Signals and Blood Pressure by a Plastic Optical Fiber FBG Sensor. *Sensors* **2019**, *19*, 5088. [CrossRef] [PubMed]
117. Kang, H.-K.; Kang, D.-H.; Bang, H.-J.; Hong, C.-S.; Kim, C.-G. Cure Monitoring of Composite Laminates Using Fiber Optic Sensors. *Smart Mater. Struct.* **2002**, *11*, 279–287. [CrossRef]
118. Życzkowski, M.; Szustakowski, M.; Ciurapiński, W.; Uziębło-Życzkowska, B. Interferometric Fiber Optics Based Sensor for Monitoring of the Heart Activity. *Acta Phys. Pol. A* **2011**, *120*, 782–784. [CrossRef]
119. Życzkowski, M.; Uziebło-Zyczkowska, B.; Dziuda, L.; Rozanowski, K. Using Modalmetric Fiber Optic Sensors to Monitor the Activity of the Heart. In Proceedings of the Optical Fibers, Sensors, and Devices for Biomedical Diagnostics and Treatment XI, San Francisco, CA, USA, 16 February 2011; Volume 7894, pp. 1–7.
120. Mishra, V.; Singh, N.; Tiwari, U.; Kapur, P. Fiber Grating Sensors in Medicine: Current and Emerging Applications. *Sens. Actuators Phys.* **2011**, *167*, 279–290. [CrossRef]
121. Silvestri, S.; Schena, E. *Optical-Fiber Measurement Systems for Medical Applications*; Optoelectronics—Devices and Applications; IntechOpen: Rome, Italy, 2011; Volume 11.
122. Tok, S.; de Haan, K.; Tseng, D.; Usanmaz, C.F.; Ceylan Koydemir, H.; Ozcan, A. Early Detection of *E. Coli* and Total Coliform Using an Automated, Colorimetric and Fluorometric Fiber Optics-Based Device. *Lab. Chip* **2019**, *19*, 2925–2935. [CrossRef] [PubMed]
123. Chen, C.; Zhao, X.-L.; Li, Z.-H.; Zhu, Z.-G.; Qian, S.-H.; Flewitt, A. Current and Emerging Technology for Continuous Glucose Monitoring. *Sensors* **2017**, *17*, 182. [CrossRef] [PubMed]
124. Najafi, B.; Mohseni, H.; Grewal, G.S.; Talal, T.K.; Menzies, R.A.; Armstrong, D.G. An Optical-Fiber-Based Smart Textile (Smart Socks) to Manage Biomechanical Risk Factors Associated With Diabetic Foot Amputation. *J. Diabetes Sci. Technol.* **2017**, *11*, 668–677. [CrossRef] [PubMed]
125. Rifat, A.A.; Ahmed, R.; Yetisen, A.K.; Butt, H.; Sabouri, A.; Mahdiraji, G.A.; Yun, S.H.; Adikan, F.R.M. Photonic Crystal Fiber Based Plasmonic Sensors. *Sens. Actuators B Chem.* **2017**, *243*, 311–325. [CrossRef]
126. Dass, S.; Jha, R. Microfiber-Wrapped Bi-Conical-Tapered SMF for Curvature Sensing. *IEEE Sens. J.* **2016**, *16*, 3649–3652. [CrossRef]
127. Dash, J.N.; Dass, S.; Jha, R. Photonic Crystal Fiber Microcavity Based Bend and Temperature Sensor Using Micro Fiber. *Sens. Actuators Phys.* **2016**, *244*, 24–29. [CrossRef]
128. Dass, S.; Jha, R. Micron Wire Assisted Inline Mach-Zehnder Interferometric Curvature Sensor. *IEEE Photonics Technol. Lett.* **2015**, *28*, 1. [CrossRef]
129. Yhuwana, Y.G.Y.; Apsari, R.; Yasin, M. Fiber Optic Sensor for Heart Rate Detection. *Optik* **2017**, *134*, 28–32. [CrossRef]
130. *Fiber Optic Sensors: An Introduction for Engineers and Scientists*, 2nd ed.; Available online: https://www.wiley.com/en-cl/Fiber+Optic+Sensors%3A+An+Introduction+for+Engineers+and+Scientists%2C+2nd+Edition-p-9780470126844 (accessed on 10 February 2021).

biosensors

Review

Electromagnetic Torso Scanning: A Review of Devices, Algorithms, and Systems

Sasan Ahdi Rezaeieh *, Amin Darvazehban, Azin S. Janani and Amin M. Abbosh

School of ITEE, The University of Queensland, Brisbane 4072, Australia; a.darvazehban@uq.edu.au (A.D.); azin.janani@uq.edu.au (A.S.J.); a.abbosh@uq.edu.au (A.M.A.)
* Correspondence: s.ahdirezaeieh@uq.edu.au

Abstract: The past decade has witnessed a surge into research on disruptive technologies that either challenge or complement conventional thoracic diagnostic modalities. The non-ionizing, non-invasive, compact, and low power requirements of electromagnetic (EM) techniques make them among the top contenders with varieties of proposed scanning systems, which can be used to detect wide range of thoracic illnesses. Different configurations, antenna topologies and detection or imaging algorithms are utilized in these systems. Hence, to appreciate their progress and assess their potential, a critical review of EM thoracic scanning systems is presented. Considering the numerous thoracic diseases, such as fatty liver disease, lung cancer, respiratory and heart related complications, this paper will exclusively focus on torso scanning systems, tracing the early foundation of research that studied the possibility of using EM waves to detect thoracic diseases besides exploring recent progresses. The advantages and disadvantages of proposed systems and future possibilities are thoroughly discussed.

Keywords: torso scanning; antennas; processing algorithms; electromagnetic imaging

Citation: Ahdi Rezaeieh, S.; Darvazehban, A.; Janani, A.S.; Abbosh, A.M. Electromagnetic Torso Scanning: A Review of Devices, Algorithms, and Systems. *Biosensors* **2021**, *11*, 135. https://doi.org/10.3390/bios11050135

Received: 8 April 2021
Accepted: 23 April 2021
Published: 27 April 2021

1. Introduction

The urge and curiosity of human beings toward understanding diseases and developing tools to diagnose them can be backdated to ancient times and has been an ever-developing part of science throughout past centuries. However, it was the invention of the X-ray by Wilhelm Rontgen in the 19th century that created a new direction in medical science. It provided a third eye to the medical staff that could confirm or reject their hypothetical diagnosis. This advancement has changed the course of treatments and raised medical standards significantly. Hence, huge investments were made to enhance this technology and improve the quality of the obtained images, leading to the invention of Computed Tomography (CT scan) in the second half of 20th century. However, despite all these innovations, both systems came with an undesirable caveat: they use ionizing radiation. Efforts by numerous researchers resulted in the invention of magnetic resonance imaging (MRI) that utilizes a combination of a strong magnetic field and electromagnetic (EM) waves to map the changes inside human body. MRI is accepted as the gold standard among medical imaging devices and its image quality has significantly improved, thanks to advancements in imaging algorithms and coil fabrication technology. Consequently, MRI technology comes with a huge price tag and highly shielding requirements to contain the strong magnetic fields. Hence, it is not suitable for rapid onsite diagnosis and frequent monitoring, besides limiting its accessibility to large medical centers.

Motivated by the aforementioned limitations, researchers have been investigating alternative or complementary techniques that are non-invasive, safe, low cost and portable. EM techniques are among the top candidates that have been widely studied. The basis of using those techniques is the fact that characteristics of EM waves, such as phase and magnitude, are altered by the dielectric properties of biological tissues [1]. For instance, a cancerous cell has higher fluid content compared to a healthy one [2]. This results in a change in the response of EM signals, and this change can be potentially exploited for detection purposes. This review focusses on application of EM scanning systems for thoracic diseases, as one of the main contributors to mortality rates in the world [3]. Fluid accumulation inside (pulmonary edema) and around (pleural effusion) the lungs is the common symptom for various diseases such as heart failure, lung cancer, breast cancer and more recently COVID-19 [4–6]. Hence, the ability to detect accumulated fluid at early stages can potentially lead to early diagnosis of the underlying diseases. The possibility of detecting lung fluid was first proposed by Susskind in 1973 [7], where the use of a combination of a cathode ray tube and an antenna was proposed as a scanning platform. The first studies on the application of microwaves to detect pulmonary edema were performed by Pedersen [8,9]. These studies modeled human body as a load and calculated reflection coefficient based on the changes inside the torso. Hence, those studies concluded that any changes inside the torso affect the magnitude and phase of EM reflection and transmission coefficients, although only limited tests were performed on the magnitude of those signals. To further improve these outcomes, two methods were later introduced in 1983 by Iskandar et al. [10] to measure the variations in lung water. The first method analyzed the phase of transmission coefficient of a microwave applicator [11], whereas the second method employed the radiometry concept that monitors changes in microwave emission levels [12]. To provide a more robust detection process, dielectric properties estimation methods were used to detect pathological changes in tissues inside lungs [13]. It follows the same logic that accumulated fluid inside or around lungs alters the average permittivity as experienced by the sensing antennas.

While all these methods have advanced the field, they all have limitation in that detection is performed based on the assumption of known dielectric properties of the healthy status of the scanned subject [8–11,13], which is difficult to achieve in practice. Accordingly, different approaches such as the combination of microwave transmission with X-ray scanning were investigated. The method in [14] estimated the amount of water content inside lungs using a method of moments. This study was an important step in moving towards a more sophisticated analysis of the torso's internal tissues. This view was further advanced by the introduction of radar-based microwave imaging [15–21] and tomography methods [22,23], where the scattered fields from the abnormal tissue inside the torso is calculated using forward and backward processing methods. Considering the higher water content, in the case of edemas and cancerous tissues, the reflected/transmitted signals from those pathologies create a strong scattered field or high contrast permittivity map that can be shown as two- or three-dimensional images. The advances in computational methods and artificial intelligence open new horizons in research that were not feasible before. For example, a statistical analysis was adopted to predict changes in fat levels inside the liver [24]. This method is based on the high symmetry between the right and left sides of the body and analysis of the correlation between signals from these areas at several frequency points. In another method, the feasibility of a supervised decent method [25] for torso imaging was adopted to estimate the torso's structural information, paving the way for real time imaging.

To ascertain the importance of the abovementioned advances, the present review focuses on torso scanning systems aimed at detecting thoracic diseases and thus does not include vital sign monitoring applications. Section 2 of the paper reviews electromagnetic scanning systems and discusses their operation principles. It then investigates different data acquisition methods and their role in complexity and accuracy of the system. Moreover, it reviews the most utilized scanning platforms for torso scanning systems, their capabilities and limitations. The safety considerations for torso scanning systems are thoroughly investigated in Section 3. The safely limits are investigated using specific absorption rate (SAR), followed by a discussion on range of SAR values for current EM torso scanners. Section 4 presents an overview of design criteria for torso scanning systems and provides a detailed review of two different categories of on-body matched and free space antennas. The advantages and disadvantages of each subcategory in terms of penetration, impedance matching, size and fabrication complexity are investigated. Section 5 classifies the utilized algorithms in these systems into three subcategories of detection only, classification and detection and localization (imaging). This section concludes that there is a compromise between accuracy of the scanning system, its complexity, and practicality. Each section is accompanied by a comparison table that highlights the pros and cons of each subcategory. Section 6 offers conclusions based on the discussions provided in previous sections and provides thoughts for future development of these systems.

2. Electromagnetic Scanning Systems

An electromagnetic scanning platform includes two main elements: (1) hardware and (2) software. In the hardware unit, antennas are used to transmit signals towards torso (imaging domain) and receive the reflected or transmitted signals that are then recorded using a vector network analyzer (VNA), or any proper EM transceiver, and stored in a computational tool that contains the detection algorithm (software). The detection algorithm analyzes different aspects of the signal and either form an image or give estimation of the pathology. While the next sections of the paper will go through details of utilized elements and methods, this section focuses on the overall operating mechanism and formation of EM scanning systems.

2.1. Data Acquisition Methods

Data acquisition methods can be broadly categorized as mono-static and multi-static ones. In a mono-static system the same antenna is utilized as the transmitter and receiver. Therefore, these systems can only analyze the reflection coefficient signals [26–28]. Since the system has no switching requirement, its implementation is simple. However, due to the lack of transmission coefficient data, the system has less degree of freedom to compensate for the effect of noise on detection decision. Additionally, the system is less sensitive to changes in the dielectric properties of deeper targets. Hence, this method is more suitable for surface or subsurface abnormalities [29]. In the multi-static system, an array of antennas, at least two, are used to transmit and receive signals [24,30–32]. This process continues sequentially until all the antennas in the array have received signals from each other. Hence, more data in the form of transmission coefficient signals between different channels is obtained compared to a mono-static method. This enables higher detection accuracy and enhances deep target detection. However, the system requires a switching network, which introduces additional insertion loss to signals plus increasing the scanning time [33,34].

2.2. Scanning Platform

Two main factors define the configuration of the scanning platform: (1) requirements of the detection algorithm and (2) size of the antennas. Detection algorithms are divided between the ones that focus on detection only and the ones that aim at detection and localization of the malignancy. The size of the antenna mainly depends on the operating frequency band and this could be a limiting factor when operating at low microwave frequencies. Generally, algorithms that only aim at making a binary decision of presence or absence of the malignancy, e.g., water accumulation, require simpler setups. As seen from Figure 1, the system that is used in early studies [31] follow this approach and use two applicators on either side of the torso to capture and utilize transmitted signal. A proper detector is used to measure the phase of the transmitted signal, which is then recorded using a VNA.

The second category of algorithms requires more sophisticated scanning platforms with higher number of antennas. To provide a statically quantifiable data or map the location of the malignancy, several scans from different angles and positions around the torso are required. To fulfill the requirements of these algorithms, three main categories of scanning platforms are designed: (1) linear platforms, (2) circular platforms and (3) quasi-circular platforms.

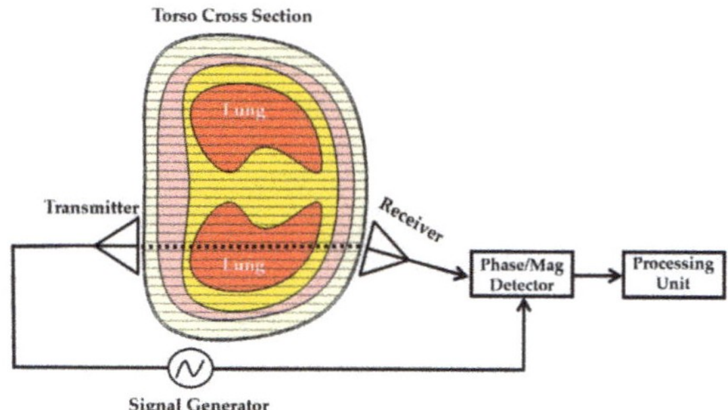

Figure 1. Configuration of an early torso scanner setup (concept taken from [31]).

2.2.1. Linear Platforms

Linear platforms are designed to scan the rear side of the torso and are generally used with algorithms that utilize differential detection approach [18,26]. In these systems an antenna or an array of antennas scan the right and left sides of the torso and their location is mechanically displaced along the torso [18,35]. Stepper motors are used to perform the movement with small, around 1 mm, steps at each scan. To simplify the scanning setup, a mono-static data acquisition technique can also be utilized in these systems. An example of these systems is depicted in Figure 2, where two antennas are located side by side on a lever to eliminate positioning errors while being displaced up and down. The main advantage of this scanning system is that high number of scans can be performed, and hence the obtained image could provide an accurate estimation of the location of the malignancy. However, these systems come at the cost of more complicated setup and slower scanning process.

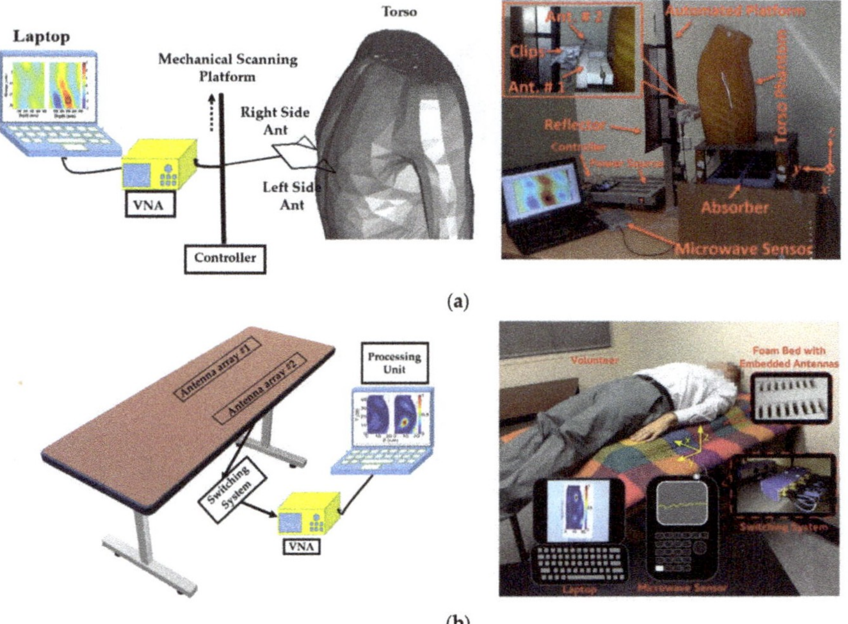

Figure 2. Linear scanning platform. (**a**) Dynamic system, schematic view (left) and system view (right). Reproduced with permission from [35]. Copyright 2014 IEEE. (**b**) Static system, schematic view (left) and system view (right). Reproduced with permission from [18] (open access).

The second types of linear systems are the static ones, where antenna elements form an array and are fixed in a position along the torso. Examples of the systems are presented in [18,34], where two arrays of antennas are embedded in a foam or a bed for a patient to lay on when the scan is performed. These platforms remove the complications with the movable systems at the cost of using lower number of antennas due to mutual coupling considerations. To compensate for reduced antenna numbers and to maintain the accuracy of the system, a multi-static data acquisition method is utilized. Hence, a switching network is added as part of the system, and this in turn increases the overall cost of the scanning system (See Figure 2b)

The linear scanning systems are most suitable for applications where scans of the upper sides of the torso are required. This facilitates a uniform scan for all male and female subjects and the obtained signals are not convoluted by the presence of breast. However, due to limited power allowed in EM scans, these systems are less sensitive to deep malignancies or the ones that are far away from the antennas.

2.2.2. Circular Platforms

To alleviate the problem of low accuracy for deep targets, circular platforms are proposed in which the antennas are located around the torso [36,37]. An example of a circular platform configuration is depicted in Figure 3. The system consists of an array of 16-element antennas, a VNA and a laptop as the processing unit. A monostatic data acquisition method is performed where each of the antennas are used to both transmit and receive backscattered signals. These setups are used for imaging algorithms that require obtaining information from all angles around the object. This is necessary to produce a scattering profile or map dielectric properties across the imaging domain, e.g., torso. Three dimensional images can also be obtained by locating the circular array on a moving flange [29] and scanning different heights across the torso. Both mono-static and multi-static data acquisition techniques can be used to obtain the scattered signals. The main consideration in these systems is the distance of the antenna from the torso, due to semi-elliptical shape of the body. Considering the significant variations between torso sizes in different individuals, the designed systems are generally not fit for every single case. While a certain degree of movement for antenna locations is allowed within the system, a fixed distance between the antenna and body cannot be guaranteed/achieved for all scanning cases, hence, creating difficulties for the imaging algorithm in the form of ghost targets due to stronger reflections at the skin/air boundary. This is addressed by using strict calibration measures before each test, besides filtering techniques that add to the complexity of the system. The circular system setup is best suited for applications where information regarding the location, size and severity of the malignancy is required.

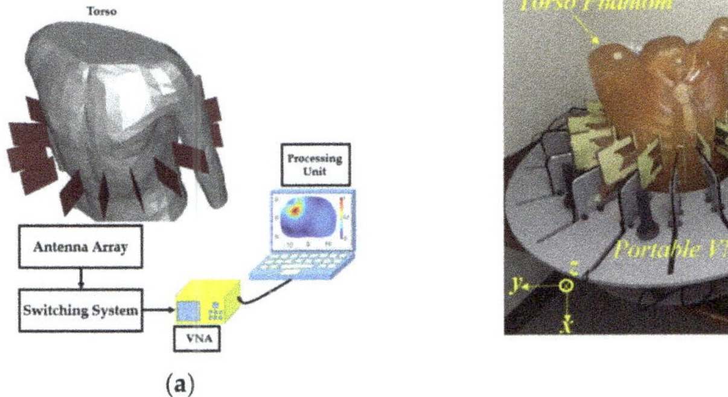

Figure 3. Antenna array platform for scanning the torso area: (**a**) schematic view; (**b**) system view. Reproduced with permission from [36]. Copyright 2016 IEEE.

2.2.3. Quasi-Circular Platforms

While circular systems provide a convenient setup for scanning, their design can be complex due to high number of antennas and switching network that ensures timely scanning of the patient. Therefore, a simpler version of these systems has also been utilized, in which the system is comprised of a single antenna and the subject is located on a turning platform [19]. In this system the subject is rotated at desired intervals, e.g., every five degrees, and the antenna constantly scans the patient until a full scan is performed. This system can only operate in mono-static data acquisition mode and the imaging algorithm is limited to the use of reflected signals only. However, a higher number of scans/data points are obtained that can potentially compensate for the lack of transmission coefficient data. The configuration of the proposed system is depicted in Figure 4.

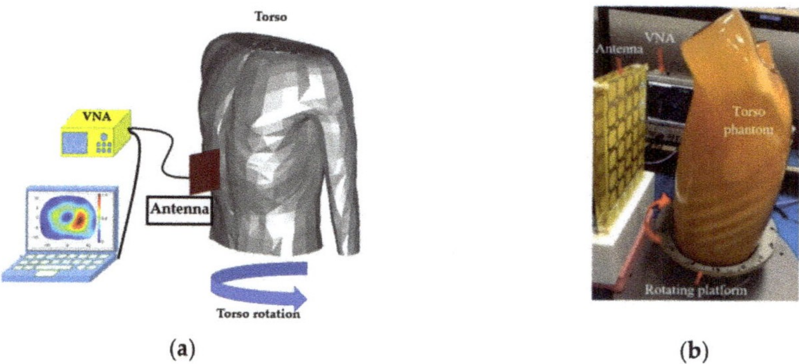

(a)   (b)

Figure 4. Quasi circular platform: (**a**) fixed beam antenna (concept taken from [19]). (**b**) Pattern reconfigurable antenna. Reproduced with permission from [38] (Copyright 2019 IEEE).

Additionally, three dimensional scans can also be obtained using pattern reconfigurable antennas [38]. An example of a quasi-circular platform using pattern reconfigurable antenna is shown in Figure 4. The schematic of pattern reconfigurable antenna for torso scanning is depicted in Figure 5.

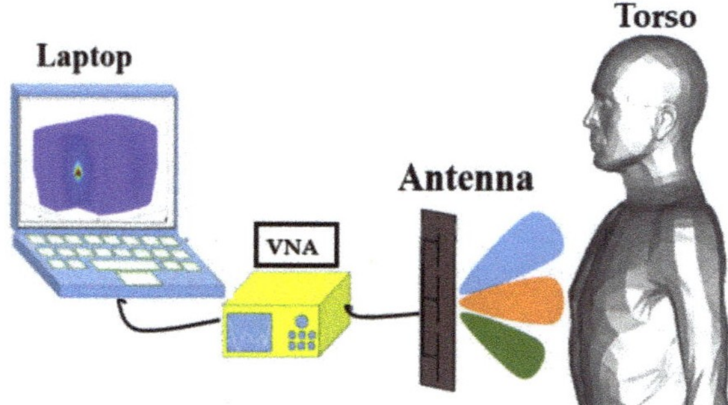

Figure 5. Schematic of pattern reconfigurable antenna for torso scanners.

2.2.4. Wearable Platforms

To overcome the complications of finding the exact location of the antenna with respect to the body, wearable systems can be used [13,24,39,40]. As shown in Figure 6, in these systems the antenna array is located on the body, and depending on the utilized detection algorithm, the antennas can be located locally adjacent to the subject [40] or surround the circumference of the imaged body [24]. These systems can utilize multi or mono static data acquisition methods and can accommodate larger number of antennas as antenna size can be significantly reduced due to the body loading on the antenna. This arrangement provides the scanning system with larger number of data points that can be collected during the scan.

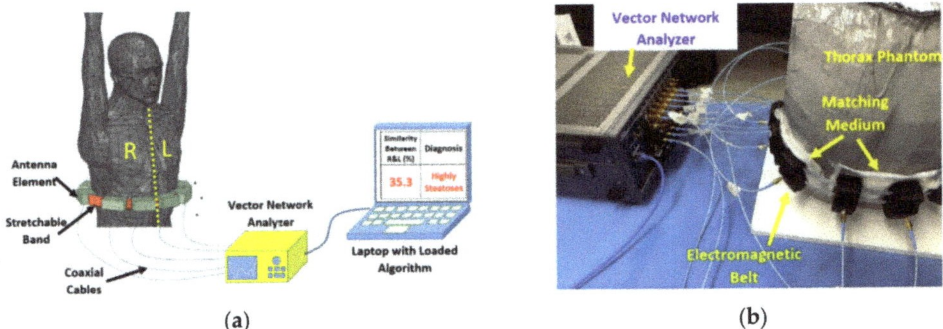

Figure 6. Wearable platform: (**a**) Schematic view; (**b**) system view. Reproduced with permission from [24] (open access).

3. Safety Considerations

Electromagnetic medical scanning systems should follow the safety regulations that are defined by different government bodies such as Federal Communications Commission (FCC) [41] in United States and European Council (EC) [42] in Europe. The safety limit is specified using specific absorption rate (SAR), which is the amount of energy stored in human body during each exposure. These limits vary depending on different jurisdictions and are generally defined as Watts per kilogram. For instance, FCC limits maximum exposure at 1.6 W/kg, whereas EC allows exposures up-to 2 W/kg.

Two main factors affect the obtained SAR value in an EM scanning system; (1) the operating frequency band and (2) the distance of the antenna from body [43]. The operating frequency of the antenna can significantly affect the obtained SAR value as it has an inverse relation with the operating wavelength. Consequently, lower microwave frequencies have deeper penetration that result in higher SAR values [44]. Similarly, SAR value has an inverse relation with the distance of the antenna from human body. This is attributed to a wider distribution of EM field at longer distances compared to the focused distribution at closer distances [43]. Studying torso scanning systems reveals that using a 1 mW (0 dBm) transmitted power results in SAR values between 0.004–0.04 W/kg at 0.6–0.9 GHz [18,24,29,33,34]. Lower values are obtained when antennas are located farther from body [29,33] and higher values are attributed to antennas closer to body [18,24,34].

4. Antenna Designs

The effectiveness of EM scanning system for the detection of different types of diseases highly depends on performance of the utilized antennas. A review of some important antennas that are utilized for medical EM scanning systems, their operation principles and performance requirements are presented in this section.

4.1. Antenna Design Criteria

The following main requirements should be considered for designing antennas for EM scanning systems: operating frequency, bandwidth, and radiation characteristics, such as directivity, penetration level and efficiency.

The first step in designing antennas for EM scanning systems is to define the optimum frequency bandwidth for the scanning system. Human tissues have frequency-dependent dielectric constant (relative permittivity) that generally decreases with increasing the operating frequency. Moreover, human tissue is a lossy medium with losses that increase with frequency [45]. Hence, increasing the operating frequency in general decreases signal penetration into human tissues. Studying the literature reveals that historically low microwave frequencies at 0.9 GHz have been utilized for torso scanning applications [8–11,30–32] that is consistent with signal loss considerations. However, to define a range for wideband torso imaging/detection applications, a study is conducted in [46], where the equivalent circuit model of an average human torso is utilized. The variations of signal magnitude at the center of the lungs for different frequencies are calculated (Figure 7). As seen in Figure 7b the attenuation of the penetrated signal increases significantly at frequencies above 1.5 GHz. Moreover, there is a direct relation between antenna's physical size and operating wavelength. Microwave frequencies below 0.5 GHz lead to physically large antennas, which makes the EM scanning system bulky and hinders its portability. Additionally, decreasing the operating frequency adversely affects resolution of the obtained images [47]. Considering all these factors, recent studies indicate that using the operation bandwidth of 0.5–1.5 GHz provides the best compromise between the signal penetration into human torso, image resolution, and antenna size [15,30,48].

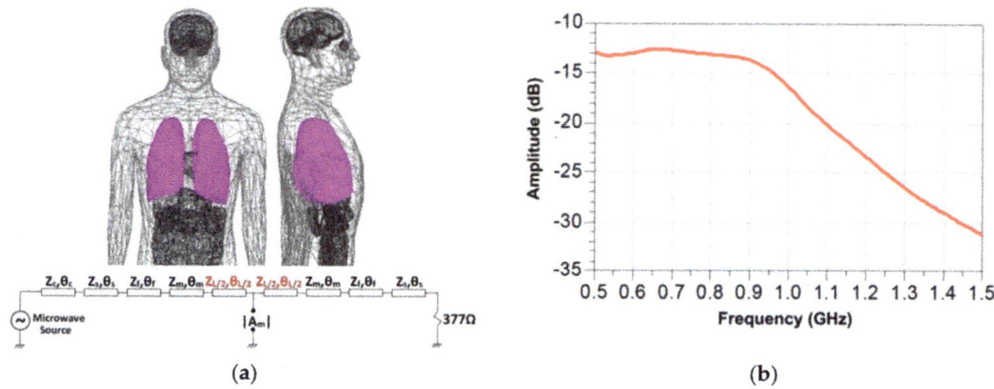

Figure 7. Signal penetration into human torso: (**a**) lumped element model of human torso; (**b**) variation of signal magnitude at the center of human lungs with frequency. Reproduced with permission from [46]. Copyright 2013 IEEE.

Different studies reveal that the accuracy of radar-based imaging algorithm is directly related to the utilized bandwidth of the antenna. The study in [49] reveals that in radar-based EMI systems, the lack of discrete observation points can be compensated using different scattered profiles from frequency samples across a wideband signal. Hence, wideband signals are preferred in EM scanning systems [18–20,50].

The other important requirement of antenna design for EM scanning systems is focused beam radiation in near field and/or far-field. In EM imaging systems, unidirectional radiation is preferred to reduce the adverse effects of environmental noise and scattered fields on signal-to-noise ratio, hence providing better detection results [51,52]. The effects of the focused beam antennas on the reconstructed images in EMI systems are thoroughly investigated in [51]. As seen from Figure 8, unidirectional focused beam antennas can reduce the adverse effects of undesired scatterers, such as surrounding organs in the imaging domain. A performance comparison of the unidirectional and omnidirectional antennas for the EMI system is presented in [52]. The results indicate that the unidirectional antennas have significantly better performance in reducing the artifacts that result in higher resolution/accuracy images.

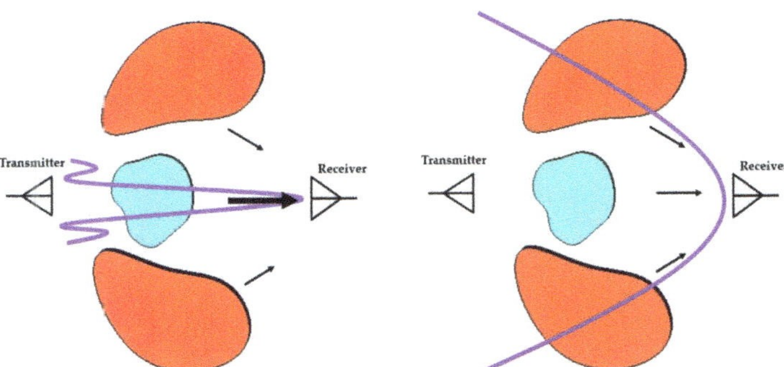

Figure 8. EM scanning system using focused beam antennas (left) and non-focused beam antennas (concept taken from [51]).

4.2. Antenna Categories

Investigating EM scanning systems reveals that two broad categories of radiators are used to scan the human torso; (1) on-body matched antennas [24,30,31,53,54], and (2) free-space antennas [55–60]. Different methods were proposed to design these antennas that are detailed in this section.

4.2.1. On-Body Matched Antennas

An on-body matched antenna is designed in the presence of human body model and its performance is optimized considering the frequency-dependent characteristic of investigated body area, e.g., torso. This ensures that the radiated power by the antenna is directly penetrated inside the body and no reflection occurs in the air-body boundary. Because these antennas are located on-body, their performance is not measured using conventional measurement concepts such as directivity or far-field radiation patterns. Instead, their success is measured by the intensity of electromagnetic field that is induced/directed inside investigated area. The EM field intensity can be increased by optimum impedance matching and directing the radiated EM signal towards the body. Considering the frequency dispersive properties of human tissues, wideband impedance matching is a challenging issue.

The most straight forward method to achieve impedance matching is to terminate the antenna with a resistor. The design of these antennas can be traced back to early studies where an electromagnetic coupler was used as microwave sensors to transmit electromagnetic signals into the torso and receive reflected/transmitted signals [30–32,55]. These couplers are designed based on co-planar waveguide structures. A strip-line impedance matching transition between coaxial cable and coplanar waveguide is implemented to increase their operating bandwidth as depicted in Figure 9a [30]. These structures are terminated by resistors to provide impedance matching to human body. However, the addition of the resistor decreases the radiation efficiency of this antenna/coupler significantly, hence, it reduces their practicality for deep target detection.

To overcome the limitations in obtaining matching with resistor termination, different methods have been developed. An L-shaped monopole antenna is proposed to reduce the size of the antenna by increasing its electrical length [40]. The configuration of the antenna is depicted in Figure 9b. As seen, a meandered shorted stub and open-ended stub are used to adjust the two main operating modes of the antenna. To enhance the bandwidth of the antenna, a triangular patch is added to the L-shaped monopole. A combination of a rectangular ring-shaped monopole and a meandered patch is proposed in

Two sets of L-shaped and I-shaped slots are etched to the ground plane to create new resonances at higher frequencies and improve the operating bandwidth [53].

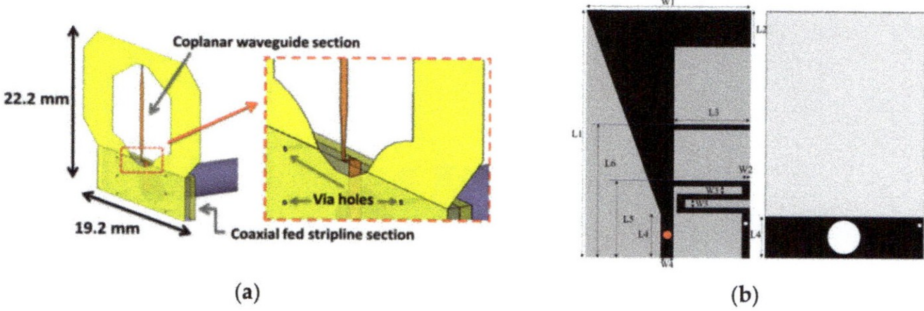

Figure 9. On-body matched antenna: (**a**) coplanar waveguide-type antenna (reproduced with permission from [30], Copyright 2014 IEEE); (**b**) L-shaped monopole antenna (reproduced with permission from [40], Copyright 2020 IEEE).

Despite their success in achieving wide operating bandwidth, the utilized antennas possess monopole-type radiation characteristics. Consequently, while part of the accepted power is matched to the torso, the rest of the power is dissipated from the back of the antenna. To address this problem, hybrid loop-dipole structure is proposed in [24], where the theory of complimentary antennas is utilized (Figure 10a). This method utilizes the well-known fact [61–64] that if the loop and dipole antennas are excited simultaneously with equal phase and magnitude, the resulting radiation pattern would be unidirectional with suppressed back lobe radiation, therefore, enhancing signal penetration inside torso by almost two-fold compared to a monopole structure [24]. The electric field penetration of the proposed design is depicted in Figure 10b.

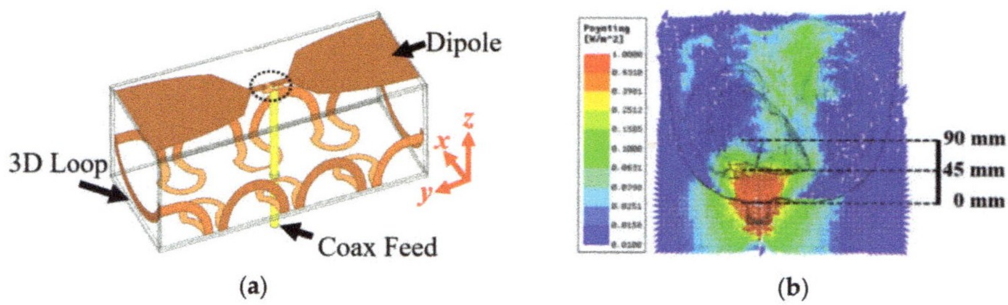

Figure 10. Hybrid loop-dipole antenna: (**a**) configuration of the antenna; (**b**) signal penetration. Reproduced with permission from [24] (open access).

Most radar-based imaging algorithms assume that the incident wave is a planar wave. This assumption is not fulfilled when using conventional body matched antennas that possess spherical wave-front for EM inside human torso. Consequently, the obtained images using these antennas are accompanied by errors regarding location of the malignancy, its size and formation of ghost targets due to the plane wave assumption of utilized algorithms. To address this problem, a body matched graded index (GRIN) lens antenna is proposed in [54] and depicted in Figure 11. This lens, which is based on the theory of multi-layer structures, transforms the spherical wave from its exciting source (slot antenna) to a planar wave inside the torso as shown in Figure 11b. This is achieved through transitioning wave from air that has a low permittivity of 1 to that of an average torso, 45, through gradual increase in permittivity in each lens layer. To fabricate different dielectric layers, a combination of low permittivity plastic and water is utilized. To obtain the desired dielectric values, a host structure that has different hole sizes is 3-D printed using the plastic material, and then filled with water. The obtained results reveal that the antenna achieves a strong penetration (more than 6 dB) compared to the body matched antenna (see Figure 12).

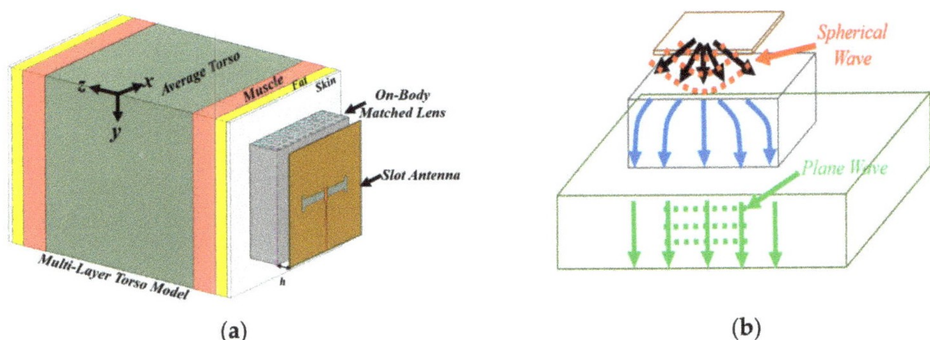

Figure 11. On-body matched GRIN lens: (**a**) configuration of the antenna; (**b**) spherical wave to plane wave transformer. Reproduced with permission from [54], Copyright 2021 IEEE.

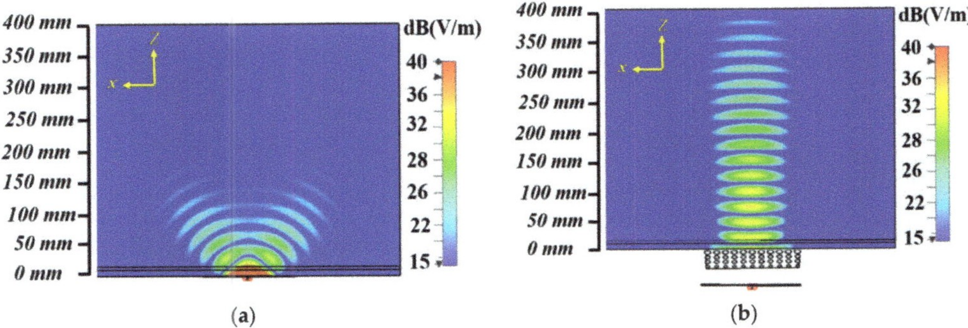

Figure 12. Electric field comparison: (**a**) on-body matched slot'; (**b**) on-body matched GRIN lens. Reproduced with permission from [54], Copyright 2021 IEEE.

4.2.2. Free-Space Antennas

To accommodate the requirements of different detection algorithms, free-space antennas were proposed and widely used to build alternative platforms. Free-space antennas are designed without the presence of body and their radiation properties are characterized using directivity, gain and radiation patterns. The detection is performed using the differences in the reflection and transmission coefficients in free space and in front of human body. These antennas strive to have compact sizes, wide operating bandwidth, and high gain/directivity. Considering that torso scanning systems operate at low microwave frequencies, obtaining all these merits is a challenging task. Hence, there will always be a compromise that is defined by the scanning system. This section summarizes the proposed designs.

Vivaldi antennas have been widely used in EM scanning systems due to their wideband operation and high radiating gain [28,55,56,65–67]. These antennas are generally comprised of a strip feed that is magnetically coupled to a flared ground plane. Corrugation methods [67,68] and fractal leaf arm techniques [56,66] are applied on the flared ground structure to increase the electrical length of the antenna for a broad bandwidth and high gain. An example of a Vivaldi antenna with fractal leaf arm is depicted in Figure 13a. Vivaldi antennas have been investigated for pulmonary edema detection [55] as well as lung tumor detection [56].

Owing to their unidirectional radiation pattern, patch antennas are widely used in some EM scanning systems [42,43]. In [69,70], dual patch antennas are used as electromagnetic sensors for pneumothorax diagnosis applications. Monopole antennas are used for lung tumor detection applications [44,45]. To improve signal penetration, the monopole antenna structure is loaded with a cavity back to achieve unidirectional radiation and hence increase the signal penetration into the human chest [71]. An example of cavity-backed elliptical monopole is illustrated in Figure 13b. Similar to on-body matched antennas, a combination of a loop and dipole can be used to increase directivity of the antenna and reduce its back radiation [24,36,57].

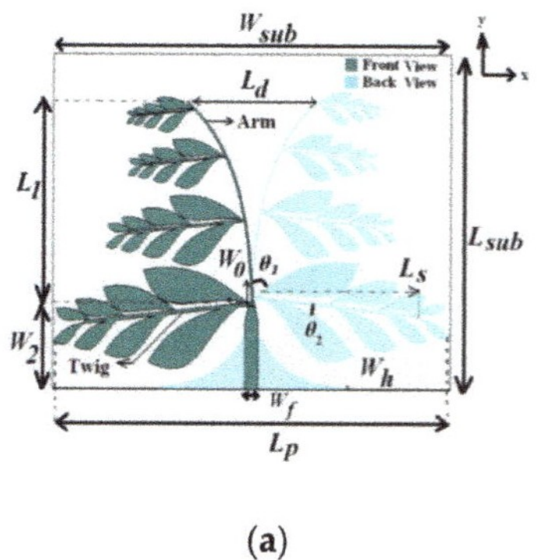

(a)

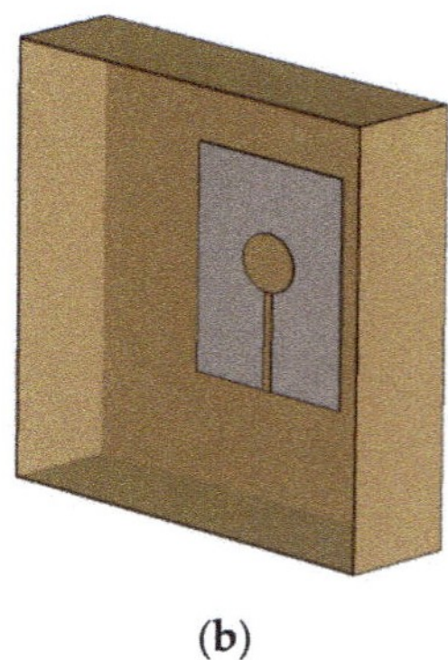

(b)

Figure 13. Free-space antenna: (**a**) fractal leaf Vivaldi antenna (reproduced with permission from [66], Copyright 2017 IEEE); (**b**) cavity-backed elliptical monopole antenna (concept taken from [71]).

To overcome the size constraints that are imposed by the limited torso area, folding techniques are used in EM scanning antennas to reduce the physical size of antennas and increase the operating bandwidth and antenna's directivity [26,58,72,73]. The process involves folding the edges of a planar antenna to form three dimensional structures. This method reduces the back radiation of antenna by changing the current flow alongside the antenna's edge which eventually reduces its size. A three-dimensional folded loop-monopole structure for EM scanning systems is presented in [57]. The loop-dipole composite is first designed based on a planar structure and then folded over an optimal folding line to increase the antenna's directivity and reduce its size. Another wideband folded antenna operating in the frequency bandwidth of 0.77–1 GHz for the early-stage detection of congestive heart failure (CHF) is presented in [26].

Metamaterial techniques are also applied to reduce the size of antennas and improve directivity at low microwave frequencies. The structures are formed by applying series capacitance and/or shunt inductance to the host antenna to tune the resonance frequency. For example, mu-negative (MNG) metamaterial unit-cells were applied to a conventional square loop antenna in [60,74] to lower the resonance frequency of the antenna (Figure 14a). Since this resonance is independent of the antenna size, the overall size of the antenna is miniaturized by more than 50%. Additionally, it is shown that by non-periodic distribution of unit-cells, a unidirectional radiation pattern can be obtained. MNG unit-cell loading is also used to enhance the operating bandwidth of directional Yagi-antennas trough excitation of mu zero resonance below existing resonance of the Yagi antenna [75,76]. An example of MNG loaded reflector Yagi-antenna is illustrated in Figure 14b.

As stated before, certain detection algorithms require scanning different locations along the torso to perform comparative detection decisions. While mechanical movement is used to scan high number of regions, certain comparative algorithms can provide detection decisions using low number of torso regions, such as three in [29]. To eliminate the complications of mechanical movements, these systems can be fabricated using pattern reconfigurable antennas to electronically scan the upper, middle and lower parts of the torso [38,54,59,77–79]. A wideband reconfigurable loop antenna operating at 0.8–1.15 GHz for torso scanners is presented in [77]. Capacitive gaps are created along the loop arms to form virtual dipole arrays, which leads to unidirectional radiation with a peak gain of 2.1 dBi. Changing the location of gaps changes the direction of the dipole array, which leads to changes in the direction of the beam. Another pattern reconfigurable loop-dipole antenna with the capability of scanning the upper, middle and lower parts of an average human torso for pleural effusion detection is presented in [78]. The antenna consists of a one-wavelength loop, a have-wavelength bow-tie dipole, and two parasitic directors. The combination of loop-dipole mode increases the antenna's directivity. The antenna operates in a wide fractional bandwidth of 55% at 0.8–1.4 GHz with a peak gain of 5 dBi. The beam steering is achieved using parasitic directors to alter the current distribution on the loop and enable beam switching in different directions to scan the whole chest area.

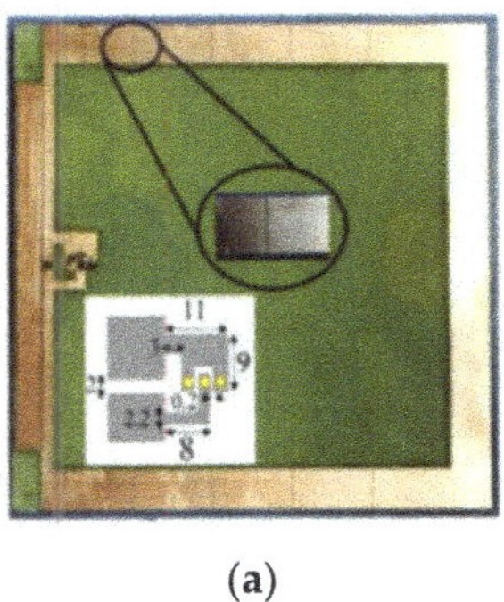

(a)

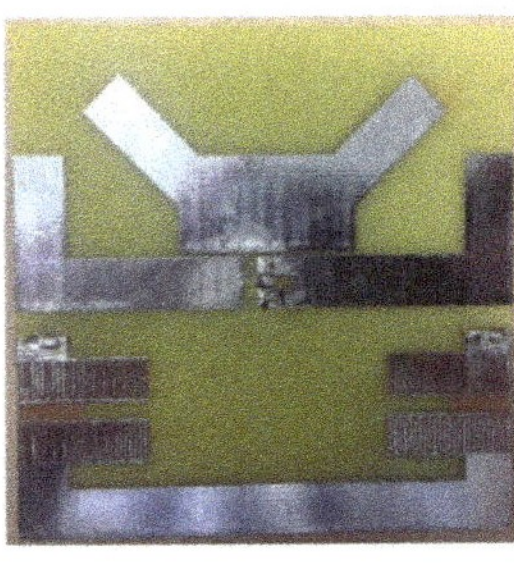

(b)

Figure 14. MNG loaded antennas: (**a**) MNG loaded loop antenna (reproduced with permission from [60], Copyright 2016 IEEE); (**b**) MNG loaded reflector Yagi-antenna. Reproduced with permission from [75]. Copyright 2017 IEEE.

To satisfy the plane wave radiation assumption inside the imaging domain for radar-based imaging applications [80], reconfigurable metasurface lenses are proposed to achieve near field beam focused plane wave radiation inside the imaging domain [38,54,59]. Metasurface structures are built using an array of unit-cells that are distributed periodically and illuminated using a source antenna, e.g., a slot antenna. These unit-cells collimate the incident field into a focused transmitted beam. A pattern of reconfigurable metasurface antennas based on the offsetting technique is presented in [38]. Three half-wavelength microstrip-fed slots operating in the frequency band of 0.9–1.2 GHz radiate a metasurface layer. Beam steering is performed based on the excitation of each metasurface unit cell with different phase delays by changing the activated slot.

Pattern reconfigurability in metasurfaces can also achieve using programable unit cells by changing the characteristics of unit cells. A programable pattern reconfigurable metasurface for pulmonary edema detection is proposed in [59], where a metasurface layer containing 5×5 programable square ring resonator as the superstrate layer on an H-shape slot radiator is designed. Four PIN diodes are embedded in each cell to alter the electric field intensity within the metasurface layer and consequently switch the high intensify electric field in different directions inside the human torso.

Figure 15 summarizes and compares various types of antennas used in EM scanning systems to detect or localize any abnormality. Based on the requirements of different imaging/detecting algorithms and the required level of signal penetration into the human torso, a suitable antenna type is selected.

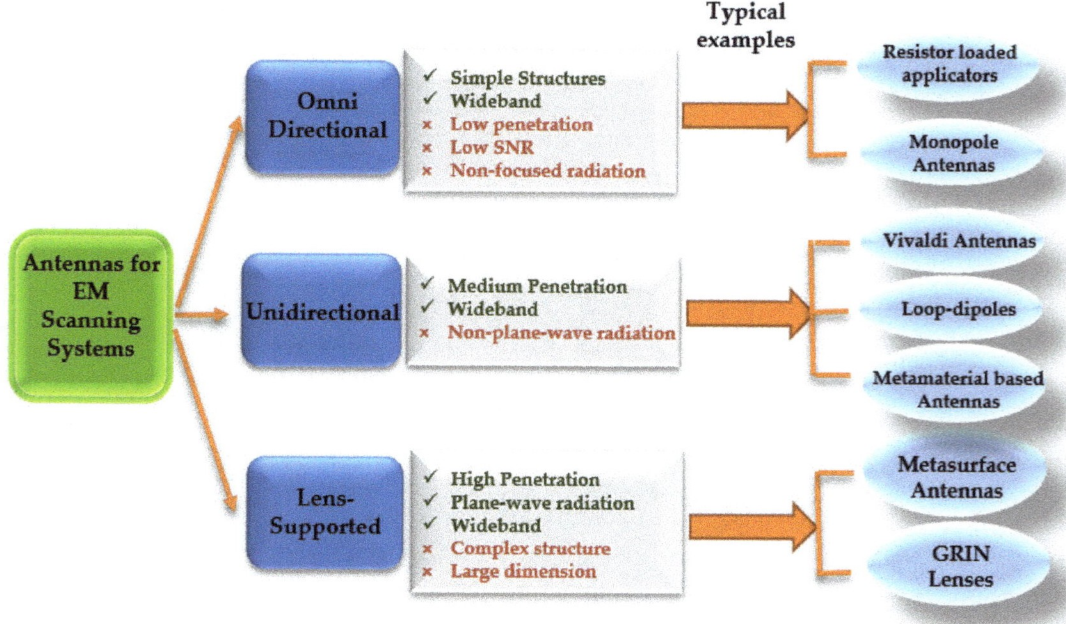

Figure 15. Characteristics of different types of antennas used in EM thoracic scanning systems.

5. Microwave Detection Techniques

EM torso scanners utilize the high dielectric contrast between healthy and diseased tissues for detection purposes. Generally, the processing unit exploits any changes in the phase and/or magnitude of transmitted EM signals for abnormality detection or localization. The variation in dielectric properties of tissues along the wave's propagation path alters the wave speed that results in the changes in phase and/or magnitude of the transmitted wave.

These changes can be used to detect/classify the abnormality and/or localize it by creating an image from the investigated domain. This image can be created either by calculating the scatter fields inside the domain (radar-based imaging) or by calculating dielectric properties of the tissues (tomography). Based on processing outcome, EM processing techniques can be classified into three main categories: (1) detection only methods, (2) detection and classification methods, and (3) detection and localization (imaging) methods. Figure 16 presents an overview of the current microwave techniques. In this section, each technique applied to the detection of different thoracic diseases is explained in detail. Then, the advantages and disadvantages of each method are discussed and compared in terms of practicality, computational time and accuracy.

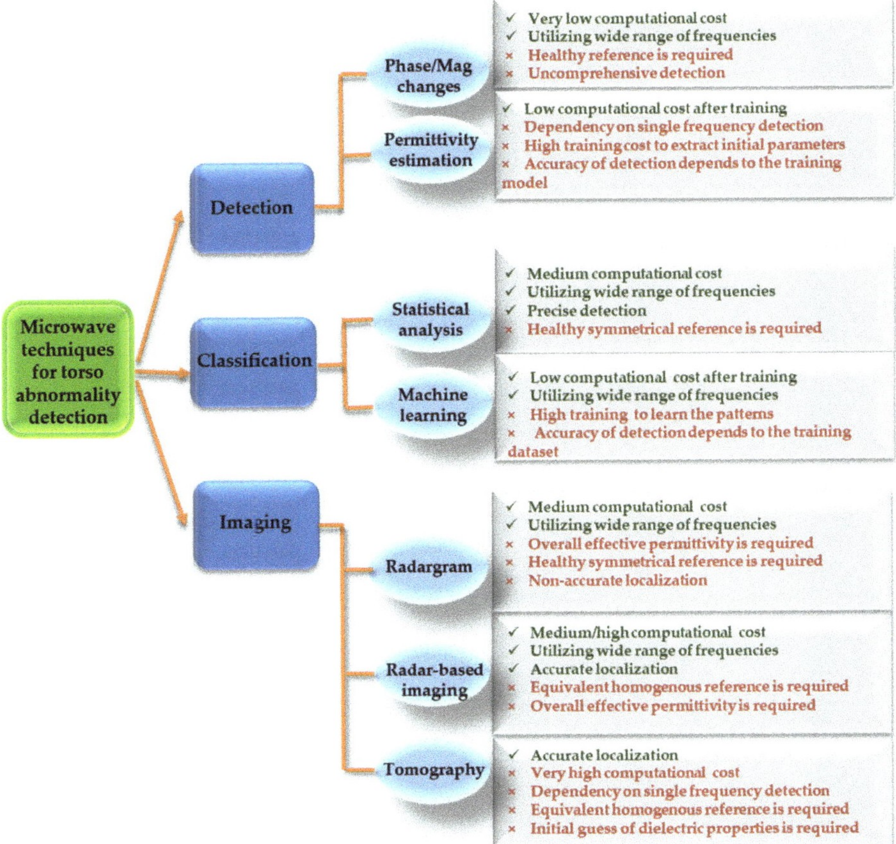

Figure 16. An overview of EM thoracic abnormality detection techniques.

5.1. Detection Only Methods

These methods aim at detecting any abnormality within the torso without providing any information on the location of that abnormality. Hence, their computational time is lower than localization techniques. These techniques are more suitable in disease detection where the location of abnormality is not important such as, hepatic steatosis or bronchial asthma detection. In this section, different methods utilized in malignancy detection are reviewed. These methods can be classified into two main categories based on their detection approach. First group exploits the change in magnitude and/or phase of electromagnetic signals. Second group estimates the overall effective permittivity of the medium.

5.1.1. Phase/Magnitude Changes

Some techniques compare the magnitude or phase of the propagation coefficient between control and diseased group to achieve a distinguishable trend [9,70,71,81–85]. These techniques are mostly utilized in linear scanning platforms operating in a mono-static or bi-static (multi-static with use of two antennas) data acquisition mode. The assessment of low and high frequency regions of reflection or transmission coefficients results in a contrast between measured healthy and unhealthy signals.

In [82], propagation coefficients of upper and lower parts of the lungs are compared between 12 patients with confirmed diagnosis of brachial asthma and a healthy controlled group comprised of 10 individuals with appropriate age and sex, matched with patients by morphological characteristics of the chest. Analyses of measured signals from patient group shows that the signals experience significantly lower attenuation compared to the measured signals from healthy controlled group in the bandwidth from 0.9 to 1.5 GHz when propagating through the upper part of the lungs (see Figure 17a). However, scanning of lower parts of the lungs show maximum different between healthy and controlled group at higher frequency bandwidth from 1.2 to 2 GHz, (see Figure 17b). Consequently, monitoring of propagation coefficient at these frequency regions provides higher accuracy of disease detection.

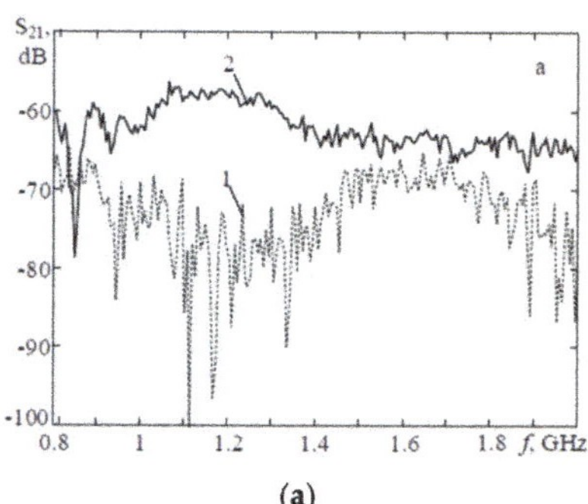
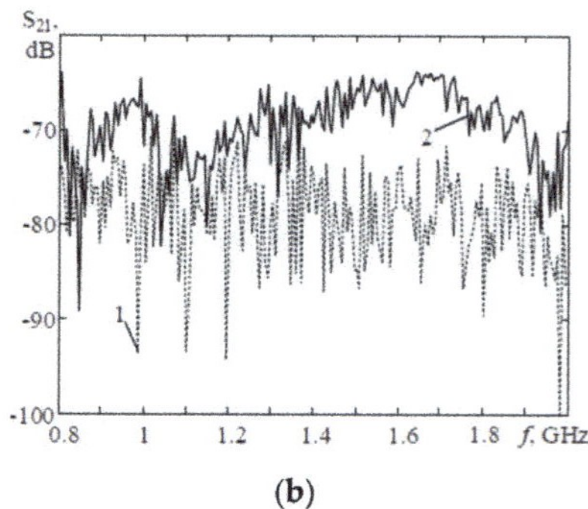

(a)   (b)

Figure 17. Propagation coefficient in a brachial asthma subject (solid line) compared to a healthy case (dashed line): (**a**) scan of upper part of lungs; (**b**) scan of lower part of lungs. Reproduced with permission from [82]. Copyright 2017. IEEE.

In [81], the phase of transmitted signals at a single frequency is used to create a phase diagram of the chest to detect any inhomogeneity within lungs. In [83], the transmission coefficient is converted into a corresponding voltage at the output and the diagram of the output voltage is created to detect any inhomogeneity inside the chest area. It is also shown that the maximum probability of detection with more precise borders takes place at higher frequencies.

The magnitude of transmission coefficients is also exploited to detect pneumothorax [70] or pulmonary edema [9]. It is shown that the maximum difference between transmission coefficients in healthy and pneumothorax scenarios occurs at frequency of 2.3 GHz and is about 7.1 dB, (see Figure 18). The existence of the trapped air in the chest cavity due to the disease results in changes in electric field distribution at this frequency. Additionally, it is shown that the magnitude of transmission coefficients increases in the bandwidth 800–955 MHz as the result of increase in lung water volume [9] (see Figure 19).

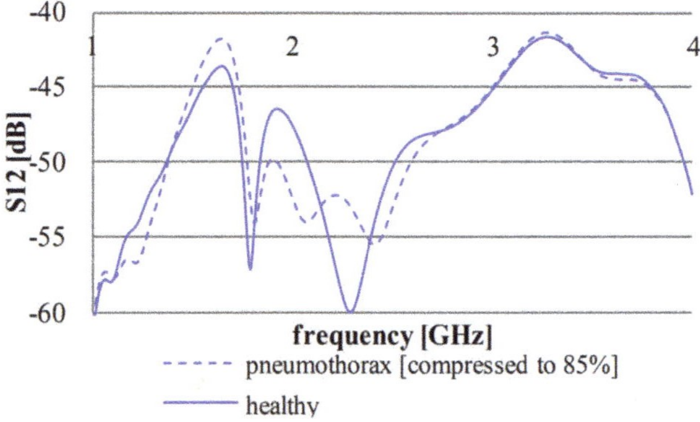

Figure 18. Magnitude of transmission coefficient in healthy and pneumothorax scenarios. Reproduced with permission from [70]. Copyright 2014 IEEE.

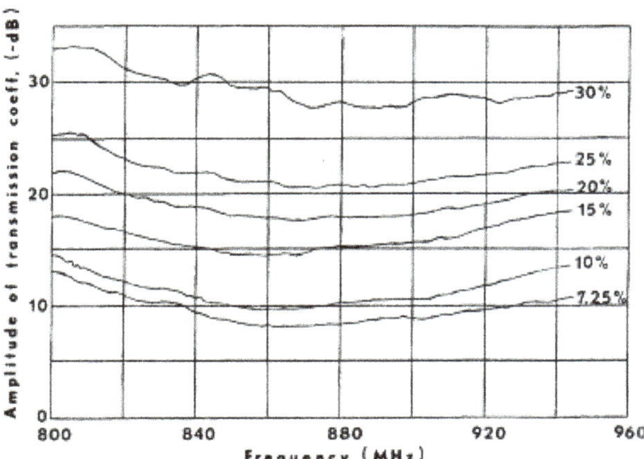

Figure 19. Increase in magnitude of transmission coefficients with increased percentage of accumulated fluid in lungs. Reproduced with permission from [9]. Copyright 1978 IEEE.

Furthermore, changes in the magnitude of reflected signals is exploited to detect the presence of tumors in lungs [84]. Differences between the measured signals with and without presence of tumor are used for detection purposes The differences between healthy and unhealthy signals were mostly found at frequencies below 6 GHz and changing the size of the tumor in the lung, creates a shift in the magnitude of reflection coefficients [85].

Although phase or magnitude of electromagnetic signals might differ in unhealthy torso due to the existence of any abnormalities or disease, this difference cannot lead to a comprehensive and accurate detection of the disease. For example, as realized from the results in [82], the frequency at which maximum contrast occurs might vary depending on the used sensors (antennas) and size of the torso. Hence, this contrast is not always an indicator of the disease or abnormality.

5.1.2. Effective Permittivity Estimation

Some approaches try to estimate the overall effective permittivity of the torso to detect changes in permittivity due to presence of an abnormality. These methods usually model the overall effective permittivity of torso as a function of transmission coefficients. Then, the weight parameters for the model are extracted using training process [13,86]. This model can be linear [13] or non-linear based on spatial statistical technique [86]. In the spatial technique, the effective permittivity of torso from the receiver perspective is modeled using a variogram. This is related to spatial dependence of each signal, a vector of quadric regression function, and a vector of regression coefficients. Variogram is calculated for each receiver using the measured S-parameters at that receiver due to transmission from antennas located at a determined neighboring antenna. To estimate the effective permittivity at each receiver, the best unbiased estimation regression coefficients is obtained by training the model using training samples. These training samples are generated by measured S-parameters of the homogenous equivalent medium with different known permittivity values. After finding the regression coefficients vector by training, it can be used to estimate the effective permittivity of any test medium.

In [86], the spatial statistical technique is used to estimate the effective permittivity and conductivity from the viewpoint of each antenna. To do so, multi-static signals are collected in a circular platform for healthy lung and lung with a tumor. The estimated permittivity and conductivity values in unhealthy scenario are slightly higher than the healthy one due to existence of tumor (see Figure 20). Although, it cannot directly interpret the health status of the lung by these estimated values, they can be used as prior information in radar-based imaging to improve the quality of the image in localizing the tumor. In [13], a wearable health monitoring sensor for detection of pulmonary edema is proposed. The 16-port sensor is placed on the human chest to detect any lung abnormalities by estimating the effective permittivity of lung (see Figure 21). Therefore, the effective permittivity of lung at a single frequency is expressed as a weighted summation of S-parameters measured at each port. The parameters of the weight vector are set based on training the model by assigning different dielectric characteristic to the inner tissue (lungs). Using multiple ports mitigate the effect of the outer layer (skin, fat, and muscle) on lung's permittivity and allow to characterize the inner layer tissue. The validation of this technique on measurements of the healthy and edema lung shows less than 11% error in the calculated permittivity of lung compared to the measured value. It is observed that the unhealthy lung has a higher permittivity value than the healthy lung due to accumulated water inside the unhealthy lung.

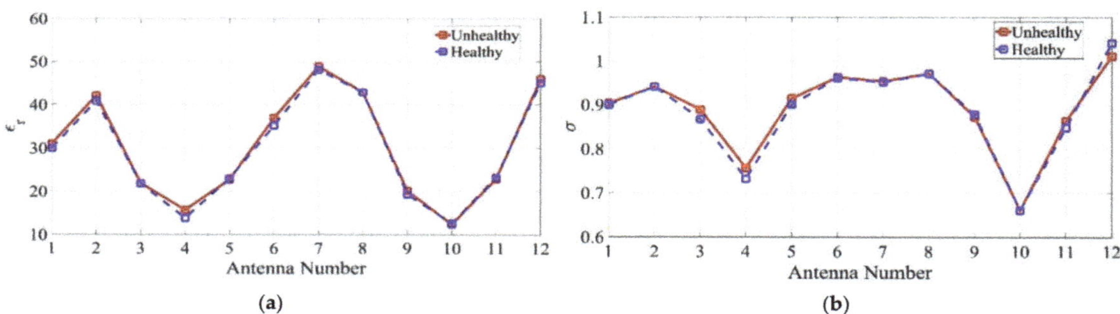

Figure 20. (**a**) Estimated permittivity and (**b**) conductivity from viewpoint of various antennas around torso for healthy and unhealthy lung cases. Reproduced with permission from [86]. Copyright 2017 IEEE.

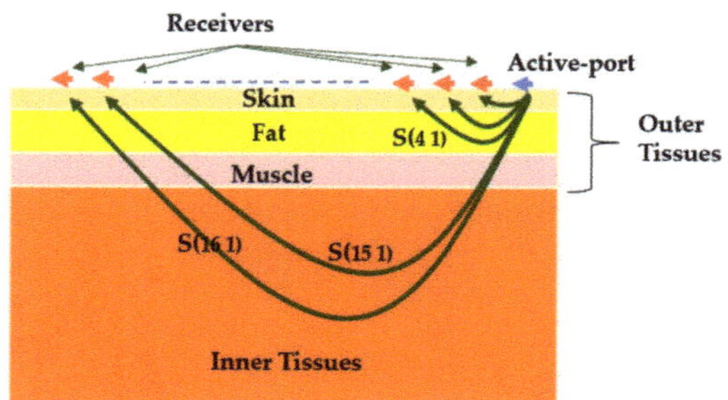

Figure 21. Permittivity estimation of inner layer (lung) by modeling it as a weighted summation of S-parameters (concept taken from [13]).

Permittivity estimation methods reveal promising results in detecting lung abnormalities. However, the performance is hindered by training requirement to determine coefficient vectors and achieve accurate result. Hence, they may not achieve reliable results when the test models are considerably different from the training model, which is the case in clinical application.

5.2. Detection and Classification Methods

These methods aim at classifying and labelling the collected data as healthy or unhealthy based on their underlying pattern and characteristics. They can be classified into two main categories: the first group exploits symmetry of the torso in a statistical approach, whereas the second group utilizes artificial intelligence techniques to differentiate between healthy and unhealthy torso.

5.2.1. Statistical Analysis

Multivariate energy statistics method is another detection technique that operates based on the assumption that organs within the left and right sides of torso have close dielectric properties [45]. Hence, the changes in the dielectric properties of unhealthy tissue enhance the contrast with surrounding tissues. Therefore, the similarity of electromagnetic signals collected from left and right sides of torso is reduced.

In [24], a wearable electromagnetic belt is used to detect hepatic steatosis using multivariate energy statistics method. Any changes in the dielectric property of liver due to excess fat in hepatic steatosis increase the contrast between the liver and surrounding tissues. The method calculates the distance correlation of the measured transmission coefficients between symmetric paths from the left and right sides of torso. The results indicate a peak measured dissimilarity of 15.1% between transmission coefficients of left and right sides of the torso in steatotic liver, which is much higher than the healthy case [24]. Hence, healthy and steatotic liver can be classified based on the left/right permittivity contrast after determining a threshold for healthy liver.

The limitation of this technique is the requirement of almost symmetrical setup to achieve a reliable classification. Additionally, a sufficient number of healthy signals is required in order to define the threshold and set the boundary between healthy and unhealthy signals.

5.2.2. Machine Learning

Supervised machine learning is a form of artificial intelligence technique, which can be used in classifying healthy and unhealthy cases. This method requires a set of training signals to learn the characteristics of input signals. Then, the trained model is validated on test signals. Supervised machine learning framework is utilized to learn an inferring model for hepatis steatosis from the data collected by an antenna operating across 0.4–1 GHz bandwidth in a mono-static mode [27]. Data are collected from a simulated numerical torso by changing the permittivity of the liver from 30–60 within the frequency range of the used antenna. Real and imaginary parts, magnitude and phase of the collected S-parameters are used as inputs for the supervised classifier. Labels of healthy and unhealthy liver are assigned based on the permittivity of liver. Learning is performed using different classification techniques and leave-one-out-cross-validation is used to evaluate the classification performance. The results indicate that this system can detect hepatic steatosis with accuracy of more than 97% for the simulated torso model.

The main drawback of this method is the requirement for enough number of training signals from various stages of disease. The accuracy of this method is highly dependent on the training data base. Insufficient and non-comprehensive training set results in misclassification.

5.3. Detection and Localization Methods

These methods aim at finding the location of abnormality within the torso by forming an image of the investigated domain. This is achieved either by calculating the scattered fields or by calculating dielectric map of the investigated domain. The target, which can be tumor, fluid, or fat infiltration, can be recognized by its high intensity or dielectric contrast with surrounding tissues. Based on the utilized approach in creating the image, the localization methods can be categorized into three main groups: (1) radargram, (2) radar-based imaging, and (3) tomography. In this section, an overview of each approach is presented. All these imaging techniques require an equivalent homogenous medium for detection and localization. The difference of measured signals with and without of the investigated domain is considered in the calculations.

5.3.1. Radargram

This technique provides a two-dimension visualization of torso by analyzing the reflection coefficients [26,56,58,72]. This method transfers measured signals to time domain using an inverse Fourier transform. Then, using the wave speed in the medium the time domain signal is scaled to show the intensity of the reflected signal within an investigated depth. Finally, each of the measurement are overlaid to create the radargram (image). Usually, the average permittivity of torso tissues is used to define the propagation speed. Hence, the target is highlighted in the image due to its variant permittivity. In [26], radargram technique is used for early detection of congestive heart failure due to fluid accumulation inside lungs. The rear side of torso is scanned using one antenna in a linear platform and a mono-static data acquisition mode. The symmetry of left and right lungs is exploited to detect and localize the target (see Figure 22).

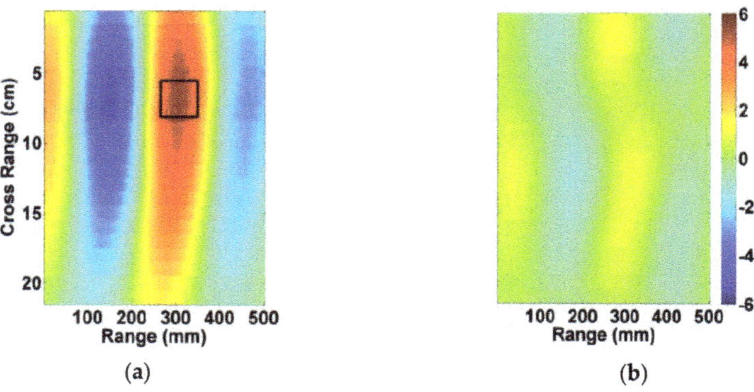

Figure 22. Radargram of (**a**) unhealthy and (**b**) healthy lungs. The square shows the location of 10 mL water inside the lung. Reproduced with permission from [26] (open access).

In [56], time domain reflectometer (TDR) data, obtained in a linear platform, are used to create an image of the scanned area. The imaging algorithm consists of pre-processing and quadratic envelop detection. The pre-processing step reduces background noise and clutters using absolute function and Shannon energy, which calculates the average energy spectrum of the signal. Then, local maximums of the signal are derived using an envelope function to obtain shape of the tumor. Finally, the fourth order quadradic function is applied to enhance the edge detection of tumor. Figure 23 shows the created image of lungs with three tumors in right and left lungs.

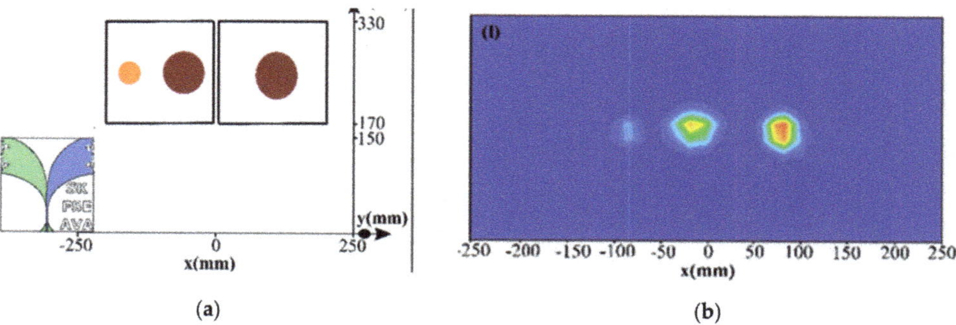

Figure 23. (**a**) Simulated lungs with tumors; (**b**) created image by TDR data analyzing. Reproduced with permission from [56]. Copyright 2020 IEEE.

The main limitation of this technique is its insensitivity to deeper malignant tissues that are far away from the antenna. Hence, its accuracy in localization small and deep targets is low. Additionally, it needs the average effective permittivity of tissue to define the propagation speed. This limits the suitability of the technique in clinical applications.

5.3.2. Radar-Based Imaging

Radar-based imaging techniques can be classified as time-domain confocal microwave imaging [20,21,50] and fast frequency-based radar imaging [18,19,29,38,77,87]. Both approaches require a calibration step when the free-space antennas are utilized to scan the torso. This step is necessary to reduce the significant reflections from air-skin interface which mask the desired target reflections. To calibrate the measured signals, the average of all measured signals is subtracted from each measured signal. As antennas are located at the same distance from skin, the effect of air-skin reflections is almost similar in all measurements. Calibrated signals are then used in further calculations to localize and detect the malignant tissue.

Time-domain confocal microwave imaging has broadly been used in head and breast imaging and demonstrates successful results in tumor or torso fluid detection [20,21,50]. The method uses an ultra-wideband signal to illuminate the imaging object, whereas the received signals in time domain are used to create a map of scattered fields inside the investigated domain. Assuming the focal points inside the imaging domain as point scatterers, the received signals are delayed depending on the wave traveled distances from each transmitter to the individual point scatterers and the receiver. The sum of the delayed signals at all the focal points are then used to calculate the scattered energy and obtain the final image.

In [20], the human torso is modeled as a $100 \times 100 \times 70$ mm block in which different layers (skin, fat, muscle, bone, and lung) are presented with their dispersive dielectric properties. The difference between reflection coefficients with and without the presence of tumor is used for confocal imaging. A smoothing process is performed on signals to improve the quality of the resultant image by applying a filter which provides nonparametric smoothing of signals peaks and filter the noise. Using this filter, the detection and localization of the tumor is enhanced (see Figure 24). In [21], confocal imaging of human chest model using the reflection coefficients reveals better tumor detection and localization during exhale.

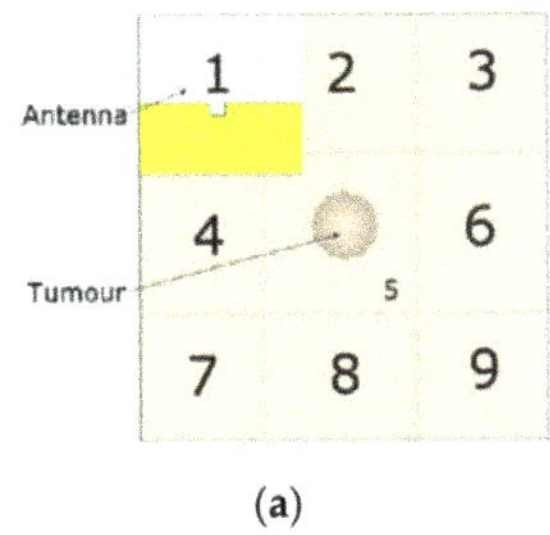

(a)

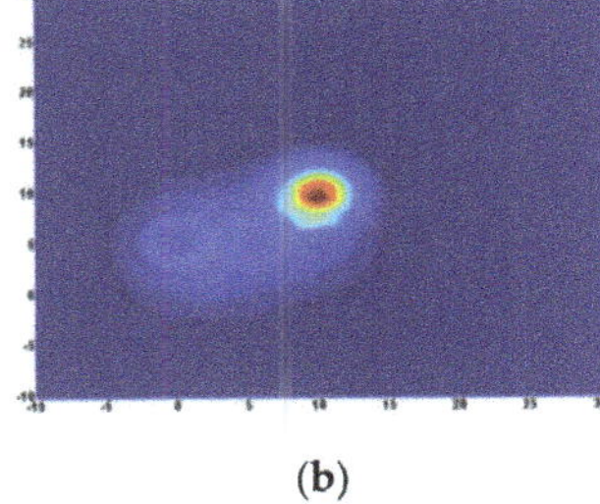

(b)

Figure 24. (a) Front view of the locations of the antenna and tumor; (b) created image with confocal imaging after smoothing. Reproduced with permission from [20]. Copyright 2014 IEEE.

Fast frequency-based radar imaging is used in several microwave systems to detect and localize tumor [19] or fluid [18,29,38,77,87] inside torso. This method first calibrates measured signals, which are the difference in S-parameters between presence and absence of torso, using average subtraction techniques to remove artifacts at each frequency step. Then, the calibrated data are multiplied by a back-propagation Green's function and summed over all antennas' positions to calculate intensity of electromagnetic fields at each scatter position inside torso. The Green's function is modeled as the multiplication of first order first kind Bessel function and an exponential term. Both Bessel and exponential functions are dependent to the wave path distance and wave number.

In [18,19], fast-frequency radar based imaging is performed to detect and localize abnormality inside lungs. Figure 25 present the detection and localization of accumulated fluid in lung using this technique.

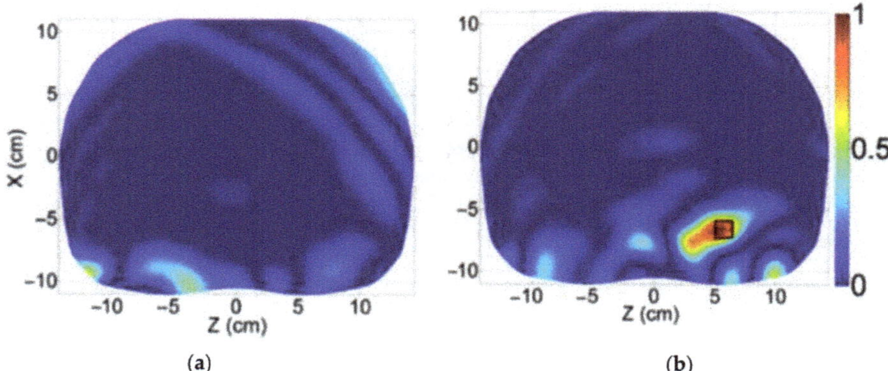

(a)   (b)

Figure 25. Reconstructed image of a (**a**) healthy lung, and (**b**) edema lung with accumulated fluid using fast-frequency radar-based imaging method. Reproduced with permission from [18] (open access).

In [33], fast-frequency imaging in conjunction with slice interpolation technique is used to create a 3D visualization of torso. To do this, different slices along torso are scanned with 1 mm resolution using a mechanically moveable array of antennas located on a flange. This system could accurately determine the location and volume of the embedded water in a phantom lung (See Figure 26).

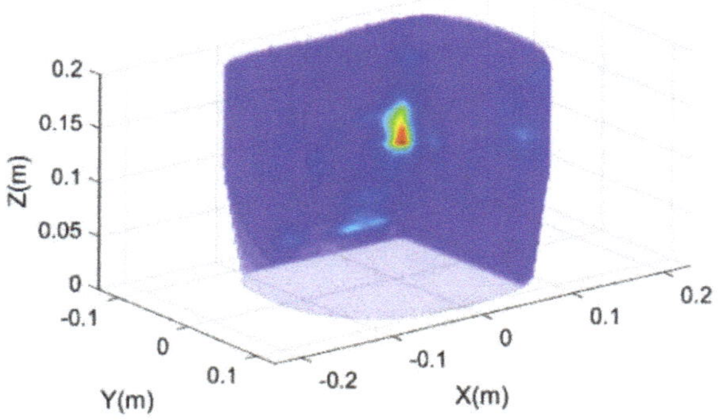

Figure 26. Three-dimensional (3D) image of the torso with accumulated water in the lung. Reproduced with permission from [33] (open access).

Radar-based imaging techniques need a prior information about the effective permittivity value of the medium. This value is usually set based on the average permittivity of tissues at mid-point frequency. However, in clinical application the average effective permittivity may be variant in different individuals who have different body shapes and habitus. Hence, effective permittivity estimation technique is usually needed to estimate the overall effective permittivity of the medium prior to imaging.

5.3.3. Tomography

Microwave tomography is based on the fact that biological tissues have different dielectric properties at microwave frequencies and thus they can be imaged by solving an inverse and forward scattering problem. In this method, the measured electric fields are used as the input of an inverse problem derived from Maxwell's equations. The non-linear invers problem is ill-posed as there is more than one solution to satisfy the equations. Hence, additional information and assumption are required to reach the correct solution. Optimization-based methods [88–90], born iterative methods [91,92], and Newton-based methods [93–95] are mostly exploited to solve the inverse problem in an efficient way in at iteration mode. At each iteration, the measured field and calculated field with current dielectric distribution are compared and the error is estimated. The solution is found when the calculated error is reached to a predefined threshold. This technique requires a priori information of dielectric properties of investigated domain to achieve reliable results. Torso microwave tomography is not as a common as breast and head tomography. This is due to the large size of the human torso, which increases the number of parameters in calculations, resulting in high computational costs and less probability of a successful convergence of the solution.

Microwave tomography is utilized to detect and localize lung cancer [96,97] or create an image of intact canine or swan heart [23,98]. In [97], the method is applied to two-dimensional computer-simulated model of the chest with three high contrast inhomogeneities which model lung tumors. The location of the tumors is successfully detected using the tomography approach. In [23], a three-dimensional gradient method is used to reconstruct a dielectric map of an excited static canine heart at frequency of 2.4 GHz. The resultant image, presented as 2-D vertical cross sections, reflects the structure complexity of the heart well, (See Figure 27). The reconstructed dielectric properties are close to the dielectric properties of myocardium and left/right ventricles are obvious in the reconstructed image.

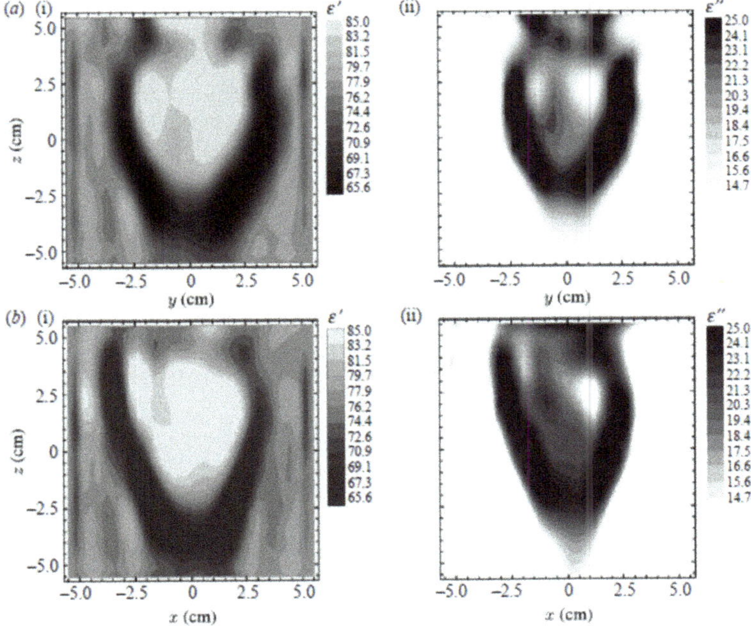

Figure 27. Reconstructed image of canine heart at 2.4 GHz at two cross-sections using microwave tomography (**a**) X = 1.5 cm, (**b**) Y = 1.5 cm. Reproduced with permission from [23]. Copyright 2000 IEEE.

Three-dimensional microwave tomography has also shown successful results in the reconstruction image of infracted canine heart [99]. The reconstructed images show both the shape of the heart and the region of infraction which is highly correlated with anatomical slices. Hence, microwave tomography might have potential for both structural and functional cardiac imaging.

Capability of the in-body microwave tomography has also been tested using a circular heterogenous phantom generated by FDTD simulations [100]. The radius of the phantom is 13 cm with two more embedded circular cylinders to create inhomogeneity. The 2-D microwave tomography at 403.5 MHz using Newton-based method showed successful result in detection and localization of the inhomogeneities within the simulated torso, (See Figure 28).

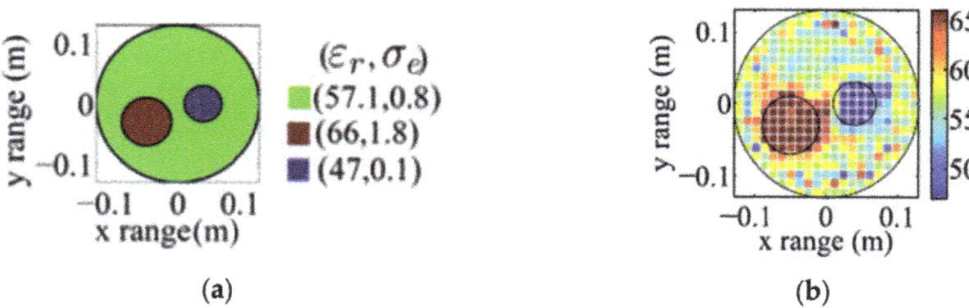

(a)

(b)

Figure 28. (**a**) Simulated circular inhomogeneous phantom; (**b**) reconstructed image of the relative permittivity using microwave tomography. Reproduced with permission from [100]. Copyright 2014 IEEE.

Practicality, computation time, and accuracy of all microwave detection techniques is compared in Figure 29. Practicality is defined based on the required system configuration and its pre-requisites. As seen from this figure, there is always a compromise between practicality, computation time and accuracy of the utilized methods. For instance, tomography methods, provide the most accurate detection decision, but they have the most impractical hardware setup and require the highest computational time. On the contrary, the detection only methods, such as magnitude/phase monitoring, has the simplest setup, but also achieves the lowest accuracy. Therefore, depending on the required accuracy and system cost, a suitable configuration might be selected.

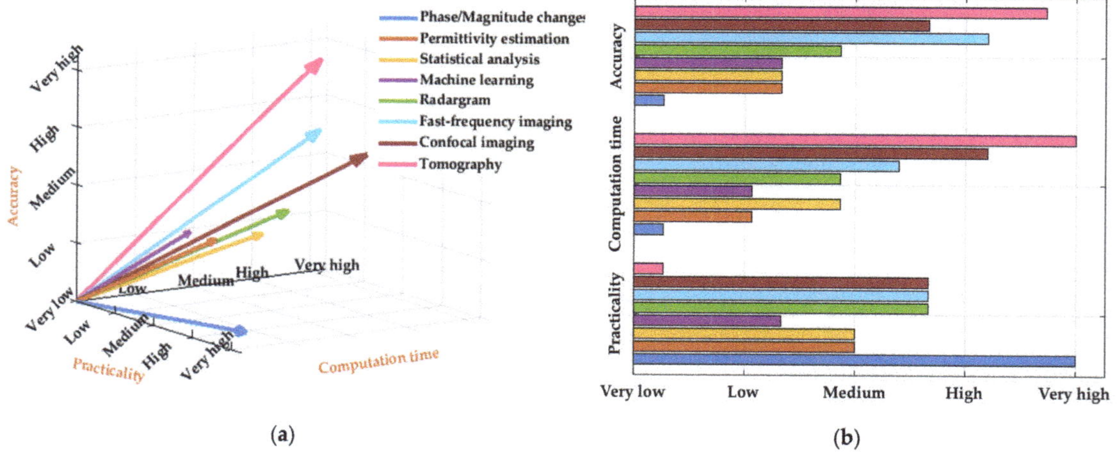

(a)

(b)

Figure 29. Comparison between EM thoracic scanning techniques based on their complexity, computation time, and accuracy; (**a**) 3-D representation; (**b**) bar graph representation.

6. Conclusions

A comprehensive review of EM torso scanning systems has been presented. Different data acquisition methods, scanning platforms, antenna types and detection/imaging algorithms are reviewed and categorized. It is shown that the idea of detecting malignancies inside torso using microwave signals has advanced significantly since its introduction in 1973. The systems have advanced from detecting the malignancy only to mapping its exact location and dimension. New detection techniques are developed thanks to significant advancements in computational capacities and incorporation of different statistical techniques and machine learning methods. This promises more accurate and real-time detection decisions that can be used as a complementary technique besides conventional imaging devices. Additionally, it strengthens the confidence in EM torso scanning systems as powerful tools for monitoring purposes and onsite diagnosis, which are the major limitations of existing imaging systems. As can be realized from Table 1, the higher accuracy comes with increased system's complexity and computational cost. This creates an exciting challenge and the future research roadmap, where solutions are needed to maintain accuracy of EM systems while simplifying their design and computational cost. While the future is unknown, it is guaranteed that only systems that can satisfy these two factors will have the competing edge in clinical settings.

Table 1. Summary of available torso scanning systems.

Ref.	System Configuration	Antenna	Algorithm	Advantages	Disadvantages
[34]	Linear array of antenna in multi-static data acquisition mode	Unidirectional wideband free space 3-D loop-monopole antenna	Fast frequency imaging	•High accuracy of detection and localization •High practicality •Medium computation time	•Requirement of healthy symmetrical reference •Requirement of average permittivity of tissues •Medium penetration
[26,35]	Linear array of antenna in mono-static data acquisition mode	Unidirectional wideband free space folded antenna	Radargram	•High practicality •Medium computation time	•Requirement of healthy symmetrical reference •Requirement of average permittivity of tissues •Not suitable for deep target detection •Medium penetration
[27]	Mono-static data acquisition mode	Unidirectional wideband on-body matched waveguide antenna	Machine learning	•Low computation time after training •Simple structure	•Requirement of training •Low practicality •Not suitable for deep target detection
[29]	Circular array of antenna in multi-static data acquisition mode	Unidirectional wideband free space metamaterial unit-cell loaded Yagi-antenna	Fast frequency imaging	•High accuracy of detection and localization •High practicality Medium computation time	•Requirement of average permittivity of tissues •Medium penetratio
[31,32,70]	Bi-static data acquisition mode	Omni directional narrowband on-body matched antenna	Phase/Mag changes	•High practicality •Low computation time	•Low accuracy •Low penetration
[33]	Circular array of antenna in multi-static data acquisition mode	Unidirectional wideband free space resonance-based reflector antenna	Fast frequency imaging	•High accuracy of •High practicality •Medium computation time	•Requirement of average permittivity of tissues •Medium penetration •Complex setup
[36]	Circular array of antenna in mono-static data acquisition mode	Unidirectional wideband free space loop-dipole antenna	Fast frequency imaging	•High accuracy •High practicality •Medium computation time	•Requirement of average permittivity of tissues •Medium penetration •Not suitable for deep target detection
[38,59,87]	Quasi-circular antenna in mono-static data acquisition mode	Unidirectional wideband free space pattern reconfigurable metasurface antenna	Fast frequency imaging	•High accuracy •High practicality •Medium computation time •High penetration	•Requirement of average permittivity of tissues
[24]	Circular antenna in multi-static data acquisition mode	Unidirectional wideband on-body matched loop-dipole antenna	Statistical analyses	•Medium practicality •Medium computation time	•Requirement of a symmetric healthy part •Low-medium accuracy •Requirement of healthy threshold •Medium penetration
[86]	Circular antenna in multi-static data acquisition mode	Unidirectional wideband free space loop-dipole antenna	Permittivity estimation	•Medium practicality •Low computation time •Suitable for enhancing radar-based imaging	•Requirement of training •Low-medium accuracy •Medium penetration

Table 1. *Cont.*

Ref.	System Configuration	Antenna	Algorithm	Advantages	Disadvantages
[21,56]	Linear scanning monostatic data acquisition mode	Unidirectional wideband free space Vivaldi antenna	Radargram	•High practicality •Low computation time	•Not suitable for deep target detection •Medium penetration
[21]	Linear scanning mono-static data acquisition mode	UWB Antenna Free Space Antenna	Confocal imaging	•High practicality •Low computation time	•Low accuracy •Requirement of average permittivity of tissues •Not suitable for deep target detection •Low penetration
[23,98,101]	Quasi-circular antenna in multi-static data acquisition mode	Dielectric Loaded Waveguide	Tomography	•High accuracy •High penetration	•Low practicality •Very high computation time

Funding: This research is funded by d Cooperative Research Centers Projects, CRC-P-516.

Institutional Review Board Statement: Not Applicable.

Informed Consent Statement: Not Applicable.

Data Availability Statement: Not Applicable.

Conflicts of Interest: The authors declare no conflict of interest.

References

1. Johnson, C.C.; Guy, A.W. Nonionizing electromagnetic wave effects in biological materials and systems. *Proc. IEEE* **1972**, *60*, 692–718. [CrossRef]
2. Ross, K.; Gordon, R.E. Water in malignant tissue, measured by cell refractometry and nuclear magnetic resonance. *J. Microsc.* **1982**, *128*, 7–21. [CrossRef] [PubMed]
3. Khaltaev, N.; Axelrod, S. Chronic respiratory diseases global mortality trends, treatment guidelines, life style modifications, and air pollution: Preliminary analysis. *J. Thorac. Dis.* **2019**, *11*, 2643. [CrossRef] [PubMed]
4. Gluecker, T.; Capasso, P.; Schnyder, P.; Gudinchet, F.; Schaller, M.-D.; Revelly, J.-P.; Chiolero, R.; Vock, P.; Wicky, S. Clinical and Radiologic Features of Pulmonary Edema. *Radiographics* **1999**, *19*, 1507–1531. [CrossRef] [PubMed]
5. Spencer, H.; Pratt, P.C. Pathology of the Lung. *Ann. Surg.* **1980**, *192*, 135. [CrossRef]
6. U.R., A.; Verma, K. Pulmonary Edema in COVID19—A Neural Hypothesis. *ACS Chem. Neurosci.* **2020**, *11*, 2048–2050. [CrossRef]
7. Susskind, C. Possible use of microwaves in the management of lung disease. *Proc. IEEE* **1973**, *61*, 673–674. [CrossRef]
8. Pedersen, P.; Johnson, C.; Durney, C.; Bragg, D. An investigation of the use of microwave radiation for pulmonary diagnostics. *IEEE. Trans. Biomed. Eng.* **1976**, 410–412. [CrossRef]
9. Pedersen, P.; Johnson, C.; Durney, C.; Bragg, D. Microwave reflection and transmission measurements for pulmonary diagnosis and monitoring. *IEEE. Trans. Biomed. Eng.* **1978**, *25*, 40–48. [CrossRef]
10. Iskander, M.; Durney, C. Microwave methods of measuring changes in lung water. *J. Microw. Power* **1983**, *18*, 265–275. [CrossRef] [PubMed]
11. Iskander, M.; Durney, C.; Shoff, D.; Bragg, D. Diagnosis of pulmonary edema by a surgically noninvasive microwave technique. *Radio Sci.* **1979**, *14*, 265–269. [CrossRef]
12. Iskander, M.; Durney, C.; Grange, T.; Smith, C. Radiometric technique for measuring changes in lung water (short papers). *IEEE Trans. Microwave Theory Tech.* **1984**, *32*, 554–556. [CrossRef]
13. Salman, S.; Wang, Z.; Colebeck, E.; Kiourti, A.; Topsakal, E.; Volakis, J.L. Pulmonary Edema Monitoring Sensor With Integrated Body-Area Network for Remote Medical Sensing. *IEEE Trans. Antennas Propag.* **2014**, *62*, 2787–2794. [CrossRef]
14. Iskander, M.; Durney, C.; Bragg, D.; Ovard, B. A microwave method for estimating absolute value of average lung water. *Radio Sci.* **1982**, *17*, 111S–118S. [CrossRef]
15. Rezaeieh, S.A.; Bialkowski, K.S.; Abbosh, A.M. Folding method for bandwidth and directivity enhancement of meandered loop ultra-high frequency antenna for heart failure detection system. *IET Microw. Antennas Propag.* **2014**, *8*, 1218–1227. [CrossRef]
16. Zamani, A.; Rezaeieh, S.A.; Abbosh, A.M. Frequency domain method for early stage detection of congestive heart failure. In Proceedings of the 2014 IEEE MTT-S International Microwave Workshop Series on RF and Wireless Technologies for Biomedical and Healthcare Applications (IMWS-Bio2014), London, UK, 8–10 December 2014; pp. 1–3.
17. Rezaeieh, S.A.; Abbosh, A.; Zamani, A.; Bialkowski, K. Pleural effusion detection system using wideband slot-loaded loop antenna. *Electron. Lett.* **2015**, *51*, 1144–1146. [CrossRef]
18. Rezaeieh, S.A.; Zamani, A.; Bialkowski, K.S.; Mahmoud, A.; Abbosh, A.M. Feasibility of Using Wideband Microwave System for Non-Invasive Detection and Monitoring of Pulmonary Oedema. *Sci. Rep.* **2015**, *5*, 14047. [CrossRef]
19. Zamani, A.; Rezaeieh, S.A.; Abbosh, A.M. Lung cancer detection using frequency-domain microwave imaging. *Electron. Lett.* **2015**, *51*, 740–741. [CrossRef]

20. Babarinde, O.J.; Jamlos, M.F. UWB microwave imaging for lung tumor detection in a thorax model. In Proceedings of the 2014 IEEE Symposium on Wireless Technology and Applications (ISWTA), Kota Kinabalu, Malaysia, 28 September–1 October 2014; pp. 130–133.

21. Babarinde, O.J.; Jamlos, M.F.; Soh, P.J.; Schreurs, D.M.M.-P.; Beyer, A. Microwave imaging technique for lung tumour detection. In Proceedings of the 2016 German Microwave Conference (GeMiC), Bochum, Germany, 14–16 March 2016; pp. 100–103.

22. Caorsi, S.; Massa, A.; Pastorino, M.; Rosani, A. Microwave medical imaging: Potentialities and limitations of a stochastic optimization technique. *IEEE Trans. Microwave Theory Tech.* **2004**, *52*, 1909–1916. [CrossRef]

23. Semenov, S.Y.; Bulyshev, A.E.; Souvorov, A.E.; Nazarov, A.G.; Sizov, Y.E.; Svenson, R.H.; Posukh, V.G.; Pavlovsky, A.; Repin, P.N.; Tatsis, G.P. Three-dimensional microwave tomography: Experimental imaging of phantoms and biological objects. *IEEE Trans. Microw. Theory Techn.* **2000**, *48*, 1071–1074. [CrossRef]

24. Rezaeieh, S.A.; Brankovic, A.; Janani, A.S.; Mohammed, B.; Darvazehban, A.; Zamani, A.; Macdonald, G.A.; Abbosh, A.M. Wearable Electromagnetic Belt for Steatotic Liver Detection Using Multivariate Energy Statistics. *IEEE Access* **2020**, *8*, 201847–201860. [CrossRef]

25. Zhang, H.; Li, M.; Yang, F.; Xu, S.; Yin, Y.; Zhou, H.; Yang, Y.; Zeng, S.; Shao, J. A Feasibility Study of 2-D Microwave Thorax Imaging Based on the Supervised Descent Method. *Electron. Lett.* **2021**, *10*, 352.

26. Rezaeieh, S.A.; Bialkowski, K.S.; Abbosh, A.M. Microwave System for the Early Stage Detection of Congestive Heart Failure. *IEEE Access* **2014**, *2*, 921–929. [CrossRef]

27. Brankovic, A.; Zamani, A.; Abbosh, A. Electromagnetic based fatty liver detection using machine learning. In Proceedings of the 2019 13th European Conference on Antennas and Propagation (EuCAP), Krakow, Poland, 31 March–5 April 2019; pp. 1–3.

28. Fear, E.C.; Bourqui, J.; Curtis, C.; Mew, D.; Docktor, B.; Romano, C. Microwave Breast Imaging With a Monostatic Radar-Based System: A Study of Application to Patients. *IEEE Trans. Microwave Theory Tech.* **2013**, *61*, 2119–2128. [CrossRef]

29. Ahdi Rezaeieh, S.; Zamani, A.; Bialkowski, K.S.; Abbosh, A.M. Novel Microwave Torso Scanner for Thoracic Fluid Accumulation Diagnosis and Monitoring. *Sci. Rep.* **2017**, *7*, 304. [CrossRef]

30. Celik, N.; Gagarin, R.; Huang, G.C.; Iskander, M.F.; Berg, B.W. Microwave Stethoscope: Development and Benchmarking of a Vital Signs Sensor Using Computer-Controlled Phantoms and Human Studies. *IEEE. Trans. Biomed. Eng.* **2014**, *61*, 2341–2349. [CrossRef] [PubMed]

31. Celik, N.; Gagarin, R.; Youn, H.; Iskander, M.F. A Noninvasive Microwave Sensor and Signal Processing Technique for Continuous Monitoring of Vital Signs. *IEEE Antennas Wirel. Propag. Lett.* **2011**, *10*, 286–289. [CrossRef]

32. Perron, R.R.; Huang, G.C.; Iskander, M.F. Textile electromagnetic coupler for monitoring vital signs and changes in lung water content. *IEEE Antennas Wirel. Propag. Lett.* **2014**, *14*, 151–154. [CrossRef]

33. Ahdi Rezaeieh, S.; Zamani, A.; Bialkowski, K.S.; Macdonald, G.A.; Abbosh, A.M. Three-Dimensional Electromagnetic Torso Scanner. *Sensors* **2019**, *19*, 1015. [CrossRef]

34. Rezaeieh, S.A.; Zamani, A.; Bialkowski, K.S.; Abbosh, A.M. Foam Embedded Wideband Antenna Array for Early Congestive Heart Failure Detection With Tests Using Artificial Phantom With Animal Organs. *IEEE Trans. Antennas Propag.* **2015**, *63*, 5138–5143. [CrossRef]

35. Rezaeieh, S.A.; Bialkowski, K.; Abbosh, A. Three-dimensional open-ended slot antenna for heart failure detection system employing differential technique. *IEEE Antennas Wirel. Propag. Lett.* **2014**, *13*, 1753–1756. [CrossRef]

36. Rezaeieh, S.A.; Bialkowski, K.S.; Zamani, A.; Abbosh, A.M. Loop-Dipole Composite Antenna for Wideband Microwave-Based Medical Diagnostic Systems With Verification on Pulmonary Edema Detection. *IEEE Antennas Wirel. Propag. Lett.* **2016**, *15*, 838–841. [CrossRef]

37. Zhang, S.; Xu, G.; Zhang, X.; Zhang, B.; Wang, H.; Xu, Y.; Yin, N.; Li, Y.; Yan, W. Computation of a 3-D Model for Lung Imaging With Electrical Impedance Tomography. *IEEE Trans. Magn.* **2012**, *48*, 651–654. [CrossRef]

38. Darvazehban, A.; Rezaeieh, S.A.; Zamani, A.; Abbosh, A.M. Pattern Reconfigurable Metasurface Antenna for Electromagnetic Torso Imaging. *IEEE Trans. Antennas Propag.* **2019**, *67*, 5453–5462. [CrossRef]

39. Blauert, J.; Kiourti, A. Bio-Matched Antennas With Flare Extensions for Reduced Low Frequency Cutoff. *IEEE Open J. Antennas Propag.* **2020**, *1*, 136–141. [CrossRef]

40. Zhang, H.; Li, M.; Yang, F.; Xu, S.; Zhou, H.; Yang, Y.; Chen, L. A Low-Profile Compact Dual-Band L-Shape Monopole Antenna for Microwave Thorax Monitoring. *IEEE Antennas Wirel. Propag. Lett.* **2020**, *19*, 448–452. [CrossRef]

41. Specific Absorption Rate (SAR) for Cellular Telephones. Available online: https://www.fcc.gov/general/specific-absorption-rate-sar-cellular-telephones (accessed on 1 April 2021).

42. On the Limitation of Exposure of the General Public to Electromagnetic Fields (0 Hz to 300 GHz). Available online: https://eur-lex.europa.eu/legal-content/EN/TXT/PDF/?uri=CELEX:31999H0519&from=EN (accessed on 1 April 2021).

43. Rezaeieh, S.A.; Sorbello, K.; Abbosh, A. Specific absorption rate in human torso from microwave-based heart failure detection systems. In Proceedings of the 2014 IEEE Antennas and Propagation Society International Symposium (APSURSI), Memphis, TN, USA, 6–11 July 2014; pp. 518–519.

44. Alqadami, A.S.M.; Nguyen-Trong, N.; Mohammed, B.; Stancombe, A.E.; Heitzmann, M.T.; Abbosh, A. Compact Unidirectional Conformal Antenna Based on Flexible High-Permittivity Custom-Made Substrate for Wearable Wideband Electromagnetic Head Imaging System. *IEEE Trans. Antennas Propag.* **2020**, *68*, 183–194. [CrossRef]

45. Tissue properties database, The Foundation for Research on Information Technologies in Society (IT'IS). Available online: https://itis.swiss/virtual-population/tissue-properties/database/tissue-frequency-chart/ (accessed on 1 April 2021).

46. Rezaeieh, S.A.; Tan, Y.-Q.; Abbosh, A.M.; Antoniades, M.A. Equivalent circuit model for finding the optimum frequency range for the detection of heart failure using microwave systems. In Proceedings of the 2013 IEEE Antennas and Propagation Society International Symposium (APSURSI), Lake Buena Vista, FL, USA, 7–13 July 2013; pp. 2059–2060.

47. Mobashsher, A.T.; Abbosh, A. On-site rapid diagnosis of intracranial hematoma using portable multi-slice microwave imaging system. *Sci. Rep.* **2016**, *6*, 1–17. [CrossRef]

48. Rezaeieh, S.A.; Zamani, A.; Bialkowski, K.S.; Abbosh, A.M. Unidirectional Slot-Loaded Loop Antenna With Wideband Performance and Compact Size for Congestive Heart Failure Detection. *IEEE Trans. Antennas Propag.* **2015**, *63*, 4557–4562. [CrossRef]

49. Zamani, A.; Abbosh, A.M.; Mobashsher, A.T. Fast Frequency-Based Multistatic Microwave Imaging Algorithm With Application to Brain Injury Detection. *IEEE Trans. Microwave Theory Tech.* **2016**, *64*, 653–662. [CrossRef]

50. Babarinde, O.J.; Jamlos, M.F.; Ramli, N.B. UWB microwave imaging of the lungs: A review. In Proceedings of the 2014 IEEE 2nd International Symposium on Telecommunication Technologies (ISTT), Langkawi Island, Malaysia, 24–26 November 2014; pp. 188–193.

51. Bayat, N.; Mojabi, P. On the use of focused incident near-field beams in microwave imaging. *Sensors* **2018**, *18*, 3127. [CrossRef] [PubMed]

52. Mobashsher, A.T.; Abbosh, A.M. Performance comparison of directional and omnidirectional ultra-wideband antennas in near-field microwave head imaging systems. In Proceedings of the 2016 International Conference on Electromagnetics in Advanced Applications (ICEAA), Queensland, Australia, 19–23 September 2016; pp. 804–807.

53. Kumar, A.; Badhai, R.K. A novel compact printed wideband on-body monopole antenna for the diagnosis of heart failure detection. *Int. J. Control Theory Appl.* **2017**, *10*, 207–217.

54. Rezaeieh, S.A.; Darvazehban, A.; Khosravi-Farsani, M.; Abbosh, A. Body-Matched Gradient Index Lens Antenna for Electromagnetic Torso Scanner. *IEEE Trans. Antennas Propag.* **2021**, 1. [CrossRef]

55. Moll, J.; Vrba, J.; Merunka, I.; Fiser, O.; Krozer, V. Non-invasive microwave lung water monitoring: Feasibility study. In Proceedings of the 2015 9th European Conference on Antennas and Propagation (EuCAP), Lisbon, Portugal, 13–17 April 2015; pp. 1–4.

56. Figueredo, R.E.; De Oliveira, A.M.; Nurhayati, N.; Neto, A.M.D.O.; Nogueira, I.C.; Justo, J.F.; Perotoni, M.B.; De Carvalho, A. A vivaldi antenna palm tree class with koch square fractal slot edge for near-field microwave biomedical imaging applications. In Proceedings of the 2020 Third International Conference on Vocational Education and Electrical Engineering (ICVEE), Surabaya, Indonesia, 3–4 October 2020; pp. 1–6.

57. Rezaeieh, S.A.; Abbosh, A.M. Compact Planar Loop–Dipole Composite Antenna With Director for Bandwidth Enhancement and Back Radiation Suppression. *IEEE Trans. Antennas Propag.* **2016**, *64*, 3723–3728. [CrossRef]

58. Rezaeieh, S.A.; Abbosh, A.; Wang, Y. Wideband unidirectional antenna of folded structure in microwave system for early detection of congestive heart failure. *IEEE Trans. Antennas Propag.* **2014**, *62*, 5375–5381. [CrossRef]

59. Darvazehban, A.; Rezaeieh, S.A.; Abbosh, A.M. Programmable Metasurface Antenna for Electromagnetic Torso Scanning. *IEEE Access* **2020**, *8*, 166801–166812. [CrossRef]

60. Rezaeieh, S.A.; Antoniades, M.A.; Abbosh, A.M. Compact Wideband Loop Antenna Partially Loaded With Mu-Negative Metamaterial Unit Cells for Directivity Enhancement. *IEEE Antennas Wirel. Propag. Lett.* **2016**, *15*, 1893–1896. [CrossRef]

61. Li, M.; Luk, K. A Differential-Fed Magneto-Electric Dipole Antenna for UWB Applications. *IEEE Trans. Antennas Propag.* **2013**, *61*, 92–99. [CrossRef]

62. Ge, L.; Luk, K.M. A Magneto-Electric Dipole for Unidirectional UWB Communications. *IEEE Trans. Antennas Propag.* **2013**, *61*, 5762–5765. [CrossRef]

63. Yan, S.; Soh, P.J.; Vandenbosch, G.A.E. Wearable Dual-Band Magneto-Electric Dipole Antenna for WBAN/WLAN Applications. *IEEE Trans. Antennas Propag.* **2015**, *63*, 4165–4169. [CrossRef]

64. Lu, W.; Liu, G.; Tong, K.F.; Zhu, H. Dual-Band Loop-Dipole Composite Unidirectional Antenna for Broadband Wireless Communications. *IEEE Trans. Antennas Propag.* **2014**, *62*, 2860–2866. [CrossRef]

65. Islam, M.T.; Mahmud, M.Z.; Misran, N.; Takada, J.; Cho, M. Microwave Breast Phantom Measurement System With Compact Side Slotted Directional Antenna. *IEEE Access* **2017**, *5*, 5321–5330. [CrossRef]

66. Biswas, B.; Ghatak, R.; Poddar, D.R. A Fern Fractal Leaf Inspired Wideband Antipodal Vivaldi Antenna for Microwave Imaging System. *IEEE Trans. Antennas Propag.* **2017**, *65*, 6126–6129. [CrossRef]

67. Wang, Z.; Zhang, H. Improvements in a high gain UWB antenna with corrugated edges. *Prog. Electromagn. Res.* **2009**, *6*, 159–166. [CrossRef]

68. Manoochehri, O.; Farzami, F.; Erricolo, D.; Chen, P.-Y.; Darvazehban, A.; Shamim, A.; Bagci, H. Design of a corrugated antipodal Vivaldi antenna with stable pattern. In Proceedings of the 2019 United States National Committee of URSI National Radio Science Meeting (USNC-URSI NRSM), Boulder, CO, USA, 9–12 January 2019; pp. 1–2.

69. Christopoulou, M.; Koulouridis, S. Inter-subject variability evaluation towards a robust microwave sensor for pneumothorax diagnosis. *Prog. Electromagn. Res.* **2015**, *42*, 61–70. [CrossRef]

70. Christopoulou, M.I.; Koulouridis, S.D. Dual patch antenna sensor for pneumothorax diagnosis: Sensitivity and performance study. In Proceedings of the 2014 36th Annual International Conference of the IEEE Engineering in Medicine and Biology Society, Chicago, IL, USA, 26–30 August 2014; Volume 2014, pp. 4827–4830.

71. Ameer, W.; Awan, D.; Bashir, S.; Waheed, A. Use of directional UWB antenna for lung tumour detection. In Proceedings of the 2019 2nd International Conference on Advancements in Computational Sciences (ICACS), Lahore, Pakistan, 18–20 February 2019; pp. 1–5.

72. Rezaeieh, S.A.; Abbosh, A. Wideband and unidirectional folded antenna for heart failure detection system. *IEEE Antennas Wirel. Propag. Lett.* **2014**, *13*, 844–847. [CrossRef]

73. Zhang, H.; Chen, X.; Li, M.; Yang, F.; Xu, S. A compact dual-band folded-cavity antenna for microwave biomedical imaging applications. In Proceedings of the 2019 IEEE International Conference on Computational Electromagnetics (ICCEM), Shanghai, China, 20–22 March 2019; pp. 1–3.

74. Rezaeieh, S.A.; Antoniades, M.A.; Abbosh, A.M. Bandwidth and Directivity Enhancement of Loop Antenna by Nonperiodic Distribution of Mu-Negative Metamaterial Unit Cells. *IEEE Trans. Antennas Propag.* **2016**, *64*, 3319–3329. [CrossRef]

75. Rezaeieh, S.A.; Antoniades, M.A.; Abbosh, A.M. Miniaturization of Planar Yagi Antennas Using Mu-Negative Metamaterial-Loaded Reflector. *IEEE Trans. Antennas Propag.* **2017**, *65*, 6827–6837. [CrossRef]

76. Rezaeieh, S.A.; Zamani, A.; Abbosh, A. Thoracic fluid detection and monitoring system using metamaterial loaded Yagi-antenna array. In Proceedings of the 2016 International Conference on Electromagnetics in Advanced Applications (ICEAA), Cairns, Australia, 19–23 September 2016; pp. 642–645.

77. Rezaeieh, S.A.; Zamani, A.; Abbosh, A.M. Pattern Reconfigurable Wideband Loop Antenna for Thorax Imaging. *IEEE Trans. Antennas Propag.* **2019**, *67*, 5104–5114. [CrossRef]

78. Darvazehban, A.; Rezaeieh, S.A.; Abbosh, A. Pattern-Reconfigurable Loop–Dipole Antenna for Electromagnetic Pleural Effusion Detection. *IEEE Trans. Antennas Propag.* **2020**, *68*, 5955–5964. [CrossRef]

79. Darvazehban, A.; Rezaeieh, S.A.; Manoochehri, O.; Abbosh, A.M. Two-Dimensional Pattern-Reconfigurable Cross-Slot Antenna With Inductive Reflector for Electromagnetic Torso Imaging. *IEEE Trans. Antennas Propag.* **2020**, *68*, 703–711. [CrossRef]

80. Bond, E.J.; Xu, L.; Hagness, S.C.; Veen, B.D.V. Microwave imaging via space-time beamforming for early detection of breast cancer. *IEEE Trans. Antennas Propag.* **2003**, *51*, 1690–1705. [CrossRef]

81. Dem'yanenko Alexander, V. Investigation on resolution of applicator antenna for device for non-invasive bronchopulmonary diseases diagnosis. In Proceedings of the 2020 IEEE Conference of Russian Young Researchers in Electrical and Electronic Engineering (EIConRus), Moscow and St. Petersburg, Russia, 27–30 January 2020; pp. 1498–1502.

82. Semernik, I.V.; Dem'Yanenko, A.V.; Semernik, O.E.; Lebedenko, A.A. Non-invasive method for bronchopulmonary diseases diagnosis in patients of all ages based on the microwave technologies. In Proceedings of the 2017 IEEE Conference of Russian Young Researchers in Electrical and Electronic Engineering (EIConRus), Moscow and St. Petersburg, Russia, 1–3 February 2017; pp. 78–81.

83. Semernik, I.V.; Dem'Yanenko, A.V.; Topalov, F.S.; Samonova, C.V.; Semernik, O.E.; Lebedenko, A.A. Bronchial asthma diagnosis device based on microwave technologies. In Proceedings of the 2017 Radiation and Scattering of Electromagnetic Waves (RSEMW), Divnomorskoe, Russia, 26–30 June 2017; pp. 275–278.

84. Camacho, L.M.; Tjuatja, S. FDTD simulation of microwave scattering from a lung tumor. In Proceedings of the 2005 IEEE Antennas and Propagation Society International Symposium, Washington, DC, USA, 3–8 July 2005; Volume 3, pp. 815–818.

85. Alhawari, A. Lung tumour detection using ultra-wideband microwave imaging approach. *J. Fundam. Appl. Sci.* **2018**, *10*, 222–234.

86. Zamani, A.; Abbosh, A.M. Estimation of frequency dispersive complex permittivity seen by each antenna for enhanced multistatic radar medical imaging. *IEEE Trans. Antennas Propag.* **2017**, *65*, 3702–3711. [CrossRef]

87. Zamani, A.; Darvazehban, A.; Rezaeieh, S.A.; Abbosh, A. Three-dimensional electromagnetic torso imaging using reconfigurable antennas. In Proceedings of the 2019 13th European Conference on Antennas and Propagation (EuCAP), Krakow, Poland, 31 March–5 April 2019; pp. 1–3.

88. Guo, L.; Abbosh, A.M. Optimization-based confocal microwave imaging in medical applications. *IEEE Trans. Antennas Propag.* **2015**, *63*, 3531–3539. [CrossRef]

89. Rocca, P.; Benedetti, M.; Donelli, M.; Franceschini, D.; Massa, A. Evolutionary optimization as applied to inverse scattering problems. *Inverse Probl.* **2009**, *25*, 1–41. [CrossRef]

90. Caorsi, S.; Massa, A.; Pastorino, M. A computational technique based on a real-coded genetic algorithm for microwave imaging purposes. *IEEE Trans. Geosci. Remote Sens.* **2000**, *38*, 1697–1708. [CrossRef]

91. Chew, W.C.; Wang, Y.-M. Reconstruction of two-dimensional permittivity distribution using the distorted Born iterative method. *IEEE Trans. Med. Imag.* **1990**, *9*, 218–225. [CrossRef]

92. Guo, L.; Abbosh, A.M. Microwave imaging of nonsparse domains using Born iterative method with wavelet transform and block sparse Bayesian learning. *IEEE Trans. Antennas Propag.* **2015**, *63*, 4877–4888. [CrossRef]

93. Souvorov, A.E.; Bulyshev, A.E.; Semenov, S.Y.; Svenson, R.H.; Nazarov, A.G.; Sizov, Y.E.; Tatsis, G.P. Microwave tomography: A two-dimensional Newton iterative scheme. *IEEE Trans. Microw. Theory Tech.* **1998**, *46*, 1654–1659. [CrossRef]

94. De Zaeytijd, J.; Franchois, A.; Eyraud, C.; Geffrin, J.-M. Full-wave three-dimensional microwave imaging with a regularized Gauss–Newton method—Theory and experiment. *IEEE Trans. Antennas Propag.* **2007**, *55*, 3279–3292. [CrossRef]

95. Harada, H.; Wall, D.J.; Takenaka, T.; Tanaka, M. Conjugate gradient method applied to inverse scattering problem. *IEEE Trans. Antennas Propag.* **1995**, *43*, 784–792. [CrossRef]
96. Semenov, S. Microwave tomography: Review of the progress towards clinical applications. *Philos Trans. A Math. Phys. Eng. Sci.* **2009**, *367*, 3021–3042. [CrossRef] [PubMed]
97. Semenov, S.; Pham, N.; Egot-Lemaire, S. Ferroelectric nanoparticles for contrast enhancement microwave tomography: Feasibility assessment for detection of lung cancer. In Proceedings of the 1st World Congress on Electroporation and Pulsed Electric Fields in Biology, Medicine and Food & Environmental Technologies, Portorož, Slovenia, 6–10 September 2015; Volume 25, pp. 311–313.
98. Semenov, S.; Posukh, V.; Bulyshev, A.; Williams, T.; Sizov, Y.; Repin, P.; Souvorov, A.; Nazarov, A. Microwave tomographic imaging of the heart in intact swine. *J. Electromagn. Waves Appl.* **2006**, *20*, 873–890. [CrossRef]
99. Semenov, S.Y.; Bulyshev, A.E.; Posukh, V.G.; Sizov, Y.E.; Williams, T.C.; Souvorov, A.E. Microwave tomography for detection/imaging of myocardial infarction. I. Excised canine hearts. *Ann. Biomed. Eng.* **2003**, *31*, 262–270. [CrossRef] [PubMed]
100. Chandra, R.; Johansson, A.J.; Gustafsson, M.; Tufvesson, F. A microwave imaging-based technique to localize an in-body RF source for biomedical applications. *IEEE. Trans. Biomed. Eng.* **2014**, *62*, 1231–1241. [CrossRef]
101. Semenov, S.Y.; Svenson, R.H.; Bulyshev, A.E.; Souvorov, A.E.; Nazarov, A.G.; Sizov, Y.E.; Pavlovsky, A.V.; Borisov, V.Y.; Voinov, B.A.; Simonova, G.I. Three-dimensional microwave tomography: Experimental prototype of the system and vector Born reconstruction method. *IEEE. Trans. Biomed. Eng.* **1999**, *46*, 937–946. [CrossRef] [PubMed]

biosensors

Review

Paving the Way for a Green Transition in the Design of Sensors and Biosensors for the Detection of Volatile Organic Compounds (VOCs)

Camilla Maria Cova [1], Esther Rincón [2], Eduardo Espinosa [2], Luis Serrano [2] and Alessio Zuliani [1,*]

[1] Department of Chemistry, University of Florence and CSGI, Via della Lastruccia 3, 50019 Sesto Fiorentino, FI, Italy; cova@csgi.unifi.it

[2] BioPren Group, Inorganic Chemistry and Chemical Engineering Department, Faculty of Sciences, University of Cordoba, 14014 Cordoba, Spain; b32rirue@uco.es (E.R.); eduardo.espinosa@uco.es (E.E.); iq3secal@uco.es (L.S.)

* Correspondence: zuliani@csgi.unifi.it

Abstract: The efficient and selective detection of volatile organic compounds (VOCs) provides key information for various purposes ranging from the toxicological analysis of indoor/outdoor environments to the diagnosis of diseases or to the investigation of biological processes. In the last decade, different sensors and biosensors providing reliable, rapid, and economic responses in the detection of VOCs have been successfully conceived and applied in numerous practical cases; however, the global necessity of a sustainable development, has driven the design of devices for the detection of VOCs to greener methods. In this review, the most recent and innovative VOC sensors and biosensors with sustainable features are presented. The sensors are grouped into three of the main industrial sectors of daily life, including environmental analysis, highly important for toxicity issues, food packaging tools, especially aimed at avoiding the spoilage of meat and fish, and the diagnosis of diseases, crucial for the early detection of relevant pathological conditions such as cancer and diabetes. The research outcomes presented in the review underly the necessity of preparing sensors with higher efficiency, lower detection limits, improved selectivity, and enhanced sustainable characteristics to fully address the sustainable manufacturing of VOC sensors and biosensors.

Keywords: biosensors; VOCs; environmental; packaging; diagnostic; pollution

1. Introduction

The United States Environmental Agency (EPA) and the European Environmental Agency (EEA) define as a volatile organic compound (VOC) any organic substance that under normal conditions is gaseous or can vaporize in the atmosphere [1,2]. Although this general description helps in easily recognizing a volatile organic compound, it is too rough and is not unequivocal in identifying VOCs. Therefore, different national and international regulations have proposed more standardized definitions according to selected physico-chemical properties of the considered chemicals. Among all, the EU Council Directive 1999/13/EC (and successive amendments and corrections) indicates as a VOC "any organic compound having at 20 °C a vapor pressure of 0.01 kPa or more or having a corresponding volatility under the particular conditions of use" [3]. Additionally, the quite dated—although still highly cited in the literature [4]—1989 World Health Organization's (WHO) definition classifies as a VOC any organic chemical having a boiling point up to 250 °C measured at a standard atmospheric pressure of 101.3 kPa. Based on this definition, the WHO subdivided VOCs into different classes: very volatile organic compounds, VVOCs, having boiling points ranging from <0 °C to 50–100 °C, such as propane (C_3H_8), butane (C_4H_{10}), methyl chloride (CH_3Cl); and volatile organic compounds, VOCs, with boiling points in the range from 50–100 °C to 240–260 °C, including substances such as

Citation: Cova, C.M.; Rincón, E.; Espinosa, E.; Serrano, L.; Zuliani, A. Paving the Way for a Green Transition in the Design of Sensors and Biosensors for the Detection of Volatile Organic Compounds (VOCs). *Biosensors* **2022**, *12*, 51. https://doi.org/10.3390/bios12020051

Received: 27 December 2021
Accepted: 18 January 2022
Published: 19 January 2022

Publisher's Note: MDPI stays neutral with regard to jurisdictional claims in published maps and institutional affiliations.

formaldehyde (CH_2O), limonene ($C_{10}H_{16}$), and ethanol (C_2H_5OH). The WHO also defined an additional category of semi-volatile organic compounds, SVOCs, including substances having boiling points ranging from 240–260 °C to 380–400 °C, such as some pesticides like dichlorodiphenyltrichloroethane (DDT), chlordane or some plasticizers like phthalates [5].

Without going deeper into the merits of the diverse definitions of VOCs, schematically summarized into Figure 1, it is quite glaring that all of them align in proving the abundance of organic chemicals identifiable as volatile in many different types of environments.

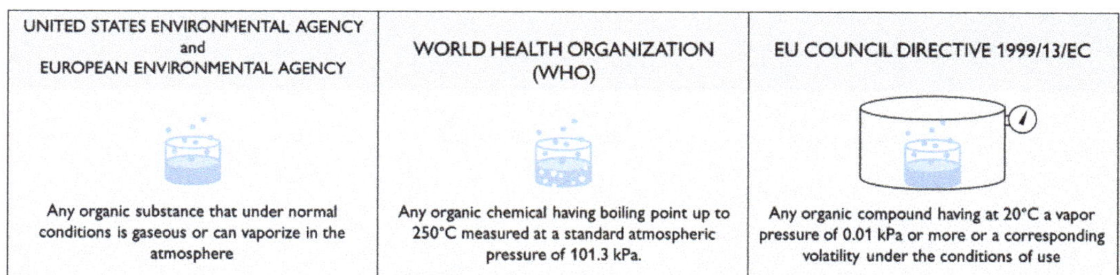

Figure 1. Definition of VOCs according to the United States Environmental Agency (EPA) and the European Environmental Agency (EEA), the World Health Organization (WHO) and the EU Council Directive 1999/13/EC.

Benzene (C_6H_6), toluene (C_7H_8), ethyl benzene (C_8H_{10}), ortho-, meta- and para-xylene, (known as BTEX) (C_8H_{10}), acetone (C_3H_6O), styrene (C_8H_8), and benzyl alcohol (C_7H_8O), are just a few examples of commonly known organic substances having vapor pressure values higher than 0.01 kPa at 20 °C and/or boiling points below 250 °C, that must therefore be considered as VOCs. These substances may be found in ordinary home indoor sites, and in other countless indoor and outdoor environments (and microenvironments) such as those located in industries [6], commercial places [7], hospitals [8], schools [9], etc. For example, among the most diffused VOCs in homes, during the analysis of the inner air of 5000 houses in Japan, acetaldehyde (C_2H_4O), toluene, and formaldehyde were found to be the most abundant VOCs [10]. In another study, the analysis of the inner air in art and craft rooms as well as in common class rooms in a primary school showed mainly the presence of benzyl alcohol, styrene, toluene, ethylbenzene (C_8H_{10}), and xylene [11].

In general, VOCs may be emitted from countless sources, such as furnishing items, building materials, lavatory and laundry products, and biological matter (such as food), etc. [12,13]. For instance, the presence has been observed of a considerably high amount of toxic formaldehyde in a sealed room containing commonly employed, medium-density fiberboards [14], and a sensibly increased concentration of toluene was proved in kitchens during dishwasher washing cycles [15].

Different environments imply the presence of different VOCs, and which varieties and their corresponding concentrations are not only determined and influenced by the materials from which they are emitted, but also from the atmospheric conditions, such as temperature or relative humidity [16], the presence of other materials which may act as adsorbers of VOCs [17,18], the rate of air flux/ventilation [19], and the presence and intensity of visible light/UV irradiation [20], etc. Thus, it is not possible to tabulate general average concentration values of VOCs in the function of similar environments; however, based on numerous studies reported in the literature, it is achievable to draw up lists of VOCs more likely emitted from specific sources and materials in determined situations [21,22]. For example, besides the recognizable emission of VOCs in chemical industries traceable to the mere pure substances [23], it is well known that cellulosic materials such as wood or paper emit acetic and formic acid due to the hydrolysis of acetyl group esters in hemicellulose [24]. Additionally, a large number of polymeric materials used in consumer goods such as furnishings [25], artificial leather or building materials, emit certain VOCs. Bis(2-ethylhexyl)phthalate (DEHP), a plasticizer with significant health concerns, is emitted from poly(vinyl chloride) (PVC) [26], while styrene, recognized as cancerogenic, is emitted from degraded polystyrene (PS) [27].

Cities and high-traffic areas are especially polluted by VOCs emitted from the use of motor vehicle fuels (considering both fuel evaporation and exhaust gas) [28,29], such as toluene, benzene or heptane (C_7H_{16}) [30].

Some specific VOCs are also emitted during food processing as well as during food degradation while 1-butanol ($C_4H_{10}O$), 1-hexanol ($C_6H_{14}O$), 2-ethyl-hexanol ($C_8H_{18}O$) and some volatile fatty acids, such as butyric ($C_4H_8O_2$), valeric ($C_5H_{10}O_2$) or caproic ($C_6H_{12}O_2$) acids, are produced during the spoilage of meat, fish, or fruit, or more generally during the decomposition, i.e., anaerobic digestion, of biomass [31].

Many plants and flowers also emit specific VOCs. Actually, phytogenic volatile organic compounds (PVOCs) represent the most abundant VOCs present in the atmosphere [32]. What we recognize as natural perfumes and fragrances capable of stimulating our senses causing an upsurge of sensations and feelings, are nothing but VOCs. For example, cinnamyl alcohol ($C_9H_{10}O$), having an intense smell of sweet hyacinth with balsamic and spicy notes, is a VOC found in cinnamon leaves and flowers [33]. Citronellol ($C_{10}H_{20}O$), smelling rosy, sweet and of citrus, is a monoterpenoid VOC principally found in roses and pelargonium flowers [34]. These substances are mainly released by flowers to attract pollinators, while other natural VOCs, such as isoprenoids, are naturally released by plants to improve resistance in response to abiotic stresses [35,36].

Many other biological and microbiological processes also imply the release of characteristic VOCs [37,38]. Among them, VOCs emitted by microorganisms (i.e., bacteria, archaea, fungi, and protists) are specifically classified as Microbial Volatile Organic Compounds (MVOCs) and comprise a large variety of chemicals such as fatty acids and their derivatives, nitrogen- and sulfur-containing compounds, aromatics and terpenoids [39,40]. Other VOCs are emitted in the biological processes occurring in human bodies [41–43]. For example, it has been observed that breath samples from breast cancer patients contain a unique combination of hydrocarbons, such as alkanes and monomethylated alkanes [44,45].

Hundreds of different VOCs are thus diffused and present in an infinite number of environments whether deriving from degradation processes, biological processes, natural events, or human activities such as industrial productions, transportation, etc. Consequently, the detection and quantification of VOCs are tactical to investigate the interactions of the volatile chemicals with the surrounding environments as well as to determine and study the emission sources. Table 1 reports some of the most common VOCs and their typical emission sources.

Table 1. Common VOCs and associated emission sources.

VOC	Typical Emission Sources
Propane	Gas grills; gas heaters
Butane	Gas grills; gas heaters; gas torches; end-life fridges, and freezers
Methyl chloride	Solvents; fire extinguishers
Formaldehyde	Plastic furniture items; fiberboards
Toluene	Paints; solvents
Acetone	Solvents; wallpaper and furniture polish
Isopropyl alcohol	Solvents; disinfecting solutions
Carbon Tetrachloride	Fire extinguishers; cleaning products
Carbon disulphide	Volcanic eruptions; marshes
Vinyl chloride	PVC pipes, wire, cable coatings, and textiles; burnt tobacco
Benzene	Fuels
Styrene	Polystyrene objects, rigid panels, and furnishings
Acetic acid	Cellulosic materials such as wood and paper
Isoprenoids	Plants

Classic methods for the analysis of VOCs are gas and liquid chromatography (GC, LC, HPCL, etc.), whether coupled with other techniques such as mass spectroscopy (MS), time of flight (TOF), thermal desorption (TD), or olfactometric detection (e.g., GC-O), etc. [46–48]. Other techniques include, for example, selected-ion flow-tube mass spectrometry (SIFT-MS) or proton-transfer-reaction mass spectrometry (PTR-MS) [49,50]. The analysis of VOCs may be carried out directly injecting the air to be analyzed into the instrument (e.g., headspace analysis) or by firstly adsorbing VOCs on passive or active samplers thus desorbing them in the selected mobile phase for analysis (such as in the case of ion chromatography). These techniques are certainly highly sensitive and efficient but are expensive and energy/time-consuming. In most of the cases they are also not portable, with important drawbacks [51], while the few commercially available portable tools for VOCs analysis are poorly efficient, have high LOD and are not selective to specific VOCs, such as in the case of a photoionization detector (PID) [52].

In the past decades the literature has reported novel VOC sensors and biosensors designed for solving these issues with remarkable results, as reported in different reviews and research papers [53–57]. In general, VOC sensors are devices capable of registering electrical, photophysical, mechanical, or biological changes, after the interaction with specific volatile compounds. These changes are converted into signals, of which the intensity normally depends on analyte concentrations, or analyte chemical and physical characteristics [58]. Among all sensors, the subclass of biosensors indicates sensors containing a biological recognition element, whether that be enzymes, proteins, antibodies, nucleic acids, cells, tissues or receptors, that interact with the VOCs [59–62].

VOC sensors and biosensors have emerged as alternatives to classic analytical tools mainly due to their faster response, cheaper analysis, and portable characteristics, while other features include enhanced selectivity, lower power consumption, or more rapid recovery times. VOC sensors and biosensors have been successfully employed in a large number of applications in food safety analysis, environmental monitoring, clinical analysis and medical diagnosis [63–66]; however, it must be highlighted that the majority of sensors and biosensors reported in previous years were developed without, or by poorly considering any green and sustainable characteristics of the final devices or of the production processes.

Recently, and more specifically in the last couple of years, different national and international policies have started firmly pushing for a sustainable development and a green transition [67–73]. For example, the European Green Deal aims at "making Europe climate neutral by 2050, by boosting the economy through green technology, by creating sustainable industry and transport, and by cutting pollution" [74]. All these policies directly influences any sort of R&D and R&I activity [75–78], including the design of novel VOC sensors and biosensors [79].

From this perspective, a review on the most innovative VOC sensors and biosensors recently developed with environmentally friendly and sustainable characteristics is herein reported, integrating the current reviews present in the literature in the field of VOC sensors and biosensors [80–87]. The review highlights recent trends in the research of green approaches to substitute and replace classic poorly sustainable sensors, in line and accordance with the most recent environmental policies and researchers' ethical spirt of sustainable growth. These approaches include manufacturing processes carried out using biomass and waste derived materials, the use of abundant elements in place of rare metals, the design of low energy consuming methods or the exploitation of biological activities, exploiting innovative technologies such as printed electronics, nanotechnology, silicon photonics, or biotechnology [88–90].

The sensors and biosensors herein reported include tools for the direct analysis of air, as well as systems for the detection of VOCs adsorbed and redispersed—using the already cited passive or active samplers—in aqueous solutions (such as electrochemical devices).

The article is presented in a logical form to be informative and pedagogic for anyone looking for a deeper understanding of the topic. The review is divided into three different sections presenting VOC sensors and biosensors in the function of highly captivating applications, including environmental analysis, intelligent food packaging design, and medical diagnosis, making the manuscript attractive for both readers having expertise in the field but also for anyone with no specific knowledge who wants to explore the matter.

In detail, Section 1 includes sensors and biosensors for environmental analysis, especially focusing on VOCs found in common indoor environments. Section 2 describes VOC sensors and biosensors for food packaging applications, where the detection of VOCs is crucial to understanding the freshness of food and the presence of possible active degradation processes. Section 3 is focused on sensors and biosensors for medical uses, of which the applicability can lead to diagnosing diseases easily and quickly. Each Section firstly discusses the most important VOCs found in the specific field and related challenges, thus, the most recent works on the preparation of sensors and biosensors with green characteristics are reported. The conclusion describes perspectives and challenges for future developments. Figure 2 summarizes the sensors and biosensors for specific VOCs' detection described in each section.

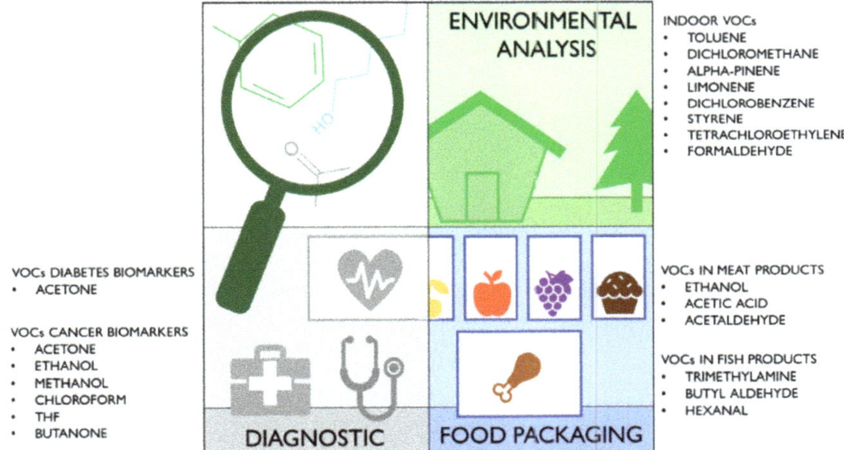

Figure 2. Field of applications of the novel sustainable sensors and biosensors and most relevant analyte VOCs reported in the present review.

2. Section A: Environmental Analysis

The environmental analysis of VOCs aims at the detection and quantification of organic compounds that might involve any biological interaction, including human health issues, plant defense mechanisms, animal toxicity concerns, etc. Without considering particular environments or situations, such as the analysis of gas leaching in pipes or in reactors (which can be undertaken, due to extremely high concentrations of VOCs, using low sensitive sensors and tools), the environmental analysis of VOCs is generally related to the selective detection of common indoor pollutants at low concentrations. Many VOCs are indeed classified as toxic and might cause asthma and other respiratory symptoms/diseases, headaches, nausea, or more severe problems such as convulsions and comas [91]. Some VOCs are also recognized as carcinogenic, especially targeting the liver, kidneys, brain, and nervous system [92]. Therefore, the analysis of VOCs in indoor environments is crucial to determine eventual chronic exposition to toxic chemicals and to avoid severe health issues. In this view, the development of sustainable sensors and biosensors for indoor pollutants has gained much interest especially addressing the current directives of sustainable R&D.

It has been calculated that normally a person spends almost 80% of its life in indoor environments. Thus, a special focus of environmental analysis is the determination and quantification of VOCs in spaces generally occupied during a day such as homes, offices, schools, classrooms, vehicles, and stores [93,94]. VOCs found in these environments are mainly emitted from sources such as construction materials, furnishing, paints, glues, heating appliances, tobacco smoke, cooking, and cleaning products [95,96]. Due to the impossibility of tabulating general concentration values in indoor environments, Table 2 reports the most diffused VOCs in houses and in a primary school and their maximum concentrations.

Table 2. Some of the most abundant VOCs normally found in indoor environments such as houses and schools.

VOCs	Maximum Concentration ($\mu g\ m^{-3}$)		
	Houses According to Héroux et al. [97] *	Houses According to Yamazaki et al. [10] **	Primary School [11]
Toluene	436	530	117
Dichloromethane	1687	/	/
α-pinene	801	/	506
Limonene	329	/	/
Dichlorobenzene	287	4900	/
Tetrachloroethylene	179	/	/
Styrene	14	2000	369
Formaldehyde	/	100	/
Acetaldehyde	/	150	/
Cumene	46	/	/
Ethylbenzene	20	590	196
Hexane	39	/	/
Naphthalene	23	/	/
n-decane	203	/	/
Xylene	77	310	153

* Houses located in Quebec, Canada, ** Houses located in different cities in Japan.

Among all VOCs present in these types of environments, researchers' efforts of recent years have specifically focused on the development of greener and more sustainable sensors and biosensors especially aimed at the detection of toluene, dichloromethane, limonene, dichlorobenzene, styrene, tetrachloroethylene, and formaldehyde.

2.1. Detection of Toluene

Toluene (C_7H_8) is an aromatic compound used in the manufacturing of many goods such as foams for furniture and insulation materials, coatings, or shoes. It has a time weighted average (TWA) of 20 ppm (8 h) and its vapor might irritate the skin, eyes, and the mucous membranes of the throat, possibly causing headache, vertigo, or fatigue [98].

Wang et al. [99] prepared an inexpensive sensor for the detection of toluene based on Fe, one of the most abundant elements in the Earth's crust, and Ni, a metal having important recyclability properties [75,100]. The sensor, in the form of mesoporous $NiFe_2O_4$, was synthesized through a solvent-free simple method producing limited quantities of waste. The sensor had a framework thickness ranging from 8.5 to 5 nm and a specific surface area ranging from 134 to 216 $m^2\ g^{-1}$. During the testing for gas detection, it was proved that the mesoporous $NiFe_2O_4$ with both an ultrathin framework and large specific surface area could detect toluene in concentrations ranging up to 1000 ppb, showing that the response, selectivity, and stability were remarkably enhanced with respect to commonly employed NiFe-based sensors.

In previous years, different lanthanide complexes have been reported as simple, sensitive, and inexpensive analytical tools for the determination of many organic solvents, metal ions and in general gases due to their structural and unique luminescent properties. Very recently, they have been also proved to be usable as sustainable sensors for the specific detection of toluene [101]. In details a new sequence of lanthanide metal-organic frameworks (LnMOFs) was prepared though a simple and inexpensive solvothermal reaction, using lanthanide (III) nitrates, methylmalonic acid as the ligand and 1,10-phenanthroline as the capping agent. The luminescence analysis of LnMOFs in the presence of different organic solvents, showed an evident and marked response though the detection of toluene, proving the possible use of LnMOFs as a highly selective luminous sensor for this type of VOC.

Some environmentally friendly carbon dots have been also proposed as possible sensors for organic compounds' detection. For example, Dong et al. recently reported the preparation of nitrogen and sulfur doped carbon dots as sensors for toluene [102]. Importantly, the materials were prepared using citric acid as the carbon source, sensibly improving the sustainability of the synthetic process, considering that citric acid might be produced by yeasts via biomass valorization [103].

A few years ago, the possibility was proved of preparing a fiber optic enzymatic biosensor featuring cost-effective, real time, continuous, and in situ measurements of toluene. A sensor was prepared using toluene ortho-monooxygenase (TOM) as the biological recognition element, and an optical fiber coated with an oxygen-sensitive ruthenium phosphorescent dye as the transducer [104]. The detection of toluene was carried out based on the enzymatic reaction catalyzed by TOM, which resulted in the consumption of oxygen and, consequently, changes in the phosphorescence intensity.

2.2. Detection of Dichloromethane

Dichloromethane (DCM) (CH_2Cl_2) is largely used in industry due to its high volatility and ability to dissolve many chemicals and it is used to produce paint removers or adhesives, among others. DCM has a TWA (8 h) of 50 ppm, and its hazardous properties include the irritation of skin and mucous membranes and the cause of headache, vertigo, nausea, vomiting and anemia. It has been classified as likely to be carcinogenic [98].

In the last decade, the quartz crystal microbalance (QCM) technique combined with a surface plasmon resonance (SPR) system using Langmuir–Blodgett (LB) thin films have emerged for the detection of VOCs due to the high sensitivity and reliability of the methodology combined with low experimental costs and limited environmental impact. Durmaz et al. exploited these features to prepare a sensitive LB film coated QCM sensor for the detection of DCM [105]. In detail, a calix[4]arene-dithiourea receptor, denoted "C[4]-DT", was used to form a thin film over quartz crystals for QCM measurements. As shown in Figure 3, the so-prepared C[4]-DT LB film-coated QCM sensor was used for the detection of several VOCs.

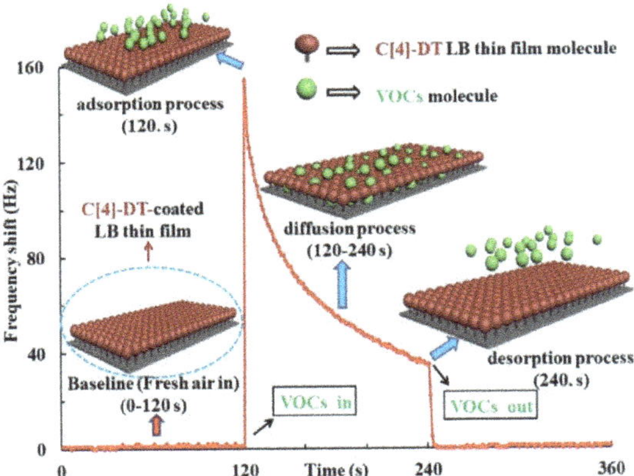

Figure 3. Schematic representation of the interaction of VOCs with the calix[4]arene-dithiourea receptor. Reprinted with permission from ref. [105]. Copyright 2021 Wiley.

The system showed a specifically selective response to the DCM rather than other vapors with a limit of quantification of 0.5 ppm. Additionally, the sensor was proved to have a good reproducibility, rapid response time, and excellent full recovery.

Based on the fact that electrochemical methods for the detection of toxic chemicals are particularly highly sensitive, economic, and portable, Shink et al. proposed an environmentally friendly electrode for the detection of DCM based on a zinc oxide modified disposable screen printed electrode (SPE) [106]. In detail, the authors developed a synthetic methodology to produce hexagonal zinc oxide (ZnO) nanopyramids (NPys), of which the morphology could remarkably improve the performance of the sensor. ZnO NPys were synthesized by a simple and fast hydrothermal procedure using zinc acetate as the precursor and oleylamine as the surfactant. As illustrated in Figure 4, the sensor showed good behavior in the detection of DMC through a series of cyclovoltammetric (CV) analysis.

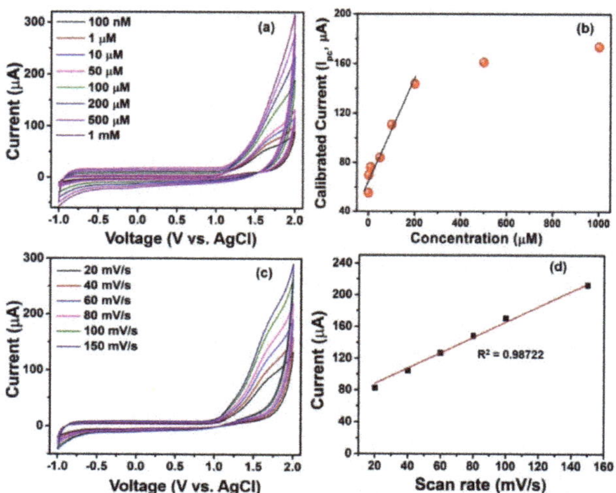

Figure 4. (**a**) CV curves obtained varying the DCM concentrations from 100 nM to 1 mM, (**b**) calibrated current of cathodic peak versus concentration of DCM chemical, (**c**) CV measurements at various scan rates and (**d**) calibrated current at cathodic peak versus scan rate of the hexagonal ZnO NPys modified electrode. Reprinted with permission from ref. [106]. Copyright 2019 Elsevier.

The modified disposable SPE chemical sensor showed a good sensing behavior for the detection of DCM with high sensitivity, a limit of detection of 17.3 µM and an excellent linearity in the range of ~100 nM to 200 µM.

More recently, another study reported the preparation of a highly sensitive sensor for the detection of DCM based on upconverting nanoparticles (UCNPs) [107]. UCNPs are nanoparticles capable of converting low energy incident photons into emitted photons with higher energy, and have particularly emerged for background-free imaging, biological detection, temperature sensing, and many other applications. The key feature of UCNPs is the possibility of preparing sensors with a high sensitivity and a low detection limit along with the important advantage of low energy consumption. The sensor for the detection of DCM was specifically prepared in the form of $NaGdF_4$:Yb,Er@$NaYF_4$:Yb active core@shell upconverting nanoparticles (UCNPs) by depositing UCNPs on porous anodic alumina oxide templates supported by glass slides, forming a thin film-like gas sensor. The nanoporous fluorescent sensor was capable of detecting dichloromethane with a detection limit of 2.9 ppm at room temperature.

Different DCM bacteria destructors have also been proved to be suitable for the preparation of sustainable sensors for DCM detection. In detail, *ethylobacteria-Methylobacterium dichloromethanicum DM4*, *Methylobacterium extorquens DM17*, *Methylopila helvetica DM6*, and *Ancylobacter dichloromethanicus DM16* immobilized on membranes fixed on a pH-sensitive transistor, could interact with DCM leading to a change in the output signal of the transistor [108].

2.3. Detection of Limonene and α-Pinene

α-pinene ($C_{10}H_{16}$) and limonene ($C_{10}H_{16}$) are natural substances mainly found in the oils of coniferous trees (α-pinene) and citrus fruit peels (limonene). α-pinene is principally used to produce perfumes and fragrances and has a TWA (8 h) of 20 ppm. At low concentrations it has therapeutics properties [109], while at high concentration it may cause allergic reactions, and could be highly toxic.

Limonene has a TWA of 30 ppm and its quite safe for human uses although it may cause allergic reactions and toxicity issues by inhalation at high concentrations. Limonene is used as solvent, fragrance, and insecticide [98].

In a similar manner to the detection of DCM, quartz crystal microbalance (QCM) techniques were also exploited for the detection of limonene and α-pinene. In detail, a sensor for the detection of limonene was prepared using a QCM chip as the sensor transducer and ethyl cellulose as the sensing material [110]. The use of ethyl cellulose (EC) is of particular interest since EC is derived from cellulose, i.e., the most renewable natural polymer on Earth [111]. The sensor was specifically proved to detect limonene up to 6000 mg m^{-3}, with a limit of detection (LOD) of 300 mg m^{-3}. The sensor was also demonstrated to be stable and efficient since it could be used for up to five cycles and for a month before observing significant losses of activity.

On the other hand, the detection of α-pinene is quite complicated and few works have reported the successful design of novel sustainable sensors, making the research highly challenging. Among the few outstanding examples, a sensor for the detection of α-pinene was prepared by manufacturing a highly selective molecularly imprinted polymer (MIP) layer combined with an interdigitated electrode (IDE) as a sensor. Importantly, the IED was prepared using methacrylic acid (MAA) as the sensing material [112]. The sensor was proved to be remarkably selective and efficient. Significantly, considering that it has been recently demonstrated that it is possible to produce MAA from biomass-derived glucose, the manufacturing of this sensor can be considered potentially sustainable, as summarized in Figure 5 [113].

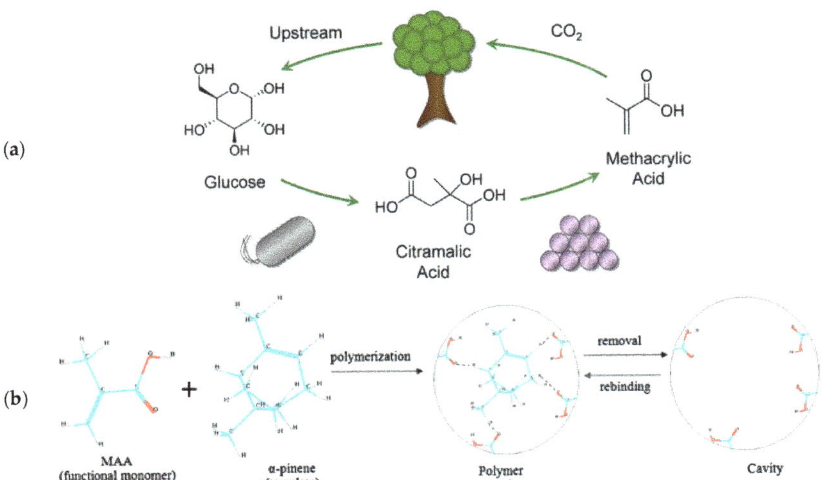

Figure 5. (**a**) Scheme of the hybrid fermentation and thermocatalysis to produce methacrylic acid (MAA) from glucose. Reprinted with permission from ref. [113]. Copyright 2021 American Chemical Society. (**b**) Illustration of molecular imprint polymer (MIP) concept made of MAA. Reprinted with permission from ref. [112]. Copyright 2013 Elsevier.

2.4. Detection of Dichlorobenzene

Dichlorobenzene (DCB) ($C_6H_4Cl_2$) in its three different isomeric forms (1,2; 1,4 and 1,3) is used in space deodorants, fumigants, insecticides, and herbicides as well as in the synthesis of dyes and resins. The lower value of TWA (8 h) of DCB (corresponding to 1,4-dichlorobenzene) is 25 ppm. Inhalation of the vapor of DCB results in irritation to the eyes, skin, and throat. DCB has also the potential to cause cancer [98].

A few years ago, Chao et al. demonstrated the possibility of producing mesoporous molecular sieves MCM-41 from coal fly ash at room temperature via a green and efficient reaction [114]. MCM-41 is a widely used material with applications in catalysis, separation processes, and adsorption of gases and liquid. This last feature was specifically exploited by Rahman et al. to design a simple, inexpensive, potentially sustainable, consistent, portable, and reliable chemical sensor for 1,2-dichlorobenzene detection [115]. The sensor was fabricated by depositing a thin layer of MCM-41 on a glassy carbon electrode (GCE). The sensor, used through an electrochemical approach, showed good sensitivity and a short response time of 14.0 s, while the linear dynamic range and the detection limit were reported as 0.089 nM to 8.9 mM and 13.0 pM, respectively.

2.5. Detection of Styrene

Styrene (C_8H_8) is extensively used in the manufacturing of numerous polymers and copolymers such as polystyrene, acrylonitrile-butadiene-styrene (ABS), styrene-butadiene latex, for the fabrication of different goods including foam packaging, toys, shoes, and furnishings. Styrene has a TWA (8 h) of 20 ppm, and its vapor irritates the eyes and mucous membranes. The inhalation of high concentrations of styrene can cause polyneuritis. It is also reasonably anticipated to be a human carcinogen [98].

Recently, Bi et al. developed a Terbium-based metal-organic frameworks (MOF) for the efficient detection of styrene. Td-MOF (Tb^{3+}) was prepared based on an innovative, facile, and low-energy consuming (at room temperature) method [116]. Td-MOF was thus homogeneously embedded into a PVA film and deposited on silica gel sheets, forming a luminescent vapor sensor film for styrene detection. A sequence of photoluminescence (PL) tests demonstrated that Tb-MOFs showed a significant response rate and high sensitivity to styrene vapor. In addition, as shown in Figure 6a, time-dependent fluorescence quenching indicated that the emission of the film was immediately quenched by exposure to styrene vapor (in only 30 s), and the intensity remained unchanged over time, proving an excellent sensitivity performance. Recyclable tests, i.e., by carrying out experiments followed by a drying procedure in an oven, also proved the good reversibility and reusability of the Td-MOF, as illustrated in Figure 6b.

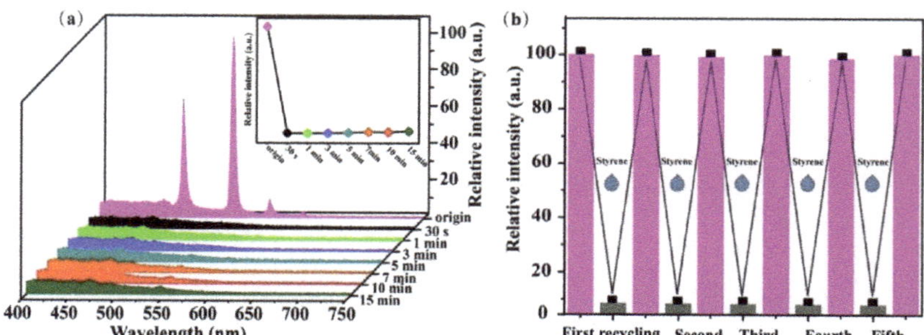

Figure 6. (a) Time-dependent emission spectra of the Tb-MOFs film responded to 20 μL styrene vapor; (b) Emission intensity of five recyclable experiments of sensing styrene, Reprinted with permission from ref. [116]. Copyright 2020 Elsevier.

A few years ago the possible utilization of bacteria for the preparation of biosensors for styrene detection was also demonstrated, such as in the case of a biosensor based on the regulation system of the styrene catabolic pathway present in the *Pseudomonas* sp. strain Y2 [117]; however, this type of approach has not been followed up in recent years, although it has tremendous potentialities.

2.6. Detection of Tetrachloroethylene

Tetrachloroethylene (C_2Cl_4) is principally used as a chemical intermediate and as a solvent in the textile and metal industries. Tetrachloroethylene has a TWA (8 h) of 25 ppm and the exposure to its vapors can cause eye irritation, narcotic action, vertigo, nausea, and headache. Tetrachloroethylene is also suspected to cause cancer [98].

A ZnO-based sensor capable of detecting tetrachloroethylene was recently proposed by Zhao et al. [118]. In detail, the researchers developed a new method for the chip-level pyrolysis of as-grown zeolitic imidazolate framework films to hierarchical and structured ZnO sheets composed of interpenetrated nanometer particles. The tunable introduction of interpenetrated particles generated adjustable oxygen vacancies, modifying the electronic structure of the sensing materials. As a result, the sensors showed improved diffusion, penetration, and adsorption of the relevant gases, resulting in enhanced sensitivity and a shortened response time toward the detection of different VOCs at the ppb-level, including tetrachloroethylene. The facile synthetic approach using a largely available material, i.e., ZnO, made the novel sensor a good candidate for sustainable scaled-up productions and commercialization.

2.7. Detection of Formaldehyde

Formaldehyde (CH_2O) is used in the manufacturing of many different products including adhesives, abrasive materials, insulating materials, coatings, and polyacetal plastics-based materials. In indoor environments it is mostly emitted from building materials. Formaldehyde is a highly toxic chemical with a TWA (8 h) of 0.1 ppm. The inhalation of formaldehyde irritates the mucous membranes, while chronic symptoms include renal and hepatic damage. It is considered cancerogenic [98].

Recently, Lee et al. reported the manufacturing of a monolithic flexible sensor for the detection of formaldehyde at the ppb-level [119]. The sensor was produced by depositing a TiO_2 sensing film on a polyethylene terephthalate substrate and by covering the film with an overlayer of molecular sieving a ZIF-7/polyether block amide (mixed matrix membrane, MMM). The sensor was designed to selectively detect formaldehyde by a sensing photoactivation at room temperature. The sensor showed ultrahigh selectivity (response ratio > 50) and response (resistance ratio > 1100) to the exposure at only 5 ppm of formaldehyde. Figure 7 illustrates the selectivity toward the detection of formaldehyde of the novel MMM/TiO_2 sensor also in the presence of ethanol (normally sensibly affecting the detection of formaldehyde).

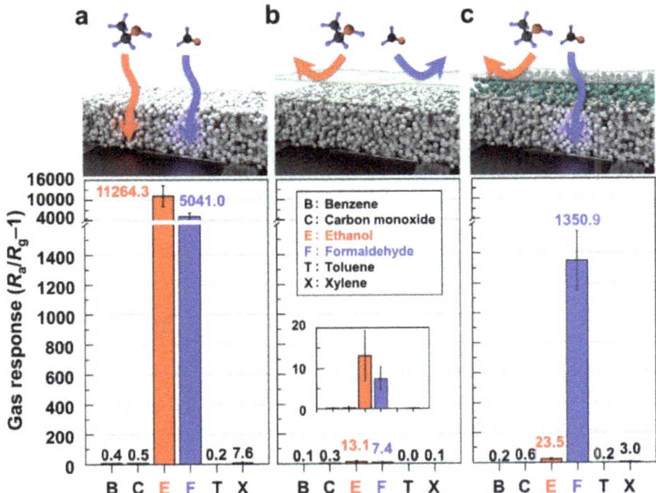

Figure 7. (**a**) Gas responses of a bare TiO_2, (**b**) Pure PEBA/TiO_2, and (**c**) 5MMM/TiO_2 sensors exposed to 5 ppm benzene, carbon dioxide, ethanol, formaldehyde, toluene, and p-xylene at 23 °C under UV illumination (λ: 365 nm). Error bars represent SD of the mean. Reprinted with permission from ref. [119]. Copyright 2021 Springer Nature.

A high-performance formaldehyde sensor was prepared by a surface micro-fabrication technique depositing a LaFeO$_3$ (LFO) thin film on a silica substrate [120]. The sensing performances demonstrated that the novel formaldehyde sensors had a remarkable sensitive response and low detection limit toward the ppb-level. In detail, the sensor exhibited a detection limit of 50 ppb and outstanding replicability with a maximum drift of the baseline resistance from different batches of the sensor gas sensors of only 5.4%, and the maximum drift of the response value of 6.5%. In addition, the response values of the sensors remained stable for up to 18 days, with an absolute deviation of response value of approximatively 0.04.

Other recent sustainable approaches for the preparation of sensors for formaldehyde detection include the use of largely available and inexpensive materials such as tin and zinc [121–128], the second most abundant element in the Earth's crusts, i.e., silicon [129,130], the use of biomass-derived materials, such as bacterial cellulose [131] or egg-white [132].

A biosensor based on formaldehyde dehydrogenase and chitosan has also been recently reported [133]. The sensor was prepared through a low-cost inkjet printing technology by depositing a polyion-complex of FDH and chitosan on an electrode connected with an organic field-effect transistor. The biosensor could detect formaldehyde with an LOD of 3.1 µM in aqueous solution.

3. Section B: Food Packaging

The demands of the users (food producers, food processors, logistic operators, distributors, and consumers) in the food industry sector are increasing in terms of food safety, quality, and traceability [134]. Throughout the food chain (production, storage, transport, and sale) there are a wide variety of factors (microorganisms, enzymes, temperature, etc.), that can corrupt food products and reduce their shelf life. This is the reason why, in particular, food packaging plays a key role in maintaining the quality of food as well as preserving it from contamination [135]. Traditional packaging systems merely isolate food from the external environment without providing information on the freshness or condition of the food beyond the expiration date. Thus, it is constantly necessary to innovate in the field of food packaging, not only to reduce its environmental footprint, but also to increase its functions. In this scenario arises intelligent packaging, a new packaging technology that integrates traditional packaging systems with intelligent functionalities, including the monitoring of changes in the food product, as well as quality and safety information [136,137], by temperature, humidity, pH, and light exposure measurements [138–141], or through the detection of specific VOCs [134,142–145]. For example, 1-butanol (C$_4$H$_{10}$O), 1-hexanol (C$_6$H$_{14}$O), 2-ethyl-hexanol (C$_8$H$_{18}$O), 1-octen-3-ol (C$_8$H$_{16}$O), butanal (C$_4$H$_8$O), hexanal (C$_6$H$_{12}$O) and nonanal (C$_9$H$_{18}$O), which are indicators of freshness in food products, while other VOCs, such as fatty volatile acids, are produced during the spoilage of foods [146].

When it comes to incorporating sensing technologies into food packaging materials, the industry trend is to do so for meat or fish products [147].

3.1. VOCs Detection in Meat Products

Microbial growth, oxidation and enzymatic autolysis are the three main mechanisms of meat deterioration. During meat spoilage, proteins and lipids decompose to form new compounds that negatively affect product quality. The intrinsic factors related to meat spoilage include pH, water activity and nutrient content of the meat, while extrinsic factors include temperature and atmospheric conditions surrounding the product [148]. For example, when microbial spoilage occurs, there is a decrease in pH due to the release of lactic acid. The microbes commonly associated with this phenomenon are of the genus *Pseudomonas* and a traditional sensor/biosensor should detect specific *Pseudomonas* presence by antigen/antibody reactions or similar [149]. Since microbial spoilage may not occur homogeneously throughout the meat product and the detection of these bacteria would require the sensor to be in direct contact with the entire product, it is most desirable that the target product detected by the sensor be a gaseous by-product released into the packaging space. Under normal packaging conditions, several metabolites are formed in the packaging space including CO_2, O_2, volatile nitrogen compounds and biogenic amines. As far as this review is concerned, it should be mentioned that the most common VOCs released during meat spoilage are alcohols, phenols, ketones, acids and sulfur-containing compounds [150].

Regarding the detection of VOCs in the meat industry, the most common trends have been towards the detection of alcohols or acetic acid. This is because alcohols such as 3-methyl-1-butanol ($C_5H_{12}O$) or 1-hexanol ($C_6H_{14}O$) are indicative of *Salmonella* contamination in packaged beef, while acetic acid is an indicator of microbial population growth. Hence, Sankaran et al., elaborated olfactory bio-derived sensors mimicking insect odorant binding protein to detect them in low concentrations at room temperature. These were biosensors based on quartz crystal microbalance (QMC) with synthetic peptides. This peptide sequence acting as the sensing material was derived from the amino acid sequence of the LUSH protein from *Drosophila* odorant binding protein and can detect alcohols with estimated lower detection limits of <5 ppm [151,152]. On the other hand, in order to be able to detect acetic acid even at low concentrations (1–3 ppm), Panigrahi et al., prepared quartz crystal microbalance (QMC) sensors deposited over synthetic polypeptide [153]. Recently, Han developed a new gas sensor employing ZnO foam as the sensing material aimed at acetic acid with superior sensing performances [154].

The latest advances in the development of sensors for the detection of alcohols in packaged meat concerned the detection of ethanol (C_2H_5OH). Senapati and Sahu prepared an Au patch electrode Ag-SnO$_2$/SiO$_2$/Si metal-insulator-semiconductor capacitive gas sensor with a high sensitivity (10 ppm) for chicken meat samples [155]. The sensor was prepared using a considerably high amount of inexpensive and largely available Sn and Si, although, it is worth mentioning that the response of these sensors to ethanol is lower than to other gases such as ammonia and trimethylamine or hydrogen sulfide, as shown in Figure 8.

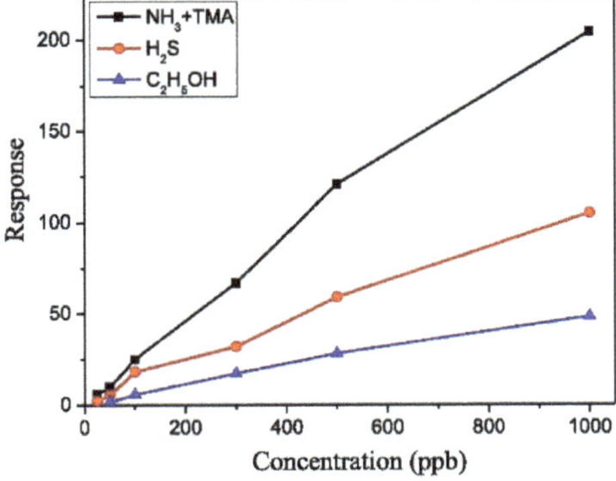

Figure 8. Response curve of Au patch electrode Ag-SnO$_2$/SiO$_2$/Si metal-insulator-semiconductor capacitive gas sensor for increasing concentrations of ammonia and trimethylamine (NH$_3$ + TMA), hydrogen sulfide (H$_2$S) and ethanol, Reprinted with permission from ref. [155]. Copyright 2020 Elsevier.

In recent years, the detection of other VOCs related to meat spoilage has also been studied. Acetaldehyde (C_2H_4O), resulting from ethanol metabolism, is one of the most important compounds to consider in sophisticated packaging systems. This compound is classified as carcinogenic, and its TWA (8 h) is 25 ppm [156]. It is therefore important to be able to detect this compound quickly and efficiently. Kim et al. fabricated a surface acoustic wave (SAW) sensor that evaluated the storage time of chicken meat (up to 15 days) as a function of increasing acetaldehyde concentration. These authors verified the feasibility of PDMS polymer composite sensors coated with a layer of the SAW device for the detection of aldehyde gas with a 0.989 coefficient of determination between the gas and storage time of chicken meat [157]. Lastly, another VOC released during the spoilage of meat products, and thus acting as a marker, is dimethyl sulfide (DMS, C_2H_6S). For its detection, Chow developed environmentally friendly chemosensors based on bimetallic donor–acceptor ensembles (BmDAE) with a selectivity toward DMS 1.0 ppm in real beef samples. This selectivity was clearly observable to the naked eye, since the chemosensor only turned pink in the presence of DMS (Figure 9a). Moreover, the chemosensor response was correlated with the microbial growth level and the storage time, as shown in Figure 9b [158].

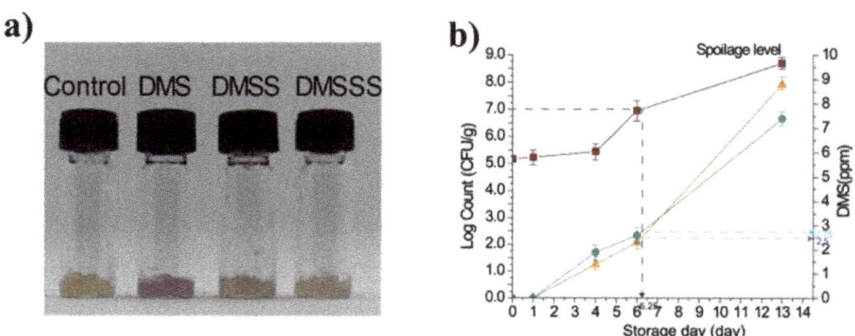

Figure 9. (**a**) Naked-eye sensing response of solid-supported chemosensor toward DMS; (**b**) changes in microbial counts (brown line) and DMS concentration measured by UV-Vis (green line) and GC-MS (orange line) for beef samples stored at 4 °C. Reprinted with permission from ref. [158]. Copyright 2019 Elsevier.

3.2. VOCs Detection in Fish Products

The consumption of fish or fish-based products is booming due to their health benefits; however, these products are extremely perishable, so it is necessary to develop non-invasive techniques that allow the freshness of the food to be known in more detail rather than just the packaging date. As with meat products, certain VOCs produced by microbial, enzymatic, or autolytic activities during fish spoilage have been identified [159]. Therefore, developing sensors for detecting these compounds is a promising approach.

One of the most characteristic VOCs released during fish spoilage is trimethylamine (TMA, C_3H_9N), a chemical produced through the decomposition of proteins, carbohydrates, and fats. Recently, Perillo and Rodríguez employed TiO_2 membrane nanotubes supported on a flexible substrate as a sensor for TMA detection. This sensor was developed using a simple electrochemical anodization and was able to detect TMA at low temperatures in a very wide detection range (40–400 ppm, Figure 10a) [160]. Importantly, TiO_2 is a largely available oxide with a very low impact on human health. Other types of sensors that can be used in the detection of TMA in canned fish are those reported by Yang et al. In this case, the authors employed α-Fe_2O_3 snowflake-like hierarchical architectures as a TMA gas sensor. The sensors showed an ultra-fast response of 0.9 and 1.5 s for response time and recovery time, respectively, for TMA and other testing gases such as ethanol, acetone, toluene, methanol and ammonia with a sensitivity of 100 ppm, as illustrated in Figure 10b [161]. Along the same lines, Liu et al., (2020) incorporated α-Fe_2O_3 nanoparticles in thick films for the detection of TMA in fish. These sensors showed very good selectivity and high sensitivity for TMA with a minimum detection of 1 ppm, as illustrated in Figure 10c [162]. This same metal oxide has been employed by Shen et al. for the development of α-Fe_2O_3 modified Au@Pt bimetallic hollow nanocube sensors. These sensors showed a very fast response time (5 s) towards 100 ppm TMA in *Larimichthys crocea* [163]. All these approaches followed the idea of exploiting an abundant element, i.e., Fe, of which its sustainable use has been already discussed.

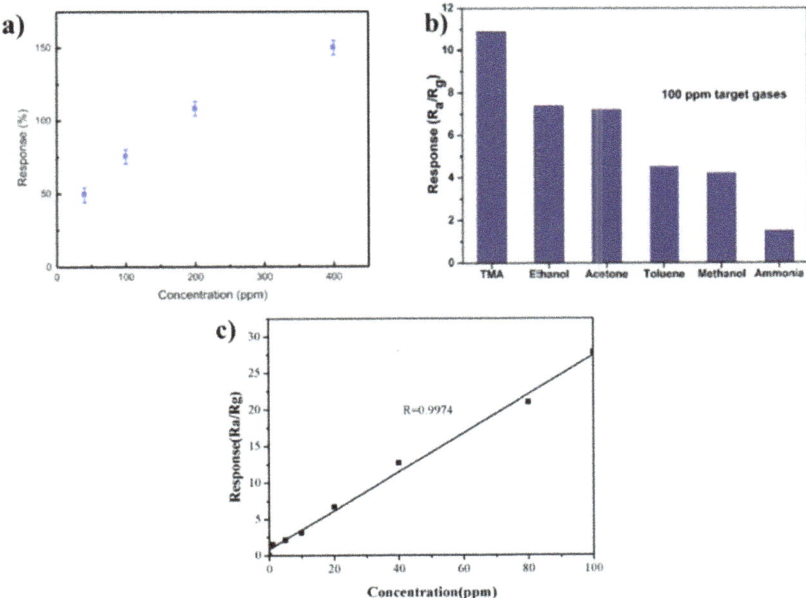

Figure 10. (**a**) TiO_2 nanotubes sensor response to increasing TMA concentrations (40–400 ppm). Reprinted with permission from ref. [160]. Copyright 2016 Elsevier. (**b**) Response of snowflake-like α-Fe_2O_3 hierarchical architectures toward 100 ppm of various testing gases Reprinted with permission from ref. [161]. Copyright 2017 Elsevier. (**c**) Response of α-Fe_2O_3 sensor to increasing concentrations of TMA gas (1–100 ppm) Reprinted with permission from ref. [162]. Copyright 2020 Frontiers Media SA.

TMA detection can be also carried out by colorimetric changes. Lv et al. laid the groundwork for the reaction mechanism of a set of colorimetric sensors that included chromogenic materials sensitive to TMA during the deterioration of packaged fresh mackerel. The authors selected six types of metalloporphyrins and tetraphenyl porphyrins (TPP) and showed that MnTPP, NiTPP and FeTPP had the best binding capacity to TMA. Thus, metal porphyrins can be employed for the construction of colorimetric sensors for TMA [164]. Meanwhile, Sun et al. developed a colorimetric printed freshness indicator for fish in modified atmosphere packaging (MAP) [165]. These authors prepared a printable ink based on a natural purple cabbage pigment—which can be potentially also extracted form waste cabbage [166]—carboxymethyl cellulose and glycerin, screen printed it on paper and applied it to grass carp MAP. This label darkens as the TMA content in the fish sample increases as an indicator of spoilage, as shown in Figure 11. The freshness of fish can also be measured non-destructively using fluorescent films. Lai et al. developed highly emissive amorphous tetraphenylethylene (TPEBA) nanoparticles capable of detecting TMA with a detection limit of 0.89 ppm in butterfish [167]. Finally, the most recent advance in the detection of TMA in fish has been the one proposed by Praoboon et al. [168]. The authors developed a paper-based electrochemiluminescence device for the estimation of TMA concentration in freshwater and marine fish samples (red tilapia, yellow tail, salmon, tuna, and catfish). The key to these sensors lay in the fast response they provided (2 min) for a TMA concentration range from 1×10^{-12} to 1×10^{-6} M.

Figure 11. Color change of printable colorimetric paper sensor during monitoring of the freshness of the grass carp within 24 h at 25 °C by Sun et al., Reprinted with permission from ref. [165]. Copyright 2021 Springer Nature.

Although to a lesser extent than the TMA, aldehydes such as hexanal ($C_6H_{12}O$), octanal ($C_8H_{16}O$) and nonanal ($C_9H_{18}O$) are also released from fish products such as grass carp or hairtail fish. In this sense, Jia et al. developed a predictive model to determine the freshness of salmon during cold storage. The authors employed electronic nose with principal component analysis (PCA) and radial basis function neural networks (RFBNN). This system allowed the detection of VOCs such as butyl aldehyde (C_4H_8O), amyl aldehyde, hexanal, heptanal ($C_7H_{14}O$), 1-propanol (C_3H_8O), and 1,2-butanone amyl alcohol, which increased proportionally with the level of salmon spoilage [169]. Lastly, Chen et al. prepared a quartz crystal microbalance (QMC) gas sensor modified with the hydrophobic amino-functionalized graphene oxide (AGO) nanocomposite for aldehydes detection in grass carp fish fillets and hairtail fillets. These sensors responded towards aldehydes within 45 ppm under 80% relative humidity during refrigerated storage at 4 °C [170].

4. Section C: Diagnostic

As estimated by the World Health Organization [171], every year 12 million global deaths (nearly 25% of total deaths) are attributable to unhealthy environments. Environmental hazards, in particular water, air, and soil pollution, causes hundreds of diseases and health problems. In addition, the WHO has pointed out that two-thirds of the total deaths related to unhealthy environments come from noncommunicable diseases (NCD) such as heart diseases, autoimmune diseases, diabetes, strokes, cancers, and others. The same institution reported that yearly about eight million people die due to the delayed diagnosis of NCD.

An effective strategy to prevent these deaths is the development of devices allowing an early diagnosis of the diseases. The accurate identification and quantification of VOCs emitted from the body can indeed provide information on health and metabolic pathological conditions. In particular, VOC sensors have gained considerable interest for the selective and continuous diagnosis of various physiological and pathological states acting as biomarkers for the identification of numerous diseases in a non-invasive way [172–175]. Indeed, the key factor of this type of analysis is the detection of VOCs in the exhaled breath of patients through simple, efficient, and inexpensive tools [176–178]. For example, some VOCs such as acetone, benzene, ethanol, and isoprene are related to specific diseases and could be used as biomarkers of diabetes, genetic disorders, infectious, cancerous, or renal diseases [75,179,180].

In recent years, scientific efforts have especially focused on the design of environmentally friendly sensors and biosensors for the sustainable diagnosis of cancer and diabetes. Moreover, some remarkable results have been also obtained in the diagnosis of asthma, chronic obstructive pulmonary disease, cystic fibrosis, liver cirrhosis and tuberculosis [181–185].

4.1. Diabetes Diagnosis

The traditional method for checking diabetes involves collecting blood samples. This type of analysis is precise and accurate but painful, expensive, and invasive. Alternatively, it has been demonstrated that diabetes can be diagnosticated in a non-invasive way by detecting different gaseous VOCs in breath samples. Indeed, the concentrations of olfactory markers of the breath in diabetic patients show significant differences compared to those of healthy patients. For example, acetone (CH_3COCH_3) is one the most studied and recognizable VOCs for diabetes diagnosis [186], considering that acetone concentration in diabetic patients is higher than 1.8 ppm [187,188].

Ma et al. [189] developed a sensor for acetone detection based on Ni, a metal having important recyclability properties, and Fe, one of the most abundant chemical elements in the Earth's crust. Porous $NiFe_2O_4$ microspheres were synthetized using an easy procedure, combining a solvothermal step with a heating annealing methodology. As proved by experimental tests, the gas sensors showed a high response to 100 ppm acetone, a low detection limit (200 ppb) and excellent reusability.

A high-performant NiO/SnO_2 acetone sensor was also prepared via a facile hydrothermal protocol [190]. The gas sensor exhibited improved performances compared to pure tin oxide and showed a fast response, low detection limit (10 ppb) and good selectivity. Similarly, a SnO_2/ZnO-based sensor able to detect acetone was recently proposed by Dong et al. [191]. In detail, an electrospinning step and a low temperature water bath method was developed for designing SnO_2/ZnO hetero nanofibers. The sensor was tested with an acetone concentration range of 1 to 100 ppm. The results demonstrated that SnO_2/ZnO materials exhibited fast response values, and a remarkable, high selectivity to acetone.

A few years ago, Zhang et al. reported a one-step route to prepare C_3N_4-SnO_2 nanocomposites with an outstanding acetone sensing performance [192]. C_3N_4 and SnO_2 are eco-friendly, economic, and easy-to-prepare materials, and the synthetic procedure reported by the researchers was simple, repeatable, and operable. The sensors exhibited about a 20 times improvement of the response sensitivity as well as remarkable selectivity, fast response and repeatability compared with pure tin oxide. The detection limit of 67 ppb was remarkably below the acetone content of diabetes patients' exhaled breath.

Recently, $ZnFe_2O_4$ has also attracted considerable interest due to its environmentally friendly characteristics, low cost, and excellent stability. Huang et al. designed $ZnFe_2O_4$ nanorods through an easy hydrothermal route [193] with a high gas response of acetone.

Another study reported the microwave-assisted synthesis of a sensor for the detection of an acetone based on a Co_3O_4/rGO nanocomposite [194]. Microwave (MW) irradiation is recognized as a time-saving heating method with remarkable environmentally friendly characteristics such as minimized heating loss and improved energy efficiency [75,195,196]. The tests showed that the materials achieved remarkable response to acetone (0.5~200 ppm) and good selectivity against the gases of hydrogen, methane, hydrogen sulphide, formaldehyde, methanol, methoxyethane and ethanol.

4.2. Cancer Diagnosis

Commonly used methodologies for cancer diagnosis implies bronchoscopy and diagnostic imaging (CT scan). These analyses entail some drawbacks such as weak sensitivity or the use of expensive tools. Moreover, bronchoscopy involves anesthesia, which is sometimes correlated with trauma and complications. In the past decade, the detection of specific VOC biomarkers has been identified as a new frontier for non-invasive cancer diagnosis [197]. In detail, VOCs such as toluene, benzene, styrene, ethanol, methanol, acetaldehyde, formaldehyde, and octanal are present in the breath of people suffering cancer [198] in concentrations higher with respect to the health subject [199].

Recently, Feller et al. presented the design of a biobased carbon nanorods VOC sensor for the effective detection of acetone, ethanol, and methanol for the early diagnosis of cancer [200]. Importantly, the device was prepared via an easy, fast and green approach through the pyrolysis of a renewable carbon source, i.e., castor oil.

Also Sahajwalla et al. have developed a new sensor with sensing performances tailored for VOC biomarker cancer detection [201]. As illustrated in Figure 12, the tool was synthetized using pristine graphene and zinc oxide nanoparticles recovered from spent Zn–C batteries.

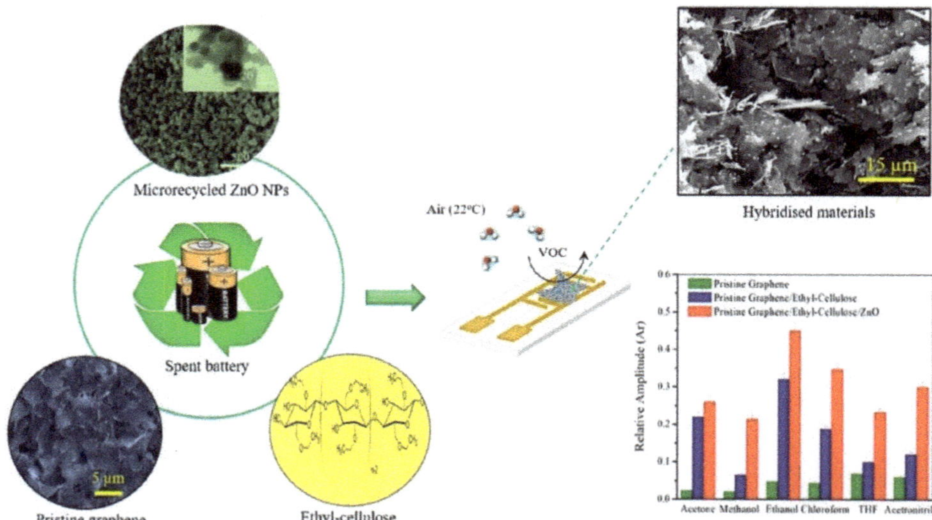

Figure 12. Schematic representation of the preparation of ZnO-based sensors for VOCs detection and cancer diagnosis using spent batteries, Reprinted with permission from ref. [201]. Copyright 2021 Elsevier.

Preliminary tests showed that the recycled ZnO nanoparticles had good selectivity along with a sensitivity towards chloroform ($CHCl_3$) and ethanol at a 5 ppm testing level, a value of concentration often found in patients suffering from cancer.

Another ZnO-based sensor has been reported for the detection of butanone (C_4H_8O), a VOC present in the breath of patients with gastric cancer [202]. In particular, a bicone-like ZnO structure was prepared through a microwave-assisted template free method. The structure showed outstanding performances in terms of selectivity, sensitivity, and detection limit (0.41 ppm).

5. Conclusions: Challenges and Opportunities

Global warning, overpopulation crisis, the decreasing availability of water, food fraud and adulteration, the overspreading of non-communicative diseases, are just some of the challenges the world is currently facing. In the most recent period, also influenced by important changes caused by the COVID-19 pandemic, society has gained more consciousness about these issues and has started asking its policy makers for relevant responses. Thus, sustainable development has become a primary necessity, not just a desirable eventuality.

Scientists have been undoubtedly among the first suggesting key strategies for a green future. In the field of analytical chemistry, researchers have specifically highlighted the importance of accessing sustainable, innovative, fast, and accurate techniques and technologies for VOCs' analysis alternatives to the traditional tools requiring expensive, long analysis, and that imply the disposal of large volumes of waste (e.g., solvents), such as mass spectrometry, adsorption/atomic emission spectroscopy or chromatography-based techniques (Table 3).

As described throughout this review, in recent years researchers have proposed novel sustainable sensors and biosensors for VOCs' detection for highly relevant applications and for the well-being of society. The monitoring of the toxicity of different environments (e.g., houses and schools), the control of the freshness and quality of foods, especially in meat and fish products, and the diagnosing of different diseases such as diabetes or cancer, are just some of the potential uses of these new devices.

Table 3. Advantages and disadvantages of classic tools and sustainable sensors and biosensors for the detection of VOCs.

	Advantages	Disadvantages
Classic methods (e.g., GC, HPLC, PTR, etc.)	High specificity; rapid separations; robust techniques	Matrix effects; high costs; higher maintenance; laborious sample preparation
Sustainable sensors and biosensors	Rapid response and recovery time; inexpensive; high sensitivity; small size; good precision; robustness	Temperature and humidity sensitive; high power consumption; short lifetime

Remarkable results have been obtained, but still there are important barriers to overcome, including optimizing the selectivity, the stability, the efficiency and the detection limit of these sensors and biosensors. For example, there are inorganic gases, pathogens, or compounds such as proteins, that can interact with the devices and interfere with their specific sensing actions, affecting selectivity. Thus, these devices are required to differentiate target substances from non-targets, showing high specificity and reducing non-specific interactions. Additionally, most of the sensors and biosensors were developed without performing a deep analysis of the production and utilization costs, which can be higher than the production and utilization costs of classic analytical tools. Finally, it must be highlighted that little effort has been given to deeper explore and investigate the end life of these sensors and biosensors, which should be considered a crucial point in the development of this type of device.

In future development, these issues can be addressed by exploiting the most recent advances in the technologies related to the different components of the sensors and biosensors. For example, the latest results in biotechnology are opening to the possibility of designing highly selective biosensors by tuning the affinity of the biological receptors to selected VOCs thanks to gene editing techniques [203,204]. Additionally, progress in microfabrication can lead to a substantial decrease in production costs, to large scale fabrication of nominally identical structures, and to the possibility of integrating different sensors and biosensors [205]. Lastly, to fully attain the sustainable characteristics needed for sustainable development, a life cycle assessment (LCA), claimed to be the best framework for assessing the potential environmental impacts of products [206], must be also determined for all sensors and biosensors before being brought to the market.

Forthcoming optimized VOC sensors and biosensors can be thus employed for the monitoring of thousands of environments and microenvironments by performing analyses at low costs and with high efficiency. This can have a tremendous impact on society, for example, by monitoring the quality of air in sensitive places such as schools and hospitals, or by making possible the massive control of food quality in the food supply chain, breaking down the food waste. The integration of the newest sensors and biosensors with innovative technologies will also potentially expand and integrate their use. For example, in combination with the Internet of Things (IOT), the sensors and biosensors can allow the real time monitoring of VOCs present in different places with communication among devices. This may result in the performing of corrective actions such as the activation of a ventilation mechanism in response to the reaching of a toxic concentration of a VOC in an environment. Additionally, integration with blockchain technology can provide information for producers, distributors and consumers about the origin, production, and traceability of food products within one portable, inexpensive, and compact device.

Author Contributions: Conceptualization, C.M.C. and A.Z.; writing—original draft preparation, C.M.C. and A.Z.; writing—review and editing, C.M.C., E.R., E.E., L.S. and A.Z. All authors have read and agreed to the published version of the manuscript.

Biosensors **2022**, *12*, 51

Funding: The authors would like to thank the Spanish Ministry of Economy, Industry and Competitiveness (Ramon y Cajal contract RYC-2015-17109) and Universidad de Córdoba (Predoctoral Grant 2019) for their financial support during this work.

Institutional Review Board Statement: Not applicable.

Data Availability Statement: Not applicable.

Conflicts of Interest: The authors declare no conflict of interest.

References

1. What Are Volatile Organic Compounds (VOCs)? Available online: https://www.epa.gov/indoor-air-quality-iaq/what-are-volatile-organic-compounds-vocs (accessed on 1 December 2021).
2. VOC. Available online: https://www.eea.europa.eu/themes/air/air-quality/resources/glossary/voc (accessed on 15 December 2021).
3. Menezes, H.C.; Amorim, L.C.A.; Cardeal, Z.L. Sampling and Analytical Methods for Determining VOC in Air by Biomonitoring Human Exposure. *Crit. Rev. Environ. Sci. Technol.* **2013**, *43*, 1–39. [CrossRef]
4. Zhang, G.X.; Feizbakhshan, M.; Zheng, S.L.; Hashisho, Z.; Sun, Z.M.; Liu, Y.Y. Effects of properties of minerals adsorbents for the adsorption and desorption of volatile organic compounds (VOC). *Appl. Clay Sci.* **2019**, *173*, 88–96. [CrossRef]
5. World Health Organization. Indoor Air Quality: Organic Pollutants. In Proceedings of the WHO Meeting, Berlin, Germany, 23–27 August 1987.
6. Hu, Y.; Li, Z.Y.; Wang, Y.T.; Wang, L.; Zhu, H.T.; Chen, L.; Guo, X.B.; An, C.X.; Jiang, Y.J.; Liu, A.Q. Emission Factors of NOx, SO2, PM and VOCs in Pharmaceuticals, Brick and Food Industries in Shanxi, China. *Aerosol Air Qual. Res.* **2019**, *19*, 1785–1797. [CrossRef]
7. Won, S.R.; Ghim, Y.S.; Kim, J.; Ryu, J.; Shim, I.K.; Lee, J. Volatile Organic Compounds in Underground Shopping Districts in Korea. *Int. J. Environ. Res. Public Health* **2021**, *18*, 5508. [CrossRef]
8. Lee, H.J.; Lee, K.H.; Kim, D.K. Evaluation and comparison of the indoor air quality in different areas of the hospital. *Medicine* **2020**, *99*, e23942. [CrossRef] [PubMed]
9. Lerner, J.E.C.; Gutierrez, M.D.; Mellado, D.; Giuliani, D.; Massolo, L.; Sanchez, E.Y.; Porta, A. Characterization and cancer risk assessment of VOCs in home and school environments in gran La Plata, Argentina. *Environ. Sci. Pollut. Res.* **2018**, *25*, 10039–10048. [CrossRef]
10. Jung, C.R.; Nishihama, Y.; Nakayama, S.F.; Tamura, K.; Isobe, T.; Michikawa, T.; Iwai-Shimada, M.; Kobayashi, Y.; Sekiyama, M.; Taniguchi, Y.; et al. Indoor air quality of 5000 households and its determinants. Part B: Volatile organic compounds and inorganic gaseous pollutants in the Japan Environment and Children's study. *Environ. Res.* **2021**, *197*, 111135. [CrossRef]
11. Inoue, M.; Mizuguchi, A.; Ueta, I.; Takahashi, K.; Saito, Y. Rapid on-site air sampling with a needle extraction device for evaluating the indoor air environment in school facilities. *Anal. Sci.* **2013**, *29*, 519–525. [CrossRef]
12. Adamova, T.; Hradecky, J.; Panek, M. Volatile Organic Compounds (VOCs) from Wood and Wood-Based Panels: Methods for Evaluation, Potential Health Risks, and Mitigation. *Polymers* **2020**, *12*, 2289. [CrossRef]
13. Ulker, O.C.; Ulker, O.; Hiziroglu, S. Volatile organic compounds (VOCs) emitted from coated furniture units. *Coatings* **2021**, *11*, 806. [CrossRef]
14. Liang, W.H.; Yang, S.; Yang, X.D. Long-Term Formaldehyde Emissions from Medium-Density Fiberboard in a Full-Scale Experimental Room: Emission Characteristics and the Effects of Temperature and Humidity. *Environ. Sci. Technol.* **2015**, *49*, 10349–10356. [CrossRef] [PubMed]
15. Rovelli, S.; Cattaneo, A.; Fazio, A.; Spinazze, A.; Borghi, F.; Campagnolo, D.; Dossi, C.; Cavallo, D.M. VOCs measurements in residential buildings: Quantification via thermal desorption and assessment of indoor concentrations in a case-study. *Atmosphere* **2019**, *10*, 57. [CrossRef]
16. Yang, F.C.; Wang, Y.; Li, H.L.; Yang, M.M.; Li, T.; Cao, F.F.; Chen, J.M.; Wang, Z. Influence of cloud/fog on atmospheric VOCs in the free troposphere: A case study at Mount Tai in Eastern China. *Aerosol Air Qual. Res.* **2017**, *17*, 2401–2412. [CrossRef]
17. Xiang, W.; Zhang, X.Y.; Chen, K.Q.; Fang, J.N.; He, F.; Hu, X.; Tsang, D.C.W.; Ok, Y.S.; Gao, B. Enhanced adsorption performance and governing mechanisms of ball-milled biochar for the removal of volatile organic compounds (VOCs). *Chem. Eng. J.* **2020**, *385*, 123842. [CrossRef]
18. Zhu, L.L.; Shen, D.K.; Luo, K.H. A critical review on VOCs adsorption by different porous materials: Species, mechanisms and modification methods. *J. Hazard. Mater.* **2020**, *389*, 122102. [CrossRef]
19. Zhu, L.; Deng, B.Q.; Guo, Y. A unified model for VOCs emission/sorption from/on building materials with and without ventilation. *Int. J. Heat Mass Transf.* **2013**, *67*, 734–740. [CrossRef]
20. Kang, I.S.; Xi, J.Y.; Hu, H.Y. Photolysis and photooxidation of typical gaseous VOCs by UV Irradiation: Removal performance and mechanisms. *Front. Environ. Sci. Eng.* **2018**, *12*, 8. [CrossRef]
21. Deng, B.Q.; Zhang, B.; Qiu, Y.F. Analytical solution of VOCs emission from wet materials with variable thickness. *Build. Environ.* **2016**, *104*, 145–151. [CrossRef]
22. Liu, Z.W.; Yan, Y.S.; Liu, T.T.; Zhao, Y.C.; Huang, Q.F.; Huang, Z.C. How to predict emissions of volatile organic compounds from solid building materials? A critical review on mass transfer models. *J. Environ. Manag.* **2022**, *302*, 114054. [CrossRef]

23. Zhang, L.; Zhu, X.Z.; Wang, Z.R.; Zhang, J.; Liu, X.; Zhao, Y. Improved speciation profiles and estimation methodology for VOCs emissions: A case study in two chemical plants in eastern China. *Environ. Pollut.* **2021**, *291*, 118192. [CrossRef] [PubMed]

24. Hunt, S.; Grau-Bove, J.; Schofield, E.; Gaisford, S. Effect of polyethylene glycol treatment on acetic acid emissions from wood. *Forests* **2021**, *12*, 1629. [CrossRef]

25. Even, M.; Girard, M.; Rich, A.; Hutzler, C.; Luch, A. Emissions of VOCs from polymer-based consumer products: From emission data of real samples to the assessment of inhalation exposure. *Front. Public Health* **2019**, *7*, 202. [CrossRef]

26. Jo, S.H.; Lee, M.H.; Kim, K.H.; Kumar, P. Characterization and flux assessment of airborne phthalates released from polyvinyl chloride consumer goods. *Environ. Res.* **2018**, *165*, 81–90. [CrossRef]

27. Noguchi, M.; Yamasaki, A. Volatile and semivolatile organic compound emissions from polymers used in commercial products during thermal degradation. *Heliyon* **2020**, *6*, e03314. [CrossRef] [PubMed]

28. Zhang, X.F.; Yin, Y.Y.; Wen, J.H.; Huang, S.L.; Han, D.M.; Chen, X.J.; Cheng, J.P. Characteristics, reactivity and source apportionment of ambient volatile organic compounds (VOCs) in a typical tourist city. *Atmos. Environ.* **2019**, *215*, 116898. [CrossRef]

29. Liu, Y.H.; Wang, H.L.; Jing, S.G.; Gao, Y.Q.; Peng, Y.R.; Lou, S.R.; Cheng, T.T.; Tao, S.K.; Li, L.; Li, Y.J.; et al. Characteristics and sources of volatile organic compounds (VOCs) in Shanghai during summer: Implications of regional transport. *Atmos. Environ.* **2019**, *215*, 116902. [CrossRef]

30. Chin, J.Y.; Batterman, S.A. VOC composition of current motor vehicle fuels and vapors, and collinearity analyses for receptor modeling. *Chemosphere* **2012**, *86*, 951–958. [CrossRef] [PubMed]

31. Tampio, E.A.; Blasco, L.; Vainio, M.M.; Kahala, M.M.; Rasi, S.E. Volatile fatty acids (VFAs) and methane from food waste and cow slurry: Comparison of biogas and VFA fermentation processes. *Glob. Chang. Biol. Bioenergy* **2019**, *11*, 72–84. [CrossRef]

32. Vivaldo, G.; Masi, E.; Taiti, C.; Caldarelli, G.; Mancuso, S. The network of plants volatile organic compounds. *Sci. Rep.* **2017**, *7*, 11050. [CrossRef]

33. Cova, C.M.; Zuliani, A.; Munoz-Batista, M.J.; Luque, R. Efficient Ru-based scrap waste automotive converter catalysts for the continuous-flow selective hydrogenation of cinnamaldehyde. *Green Chem.* **2019**, *21*, 4712–4722. [CrossRef]

34. Zuliani, A.; Cova, C.M.; Manno, R.; Sebastian, V.; Romero, A.A.; Luque, R. Continuous flow synthesis of menthol via tandem cyclisation-hydrogenation of citronellal catalysed by scrap catalytic converters. *Green Chem.* **2020**, *22*, 379–387. [CrossRef]

35. Brilli, F.; Loreto, F.; Baccelli, I. Exploiting plant volatile organic compounds (VOCs) in agriculture to improve sustainable defense strategies and productivity of crops. *Front. Plant Sci.* **2019**, *10*, 264. [CrossRef] [PubMed]

36. Boachon, B.; Lynch, J.H.; Ray, S.; Yuan, J.; Caldo, K.M.P.; Junker, R.R.; Kessler, S.A.; Morgan, J.A.; Dudareva, N. Natural fumigation as a mechanism for volatile transport between flower organs. *Nat. Chem. Biol.* **2019**, *15*, 583–588. [CrossRef]

37. Rios-Navarro, A.; Gonzalez, M.; Carazzone, C.; Ramirez, A.M.C. Learning about microbial language: Possible interactions mediated by microbial volatile organic compounds (VOCs) and relevance to understanding *Malassezia* spp. metabolism. *Metabolomics* **2021**, *17*, 39. [CrossRef]

38. D'Onofrio, C.; Knoll, W.; Pelosi, P. Aphid odorant-binding protein 9 is narrowly tuned to linear alcohols and aldehydes of sixteen carbon atoms. *Insects* **2021**, *12*, 741. [CrossRef] [PubMed]

39. Capuano, R.; Paba, E.; Mansi, A.; Marcelloni, A.M.; Chiominto, A.; Proietto, A.R.; Zampetti, E.; Macagnano, A.; Lvova, L.; Catini, A.; et al. Aspergillus species discrimination using a gas sensor array. *Sensors* **2020**, *20*, 4004. [CrossRef]

40. Weisskopf, L.; Schulz, S.; Garbeva, P. Microbial volatile organic compounds in intra-kingdom and inter-Kingdom interactions. *Nat. Rev. Microbiol.* **2021**, *19*, 391–404. [CrossRef] [PubMed]

41. Zou, Z.W.; He, J.Z.; Yang, X.D. An experimental method for measuring VOC emissions from individual human whole-body skin under controlled conditions. *Build. Environ.* **2020**, *181*, 107137. [CrossRef]

42. Janssens, E.; van Meerbeeck, J.P.; Lamote, K. Volatile organic compounds in human matrices as lung cancer biomarkers: A systematic review. *Crit. Rev. Oncol. Hematol.* **2020**, *153*, 103037. [CrossRef]

43. Course, C.; Watkins, W.J.; Muller, C.; Odd, D.; Kotecha, S.; Chakraborty, M. Volatile organic compounds as disease predictors in newborn infants: A systematic review. *J. Breath Res.* **2021**, *15*, 024002. [CrossRef] [PubMed]

44. Gouzerh, F.; Bessiere, J.M.; Ujvari, B.; Thomas, F.; Dujon, A.M.; Dormont, L. Odors and cancer: Current status and future directions. *Biochim. Et Biophys. Acta-Rev. Cancer* **2022**, *1877*, 188644. [CrossRef]

45. Xiang, L.J.; Wu, S.H.; Hua, Q.L.; Bao, C.Y.; Liu, H. Volatile organic compounds in human exhaled breath to diagnose gastrointestinal cancer: A meta-analysis. *Front. Oncol.* **2021**, *11*, 269. [CrossRef] [PubMed]

46. Corsini, L.; Castro, R.; Barroso, C.G.; Duran-Guerrero, E. Characterization by gas chromatography-olfactometry of the most odour-active compounds in Italian balsamic vinegars with geographical indication. *Food Chem.* **2019**, *272*, 702–708. [CrossRef] [PubMed]

47. Aznar, M.; Domeno, C.; Osorio, J.; Nerin, C. Release of volatile compounds from cooking plastic bags under different heating sources. *Food Packag. Shelf Life* **2020**, *26*, 100552. [CrossRef]

48. Zhou, T.; Feng, Y.Z.; Thomas-Danguin, T.; Zhao, M.M. Enhancement of saltiness perception by odorants selected from Chinese soy sauce: A gas chromatography/olfactometry-associated taste study. *Food Chem.* **2021**, *335*, 127664. [CrossRef]

49. Kumar, S.; Huang, J.Z.; Abbassi-Ghadi, N.; Spanel, P.; Smith, D.; Hanna, G.B. Selected Ion flow tube mass spectrometry analysis of exhaled breath for volatile organic compound profiling of esophago-gastric cancer. *Anal. Chem.* **2013**, *85*, 6121–6128. [CrossRef]

50. Yuan, B.; Koss, A.R.; Warneke, C.; Coggon, M.; Sekimoto, K.; de Gouw, J.A. Proton-transfer-reaction mass spectrometry: Applications in atmospheric sciences. *Chem. Rev.* **2017**, *117*, 13187–13229. [CrossRef]

51. Biasioli, F.; Yeretzian, C.; Gasperi, F.; Mark, T.D. PTR-MS monitoring of VOCs and BVOCs in food science and technology. *Trac-Trends Anal. Chem.* **2011**, *30*, 968–977. [CrossRef]

52. Spinelle, L.; Gerboles, M.; Kok, G.; Persijn, S.; Sauerwald, T. Review of portable and low-cost sensors for the ambient air monitoring of benzene and other volatile organic compounds. *Sensors* **2017**, *17*, 1520. [CrossRef]

53. Zheng, L.F.; Zhang, C.; Ma, J.; Hong, S.H.; She, Y.X.; Abd El-Aty, A.M.; He, Y.H.; Yu, H.L.; Liu, H.J.; Wang, J. Fabrication of a highly sensitive electrochemical sensor based on electropolymerized molecularly imprinted polymer hybrid nanocomposites for the determination of 4-nonylphenol in packaged milk samples. *Anal. Biochem.* **2018**, *559*, 44–50. [CrossRef]

54. Rasheed, T.; Hassan, A.A.; Kausar, F.; Sher, F.; Bilal, M.; Iqbal, H.M.N. Carbon nanotubes assisted analytical detection-Sensing/delivery cues for environmental and biomedical monitoring. *Trac-Trends Anal. Chem.* **2020**, *132*, 116066. [CrossRef]

55. Fazio, E.; Spadaro, S.; Corsaro, C.; Neri, G.; Leonardi, S.G.; Neri, F.; Lavanya, N.; Sekar, C.; Donato, N. Metal-oxide based nanomaterials: Synthesis, characterization and their applications in electrical and electrochemical sensors. *Sensors* **2021**, *21*, 2494. [CrossRef]

56. El Kazzy, M.; Weerakkody, J.S.; Hurot, C.; Mathey, R.; Buhot, A.; Scaramozzino, N.; Hou, Y.X. An overview of artificial olfaction systems with a focus on surface plasmon resonance for the analysis of volatile organic compounds. *Biosensors* **2021**, *11*, 244. [CrossRef] [PubMed]

57. Dobrzyniewski, D.; Szulczynski, B.; Dymerski, T.; Gebicki, J. Development of gas sensor array for methane reforming process monitoring. *Sensors* **2021**, *21*, 4983. [CrossRef]

58. Justino, C.I.L.; Duarte, A.C.; Rocha-Santos, T.A.P. Critical overview on the application of sensors and biosensors for clinical analysis. *Trac-Trends Anal. Chem.* **2016**, *85*, 36–60. [CrossRef] [PubMed]

59. Ratajczak, K.; Stobiecka, M. High-performance modified cellulose paper-based biosensors for medical diagnostics and early cancer screening: A concise review. *Carbohydr. Polym.* **2020**, *229*, 115463. [CrossRef] [PubMed]

60. Cesewski, E.; Johnson, B.N. Electrochemical biosensors for pathogen detection. *Biosens. Bioelectron.* **2020**, *159*, 112214. [CrossRef]

61. Majdinasab, M.; Mishra, R.K.; Tang, X.Q.; Marty, J.L. Detection of antibiotics in food: New achievements in the development of biosensors. *Trac-Trends Anal. Chem.* **2020**, *127*, 115883. [CrossRef]

62. Bhalla, N.; Pan, Y.W.; Yang, Z.G.; Payam, A.F. Opportunities and challenges for biosensors and nanoscale analytical tools for pandemics: COVID-19. *ACS Nano* **2020**, *14*, 7783–7807. [CrossRef]

63. Majhi, S.M.; Mirzaei, A.; Kim, H.W.; Kim, S.S.; Kim, T.W. Recent advances in energy-saving chemiresistive gas sensors: A review. *Nano Energy* **2021**, *79*, 105369. [CrossRef] [PubMed]

64. Jiang, Y.C.; Yin, S.; Dong, J.W.; Kaynak, O. A Review on soft sensors for monitoring, control, and optimization of industrial processes. *IEEE Sens. J.* **2021**, *21*, 12868–12881. [CrossRef]

65. Mani, V.; Beduk, T.; Khushaim, W.; Ceylan, A.E.; Timur, S.; Wolfbeis, O.S.; Salama, K.N. Electrochemical sensors targeting salivary biomarkers: A comprehensive review. *Trac-Trends Anal. Chem.* **2021**, *135*, 116164. [CrossRef]

66. Tajik, S.; Beitollahi, H.; Nejad, F.G.; Dourandish, Z.; Khalilzadeh, M.A.; Jang, H.W.; Venditti, R.A.; Varma, R.S.; Shokouhimehr, M. Recent developments in polymer nanocomposite-based electrochemical sensors for detecting environmental pollutants. *Ind. Eng. Chem. Res.* **2021**, *60*, 1112–1136. [CrossRef]

67. Nykamp, H. A transition to green buildings in Norway. *Environ. Innov. Soc. Transit.* **2017**, *24*, 83–93. [CrossRef]

68. Kemp-Benedict, E. Investing in a Green Transition. *Ecol. Econ.* **2018**, *153*, 218–236. [CrossRef]

69. Yang, Y.C.; Nie, P.Y.; Huang, J.B. The optimal strategies for clean technology to advance green transition. *Sci. Total Environ.* **2020**, *716*, 134439. [CrossRef]

70. Bogovic, N.D.; Grdic, Z.S. Transitioning to a green economy-possible effects on the croatian economy. *Sustainability* **2020**, *12*, 9342. [CrossRef]

71. Sharpe, S.A.; Martinez-Fernandez, C.M. The Implications of green employment: Making a just transition in ASEAN. *Sustainability* **2021**, *13*, 7389. [CrossRef]

72. Ohnesorge, L.; Rogge, E. Europe's green policy: Towards a climate neutral economy by way of investors' choice. *Eur. Co. Law* **2021**, *18*, 34–39.

73. Zuliani, A.; Cova, C.M. Green synthesis of heterogeneous visible-light-active photocatalysts: Recent advances. *Photochem* **2021**, *1*, 147–166. [CrossRef]

74. Green Transition. Available online: https://ec.europa.eu/reform-support/what-we-do/green-transition_en (accessed on 25 November 2021).

75. Zuliani, A.; Balu, A.M.; Luque, R. Efficient and Environmentally friendly microwave-assisted synthesis of catalytically active magnetic metallic Ni nanoparticles. *ACS Sustain. Chem. Eng.* **2017**, *5*, 11584–11587. [CrossRef]

76. Zuliani, A.; Ranjan, P.; Luque, R.; Van der Eycken, E.V. Heterogeneously catalyzed synthesis of imidazolones via cycloisomerizations of propargylic ureas using Ag and Au/Al SBA-15 systems. *ACS Sustain. Chem. Eng.* **2019**, *7*, 5568–5575. [CrossRef]

77. Cova, C.M.; Zuliani, A.; Manno, R.; Sebastian, V.; Luque, R. Scrap waste automotive converters as efficient catalysts for the continuous-flow hydrogenations of biomass derived chemicals. *Green Chem.* **2020**, *22*, 1414–1423. [CrossRef]

78. Zuliani, A.; Cano, M.; Calsolaro, F.; Puente Santiago, A.R.; Giner-Casares, J.J.; Rodríguez-Castellón, E.; Berlier, G.; Cravotto, G.; Martina, K.; Luque, R. Improving the electrocatalytic performance of sustainable Co/carbon materials for the oxygen evolution reaction by ultrasound and microwave assisted synthesis. *Sustain. Energy Fuels* **2021**, *5*, 720–731. [CrossRef]

79. Ivars-Barcelo, F.; Zuliani, A.; Fallah, M.; Mashkour, M.; Rahimnejad, M.; Luque, R. Novel applications of microbial fuel cells in sensors and biosensors. *Appl. Sci.* **2018**, *8*, 1184. [CrossRef]

80. Ghosh, R.; Gardner, J.W.; Guha, P.K. Air pollution monitoring using near room temperature resistive gas sensors: A review. *IEEE Trans. Electron Devices* **2019**, *66*, 3254–3264. [CrossRef]

81. Masikini, M.; Chowdhury, M.; Nemraoui, O. Review-metal oxides: Application in exhaled breath acetone chemiresistive sensors. *J. Electrochem. Soc.* **2020**, *167*, 037537. [CrossRef]

82. Wang, J.X.; Zhou, Q.; Peng, S.D.; Xu, L.N.; Zeng, W. Volatile organic compounds gas sensors based on molybdenum oxides: A mini review. *Front. Chem.* **2020**, *8*, 339. [CrossRef] [PubMed]

83. Li, H.Y.; Zhao, S.N.; Zang, S.Q.; Li, J. Functional metal-organic frameworks as effective sensors of gases and volatile compounds. *Chem. Soc. Rev.* **2020**, *49*, 6364–6401. [CrossRef] [PubMed]

84. Li, B.L.; Zhou, Q.; Peng, S.D.; Liao, Y.M. Recent advances of SnO_2-based sensors for detecting volatile organic compounds. *Front. Chem.* **2020**, *8*, 321. [CrossRef]

85. Tholl, D.; Hossain, O.; Weinhold, A.; Rose, U.S.R.; Wei, Q.S. Trends and applications in plant volatile sampling and analysis. *Plant J.* **2021**, *106*, 314–325. [CrossRef]

86. Zeng, H.; Zhang, G.Z.; Nagashima, K.; Takahashi, T.; Hosomi, T.; Yanagida, T. Metal-oxide nanowire molecular sensors and their promises. *Chemosensors* **2021**, *9*, 41. [CrossRef]

87. Tomic, M.; Setka, M.; Vojkuvka, L.; Vallejos, S. VOCs sensing by metal oxides, conductive polymers, and carbon-based materials. *Nanomaterials* **2021**, *11*, 552. [CrossRef]

88. Pelosi, P.; Mastrogiacomo, R.; Iovinella, I.; Tuccori, E.; Persaud, K.C. Structure and biotechnological applications of odorant-binding proteins. *Appl. Microbiol. Biotechnol.* **2014**, *98*, 61–70. [CrossRef]

89. Wasilewski, T.; Szulczynski, B.; Wojciechowski, M.; Kamysz, W.; Gebicki, J. Determination of long-chain aldehydes using a novel quartz crystal microbalance sensor based on a biomimetic peptide. *Microchem. J.* **2020**, *154*, 104509. [CrossRef]

90. Magna, G.; Muduganti, M.; Stefanelli, M.; Sivalingam, Y.; Zurlo, F.; Di Bartolomeo, E.; Catini, A.; Martinelli, E.; Paolesse, R.; Di Natale, C. Light-activated porphyrinoid-capped nanoparticles for gas sensing. *ACS Appl. Nano Mater.* **2021**, *4*, 414–424. [CrossRef]

91. Gong, Y.; Wei, Y.J.; Cheng, J.H.; Jiang, T.Y.; Chen, L.; Xu, B. Health risk assessment and personal exposure to Volatile Organic Compounds (VOCs) in metro carriages—A case study in Shanghai, China. *Sci. Total Environ.* **2017**, *574*, 1432–1438. [CrossRef] [PubMed]

92. Soni, V.; Singh, P.; Shree, V.; Goel, V. Effects of VOCs on human health. In *Air Pollution and Control. Energy, Environment, and Sustainability*; Springer: Singapore, 2018; pp. 119–142. [CrossRef]

93. Meciarova, L.; Vilcekova, S.; Burdova, E.K.; Kiselak, J. Factors effecting the total volatile organic compound (TVOC) concentrations in Slovak households. *Int. J. Environ. Res. Public Health* **2017**, *14*, 1443. [CrossRef] [PubMed]

94. Yang, S.; Yang, X.D.; Licina, D. Emissions of volatile organic compounds from interior materials of vehicles. *Build. Environ.* **2020**, *170*, 1443. [CrossRef]

95. Harb, P.; Locoge, N.; Thevenet, F. Emissions and treatment of VOCs emitted from wood-based construction materials: Impact on indoor air quality. *Chem. Eng. J.* **2018**, *354*, 641–652. [CrossRef]

96. Caron, F.; Guichard, R.; Robert, L.; Verriele, M.; Thevenet, F. Behaviour of individual VOCs in indoor environments: How ventilation affects emission from materials. *Atmos. Environ.* **2020**, *243*, 117713. [CrossRef]

97. Heroux, M.E.; Gauvin, D.; Gilbert, N.L.; Guay, M.; Dupuis, G.; Legris, M.; Levesque, B. Housing characteristics and indoor concentrations of selected volatile organic compounds (VOCs) in Quebec City, Canada. *Indoor Built Environ.* **2008**, *17*, 128–137. [CrossRef]

98. Toxicological Profiles. Available online: https://www.atsdr.cdc.gov/toxprofiledocs/index.html (accessed on 3 December 2021).

99. Lai, X.Y.; Cao, K.; Shen, G.X.; Xue, P.; Wang, D.; Hu, F.; Zhang, J.L.; Yang, Q.F.; Wang, X.Z. Ordered mesoporous $NiFe_2O_4$ with ultrathin framework for low-ppb toluene sensing. *Sci. Bull.* **2018**, *63*, 187–193. [CrossRef]

100. Cui, F.H.; Wang, G.; Yu, D.W.; Gan, X.D.; Tian, Q.H.; Guo, X.Y. Towards "zero waste" extraction of nickel from scrap nickel-based superalloy using magnesium. *J. Clean. Prod.* **2020**, *262*, 121275. [CrossRef]

101. Li, H.R.; Li, Y.; Cheng, H.L.; Yang, Q.R.; Xiong, J.H.; Ma, Y.J.; Ding, L.W.; Zeng, C.H. Lanthanide Metal-organic frameworks as luminescent sensor for toluene. *J. Inorg. Organomet. Polym. Mater.* **2020**, *30*, 2645–2653. [CrossRef]

102. Wang, J.L.; Wu, Z.L.; Chen, S.H.; Yuan, R.; Dong, L. A novel multifunctional fluorescent sensor based on N/S co-doped carbon dots for detecting Cr (VI) and toluene. *Microchem. J.* **2019**, *151*, 104246. [CrossRef]

103. Sarris, D.; Galiotou-Panayotou, M.; Koutinas, A.A.; Komaitis, M.; Papanikolaou, S. Citric acid, biomass and cellular lipid production by Yarrowia lipolytica strains cultivated on olive mill wastewater-based media. *J. Chem. Technol. Biotechnol.* **2011**, *86*, 1439–1448. [CrossRef]

104. Zhong, Z.; Fritzsche, M.; Pieper, S.B.; Wood, T.K.; Lear, K.L.; Dandy, D.S.; Reardon, K.F. Fiber optic monooxygenase biosensor for toluene concentration measurement in aqueous samples. *Biosens. Bioelectron.* **2011**, *26*, 2407–2412. [CrossRef] [PubMed]

105. Durmaz, M.; Acikbas, Y.; Bozkurt, S.; Capan, R.; Erdogan, M.; Ozkaya, C. A novel Calix 4 arene thiourea decorated with 2-(2-aminophenyl)benzothiazole moiety as highly selective chemical gas sensor for dichloromethane vapor. *ChemistrySelect* **2021**, *6*, 4670–4676. [CrossRef]

106. Kim, E.B.; Abdullah; Ameen, S.; Akhtar, M.S.; Shin, H.S. Environment-friendly and highly sensitive dichloromethane chemical sensor fabricated with ZnO nanopyramids-modified electrode. *J. Taiwan Inst. Chem. Eng.* **2019**, *102*, 143–152. [CrossRef]

107. Wang, H.Y.; Zhan, S.P.; Wu, X.F.; Wu, L.Q.; Liu, Y.X. Nanoporous fluorescent sensor based on upconversion nanoparticles for the detection of dichloromethane with high sensitivity. *RSC Adv.* **2021**, *11*, 565–571. [CrossRef]

108. Plekhanova, Y.V.; Firsova, Y.E.; Doronina, N.V.; Trotsenko, Y.A.; Reshetilov, A.N. Aerobic methylobacteria as the basis for a biosensor for dichloromethane detection. *Appl. Biochem. Microbiol.* **2013**, *49*, 188–193. [CrossRef]

109. Weston-Green, K.; Clunas, H.; Naranjo, C.J. A review of the potential use of pinene and linalool as terpene-based medicines for brain health: Discovering novel therapeutics in the flavours and fragrances of cannabis. *Front. Psychiatry* **2021**, *12*, 1309. [CrossRef]

110. Wen, T.; Sang, M.X.; Wang, M.L.; Han, L.B.; Gong, Z.L.; Tang, X.H.; Long, X.Z.; Xiong, H.; Peng, H.L. Rapid detection of d-limonene emanating from citrus infestation by *Bactrocera dorsalis* (Hendel) using a developed gas-sensing system based on QCM sensors coated with ethyl cellulose. *Sens. Actuators B Chem.* **2021**, *328*, 129048. [CrossRef]

111. Li, B.W.; Xu, C.Q.; Liu, L.; Yu, J.; Fan, Y.M. Facile and sustainable etherification of ethyl cellulose towards excellent UV blocking and fluorescence properties. *Green Chem.* **2021**, *23*, 479–489. [CrossRef]

112. Hawari, H.F.; Samsudin, N.M.; Shakaff, A.Y.M.; Wahab, Y.; Hashim, U.; Zakaria, A.; Ghani, S.A.; Ahmad, M.N. Highly selective molecular imprinted polymer (MIP) based sensor array using interdigitated electrode (IDE) platform for detection of mango ripeness. *Sens. Actuators B Chem.* **2013**, *187*, 434–444. [CrossRef]

113. Wu, Y.; Shetty, M.; Zhang, K.; Dauenhauer, P.J. Sustainable hybrid route to renewable methacrylic acid via biomass-derived citramalate. *ACS Eng. Au* **2021**. [CrossRef]

114. Hui, K.S.; Chao, C.Y.H. Synthesis of MCM-41 from coal fly ash by a green approach: Influence of synthesis pH. *J. Hazard. Mater.* **2006**, *137*, 1135–1148. [CrossRef]

115. Abu-Zied, B.M.; Alam, M.M.; Asiri, A.M.; Schwieger, W.; Rahman, M.M. Fabrication of 1,2-dichlorobenzene sensor based on mesoporous MCM-41 material. *Colloids Surf. A Physicochem. Eng. Asp.* **2019**, *562*, 161–169. [CrossRef]

116. Feng, L.; Dong, C.L.; Li, M.F.; Li, L.X.; Jiang, X.; Gao, R.; Wang, R.J.; Zhang, L.J.; Ning, Z.L.; Gao, D.J.; et al. Terbium-based metal-organic frameworks: Highly selective and fast respond sensor for styrene detection and construction of molecular logic gate. *J. Hazard. Mater.* **2020**, *388*, 121816. [CrossRef]

117. Alonso, S.; Navarro-Llorens, J.M.; Tormo, A.; Perera, J. Construction of a bacterial biosensor for styrene. *J. Biotechnol.* **2003**, *102*, 301–306. [CrossRef]

118. Yuan, H.Y.; Aljneibi, S.; Yuan, J.R.; Wang, Y.X.; Liu, H.; Fang, J.; Tang, C.H.; Yan, X.H.; Cai, H.; Gu, Y.D.; et al. ZnO nanosheets abundant in oxygen vacancies derived from metal-organic frameworks for ppb-level gas sensing. *Adv. Mater.* **2019**, *31*, e1807161. [CrossRef]

119. Jo, Y.K.; Jeong, S.Y.; Moon, Y.K.; Jo, Y.M.; Yoon, J.W.; Lee, J.H. Exclusive and ultrasensitive detection of formaldehyde at room temperature using a flexible and monolithic chemiresistive sensor. *Nat. Commun.* **2021**, *12*, 4955. [CrossRef] [PubMed]

120. Hu, J.Y.; Chen, X.Q.; Zhang, Y. Batch fabrication of formaldehyde sensors based on LaFeO$_3$ thin film with ppb-level detection limit. *Sens. Actuators B Chem.* **2021**, *349*, 130738. [CrossRef]

121. Xu, L.; Ge, M.Y.; Zhang, F.; Huang, H.J.; Sun, Y.; He, D.N. Nanostructured of SnO$_2$/NiO composite as a highly selective formaldehyde gas sensor. *J. Mater. Res.* **2020**, *35*, 3079–3090. [CrossRef]

122. Kang, Z.J.; Zhang, D.Z.; Li, T.T.; Liu, X.H.; Song, X.S. Polydopamine-modified SnO$_2$ nanofiber composite coated QCM gas sensor for high-performance formaldehyde sensing. *Sens. Actuators B Chem.* **2021**, *345*, 130299. [CrossRef]

123. Sun, Y.H.; Wang, J.; Du, H.Y.; Li, X.G.; Wang, C.; Hou, T.Y. Formaldehyde gas sensors based on SnO2/ZSM-5 zeolite composite nanofibers. *J. Alloys Compd.* **2021**, *868*, 159140. [CrossRef]

124. Chang, H.K.; Ko, D.S.; Cho, D.H.; Kim, S.; Lee, H.N.; Lee, H.S.; Kim, H.J.; Park, T.J.; Park, Y.M. Enhanced response of the photoactive gas sensor on formaldehyde using porous SnO$_2$@TiO$_2$ heterostructure driven by gas-flow thermal evaporation and atomic layer deposition. *Ceram. Int.* **2021**, *47*, 5985–5992. [CrossRef]

125. Jali, M.H.; Rahim, H.R.A.; Johari, M.A.M.; Ali, U.U.M.; Johari, S.H.; Mohamed, H.; Harun, S.W.; Yasin, M. Formaldehyde sensor with enhanced performance using microsphere resonator-coupled ZnO nanorods coated glass. *Opt. Laser Technol.* **2021**, *139*, 106853. [CrossRef]

126. Liu, J.; Chen, Y.; Zhang, H.Y. Study of highly sensitive formaldehyde sensors based on ZnO/CuO heterostructure via the sol-gel method. *Sensors* **2021**, *21*, 4685. [CrossRef]

127. Khojier, K. Preparation and investigation of Al-doped ZnO thin films as a formaldehyde sensor with extremely low detection limit and considering the effect of RH. *Mater. Sci. Semicond. Process.* **2021**, *121*, 105283. [CrossRef]

128. Huang, J.Y.; Liang, H.; Ye, J.X.; Jiang, D.T.; Sun, Y.L.; Li, X.J.; Geng, Y.F.; Wang, J.Q.; Qian, Z.F.; Du, Y. Ultrasensitive formaldehyde gas sensor based on Au-loaded ZnO nanorod arrays at low temperature. *Sens. Actuators B Chem.* **2021**, *346*, 130568. [CrossRef]

129. Zheng, Q.Y.; Zhang, H.; Liu, J.L.; Xiao, L.H.; Ao, Y.H.; Li, M. High porosity fluorescent aerogel with new molecular probes for formaldehyde gas sensors. *Microporous Mesoporous Mater.* **2021**, *325*, 111208. [CrossRef]

130. Gautam, V.; Kumar, A.; Kumar, R.; Jain, V.K.; Nagpal, S. Silicon nanowires/reduced graphene oxide nanocomposite based novel sensor platform for detection of cyclohexane and formaldehyde. *Mater. Sci. Semicond. Process.* **2021**, *123*, 105571. [CrossRef]

131. Wang, J.L.; Shang, J.H.; Guo, Y.J.; Jiang, Y.Y.; Xiong, W.K.; Li, J.S.; Yang, X.; Torun, H.; Fu, Y.Q.; Zu, X.T. Hydrophobic metal organic framework for enhancing performance of acoustic wave formaldehyde sensor based on polyethyleneimine and bacterial cellulose nanofilms. *J. Mater. Sci. Mater. Electron.* **2021**, *32*, 18551–18564. [CrossRef]

132. Padmalaya, G.; Vardhan, K.H.; Kumar, P.S.; Ali, M.A.; Chen, T.W. A disposable modified screen-printed electrode using egg white/ZnO rice structured composite as practical tool electrochemical sensor for formaldehyde detection and its comparative electrochemical study with Chitosan/ZnO nanocomposite. *Chemosphere* **2022**, *288*, 132560. [CrossRef]

133. Tsuchiya, K.; Furusawa, H.; Nomura, A.; Matsui, H.; Nihei, M.; Tokito, S. Formaldehyde detection by a combination of formaldehyde dehydrogenase and chitosan on a sensor based on an organic field-effect transistor. *Technologies* **2019**, *7*, 48. [CrossRef]

134. Vanderroost, M.; Ragaert, P.; Devlieghere, F.; De Meulenaer, B. Intelligent food packaging: The next generation. *Trends Food Sci. Technol.* **2014**, *39*, 47–62. [CrossRef]

135. Shao, P.; Liu, L.M.; Yu, J.H.; Lin, Y.; Gao, H.Y.; Chen, H.J.; Sun, P.L. An overview of intelligent freshness indicator packaging for food quality and safety monitoring. *Trends Food Sci. Technol.* **2021**, *118*, 285–296. [CrossRef]

136. Sohail, M.; Sun, D.W.; Zhu, Z.W. Recent developments in intelligent packaging for enhancing food quality and safety. *Crit. Rev. Food Sci. Nutr.* **2018**, *58*, 2650–2662. [CrossRef]

137. Azeredo, H.M.C.; Souza Correa, D. Smart choices: Mechanisms of intelligent food packaging. *Curr. Res. Food Sci.* **2021**, *4*, 932–936. [CrossRef] [PubMed]

138. Shukla, V.; Kandeepan, G.; Vishnuraj, M.R.; Soni, A. Anthocyanins based indicator sensor for intelligent packaging application. *Agric. Res.* **2016**, *5*, 205–209. [CrossRef]

139. Nguyen, L.H.; Naficy, S.; McConchie, R.; Dehghani, F.; Chandrawati, R. Polydiacetylene-based sensors to detect food spoilage at low temperatures. *J. Mater. Chem. C* **2019**, *7*, 1919–1926. [CrossRef]

140. Hashim, S.B.H.; Elrasheid Tahir, H.; Liu, L.; Zhang, J.; Zhai, X.; Ali Mahdi, A.; Nureldin Awad, F.; Hassan, M.M.; Xiaobo, Z.; Jiyong, S. Intelligent colorimetric pH sensoring packaging films based on sugarcane wax/agar integrated with butterfly pea flower extract for optical tracking of shrimp freshness. *Food Chem.* **2021**, *373*, 131514. [CrossRef] [PubMed]

141. Moustafa, H.; Morsy, M.; Ateia, M.A.; Abdel-Haleem, F.M. Ultrafast response humidity sensors based on polyvinyl chloride/graphene oxide nanocomposites for intelligent food packaging. *Sens. Actuators A-Phys.* **2021**, *331*, 112918. [CrossRef]

142. Abou Saoud, W.; Assadi, A.A.; Kane, A.; Jung, A.V.; Le Cann, P.; Gerard, A.; Bazantay, F.; Bouzaza, A.; Wolbert, D. Integrated process for the removal of indoor VOCs from food industry manufacturing: Elimination of Butane-2,3-dione and Heptan-2-one by cold plasma-photocatalysis combination. *J. Photochem. Photobiol. A Chem.* **2020**, *386*, 112071. [CrossRef]

143. Zhai, X.; Wang, X.; Zhang, J.; Yang, Z.; Sun, Y.; Li, Z.; Huang, X.; Holmes, M.; Gong, Y.; Povey, M.; et al. Extruded low density polyethylene-curcumin film: A hydrophobic ammonia sensor for intelligent food packaging. *Food Packag. Shelf Life* **2020**, *26*, 100595. [CrossRef]

144. Reji, R.P.; Marappan, G.; Sivalingam, Y.; Surya, V.J. VOCs adsorption induced surface potential changes on phthalocyanines: A combined experimental and theoretical approach towards food freshness monitoring. *Mater. Lett.* **2022**, *306*, 130945. [CrossRef]

145. Wang, A.; Stancik, C.M.; Yin, Y.; Wu, J.; Duncan, S.E. Performance of cost-effective PET packaging with light protective additives to limit photo-oxidation in UHT milk under refrigerated LED-lighted storage condition. *Food Packag. Shelf Life* **2022**, *31*, 100773. [CrossRef]

146. Preis, S.; Klauson, D.; Gregor, A. Potential of electric discharge plasma methods in abatement of volatile organic compounds originating from the food industry. *J. Environ. Manag.* **2013**, *114*, 125–138. [CrossRef]

147. Fortin, C.; Goodwin, H.L.; Thomsen, M. Consumer attitudes toward freshness indicators on perishable food products. *J. Food Distrib. Res.* **2009**, *40*, 1–15.

148. Park, Y.W.; Kim, S.M.; Lee, J.Y.; Jang, W. Application of biosensors in smart packaging. *Mol. Cell. Toxicol.* **2015**, *11*, 277–285. [CrossRef]

149. Kuswandi, B.; Damayanti, F.; Abdullah, A.; Heng, L.Y. Simple and low-cost on-package sticker sensor based on litmus paper for real-time monitoring of beef freshness. *J. Math. Fundam. Sci.* **2015**, *47*, 236–251. [CrossRef]

150. Alessandroni, L.; Caprioli, G.; Faiella, F.; Fiorini, D.; Galli, R.; Huang, X.H.; Marinelli, G.; Nzekoue, F.; Ricciutelli, M.; Scortichini, S.; et al. A shelf-life study for the evaluation of a new biopackaging to preserve the quality of organic chicken meat. *Food Chem.* **2022**, *371*, 131134. [CrossRef] [PubMed]

151. Sankaran, S.; Panigrahi, S.; Mallik, S. Odorant binding protein based biomimetic sensors for detection of alcohols associated with Salmonella contamination in packaged beef. *Biosens. Bioelectron.* **2011**, *26*, 3103–3109. [CrossRef] [PubMed]

152. Sankaran, S.; Panigrahi, S.; Mallik, S. Olfactory receptor based piezoelectric biosensors for detection of alcohols related to food safety applications. *Sens. Actuators B Chem.* **2011**, *155*, 8–18. [CrossRef]

153. Panigrahi, S.; Sankaran, S.; Mallik, S.; Gaddam, B.; Hanson, A.A. Olfactory receptor-based polypeptide sensor for acetic acid VOC detection. *Mater. Sci. Eng. C Mater. Biol. Appl.* **2012**, *32*, 1307–1313. [CrossRef] [PubMed]

154. Han, D.Q. Sol-gel autocombustion synthesis of zinc oxide foam decorated with holes and its use as acetic acid gas sensor at sub-ppm level. *Ceram. Int.* **2020**, *46*, 3304–3310. [CrossRef]

155. Senapati, M.; Sahu, P.P. Meat quality assessment using Au patch electrode Ag-SnO2/SiO2/Si MIS capacitive gas sensor at room temperature. *Food Chem.* **2020**, *324*, 126893. [CrossRef]

156. ACGIH. Available online: https://www.acgih.org/ (accessed on 20 November 2021).

157. Kim, G.; Cho, B.; Oh, S.H.; Kim, K. Feasibility study for the evaluation of chicken meat storage time using surface acoustic wave sensor. *J. Biosyst. Eng.* **2020**, *45*, 261–271. [CrossRef]
158. Chow, C.-F. Bimetallic-based food sensors for meat spoilage: Effects of the accepting metallic unit in Fe(II)-C N-M-A (M-A = Pt(II) or Au(I)) on device selectivity and sensitivity. *Food Chem.* **2019**, *300*, 125190. [CrossRef] [PubMed]
159. Odeyemi, O.A.; Burke, C.M.; Bolch, C.C.J.; Stanley, R. Seafood spoilage microbiota and associated volatile organic compounds at different storage temperatures and packaging conditions. *Int. J. Food Microbiol.* **2018**, *280*, 87–99. [CrossRef] [PubMed]
160. Perillo, P.M.; Rodriguez, D.F. Low temperature trimethylamine flexible gas sensor based on TiO_2 membrane nanotubes. *J. Alloy. Compd.* **2016**, *657*, 765–769. [CrossRef]
161. Yang, T.Y.; Du, L.Y.; Zhai, C.B.; Li, Z.F.; Zhao, Q.; Luo, Y.; Xing, D.J.; Zhang, M.Z. Ultrafast response and recovery trimethylamine sensor based on alpha-Fe_2O_3 snowflake-like hierarchical architectures. *J. Alloy. Compd.* **2017**, *718*, 396–404. [CrossRef]
162. Liu, L.K.; Fu, S.; Lv, X.; Yue, L.L.; Fan, L.; Yu, H.T.; Gao, X.L.; Zhu, W.B.; Zhang, W.; Li, X.; et al. A gas sensor with Fe_2O_3 nanospheres based on trimethylamine detection for the rapid assessment of spoilage degree in fish. *Front. Bioeng. Biotechnol.* **2020**, *8*, 567584. [CrossRef] [PubMed]
163. Shen, J.; Xu, S.; Zhao, C.; Qiao, X.; Liu, H.; Zhao, Y.; Wei, J.; Zhu, Y. Bimetallic Au@Pt nanocrystal sensitization mesoporous α-FeO_3 hollow nanocubes for highly sensitive and rapid detection of fish freshness at low temperature. *ACS Appl. Mater. Interfaces* **2021**, *13*, 57597–57608. [CrossRef]
164. Lv, R.Q.; Huang, X.Y.; Ye, W.T.; Aheto, J.H.; Xu, H.X.; Dai, C.X.; Tian, X.Y. Research on the reaction mechanism of colorimetric sensor array with characteristic volatile gases-TMA during fish storage. *J. Food Process Eng.* **2019**, *42*, e12952. [CrossRef]
165. Sun, Y.; Wen, J.W.; Chen, Z.J.; Qiu, S.B.; Wang, Y.X.; Yin, E.Q.; Li, H.B.; Liu, X.H. Non-destructive and rapid method for monitoring fish freshness of grass carp based on printable colorimetric paper sensor in modified atmosphere packaging. *Food Anal. Methods* **2021**, 1–11. [CrossRef]
166. Capello, C.; Trevisol, T.C.; Pelicioli, J.; Terrazas, M.B.; Monteiro, A.R.; Valencia, G.A. Preparation and Characterization of Colorimetric Indicator Films Based on Chitosan/Polyvinyl Alcohol and Anthocyanins from Agri-Food Wastes. *J. Polym. Environ.* **2021**, *29*, 1616–1629. [CrossRef]
167. Lai, F.Y.; Yang, J.L.; Huang, R.R.; Wang, Z.L.; Tang, J.Q.; Zhang, M.L.; Miao, R.; Fang, Y. Nondestructive evaluation of fish freshness through nanometer-thick fluorescence-based amine-sensing films. *ACS Appl. Nano Mater.* **2021**, *4*, 2575–2582. [CrossRef]
168. Praoboon, N.; Siriket, S.; Taokaenchan, N.; Kuimalee, S.; Phaisansuthichol, S.; Pookmanee, P.; Satienperakul, S. Paper-based electrochemiluminescence device for the rapid estimation of trimethylamine in fish via the quenching effect of thioglycolic acid-capped cadmium selenide quantum dots. *Food Chem.* **2022**, *366*, 130590. [CrossRef]
169. Jia, Z.X.; Shi, C.; Wang, Y.B.; Yang, X.T.; Zhang, J.R.; Ji, Z.T. Nondestructive determination of salmon fillet freshness during storage at different temperatures by electronic nose system combined with radial basis function neural networks. *Int. J. Food Sci. Technol.* **2020**, *55*, 2080–2091. [CrossRef]
170. Chen, W.; Wang, Z.H.; Gu, S.; Wang, J.; Wang, Y.W.; Wei, Z.B. Hydrophobic amino-functionalized graphene oxide nanocomposite for aldehydes detection in fish fillets. *Sens. Actuators B Chem.* **2020**, *306*, 127579. [CrossRef]
171. Public Health and Environment. Available online: https://www.who.int/data/gho/data/themes/public-health-and-environment (accessed on 3 December 2021).
172. Gardner, J.W.; Vincent, T.A. Electronic Noses for Well-Being: Breath Analysis and Energy Expenditure. *Sensors* **2016**, *16*, 947. [CrossRef]
173. Terrington, D.; Hayton, C.; Peel, A.; Fowler, S.; Wilson, A. Exhaled breath biomarkers in sarcoidosis: A systematic review. *Eur. Respir. J.* **2018**, *52*, PA3003. [CrossRef]
174. Peel, A.M.; Wilkinson, M.; Sinha, A.; Loke, Y.K.; Fowler, S.J.; Wilson, A.M. Volatile organic compounds associated with diagnosis and disease characteristics in asthma—A systematic review. *Respir. Med.* **2020**, *169*, 105984. [CrossRef]
175. Di Zazzo, L.; Magna, G.; Lucentini, M.; Stefanelli, M.; Paolesse, R.; Di Natale, C. Sensor-embedded face masks for detection of volatiles in breath: A proof of concept study. *Chemosensors* **2021**, *9*, 356. [CrossRef]
176. Li, D.S.; Shao, Y.Z.; Zhang, Q.; Qu, M.J.; Ping, J.F.; Fu, Y.Q.; Xie, J. A flexible virtual sensor array based on laser-induced graphene and MXene for detecting volatile organic compounds in human breath. *Analyst* **2021**, *146*, 5704–5713. [CrossRef] [PubMed]
177. Lee, B.M.; Eetemadi, A.; Tagkopoulos, I. Reduced Graphene Oxide-Metalloporphyrin Sensors for Human Breath Screening. *Appl. Sci.* **2021**, *11*, 11290. [CrossRef]
178. Tyagi, H.; Daulton, E.; Bannaga, A.S.; Arasaradnam, R.P.; Covington, J.A. Non-invasive detection and staging of colorectal cancer using a portable electronic nose. *Sensors* **2021**, *21*, 5440. [CrossRef]
179. Jalal, A.H.; Alam, F.; Roychoudhury, S.; Umasankar, Y.; Pala, N.; Bhansali, S. Prospects and challenges of volatile organic compound sensors in human healthcare. *ACS Sens.* **2018**, *3*, 1246–1263. [CrossRef]
180. Capuano, R.; Catini, A.; Paolesse, R.; Di Natale, C. Sensors for lung cancer diagnosis. *J. Clin. Med.* **2019**, *8*, 235. [CrossRef]
181. Sanchez-Vicente, C.; Santos, J.P.; Lozano, J.; Sayago, I.; Sanjurjo, J.L.; Azabal, A.; Ruiz-Valdepenas, S. Graphene-doped Tin oxide nanofibers and nanoribbons as gas sensors to detect biomarkers of different diseases through the breath. *Sensors* **2020**, *20*, 7223. [CrossRef] [PubMed]
182. Hanh, N.H.; Ngoc, T.M.; Duy, L.V.; Hung, C.M.; Duy, N.V.; Hoa, N.D. A comparative study on the VOCs gas sensing properties of Zn_2SnO_4 nanoparticles, hollow cubes, and hollow octahedra towards exhaled breath analysis. *Sens. Actuators B Chem.* **2021**, *343*, 130147. [CrossRef]

183. Rodriguez-Aguilar, M.; de Leon-Martinez, L.D.; Gorocica-Rosete, P.; Perez-Padilla, R.; Dominguez-Reyes, C.A.; Tenorio-Torres, J.A.; Ornelas-Rebolledo, O.; Mehta, G.; Zamora-Mendoza, B.N.; Flores-Ramirez, R. Application of chemoresistive gas sensors and chemometric analysis to differentiate the fingerprints of global volatile organic compounds from diseases. Preliminary results of COPD, lung cancer and breast cancer. *Clin. Chim. Acta* **2021**, *518*, 83–92. [CrossRef] [PubMed]

184. Zaim, O.; Diouf, A.; El Bari, N.; Lagdali, N.; Benelbarhdadi, I.; Ajana, F.Z.; Llobet, E.; Bouchikhi, B. Comparative analysis of volatile organic compounds of breath and urine for distinguishing patients with liver cirrhosis from healthy controls by using electronic nose and voltammetric electronic tongue. *Anal. Chim. Acta* **2021**, *1184*, 339028. [CrossRef] [PubMed]

185. Wu, X.Y.; Wang, H.R.; Wang, J.H.; Wang, D.Z.; Shi, L.J.; Tian, X.; Sun, J.H. VOCs gas sensor based on MOFs derived porous Au@Cr$_2$O$_3$-In$_2$O$_3$ nanorods for breath analysis. *Colloids Surf. A Physicochem. Eng. Asp.* **2022**, *632*, 127752. [CrossRef]

186. Guo, W.W.; Huang, L.L.; Zhao, B.Y.; Gao, X.; Fan, Z.H.; Liu, X.Y.; He, Y.Z.; Zhang, J. Synthesis of the ZnFe$_2$O$_4$/ZnSnO$_3$ nanocomposite and enhanced gas sensing performance to acetone. *Sens. Actuators B Chem.* **2021**, *346*, 130524. [CrossRef]

187. Yoon, J.W.; Lee, J.H. Toward breath analysis on a chip for disease diagnosis using semiconductor-based chemiresistors: Recent progress and future perspectives. *Lab Chip* **2017**, *17*, 3537–3557. [CrossRef]

188. Guo, W.W.; Huang, L.L.; Zhang, J.; He, Y.Z.; Zeng, W. Ni-doped SnO$_2$/g-C$_3$N$_4$ nanocomposite with enhanced gas sensing performance for the effective detection of acetone in diabetes diagnosis. *Sens. Actuators B Chem.* **2021**, *334*, 129666. [CrossRef]

189. Zhang, S.F.; Jiang, W.H.; Li, Y.W.; Yang, X.L.; Sun, P.; Liu, F.M.; Yan, X.; Gao, Y.; Liang, X.H.; Ma, J.; et al. Highly-sensitivity acetone sensors based on spinel-type oxide (NiFe$_2$O$_4$) through optimization of porous structure. *Sens. Actuators B Chem.* **2019**, *291*, 266–274. [CrossRef]

190. Hu, J.; Yang, J.; Wang, W.D.; Xue, Y.; Sun, Y.J.; Li, P.W.; Lian, K.; Zhang, W.D.; Chen, L.; Shi, J.; et al. Synthesis and gas sensing properties of NiO/SnO$_2$ hierarchical structures toward ppb-level acetone detection. *Mater. Res. Bull.* **2018**, *102*, 294–303. [CrossRef]

191. Du, H.Y.; Li, X.G.; Yao, P.J.; Wang, J.; Sun, Y.H.; Dong, L. Zinc oxide coated Tin oxide nanofibers for improved selective acetone sensing. *Nanomaterials* **2018**, *8*, 509. [CrossRef]

192. Hu, J.; Zou, C.; Su, Y.J.; Li, M.; Yang, Z.; Ge, M.Y.; Zhang, Y.F. One-step synthesis of 2D C$_3$N$_4$-tin oxide gas sensors for enhanced acetone vapor detection. *Sens. Actuators B Chem.* **2017**, *253*, 641–651. [CrossRef]

193. Li, L.; Tan, J.F.; Dun, M.H.; Huang, X.T. Porous ZnFe2O4 nanorods with net-worked nanostructure for highly sensor response and fast response acetone gas sensor. *Sens. Actuators B Chem.* **2017**, *248*, 85–91. [CrossRef]

194. Ao, W.Y.; Fu, J.; Mao, X.; Kang, Q.H.; Ran, C.M.; Liu, Y.; Zhang, H.D.; Gao, Z.P.; Li, J.; Liu, G.Q.; et al. Microwave assisted preparation of activated carbon from biomass: A review. *Renew. Sustain. Energy Rev.* **2018**, *92*, 958–979. [CrossRef]

195. Cova, C.M.; Zuliani, A.; Santiago, A.R.P.; Caballero, A.; Munoz-Batista, M.J.; Luque, R. Microwave-assisted preparation of Ag/Ag$_2$S carbon hybrid structures from pig bristles as efficient HER catalysts. *J. Mater. Chem. A* **2018**, *6*, 21516–21523. [CrossRef]

196. Zuliani, A.; Munoz-Batista, M.J.; Luque, R. Microwave-assisted valorization of pig bristles: Towards visible light photocatalytic chalcocite composites. *Green Chem.* **2018**, *20*, 3001–3007. [CrossRef]

197. Sun, X.H.; Shao, K.; Wang, T. Detection of volatile organic compounds (VOCs) from exhaled breath as noninvasive methods for cancer diagnosis. *Anal. Bioanal. Chem.* **2016**, *408*, 2759–2780. [CrossRef] [PubMed]

198. Hakim, M.; Broza, Y.Y.; Barash, O.; Peled, N.; Phillips, M.; Amann, A.; Haick, H. Volatile organic compounds of lung cancer and possible biochemical pathways. *Chem. Rev.* **2012**, *112*, 5949–5966. [CrossRef]

199. Behera, B.; Joshi, R.; Vishnu, G.K.A.; Bhalerao, S.; Pandya, H.J. Electronic nose: A non-invasive technology for breath analysis of diabetes and lung cancer patients. *J. Breath Res.* **2019**, *13*, 024001. [CrossRef]

200. Tripathi, K.M.; Sachan, A.; Castro, M.; Choudhary, V.; Sonkar, S.K.; Feller, J.F. Green carbon nanostructured quantum resistive sensors to detect volatile biomarkers. *Sustain. Mater. Technol.* **2018**, *16*, 1–11. [CrossRef]

201. Hassan, K.; Hossain, R.; Farzana, R.; Sahajwalla, V. Microrecycled zinc oxide nanoparticles (ZnO NP) recovered from spent Zn-C batteries for VOC detection using ZnO sensor. *Anal. Chim. Acta* **2021**, *1165*, 338563. [CrossRef]

202. Zito, C.A.; Perfecto, T.M.; Oliveira, T.N.T.; Volanti, D.P. Bicone-like ZnO structure as high-performance butanone sensor. *Mater. Lett.* **2018**, *223*, 142–145. [CrossRef]

203. Pu, J.Y.; Kentala, K.; Dickinson, B.C. Multidimensional control of Cas9 by evolved RNA polymerase-based biosensors. *ACS Chem. Biol.* **2018**, *13*, 431–437. [CrossRef] [PubMed]

204. Xie, S.Y.; Ji, Z.R.; Suo, T.Y.; Li, B.Z.; Zhang, X. Advancing sensing technology with CRISPR: From the detection of nucleic acids to a broad range of analytes e A review. *Anal. Chim. Acta* **2021**, *1185*, 338848. [CrossRef] [PubMed]

205. Baracu, A.M.; Gugoasa, L.A.D. Review-recent advances in microfabrication, design and applications of amperometric sensors and biosensors. *J. Electrochem. Soc.* **2021**, *168*, 037503. [CrossRef]

206. Kleinekorte, J.; Fleitmann, L.; Bachmann, M.; Katelhon, A.; Barbosa-Povoa, A.; von der Assen, N.; Bardow, A. Life cycle assessment for the design of chemical processes, products, and supply chains. *Annu. Rev. Chem. Biomol. Eng.* **2020**, *11*, 203–233. [CrossRef] [PubMed]

biosensors

Review

Rolling Circle Amplification as an Efficient Analytical Tool for Rapid Detection of Contaminants in Aqueous Environments

Kuankuan Zhang [1,2], **Hua Zhang** [1,*], **Haorui Cao** [1,2], **Yu Jiang** [1], **Kang Mao** [1,*] **and Zhugen Yang** [3]

1   State Key Laboratory of Environmental Geochemistry, Institute of Geochemistry,
     Chinese Academy of Sciences, Guiyang 550081, China; zhangkuankuan@mail.gyig.ac.cn (K.Z.);
     caohaorui@mail.gyig.ac.cn (H.C.); jiangyu@mail.gyig.ac.cn (Y.J.)
2   University of Chinese Academy of Sciences, Beijing 100049, China
3   Cranfield Water Science Institute, Cranfield University, Cranfield MK43 0AL, UK;
     Zhugen.Yang@cranfield.ac.uk
*   Correspondence: zhanghua@mail.gyig.ac.cn (H.Z.); maokang@mail.gyig.ac.cn (K.M.)

Abstract: Environmental contaminants are a global concern, and an effective strategy for remediation is to develop a rapid, on-site, and affordable monitoring method. However, this remains challenging, especially with regard to the detection of various contaminants in complex water environments. The application of molecular methods has recently attracted increasing attention; for example, rolling circle amplification (RCA) is an isothermal enzymatic process in which a short nucleic acid primer is amplified to form a long single-stranded nucleic acid using a circular template and special nucleic acid polymerases. Furthermore, this approach can be further engineered into a device for point-of-need monitoring of environmental pollutants. In this paper, we describe the fundamental principles of RCA and the advantages and disadvantages of RCA assays. Then, we discuss the recently developed RCA-based tools for environmental analysis to determine various targets, including heavy metals, organic small molecules, nucleic acids, peptides, proteins, and even microorganisms in aqueous environments. Finally, we summarize the challenges and outline strategies for the advancement of this technique for application in contaminant monitoring.

Keywords: rolling circle amplification; environmental monitoring; heavy metals; organic molecules; microorganisms

Citation: Zhang, K.; Zhang, H.; Cao, H.; Jiang, Y.; Mao, K.; Yang, Z. Rolling Circle Amplification as an Efficient Analytical Tool for Rapid Detection of Contaminants in Aqueous Environments. *Biosensors* **2021**, *11*, 352. https://doi.org/10.3390/bios 11100352

Received: 12 August 2021
Accepted: 21 September 2021
Published: 23 September 2021

Publisher's Note: MDPI stays neutral with regard to jurisdictional claims in published maps and institutional affiliations.

1. Introduction

In recent years, the discharge of contaminants from industrial and agricultural activities and urban wastewater has caused serious contamination of the aqueous system, posing a great potential threat to human health and aquatic life. These contaminants can be divided into three categories: (i) inorganic chemical substances, (ii) organic pollutants and (iii) microorganisms. These substances can cause adverse effects on the environment [1–4], for example, the disruption of hormones and the endocrine system and the induction of cytotoxicity and/or genotoxicity and carcinogenesis [5,6]. The variable composition of pollutants and their location in aqueous environments over time have resulted in increasing focus on new technologies that use cheap and real-time strategies to monitor pollutants. Most of these strategies are based on laboratory platforms, such as inductively coupled plasma mass spectrometry (ICP-MS) for the detection of heavy metal ions, liquid chromatography-tandem mass spectrometry (LC-MS) for the detection of small organic chemicals or their metabolites, and polymerase chain reaction (PCR) for the detection of nucleic acids and genetic information, which require preprocessing and frequent data sampling, which means that they are both expensive and slow. These aspects highlight the need to develop a new strategy that is more sensitive, portable, and efficient for on-site detection of pollutants composed of multiple substances [7–10].

Recently, rolling circle amplification (RCA)-based analytical methods have received increasing attention in environmental monitoring. RCA is an uncomplicated and efficient isothermal enzymatic process using unique DNA and RNA polymerases to produce long single-stranded DNA (ssDNA) and RNA [11,12]. In RCA, the polymerase will spontaneously and continuously add nucleotides to the primers that bind to the circular template, generating long ssDNA with tandem repeats of tens to hundreds of orders of magnitude. Unlike PCR, which requires a thermal cycler and thermostable DNA polymerase. RCA can be in solution, on a solid support, or in a complex biological environment at a constant temperature (room temperature to 37 °C). The ability of RCA to grow a long DNA chain on a solid support or inside a cell from one molecular binding event enables the detection of targets at a single molecule level [13–15]. In addition, an RCA product comprising repeating cyclic sequences complementary to template DNA can be customized by template design. By designing the template, the customizable DNA product includes functional sequences, including aptamers, DNAzymes, spacer domains, and restriction endonuclease sites. Of course, by hybridizing the RCA product with a complementary nucleic acid linked to a functional part including biotin [16,17], fluorophores [18,19], antibodies [20], and nanoparticles [21–24], it is easy to synthesize a multifunctional material with a variety of properties, including biorecognition and biosensing. Collectively, the properties of high-efficiency isothermal amplification, single-molecule sensitivity, versatility of structure and composition, and multivalences make RCA a powerful tool in aqueous environments [25–27]. Currently, RCA has been extensively studied to develop sensitive methods for detecting DNA, RNA, DNA methylation, single nucleotide polymorphisms, small molecules, proteins, and cells. In addition to diagnosis, RCA has also been proven to be effective for cell-free cloning and sequencing [28,29], in situ genotyping and genome-wide analysis of cells and tissues [30–34]. Recently, RCA has received widespread attention for its use in the production of DNA nanostructures such as origami, nanoribbons, nanotubes, DNA nanoscaffolds, and DNA metamaterials for periodic nanocomponents [11,35–38]. Importantly, these materials have high prospects in a wide range of applications, including environmental monitoring, drug delivery, and in vivo imaging of manufacturing electronic circuits, including DNA-based materials.

In this article, we outline the basic engineering principles for implementing RCA design. Then, we discuss the progress of research in the last five years using RCA technology to analyse various pollutants in the water environment. Taking four kinds of common analytes in the environment as examples, including heavy metal ions, organic molecules, biological macromolecules, bacteria, and other microorganisms, the application of RCA for water environment monitoring is discussed. Among the analytes, biomacromolecules are divided into nucleic acids, lipids, peptides, and proteins. Finally, the contents of this paper are summarized, and the application prospects of RCA-based analytical methods in environmental monitoring are discussed.

2. Advantages and Disadvantages of the RCA Assay

2.1. Fundamentals of RCA

The RCA reaction typically requires four components: (1) DNA polymerase (e.g., Phi29 DNA polymerase), which includes an appropriate buffer; (2) a short nucleic acid primer; (3) a circular DNA template; and (4) deoxynucleotide triphosphate (dNTP) (monomer or structural unit of RCA product) [11,39–41]. In polymerases, Phi29 DNA polymerase is most commonly used because of its excellent capability and continuous strand displacement synthesis capability. Phi29 can handle topological constraints, four-way cross connections, and multiple circular DNA template complexes [39,42,43]. For RCA primers, both RNA and DNA (usually the "target" molecule to be detected) can achieve this goal. Indeed, the target DNA and RNA can be used to connect the first template mediated as a padlock probe (PLP) using RCA reaction circular template cyclizing [44,45]. The circular DNA template (usually 15–200 nucleotides (NT) in length) is a component that can be enzymatically or chemically synthesized through intramolecular phosphate and hydroxyl end groups.

Most commonly, the template is a circular DNA template mediated by enzymatic ligation (e.g., T4 DNA ligase) or the use of a special DNA ligase enzyme with a template-free connection to a synthetic CircLigase [15,46]. By designing primers and circular templates, RCA product length, sequence, composition, structure and rigidity may be appropriately adjusted, thereby becoming a highly versatile RCA technique (summarized in Table 1).

Table 1. Comparison of RCA, PCR and real-time PCR for the detection of DNA.

Features	Conventional PCR Assay	Real Time-PCR Assay	RCA Assay
Sensitivity	Sensitive	Highly sensitive	Highly sensitive
Specificity	Specific	Specific	Specific
Temperature conditions	Thermal cycle	Thermal cycle	Isothermal
Inhibition by biological samples	Yes	Yes	No
Instruments required	Thermocycler	Thermocycler	Not required
Post-assay analysis	Required	Required	Generally not required
Amplicon detection methods	Gel electrophoresis	Real-time detection/ amplification graph	Gel electrophoresis, Turbidity measurement by visual inspection or using a real-time turbidimeter; dye-based visual detection
Qualitative detection	Yes	Yes	Yes
Quantitative detection	No	Yes	Semi-quantitative
Portability	Partially	Yes	Yes
Overall assay time	3–5 h	2.5–4 h	1–1.5 h
Cost effectiveness	Less expensive	Expensive	Less expensive

2.2. Exponential RCA Amplification

One of the powerful functions of RCA is the ability to design a circular template so that the signal generated by a single binding event is exponentially amplified [47–49]. Using a plurality of primers hybridizing to the same ring can lead to amplification of a plurality of events, thereby producing a plurality of RCA products [50,51] (Figure 1). The number of primers that one circular template can accommodate depends on the length of the primers and the circle. Another method for exponential amplification of RCA uses a so-called hyperbranched RCA (HRCA) (branched or amplification) method, in which the RCA product used as a template for the second and third groups is further expanded using primers [52–55]. Note that a primer can be integrated into the hyperbranched RCA method to increase the sensitivity, especially when the target is detected at low abundance [56]. Additionally, restriction enzyme digestion followed by enzymatic ligation template-mediated, linear RCA products may be converted to a variety of cyclic products [57,58]. A second set of primers may then be used to incorporate these new cyclic products for further amplification. This "circle to circle amplification" restriction digestion process, cyclization and amplification may be repeated for additional amplification. Finally, after hybridization with a second set of circles, the RCA product may be treated with nicking enzyme to generate a plurality of primers. The hybridized primer/circular template product obtained from the nickase reaction can be directly used for the next cycle of RCA amplification.

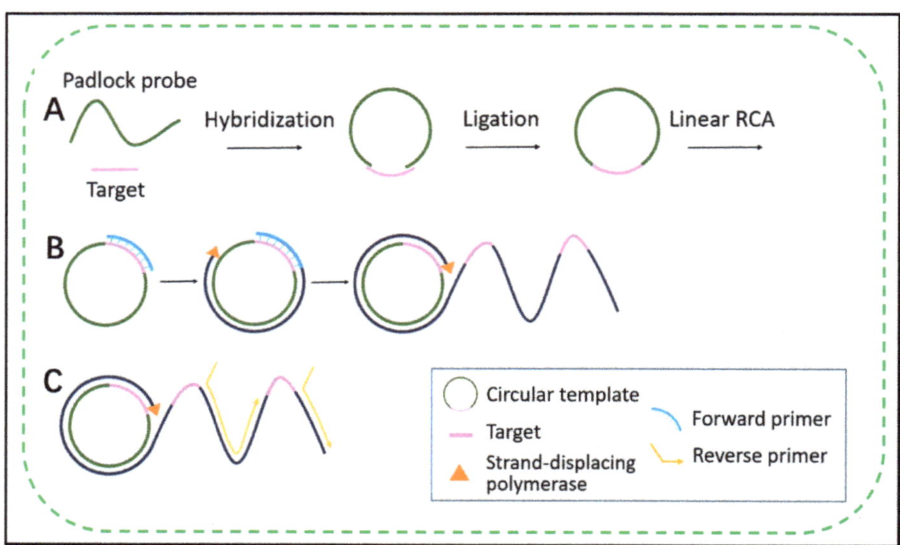

Figure 1. Schematics of RCA mechanisms. (**A**) Ligation RCA. (**B**) Linear RCA. (**C**) Hyperbranched RCA. Adapted with permission from ref. [50].

2.3. Detection of the RCA Product

A variety of signal reading technologies can be used to monitor and detect RCA processes and products. The most common RCA product analysis was carried out by gel electrophoresis. Furthermore, during RCA fluorophore dNTP coupling, fluorescent dye incorporated into the RCA product, bound by a fluorophore, or a complementary strand, can be easily observed using fluorescence-based techniques, including fluorescence spectroscopy, microscopy and flow cytometry [26,59,60]. Combining RCA with molecules such as modified AuNPs, quantum dots, or magnetic beads, it is easy to achieve visualization of RCA products [18,22,61,62]. For instance, RCA products can trigger the assembly of AuNPs for colorimetric and spectral visualization. RCA products can also be combined with magnetic beads to generate diffraction signals.

An electrochemical signal can also be generated by QD hybridization of the RCA products followed by dissolution to achieve high sensitivity [63]. Another method of detecting an electrochemical signal generated by an RCA product comprises inserting the DNA into an organic molecule (e.g., methylene blue), which is inserted into the RCA product [64]. Molecular beacons [65,66] and DNA-intercalating dyes (e.g., SYBR green [67,68]) have also been widely used to detect RCA products, which is also important since real-time monitoring of the reaction in the absence of RCA DNA products results in minimal background fluorescence. In addition to the fluorescence signal, there are some intercalating dyes, such as 3,3-diethylthiadicarbocyanine iodide (DiSC2(5)), that, when combined with the duplex between DNA and peptide nucleic acid (PNA), can be used to produce a colour change from blue to purple [69]. Then, the colour change signal can achieve the goals of the naked eye for detection, which is particularly suitable for real-time diagnostic applications. Furthermore, HRP (horseradish peroxidase) was immobilized on RCA products through biotin modification of DNA to realize visual detection based on colorimetry [70]. Interestingly, the DNAzyme sequence that mimics HRP can catalyse the oxidation of 2,20-azidobis (3-ethylbenzothiazolin-6-sulfonic acid) and generate a blue-green colorimetric signal. It can also be incorporated into RCA products [71]. This "dual-amplification system" (i.e., the RCA and has multiple converting enzyme DNA) enables real-time supersensitive colorimetric detection of target molecules. Finally, the RCA product can be detected by bioluminescence. In this case, the RCA reaction generates a large amount of pyrophosphate that can be used as an adenyl transferase substrate to produce ATP. Then, firefly luciferase ATP acts as a cofactor to produce a bioluminescent signal [72,73].

RCA can not only achieve signal amplification of target nucleic acids through amplification but also has flexible and diverse visualization methods; therefore, it has great potential for application in nucleic acid detection. The advantages of RCA include the following: (1) high sensitivity: RCA has strong amplification ability, the efficiency of exponential RCA can reach 10^9 fold, and it has the potential to detect single copies; (2) high sequence specificity: it can distinguish single nucleotide polymorphisms; (3) the amplified product can be directly used for sequencing after phosphorylation treatment; (4) high throughput: RCA can form a closed circular sequence on the target, ensuring that the signal generated by RCA is concentrated at one point, thereby achieving in situ amplification and slide amplification. However, there are still some shortcomings in the development of the RCA method: (1) the padlock probe is often close to 100 bp, and therefore, the synthesis cost is relatively high; (2) background interference is a problem during signal detection.

3. RCA Assay for the Detection of Targets in Aqueous Environments

Over the years, a number of RCA assays have been developed for the sensitive and specific detection of various targets, including heavy metals, organic small molecules, nucleic acids, peptides and proteins, and microorganisms in aqueous environments (listed in Table 2).

Table 2. Overview of RCA assays for the detection of targets in aqueous environments.

Targets		Detection Signal	Detection Range	LOD	Reference
Heavy metal ions	Hg^{2+}	Fluorescence	0.42 pM–42.5 nM	0.14 pM	[74]
Heavy metal ions	Hg^{2+}	Electrochemical	0.2 pM–100 nM	0.097 pM	[75]
Heavy metal ions	Hg^{2+}	Fluorescence	0–20 nM	200 pM	[76]
Heavy metal ions	Hg^{2+}	ECL	0.1 pM–0.1 µM	33 fM	[27]
Heavy metal ions	Hg^{2+}	Electrochemical	1 pM–1 µM	0.684 pM	[77]
Heavy metal ions	Hg^{2+}	Colorimetry	2.5–100 nM	1.6 nM	[78]
Heavy metal ions	Hg^{2+}	Colorimetry	0–14 µg L^{-1}	3.3 µg L^{-1}	[79]
Heavy metal ions	Pb^{2+}	Fluorescence	1.0–100 nM	1 nM	[80]
Heavy metal ions	Pb^{2+}	pH values	1.0–100 nM	0.91 nM	[81]
Heavy metal ions	Pb^{2+}	Fluorescence	0.1–50 nM	0.03 nM	[60]
Heavy metal ions	$UO_2{}^{2+}$	Colorimetry	0.02–15 ng mL^{-1}	1.0 pg mL^{-1}	[82]
Organic small molecules	Bisphenol A (BPA)	Fluorescence	1 nM–0.1 fM	5.4×10^{-17} M	[83]
Nucleic acids	miRNA	Fluorescence	50–500 fM	25 fM	[84]
Nucleic acids	miRNA	Fluorescence	10–10^6 fM	20 fM	[85]
Nucleic acids	R6G	Fluorescence	10^{-16}–10^{-11} M	8.7×10^{-18} M	[86]
Nucleic acids	gene point mutation	Fluorescence		1 µM	[87]
Peptides and proteins	microcystin-LR	Electrochemical	0.01–50 µg L^{-1}	0.007 µg L^{-1}	[88]
Peptides and proteins	glutamate dehydrogenase (GDH)	Fluorescence	10–100 nM	3 nM	[89]
Microorganisms	*Karenia mikimotoi*	Lateral flow assay	1–1000 cells mL^{-1}	0.1 cell mL^{-1}	[90]
Microorganisms	*Karenia mikimotoi*	Colorimetry	1–1000 cells mL^{-1}	1 cell mL^{-1}	[91]
Microorganisms	Harmful algal blooms (HABs)	Colorimetry	0.1–1000 cells mL^{-1}	0.1 cell mL^{-1}	[92]
Microorganisms	Exophiala	Electrophoresis	-	single-nucleotide level	[93]
Microorganisms	16S rDNA	THz absorption	10^{-10}–10^{-7} M	0.6×10^{-10} M	[94]
Microorganisms	*Chattonella marina*	Fluorescence	10–10^5 cells mL^{-1}	10 cell mL^{-1}	[95]
Microorganisms	circular ssDNA viruses	Whole-genome sequencing	-		[96]

Table 2. *Cont.*

	Targets	Detection Signal	Detection Range	LOD	Reference
Microorganisms	*Amphidinium carterae*	Electrophoresis	100 ng mL^{-1}–1 fg mL^{-1}	281 copies	[97]
Microorganisms	bacterial DNA sequences	Optical (laser)	-	one bacterial DNA sequence	[98]
Living bacteria	*Salmonella typhimurium*	Current	20–2 × 10^8 CFU mL^{-1}	16 CFU mL^{-1}	[99]
Other targets	ATP	Droplet motion	50 pM–5 mM	5 nM	[100]

3.1. RCA Assay for Heavy Metal Ions

The basic definition of heavy metal elements refers to any metal element that has a relatively high density and is poisonous or toxic even at low concentrations, such as lead (Pb), cadmium (Cd), mercury (Hg), chromium (Cr), and arsenic (As). As the chemical properties of arsenic are similar to those of heavy metals, arsenic is also grouped with heavy metals [101]. Heavy metal pollution has gradually developed as an environmental problem affecting human health in many countries [102–104]. Therefore, the development of sensitive and selective heavy metal ion detection methods is imperative to preserve the environment and protect human health.

Traditional methods developed for heavy metal ion detection include chromatographic, spectroscopic, and electrochemical methods. These techniques have the advantages of high accuracy and sensitivity; however, expensive and complicated instruments, complicated sample preparation, and well-trained operators are all indispensable, which means they cannot meet the requirements of portability and ease of use. Recently, novel RCA-based methods have shown great potential in heavy metal ion detection in aqueous environments due to their advantages of low cost and easy operation. Thus, the following sections review some of the research efforts in this area in recent years [105].

3.1.1. Mercury (Hg)

In recent decades, mercury pollution has been commonly found in water, food, cosmetics, the atmosphere, and human health and poses a serious threat to the economy [106]. Contaminants of mercury exist in different forms in nature, such as elemental mercury, $HgCl_2$, Hg_2Cl_2, methyl mercury (CH_3Hg), and $Hg(NH_2)Cl$. These molecules can be ingested, absorbed into the body (through the skin) and inhaled and accumulate in vital organs and tissues, leading to organ dysfunction and irreversible damage to the nervous system. Therefore, the World Health Organization (WHO) has determined the maximum allowable mercury content in different samples to control the harm caused by mercury. Fast, simple, and cost-effective development of in situ testing methods will facilitate the management of heavy metal pollution and mitigation [107].

The colorimetric detection that converts density information into colour changes can be directly interpreted by the naked eye terminal. Due to the low cost, portability, and ease of operation of the colorimetric method, it has been widely and routinely used for the detection of various targets, such as DNA [108,109], proteins [110], cells [111–113], and heavy metal ions [114]. The combination of RCA and colorimetric assays has also been used to analyse heavy metals in aqueous environments. Wang et al. developed an RCA detection method based on a signal enhancement Hg colorimetric aptasensor [78]. As Hg^{2+} poses a serious threat to public health and food safety, the technology for sensitive detection of Hg^{2+} is constantly innovating. In a previous study, Lim et al. constructed an instant detection chip for the colorimetric detection of inorganic Hg^{2+} based on microfluidics that was portable and easy to operate [79]. Manufactured by a three-dimensional printing technique, a disposable chip comprising DNAzyme RCA was generated. A colour change caused by the enzymatic reaction between DNAzymes and the peroxidase substrate 2,2′-azino-bis(3-ethylbenzthiazoline-6-sulfonic acid) (ABTS) was measured using a portable spectrophotometer (Figure 2A). In the "turn-off" type RCA reaction, the interaction of thymine with Hg^{2+} prevents the annealing of the T-rich primer that initiates the RCA reaction. Therefore, depending on the Hg^{2+} concentration, the number of amplified DNases that cause colour changes is reduced. The colorimetric signal is enhanced by amplifying double-repeat DNAzymes from a circular DNA template. The chip detects Hg^{2+} in tap water samples with a high sensitivity of 3.6 µg L^{-1}. Compared with conventional analytical instruments, it has higher selectivity, precision, and reproducibility. This low-cost, easy-to-use platform can reduce the risk of accidental poisoning by Hg^{2+}.

Similarly, Wu et al. recently fabricated a colorimetric aptamer sensor based on RCA to detect Hg^{2+} which possesses an even lower detection limit [78]. First, the aptamer hybridized with its complementary strand ($cDNA_1$) is fixed on the microtiter plate, and the complementary strand ($cDNA_1$) is connected to the primer at the same time to trigger the RCA reaction of the circular template. A successful RCA process will result in the formation of long ssDNA strands on the microtiter plate, resulting in DNA fragments that hybridize with $cDNA_2$ from many organisms. The avidin/biotin binding between avi-HRP and bio-$cDNA_2$ increases the amount of labelled HRP. By adding TMB-H_2O_2, HRP catalyses the reaction and generates a light signal. When there is a target, the situation will be completely different. Hg^{2+} preferentially binds to the aptamer to form a strong and stable T-Hg^{2+}-T complex, resulting in the release of the HRP $cDNA_1$ cluster. Therefore, the optical signal is reduced. The results show that the limit of detection (LOD) was 1.6 nM, with excellent specificity. Compared with the detection signal of the RCA-free system, the detection signal can be increased up to 18 times.

Among different detection methods, fluorescent strategies have unique properties, such as easy installation, suitable signal transduction, a wide linear range and quick response. For instance, Chen et al. reported a highly sensitive Hg^{2+} fluorescent sensor based on hyperbranched RCA [74], with a detection limit of 0.14 pM. More recently, Zhao et al. established a method that used trifunctional molecular beacon-mediated quadratic amplification for the highly sensitive and rapid detection of Hg^{2+} with a tunable dynamic range [76]. Due to its moderate sensitivity and limited dynamic range, it is challenging to analyse targets with low abundance or multiple orders of magnitude changes in concentration. Here, the authors introduced a homogeneous and fast quadratic polynomial amplification strategy by rationally designing three functional molecular beacons. This strategy not only acts as a reporter but also acts as a coupled two-stage amplification module without adding any bridges of reaction components or processes. The Hg^{2+} assay as an example and achieved high sensitivity with an LOD of 200 pM within 30 min. To create an adjustable dynamic range, isomorphisms are used to regulate target-specific binding. When the number of metal binding sites changes from one to three, the useful dynamic range (spanning 50-, 25-, and 10-fold) is used to program the signal response accordingly. In addition, the applicability of this method in river water samples has been successfully verified, and it has good recovery and reproducibility, indicating that it has great practicability in complex actual samples.

Electrochemical response signals are fast, inexpensive, and can be miniaturized for use with other portable devices, which enables the use of very few samples by nontechnical personnel to measure a target on the spot; thus, electrochemical methods have attracted increasing attention. Zhao et al. developed a novel perylene derivative with electrochemiluminescence (ECL) and applied it to Hg^{2+} detection based on a dual amplification strategy [27]. The cathodic ECL of a new covalently cross-linked perylene derivative (PTC-PEI) composed of polyethyleneimine (PEI) and perylene tetracarboxylic acid (PTCA) in an aqueous system was first studied (Figure 2B). Promising novel materials with ECL in PTC-PEI exhibit excellent physical and chemical stability and high ECL intensity, presenting an alternative way to construct an ECL sensor with improved sensitivity. Thus, this sensor was applied to construct a dual amplified "signal-on" Hg^{2+} sensor by employing nicking endonuclease (NEase)-assisted target recycling and RCA for semaphore amplification. Herein, the process is produced by RCA of a long G-rich sequence to capture large amounts of haem on the electrode surface, and then a significant amplification of ECL signals by a PTC-PEI is obtained. This sensor platform showed a detection limit as low as 33 fM with a wide linear range from 0.1 pM to 0.1 µM. Based on the dual-signal amplification strategy, the designed sensor was successfully used to directly detect real water samples from lakes using the standard addition method.

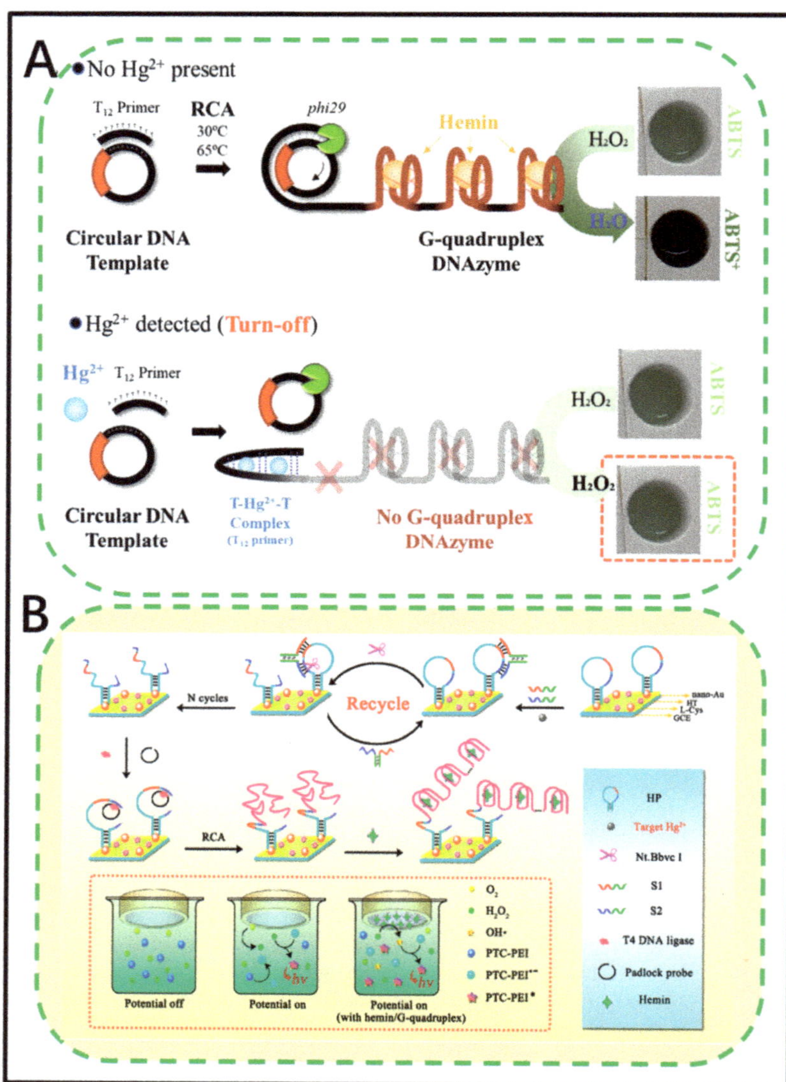

Figure 2. RCA assay for Hg detection. Adapted with permission from ref. [27,79]. (**A**) Colorimetric change of the detection chip when there are no mercury ions in the sample and when mercury ions are present. (**B**) Signal opening sensor based on novel covalently crosslinked perylene derivative (PTC-PEI) system design.

Thymine-Hg²⁺-thymine (T-Hg²⁺-T) induces DNA strand replacement and realizes the specific recognition of Hg²⁺. It is a common strategy for Hg²⁺ detection in environmental samples. Lv et al. developed an ultrasensitive electrochemical measurement method for Hg²⁺ using an efficient target conversion method [77]. First, AuNPs were uniformly coated on polystyrene magnetic microspheres as a magnetic separator, and then ssDNA D1 (rich in thymine) and S1/D2 DNA duplexes (rich in guanine S1) were used as markers. When Hg²⁺

and long ssDNA D3 (rich in thymine at the 5' end) are present in the tested sample, a stable T-Hg^{2+}-T structure between D2 and D3 is immediately formed, and S1 is changed from the S1/D2 DNA duplex, thus realizing the transformation of S1. At this time, the target Hg^{2+} is combined with the output S1. Therefore, the total amount of output S1 is proportional to the amount of input Hg^{2+}. After that, the output S1 will be used as a primer to start an RCA reaction to obtain long guanine-rich ssDNA, thus achieving further hybridization with the DNA captured on the electrode surface. Eventually, methylene blue, as an electron mediator, will interact with the ssDNA polymer through electrostatic binding to generate a detection signal. The electrochemical biosensor based on RCA has a wide linear range, good accuracy, and excellent recovery rate. These stable properties are very suitable for water sample detection. It has strong competitiveness and application in detecting Hg^{2+} in the environment.

3.1.2. Lead (Pb)

Lead ions (Pb^{2+}) are highly toxic heavy metal pollutants that are widespread in the water environment. Since lead ions bioaccumulate and have nonbiodegradable properties, even at low concentrations, lead ions may also cause nervous, reproductive, cardiovascular, and other developmental disorders [115]. Figure 3C demonstrates an ultrasensitive fluorescent assay based on an RCA-assisted multisite-strand-displacement-reaction (SDR) signal-amplification strategy [60]. The proposed strategy is not only to achieve recycling targets but also to introduce RCA by the release of the DNA enzyme. Most importantly, the RCA product is used as an initiator to provide a plurality of SDR sites, which can replace the duplex signal RCA product to effectively prevent the self-quenching probe assembly RCA product signal. Therefore, the amplification efficiency and sensitivity can be significantly improved. Using this strategy for intracellular Pb^{2+} detection, a detection limit as low as 0.03 nM and a wide linear range from 0.1 pM to 0.1 µM were obtained. In addition, the proposed strategy can be extended to determine other goals and provide a new approach for environmental analysis.

Liu et al. developed a rapid and sensitive method for Pb^{2+} detection based on a cationic conjugated polymer and an aptamer [116]. By selecting a more specific aptamer probe, the probe for Pb^{2+} recognition and combination is a single-stranded oligonucleotide labelled with fluorescein. Upon combining with Pb^{2+} with high specificity, the random coiled probe changed to a G-quadruplex with a higher charge density, which enhanced the electrostatic interactions between the oligonucleotide and the cationic conjugated polymer; thus, the two fluorophores were in close proximity, leading to a significantly increased fluorescence resonance energy transfer (FRET) signal. However, other nontarget metal ions produced much lower FRET signals because they could not combine with the probe and thus quenched the fluorescence of the conjugated polymer and fluorescein. This method was rapid, highly specific, and sensitive, and common metal ions did not influence the detection of Pb^{2+}. This FRET-based method, whose LOD was lower than the national standard for drinking water quality, provides a new simple, rapid, and efficient method for the detection of Pb^{2+} in various sources of water.

Using the device integration technique, Tang et al. designed a metal-ion-induced DNAzyme on magnetic beads for Pb^{2+} detection by using RCA, glucose oxidase, and a readout of pH changes [81]. As shown in Figure 3A, the work reported a method of measuring Pb^{2+} ions in environmental samples. A Pb^{2-}-specific DNAzyme immobilized on magnetic beads was coupled to RCA and a pH-metre-based readout. The addition of Pb^{2+} ions induced partial cleavage of the DNA enzymes on the magnetic beads. The single-stranded DNA retained on the magnetic beads was used as a primer. With the help of a circular DNA template, polymerase and dNTPs trigger the RCA reaction. This results in the formation of many oligonucleotide repeats on the magnetic beads. Subsequently, these repetitive sequences are hybridized with glucose oxidase-labelled single-stranded DNA (GOx-ssDNA) to form a long coenzyme containing tens to hundreds of GOx-ssDNA tandem repeats (Figure 3B). The linked GOx molecules oxidize glucose, which is accompanied by a decrease in local pH. This method has good reproducibility, high specificity, and acceptable accuracy. It is used to analyse spiked water samples, and the results are superior to those obtained by ICP-MS.

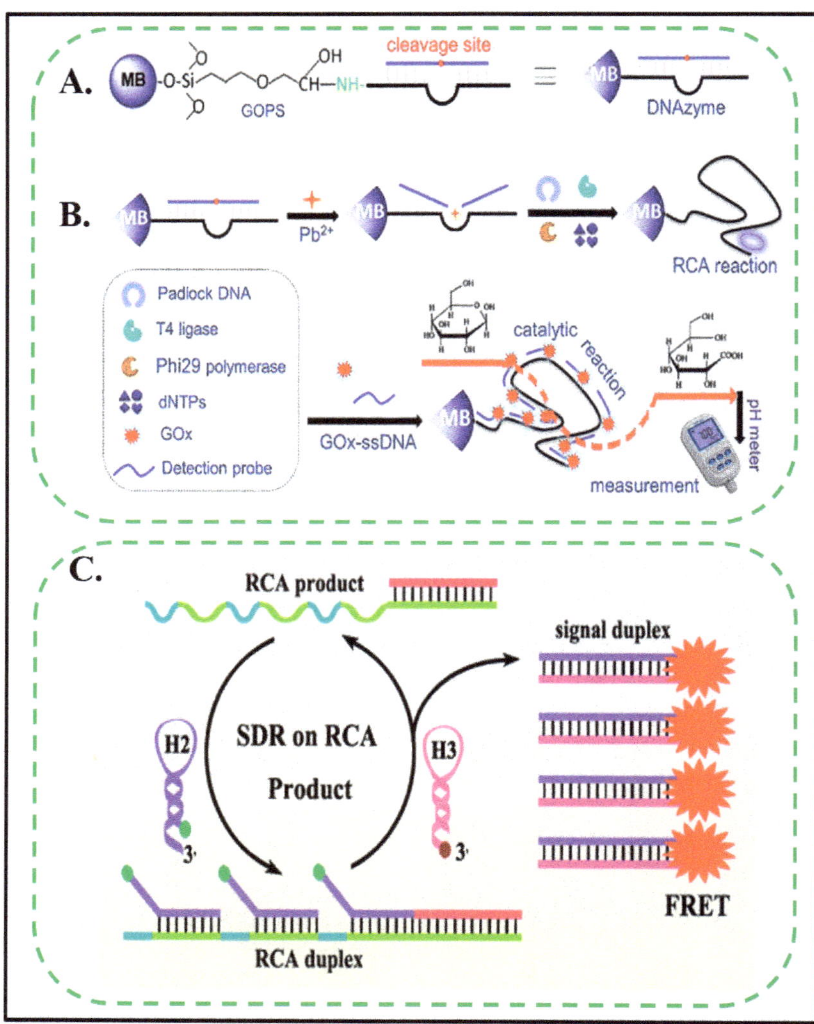

Figure 3. RCA assay for Pb^{2+} detection. Adapted with permission from ref. [60,81]. (**A,B**) Illustration of metal-ion-induced DNAzyme on magnetic beads (MB) for the detection of Pb^{2+} with rolling circle amplification (RCA) on a handheld pH metre; (**C**) Multisite-strand-displacement-reaction (SDR) signal-amplification strategy.

Rapid, portable, and efficient Pb^{2+} detection is important for monitoring environmental toxicity and evaluating human health. Lu et al. demonstrated a DNAzyme assay coupled with effective magnetic separation and RCA for the detection of Pb^{2+} with a smartphone camera [80]. In this work, a simple and low-cost homogenous fluorescence DNAzyme assay was developed for Pb^{2+} determination based on Pb^{2+}-dependent cleavage and RCA. A DNAzyme and its substrate form a double-stranded hybrid in solution, which can completely react with Pb^{2+} in the water phase. Then, DNAzyme/substrate hybrid and unreacted substrate portions with chain cleavage of the biotin-labelled biotin-streptavidin interaction avidin magnetic beads are captured and removed from the reaction solution. The rest of the substrate chain remains in solution and then acts as a primer and triggers RCA. The concentration of cleaved substrate strand Pb^{2+} concentration related, and the biotin-streptavidin-biotin separation was effective to minimize non-specific amplification. Using a smartphone camera, the fluorescence intensity was recorded and quantified after 30–90 min of amplification so that this method could be carried out with the least amount of equipment. Under the best conditions, the dynamic range is 1–100 nM, and this method has been successfully used for the detection of Pb^{2+} in spiked lake water [80].

Tsekenis et al. developed heavy metal ion detection using a capacitive micromechanical biosensor array for environmental monitoring [117]. In this work, the fabrication and evaluation of a DNAzyme-functionalized capacitive micromechanical sensor array for the detection of lead ions is proposed. In the presence of Pb^{2+}, the enzyme may catalyse DNA chain cleavage of the substrate DNA strand with ribonucleotide bases to dissociate the complex into three segments. The DNAzyme strand is laser printed and fixed on the sensor surface and hybridizes with the substrate strand. When self-cleavage occurs, the surface stress will change, which is then recorded as a change in device capacitance. The sensor can detect 10 μM Pb^{2+}, and in the reverse process, it proves the rehybridization of the immobilized catalytic chain and the substrate chain. The reaction is verified by labelling the catalytic chain with a fluorescent molecule, while the substrate chain is labelled with a quencher.

3.1.3. Other Ions

In addition to Hg^{2+} and Pb^{2+}, other ions have also been reported, such as uranyl ions (UO_2^{2+}). Chen et al. developed a visual detection method for ultratrace levels of uranyl ions using magnetic bead-based DNAzyme recognition in combination with RCA [82]. The authors describe a colorimetric method for the determination of ultratrace levels of uranyl ion in beverages and milk. The employment of DNAzyme-functionalized magnetic beads facilitates the separation and collection of the analyte from the sample matrix. The RCA strategy achieves an effect with a ratio of one UO_2^{2+} to massive amounts of HRP, which strongly improves the sensitivity. The visual detection limit is much lower than the maximum allowable level of UO_2^{2+} in drinking water as defined by the USA Environmental Protection Agency, which indicates that the method meets the requirements for simple, rapid, and on-site detection of ultratrace UO_2^{2+} in real samples.

3.2. RCA Assay for Organic Small Molecules

Organic pollutants can come from natural or anthropogenic sources, and industrial, agricultural, and domestic wastewater can be found in a wide range of these pollutants. Among various organic pollutants, bisphenol A (BPA) is a typical substance that has caused widespread concern. As the scientific understanding of bisphenol A continues to deepen, it is especially discovered that it can cause disorder and damage to the normal physiological processes of the human body. Currently, major cities have strengthened the detection and supervision of bisphenol A (BPA) in terms of food safety and drinking water safety. However, the detection of BPA relies on precision and expensive machines, such as HPLC-ICP-MS. These methods require tedious operations and long analysis times. To reduce the analysis cost and simplify the operation, researchers have developed an RCA method to detect BPA. For instance, Xia et al. creatively constructed a label-free aptamer fluorescence sensing platform based on the RCA/Exo III (Exo III) combined cascade amplification strategy, which has high selectivity and high sensitivity for BPA detection [83] (Figure 4). The first step is to design a BPA-resistant aptamer and a DNA double-stranded probe (RP) for the trigger sequence for BPA recognition and signal amplification; next, when BPA appears, it will trigger the probe to be released. On this basis, the initial amplification reaction of RCA was started. When an increasing number of RCA products appear, the RCA products will trigger a second amplification assisted by Exo III with the help of hairpin probes. To date, many G quadruplexes will be enriched in lantern-like structures.

Finally, by irradiating the G-quadruple lantern with zinc(II)-protoporphyrin IX (ZnPPIX), an enhanced fluorescence signal is generated. In the above process, RCA acts as the primary amplification, and the secondary Exo III mediates the secondary amplification. This cascade amplification gives the detection platform excellent sensitivity, and the detection limit is 5.4×10^{-17} M. The strong specificity of the anti-BPA aptamer guarantees the specificity of the platform. This kind of unlabelled fluorescent signal probe avoids the tedious labelling process, greatly reduces the design operation, and at the same time makes the cost lower. In the end, the author also carried out the measurement of real samples of the environment, and the results were reliable, demonstrating the potential application value of this method in the field of environmental detection.

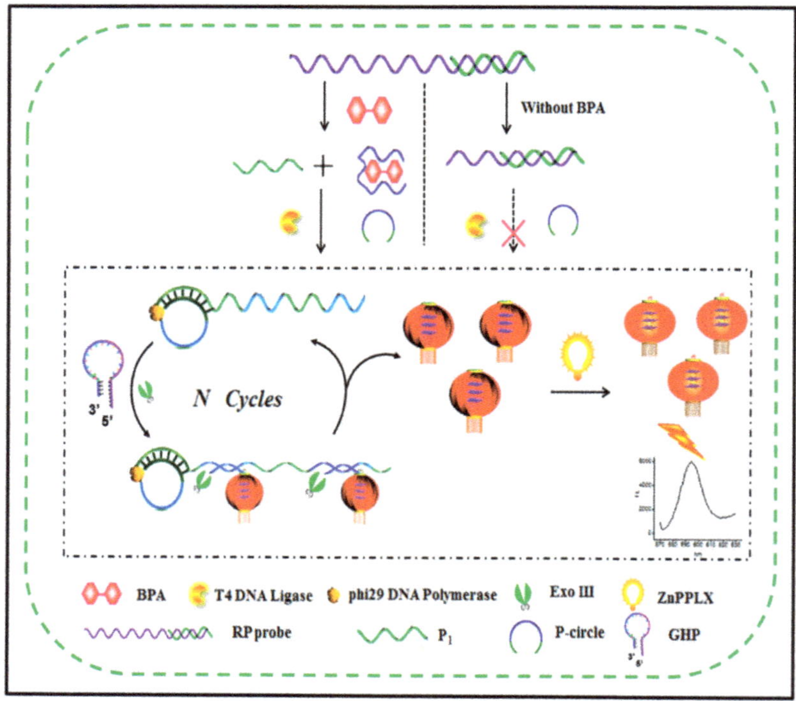

Figure 4. A label-free and sensitive fluorescent qualitative assay for bisphenol A based on RCA/exonuclease III combined cascade amplification. Adapted with permission from ref. [83].

3.3. RCA Assay for Nucleic Acids

MicroRNAs (miRNAs) are evolutionarily conserved, ~18–24-nucleotide-long noncoding RNAs that play a significant role in the control of human gene expression by posttranscriptional gene regulation or silencing. Furthermore, the abnormal expression of a single miRNA can regulate the activity of multiple genes. Previous studies have shown that changes in miRNA expression may lead to a variety of human diseases and disorders, such as cancer, cardiovascular disease, autoimmune disease, neurodegenerative disease, and liver and inflammatory diseases [118,119]. miRNAs are very stable in human peripheral blood circulation and are widely present in other body tissues and fluids, such as urine, saliva, milk, and cerebrospinal fluid. These characteristics indicate that miRNAs are potential biomarkers for diagnostic purposes. miRNAs are related to the occurrence and development of diseases and are pathologically specific; therefore, altered miRNA expression has been used for early detection and diagnosis, classification, prognosis, and predictive diagnosis [120–124].

Ma et al. developed a fast, sensitive, and highly specific label-free fluorescent quantitative biosensor for miRNA through the branched-chain RCA (BRCA) reaction [84] (Figure 5A). The target miRNA acts as a primer and can hybridize specifically to the circular DNA template. Then, RCA is initiated by Phi29 DNA polymerase, and a reverse primer complementary to the RCA product is introduced during this process to achieve isothermal BRCA. While consuming a large amount of dNTPs, it produces the same number of pyrophosphate (PPi) molecules. In this study, a simple and cheaply synthesized pyridine-based Zn(II) complex was used as a fluorescent probe for the selective detection of PPi through dNTPs. In this way, the PPi generated during the isothermal amplification process is effectively chelated to the pyridine-Zn(II) complex to form a highly fluorescent complex, pyridine-Zn(II)-PPi, whose fluorescence intensity is only comparable to the original target. The concentration of miRNA is closely related. This strategy not only achieves isothermal amplification but also allows direct monitoring of DNA polymerization byproducts. For the nonlabelled fluorescence detection of miRNA, PPi greatly simplifies the sensor procedure. This noncumbersome sensor provides a sensitive and easy-to-use platform for miRNA quantification. Significantly promote the career of miRNA as a biomarker in drug discovery, clinical diagnosis, and life science research.

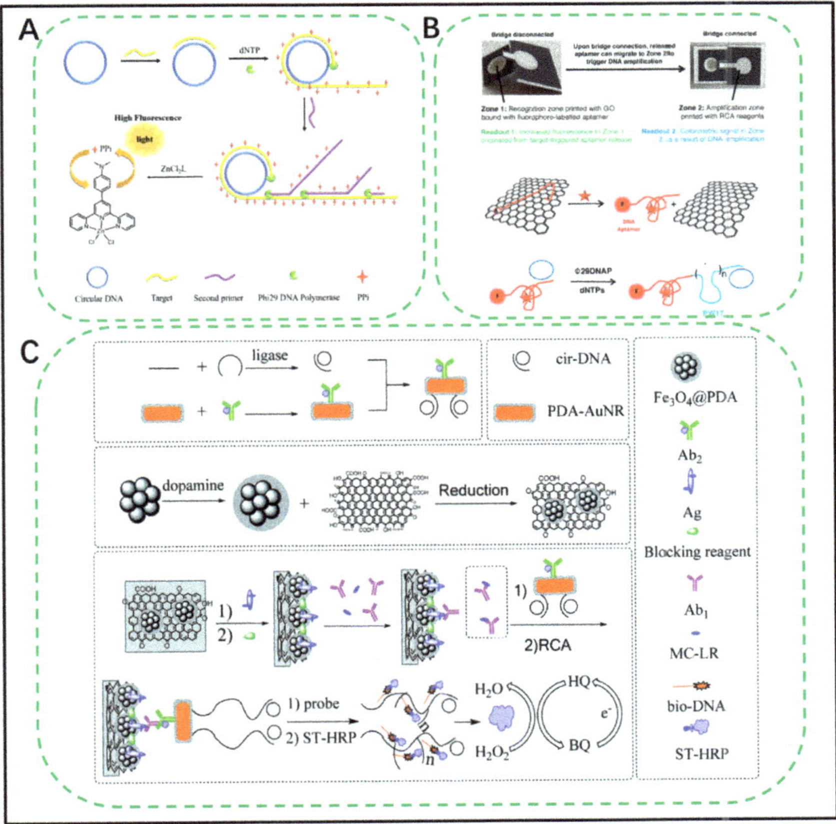

Figure 5. RCA assay for nucleic acid detection. Adapted with permission from ref. [84,88,89]. (**A**) Label-free miRNA sensor based on zinc terpyridine complex; (**B**) design of detachable paper-based biosensor with dual signal output; (**C**) construction of a dual-signal amplification immunosensing platform based on magnetic graphene. MC-LR: microcystin-leucine-arginine, PDA: polydopamine, Ab1: antigen (Ag) and antibody (primary antibody), ST-HR: streptavidin-HRP.

Zhou et al. designed a new high-throughput method to analyse the methylation pattern of individual DNA molecules [125]. High-throughput assays for methylation pattern analysis of individual DNA clones are important for research on tumour initiation, progression and transfer and chemotherapy. In this study, a new method was developed for methylation pattern analysis based on HRCA cloning and microarray technology. A library of DNA fragments with different methylation statuses was amplified from bisulfite-modified genomic DNA using PCR, and circular PCR products were then formed by ligation with a linker. HRCA was performed on streptavidin-coated beads in water-in-oil microemulsions, where the circular products were used as templates with one of the primers immobilized on the beads. Finally, the beads were immobilized on glass slides using polyacrylamide gel and hybridized with specific probes to identify the multiple C g site methylation status of each clone. This method was applied successfully to each clone methylation pattern analysis of the P16 gene promoter in 10 stomach tumour samples and 10 corresponding normal samples. The experiments showed that the method could measure the methylation pattern of each DNA clone with high sensitivity simply by counting the methylation clones.

Xu et al. developed an RCA integrated detection platform that can be used for multiple miRNA quantification by preparing a new type of porous hydrogel-encapsulated photonic crystal (PhC) barcode [85]. The development of a highly sensitive platform for the detection of multiple circulating miRNAs is very important for clinical diagnosis. The porous hydrogel shell and the hydrophilic protein scaffold are connected to each other to form an opal reverse structure in which the PhC barcode is coated. Opal anti-structure can provide homogeneous water around miRNA target reaction and RCA. The encapsulated PhC core of the barcode can provide stable diffraction peaks to encode different miRNAs and their RCAs during the detection process. In this way, the advantages of the PhC barcode and RCA are integrated. Experiments have proven that this technology shows acceptable accuracy and reproducibility for rapid quantification of low-abundance miRNA (20 fM). Therefore, the proposed porous hydrogel-encapsulated PhC barcode provides a new platform for multiple quantification of low-abundance targets in practical applications.

Xu et al. produced a sensitive nucleic acid detection platform based on superhydrophobic micropores [86]. The micropores are located on the superhydrophobic substrate. Due to the difference in wettability, ultratrace DNA molecules are enriched, which realizes the concentration of the chip, and then the fluorescence signal is amplified, which improves the detection sensitivity. Using the biosensing interface of ultrawet materials to detect ultratrace DNA through concentration has opened up a simple and cost-effective new way of thinking.

Based on the high dark phase contrast of vapour condensation, Zhang et al. developed a label-free smart device that can detect diseases related to gene mutation sites in real time [86]. A Peltier cooler and a mini PC board for image processing are the core components of the device. The workflow, in short, uses the heat of the hot end of the Peltier cooler to evaporate the fluid in the copper cavity, and then the vapour condenses on the surface of the microarray chip placed on the cold end of the cooler and further characterizes the vapour condensation relative to the microarray. The high dark phase contrast of the analytes on the chip. Used in conjunction with RCA, the device can see the change from reduced hydrophilicity to hydrophilicity caused by gene capture and DNA amplification. Analysis of lung cancer gene point mutations proved the high selectivity and multiple analysis capabilities of this inexpensive device.

3.4. RCA Assay for Peptides and Proteins

Figure 5C shows an RCA signal-enhanced immunosensor for ultrasensitive microcystin-LR (MC-LR) detection based on magnetic graphene-functionalized electrodes [88]. This novel competitive immunosensor promotes the development of the MC-LR detection field. Magnetic graphene is synthesized, characterized, and used as a substrate. Due to its large surface area and easy separation, the antigen can be immobilized on the electrode surface.

In addition, gold nanorods modified with polydopamine are modified and functionalized with secondary antibodies and circular DNA templates. Through the function of RCA, the DNA template can be replicated to generate a large number of repetitive DNA sequences. At this time, the detection probe will hybridize with the repetitive sequence; therefore, the signal has been significantly improved. Under optimal parameter conditions, the immunosensor proposed by the team has a good linear relationship between the current response in the range of 0.01–50 µg L^{-1} and the target concentration, and the detection limit is 7.0×10^{-3} µg L^{-1}. Similarly, the immunosensor has also been proven to be highly specific, and its reusability and stability are commendable. Most importantly, the proposed biosensor is used to detect MC-LR in real water samples and has a good recovery rate, indicating its application prospects in actual environmental monitoring. Hui et al. reported a paper-based multifunctional bio/nanomaterial printed sensor platform, as shown in Figure 5B [89]. The platform is divided into two reaction zones and a connecting bridge. Molecular recognition and signal amplification are realized by printing multifunctional biological/nanomaterials. When the targeted analyte appears in the first area, the fluorescently labelled nucleic acid aptamer will automatically desorb from the printed graphene oxide, thereby quickly generating an initial fluorescent signal. Then, the released aptamer flows with the wave to the second area, where it reacts with the printed reagent to initiate RCA, thereby generating DNA amplicons containing DNAzyme mimicking peroxidase, thereby generating colorimetric readings. No equipment or smartphone is required to interpret the reading. To verify the specificity and sensitivity of the sensor, the author used an adenosine triphosphate RNA aptamer (a bacterial marker) and glutamate dehydrogenase DNA aptamer for verification and successfully passed the test. Moreover, when the target is added to serum or stool samples, it can still be detected, proving the potential of this method in testing clinical samples.

3.5. RCA Assay for Microorganisms

Widespread waterborne microbial diseases have caused significant mortality and morbidity worldwide [126]. Therefore, the detection of these microorganisms or pathogens is very important. Conventional microbial pathogen detection methods require the use of artificial culture media or microscopic methods. These methods have many technical limitations, such as low detection accuracy, low sensitivity, complex sample processing and time consumption [127–129]. For instance, coliform assays are traditionally used to assess coliform bacteria in environmental samples but not to monitor the overall microorganism content.

Due to the abovementioned shortcomings of conventional methods, the development of technologies for more effective discovery of trace pathogens in dark water has received considerable attention. Recently, in the field, it has been applied in several molecular methods, such as PCR, enzyme-linked immunosorbent assay (ELISA) [130,131], and fluorescence in situ hybridization (FISH) [132], and applying policy-based biosensors to detect microorganisms in water and wastewater has become a very popular field of study [133]. The use of biosensors can identify microbial contamination in real time, while using traditional techniques, it takes several days to obtain results.

Harmful algal blooms (HABs) of toxic microalgae have received much attention worldwide because their existence has always been a threat to marine ecosystems, fisheries, and human health. The scientific community has also been developing a monitoring system that can effectively and accurately identify pathogenic algae and monitor the quality of seawater. Unfortunately, the traditional methods based on laboratory precision microscopes are too complicated and time-consuming [134,135]. Taking the coast of China as the research area, part of the large subunit rDNA (D1-D2) sequences of eight common toxic and harmful algae in the research area were cloned, and then a specific PLP consisting of universal primer binding sites and ZIP sequences was designed. Then, a sorting probe DNA array complementary to the ZIP sequence on the nylon membrane was prepared. The amplified product is labelled with biotin produced by multiple HRCAs (MHRCAs).

After heat denaturation, the MHRCA product will hybridize with the DNA array, and then spot colouring will appear. As shown in Figure 6, an MHRCA-based membrane DNA array assay (MHBMDAA) for the detection of toxic microalgae has been developed [92]. The specificity of MHBMDAA was confirmed by the double cross-reactivity test of PLP and taxonomic probes. The detection range of the MHBMDAA method in simulated samples can reach 0.1 to 1000 cell mL^{-1}, and its sensitivity is 10 times higher than that of the multiplex PCR membrane DNA array. The validity and reliability of MHBMDAA have also been verified by natural samples from the East China Sea. The results show that MHBMDAA is a sensitive and reliable detection tool for the early warning system of toxic microalgae.

Karenia mikimotoi (*K. mikimotoi*) is globally distributed, toxic, and harmful, and it easily forms water blooms. It is similar to a single spark, which explodes frequently in the global seas. To avoid endangering seafood and human health, fast, accurate, and sensitive on-site monitoring of this harmful algae is needed. Zhang et al. reported the comparative detection of *Carrenella triloba* through exponential RCA (E-RCA) and double-linked E-RCA and compared their sensitivity with traditional PCR methods [90]. Part of the large subunit rDNA (D1-D2) of *K. mikimotoi* was amplified, cloned, and then sequenced. After the obtained sequence was used for specific region comparison analysis, PLP and primers of E-RCA and dlE-RCA were designed. Through the simulation, the parameters of the E-RCA and dlE-RCA systems are optimized. The specificity test showed that both E-RCA and dlE-RCA are specific to *K. mikimotoi* bacteria. The sensitivity comparison shows that the sensitivity of E-RCA is 10 times higher than that of PCR, while the sensitivity of dlE-RCA is equivalent to that of PCR. Tests on simulated field samples show that the detection limits of the developed E-RCA and dlE-RCA methods are 1 and 10 cells, respectively. By visually observing the colour reaction and adding fluorescent SYBR green I dye to the reaction tube, it can be confirmed that E-RCA and dlE-RCA are positive. Compared with E-RCA, dlE-RCA can avoid the self-cyclization of PLP. The developed E-RCA and dlE-RCA methods are also very effective for field samples with a target cell density in the range of 10–1000 cells mL^{-1}. These results indicate that the established E-RCA and dlE-RCA detection protocols show expectations for future field applications of *K. mikimotoi* monitoring.

Zhang et al. established a method combining isothermal amplification technology and a rapid analysis method for the rapid detection of *K. mikimotoi* on site [90]. In short, it consists of two parts: HRCA isothermal amplification based on targeted nucleic acids and lateral flow dipstick (LFD) for detecting nucleic acid amplification products, namely, the HRCA-LFD sensor analysis platform, which relies on targeted DNA template PLP and LFD probes targeting PLP to detect *K. mikimotoi*. The core point is the sequenced endogenous *K. mikimotoi* spacer sequence obtained by molecular cloning and is used as the target of PLP. The analysis of on-site samples shows that HRCA-LFD analysis is suitable for samples with target cell densities ranging from 1 to 1000 cells mL^{-1}. HRCA-LFD can detect *K. mikimotoi* sensitively and reliably directly from seawater samples.

Najafzade has developed a method to accurately identify seven species of aquatic Exophiala species through RCA DNA PLPs [93]. The potential opportunistic species in the black yeast genus Exophiala are relatively high, and these opportunistic species can cause systemic or scattered infections in individuals with strong immune capabilities. Among them, the species that cause systemic diseases can generally grow at 37–40 °C, while other species lack heat tolerance, and most of them involve diseases of aquatic vertebrates, invertebrates, and most cold-blooded animals. Here, the author introduces a high-efficiency determination method that can identify and identify water-based Exophiala species without sequence restrictions. First, the author completed the sequencing and comparison of the ITS rDNA regions of seven Exophiala species and the closely related *Veronaea botryosa*. They designed a specific PLP that can be used to detect characteristic single nucleotide polymorphisms. By amplifying the DNA of the target fungus, the amplified product was observed on a 1% agarose gel to confirm the specificity of the probe-template binding and finally realize detection at the species level. During the experiment, the amount of reagents was reduced to prevent false-positive results. Simplicity, sensitivity, durability, and low cost make this PLP analysis (RCA) stand out in the diagnosis of bacterial DNA species. The application of terahertz (THz) spectroscopy in the field of sensing meets the needs of rapid and sensitive bacterial detection to a certain extent. Yang et al. developed an RCA-based THz biosensor for isothermal detection of bacterial DNA [94]. The first step is to hybridize a bacterial-specific, artificially synthesized 16S rDNA sequence with PLP, where the 5′ and 3′ ends of the PLP contain sequences that are completely complementary to the target sequence. When the target sequence is recognized and ligated, the linear PLP is circularized to form a circular PLP. Then, the capture probe (CP) immobilized on the magnetic beads plays the role of primer, and RCA starts to initialize. In the THz range, the absorbency of DNA molecules is far from that of water molecules, so the RCA product on the surface of the magnetic beads will cause the THz absorbance to decrease significantly. At this time, sensitive THz spectroscopy will detect the difference. The specificity test result is obvious, which is proven by its low signal response to interfering bacteria. The proposed strategy not only proves a new attempt to detect target bacterial DNA isothermally but also provides a general platform for sensitive and specific DNA biosensing using THz spectroscopy technology. Chen et al. used a research and development strategy that took full advantage of the sensitivity of HRCA to quickly detect *Amphidinium carterae* in environmental samples [97]. For coastal countries, the quality of marine water is decisive for the regional marine ecosystem, marine fisheries, or public health. Unfortunately, many coastal countries and regions are currently threatened by toxic microalgae, and they are becoming increasingly serious. Therefore, it is urgent and necessary to establish a large-scale water quality analysis and early warning system that can quickly, sensitively, and accurately detect toxin-producing microalgae in water bodies. In this article, the author uses HRCA to quickly detect *Amphidinium carterae*. First, the large subunit rDNA (LSU D1-D2) of *Amphidinium carterae* was sequenced to design a species-specific PLP. Then, the PLP combined with two amplification primers was used by HRCA. Of course, the HRCA sensing platform passed the specific detection experiment and passed the test with other algae. The entire operation process was controlled to be completed within 1.5 h. What is even more surprising is that the platform's repeated detection limit is one cell. During the detection process, the fluorescent dye SYBR green I can be added to the amplified product so that the positive result of HRCA can be seen through the colour reaction. HRCA provides a very useful detection tool that can accurately screen large samples of *Amphidinium carterae* and other toxic species. To further improve the sensitive, Nie et al. applied HRCA and an HRCA-based strip test (HBST) for the detection of *Chattonella marina* [95]. As mentioned above, the existence of HAB poses a threat to marine ecosystems, fisheries and human health on a global scale. How to quickly and accurately monitor pathogenic algae and seawater quality is the goal. In this research article, the author combines the two methods and proposes the use of HRCA and HBST to rapidly detect *Chattonella* in the sea. The first is to sequence the large subunit (LSU) ribosomal DNA (rDNA) characteristic region of *Candida marinus* and design a specific PLP based on the sequencing results. In this way, the entire HRCA reaction covers two amplification primers and another HBST that plays an important role. The detection procedure involves a constant-temperature HRCA reaction, paper-based hybridization, and colour development results judged by the naked eye. Verifying specificity and sensitivity is an indispensable link. After a simple and logical operation, the results show that the detection limit of HBST detection is 1 copy μL^{-1} of the *Pseudomonas marina* LSU rDNA plasmid, which is the most prominent. It is one order of magnitude higher than the detection limit of HRCA and three orders of magnitude higher than the detection limit of conventional PCR. Finally, the author also applied the scheme to simulated field samples, and the results obtained are also good. The developed HBST still has higher detection sensitivity than HRCA and traditional PCR methods. In summary, the method proposed in this study is expected to break through the monitoring and early warning dilemma caused by HAB to the global ocean system and realize the on-site, sensitive and specific detection of cryptosporidium from natural samples. At the same time, it also provides good detection cases and models for future detection of other harmful algae.

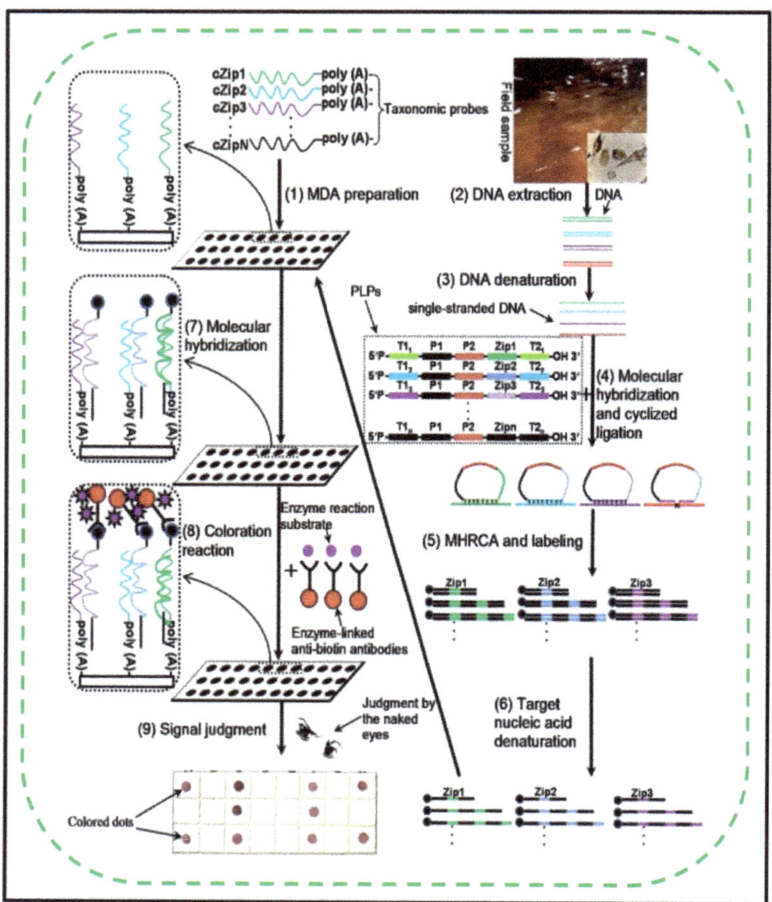

Figure 6. RCA assay for microorganism detection. Adapted with permission from ref. [92].

Pearson et al. proposed the view that "virus recombination leads to fuzzy classification" [96]. The structural composition and dynamic changes of biological communities are inevitable research fields, especially in the application of agricultural sites. Research on microbial communities with high economic attributes and related maintenance is increasingly being discussed. The wastewater treatment plant is a melting pot with multiple coexistences. First, its sources are relatively wide, including not only domestic wastewater, livestock and poultry breeding wastewater but also various environmental wastewaters with stress factors, making wastewater treatment plants a variety of virus libraries of hosts that can be infected. In addition, the dynamic changes of its internal environment cannot be ignored; that is, the phenomenon of virus recombination cannot be ignored. They used a combination of sucrose gradient size selection and RCA to isolate the full-length genome of the circular ssDNA virus from the wastewater treatment facility and sequenced it on Illumina MSeq to achieve virus collection inspection. However, compared with the relatively large dsDNA viruses that are often studied, single-stranded DNA viruses are the least known microbial pathogens, and they also face technical bottlenecks such as genome bias and difficulty in cultivation. The team studied several typical examples, including model organisms (Microviridae) for genetic and evolutionary research and agricultural pathogens (Circoviridae and Geminiviridae) that infect livestock and crops. In the end, the results of the examination of viral DNA collected at the site provided evidence for 83 unique genotype groups. On the one hand, the results show the wide diversity of the community. These groupings are genetically different from known virus types; in addition, although the expression of these genomes is similar to that of known virus families, many differences are so great that they may represent the new taxonomy group. This study demonstrates the effectiveness of this method to isolate bacteria and large viruses from ssDNA viruses and the ability to use this protocol to obtain in-depth analysis of the diversity within the group.

Bejhed et al. used magneto-optical readings to realize simple, fast and cost-effective qualitative double-stranded detection of bacterial DNA sequences [98]. Whether it is the diagnosis of infectious diseases in biomedicine, the safety control of residents' drinking water or food safety management and other health fields, the rapid, low-cost and easy-to-operate multitarget detection of pathogens has always been our unremitting pursuit. Therefore, we need an increasing number of novel bioassay methods to meet the requirements. Biological detection platforms established by magnetic and optical-magnetic bioassay methods are increasingly playing a role in developing countries due to their unique advantages. A photomagnetic method for the qualitative detection of bacterial DNA sequences has been developed, which can quickly determine whether the DNA sequence of the target bacteria appears in the sample. After two magnetic beads of different sizes are functionalized with nucleic acids, they hybridize with RCA products from two different bacteria to form the sensing platform. Among them, magnetic beads of different sizes are equipped with different oligonucleotide probes, which are only complementary to one of the RCA products. The final test results are also obtained from the same volume of samples. The measurement response is controlled by the modulation frequency of the applied magnetic field of the projected laser, which is an outstanding feature of the magneto-optical sensing platform. It is very suitable for countries and regions with low resources in the world, especially those that urgently need low-cost, large-scale screening for pathogens related to human or veterinary drugs.

In addition to the detection of microorganisms based on nucleic acids, RCA can also be used to detect living bacteria with aptamers. Ge et al. recently developed a method to detect *Salmonella typhimurium* directly [99]. The authors adopted an aptamer that hybridizes with the capture probe immobilized on the AuNP surface to recognize the target. In the presence of *Salmonella typhimurium*, the aptamer would combine with bacteria and release the primer binding site of the capture probe. This is followed by the triggering of the RCA reaction which produces various DNA fragments. Finally, the detection probe could be immobilized by the products of RCA via DNA hybridization, which could induce current change. This method has a very low detection limit of 16 CFU mL^{-1} and a broad linear range of 20 to 2×10^8 CFU mL^{-1}. This biosensor demonstrated that RCA has great potential for the direct detection of microorganisms with the help of aptamers.

4. Emerging Nanotechnology for RCA Assay

When constructing RCAs with higher performance analysis tools, in addition to new materials, several new uses of biotechnology, such as DNA integration technology and equipment, have also been developed. Among them, DNA technology has attracted much attention due to its programmed assembly and precise modelling of a strand-deforming operation. Moreover, RCAs can be used to construct various DNA machines to perform different functions via DNA technology. Additionally, engineering tools have also been widely used to make RCA more convenient and portable, such as microfluidics, paper devices and other commercial portable devices. When integrated with devices, RCA can be more easily used for point-of-care (POC) detection.

4.1. DNA Technology

4.1.1. DNA Assembly Technology

DNA has been widely used to drive nanoparticle assembly due to its programmed assembly, high selectivity, and excellent recognition ability [136]. DNA assembly technology has also been applied to RCA assays, as it can generate a long ssDNA strand that can provide binding sites and various structures for assemblies [136].

Tian et al. recently constructed a biosensor to detect SARS-CoV-2 by combining RCA and DNA assembly technology [137]. The SARS-CoV-2 outbreak began in late December 2019 and soon spread around the world, which created a demand for rapid diagnosis [138]. The authors adopted a padlock probe (PLP) for target recognition followed by RCA for signal amplification and the assembly of MNPs for signal transduction. The padlock probe will be ligated to form circular templates for the first round RCA (CT1) when the target appears. Then, the intermediate amplicons will be generated by nicking-enhanced RCA. The intermediate amplicons could anneal to circular templates for the second round RCA (CT2) to generate amplicon coils, leading to the assembly of magnetic nanoparticles that could be detected by optomagnetic sensing. The method can be finished in 100 min with a dynamic detection range of three orders of magnitude and achieved a detection limit of only 0.4 fM. This method demonstrated that DNA assembly may provide good signal transduction for RCA detection.

RCA can also be used to construct organic polymer materials via DNA assembly technology such as DNA hydrogels. Na et al. used the RCA method and microbeads to generate a DNA hydrogel for the detection of infectious viruses [139]. The primer was first immobilized on the surface of microbeads, and the dumbbell-shaped templates were then immobilized via hybridization with primers. In the presence of the target nucleic acid, the templates can be circularized, and the RCA starts, which can generate a specific dumbbell shape to form the DNA hydrogel and block the flow path. Coloured ink was adopted as a visual indicator of whether the channel was blocked. The detection time was only 15 min, and the detection limit was 0.1 pM.

4.1.2. DNA Machines

DNA machines refer to DNA molecules with several basic properties, such as the ability to perform mechanical functions accompanied by the need for fuel input, including pH and light, the generation of waste products, and energy consumption [140]. Recently, various DNA machines, such as tweezers, walkers, gears, and cranes, have been developed to perform different functions, such as drug delivery and control of the fluorescence properties of fluorophores [141,142].

An RCA assay could also be combined with a DNA machine, as it can generate DNA strands for the device. de Avila et al. designed acoustically propelled nanomotors for intracellular siRNA delivery by employing gold nanowires and the RCA method [141]. As shown in Figure 7A, the authors first utilized the RCA reaction to produce long ssDNAs made of GFP-targeting siRNA regions and noncoding spacer regions. Then, siRNA was bound to the long ssDNA strand and created alternating single-stranded and double-stranded RCA products, which could be wrapped on the positively charged surface of the gold nanomotor to form a DNA machine. This machine could be driven by ultrasound to penetrate the cell membrane. Once inside cells, the siRNA immobilized on the AuNW surface suppressed the gene-mRNA expression, making the cell fluorescence "OFF". This DNA machine is an efficient RNA delivery tool and might be a promising platform for RNA-mediated gene therapy. The development of this method suggested that RCA could not only be applied to detection but also be a promising technique for gene therapy and drug delivery.

The DNA walker is another DNA machine that has been combined with the RCA assay. Li et al. recently developed a method for the detection of *Escherichia coli* O157:H7 based on an RCA assay and a DNA walker [143]. The method can be divided into three parts, as illustrated in Figure 7B. First, DNA walker-based amplification can be instigated by the presence of the target gene, which could hybridize with BN to release DW for further hybridization between DW and TN. DNA walker-based amplification could then be started with the help of Nb. BbvC I. The DNA generated by the DNA walker could open hairpin DNA 1 (H1) to activate the RCA reaction. The long ssDNA produced by RCA could react with hairpin DNA 2 (H2) and hairpin DNA 3 (H3) to start HCR amplification. Through the combination of the DNA walker, RCA and HCR, this biosensor possesses a detection limit of 7 CFU mL^{-1} and is superior to most detection assays for *E. coli* O157:H7.

4.2. Engineering of RCA as a Portable Tool for Point-of-Use Detection

Traditional detection methods such as ICP-MS and PCR are usually based on complex and expensive instruments. Instrumental assays have various advantages, including high sensitivity, stability, and selectivity. However, they usually require long and complex pretreatment steps, costly instruments, and professional operations. Therefore, they cannot be directly applied to POC detection. In recent years, several portable assays, such as microfluidic chips and paper devices, have been established for POC detection, and some of them can be combined with RCA assays.

4.2.1. Microfluidic Chips

A microfluidic chip is an integrated platform that integrates microanalysis and pretreatment steps such as sampling, dilution, reagent addition and separation [144]. The platform possesses various advantages, including ease of control, low cost and low sample consumption, and is a promising technique for POC detection.

Heo et al. designed a valveless rotary microfluidic device based on the RCA assay that can detect multiple single-nucleotide polymorphisms simultaneously [145]. As shown in Figure 7C, the device consists of three components: a channel wafer, a resistance temperature detector (RTD) wafer with a Ti/Pt electrode pattern, and a rotating plate. The sample is loaded in the twelve ligation solution inlets, and the chamber of the top rotating plate is aligned with the radial microchannel. The padlock probe for recognition is also in the chamber. Then, the chamber was isolated by rotating the plate 7.5° for the ligation reaction and rotating again for RCA reagent injection. The RCA reaction and detection probe hybridization were conducted similarly to the ligation reaction. Finally, the results were read out via a fluorescence optical microscope. This microfluidic chip can not only achieve multiplex detection but also needs no microvalves or micropumps, simplifying the chip design and operation.

4.2.2. Paper-Based Platforms

Paper materials that are abundant, low-cost, easy to manufacture, portable, and have support over sensor devices are widely used, especially in the POC diagnostic field. Liu et al. constructed a paper device for DNA or microRNA detection based on the RCA assay [146]. The authors first adopted the wax-printing technique to produce a 96-microzone paper plate with a test zone diameter of 4 mm. The RCA primer was also printed on the test zone. Then, the RCA reagents, including the circular DNA template, phi29 DNA polymerase, dNTPs and hemin, were mixed with pullulan solution and printed into the test zone. After air-drying, the paper device was finished. The addition of the target gene will activate the RCA reaction and can be detected by the colour change with TMB and H_2O_2 because its product possesses the PW17 sequence. This device successfully simplifies the detection steps and requires no expensive instrumentation. More importantly, it achieved comparable results to the values obtained using qRT-PCR.

4.2.3. Electrochemistry Platforms

Electrochemical sensors have been widely used for point-of-care detection. Recently, the RCA method has also been applied for electrochemical sensors to detect pathogens, macromolecules, and small molecules. Huang et al. designed a biosensor based on RCA and voltametric methods to detect hepatitis B virus [147]. The method has an extraordinary sensitivity with 2.6 aM detection limit and the response is linear in the 10–700 aM range. Shen et al. established an immunoelectrochemical biosensor [148] based on the RCA method to detect human epidermal growth factor receptor 2 (HER2), and the detection limit was just 90 fg mL^{-1}. Yi et al. recently developed a versatile electrochemical platform to detect adenosine with a detection limit of 320 pM and a linear range of 1 nM–10 μM [149]. In general, these methods are similar and utilize the specificity of RCA to recognize the target and the electrochemical effect of the RCA product (many could be indirect products, such as probes immobilized on the long ssDNA generated by the RCA reaction) to transduce the chemical signal into an electrical signal. Taking Yi's study as an example, the conformation of the right probe was changed in the presence of adenosine, forming a hairpin structure. The hairpin structure can be linked to another hairpin structure generated by the left probe to produce circular DNA. Then, the RCA reaction can be triggered with primers and Phi29 polymerase [149]. The RCA product could hybridize with capture probes on the electrode surface which induced an increase in the impedance signal. This device is not only easy to operate but also flexible. Only primes need to be changed when they are used for the detection of different targets.

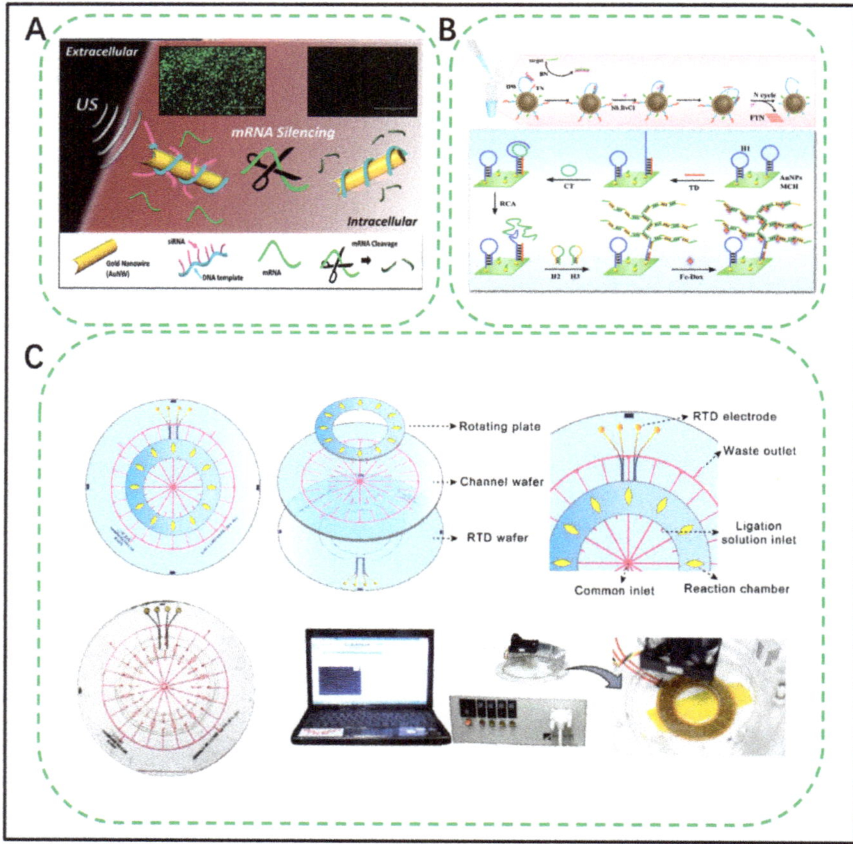

Figure 7. Emerging nanotechnology for RCA assay [141,143,145]. (**A**) Schematic of the nanomotor-based gene silencing approach; (**B**) Schematic representation of the multiple sensitizing electrochemical biosensor for the detection of *E. coli* O157:H7; (**C**) a novel rotary microfluidic device which can perform multiplex single nucleotide polymorphism typing on the mutation sites of TP53 genes.

4.2.4. Commercial Portable Device

RCA analysis is usually integrated with commercial small equipment for on-site analysis. Portable devices have rapid detection, ease of use, and low-cost characteristics and have been widely used in daily life.

The glucose metre might be one of the most successful commercial portable devices recently. Various efforts have been made to combine RCA methods with glucose metres. Jia et al. recently developed a biosensor to detect p53 DNA based on an RCA assay and glucose metre [150]. This paper first immobilized a hairpin probe on magnetic beads for recognition followed by a Padlock probe-mediated RCA step for signal amplification. Then, numerous DNA-invertase conjugations were tagged on the long ssDNA generated by the RCA assay to hydrolyse sucrose to glucose for detection by a glucose metre. This biosensor achieved a detection limit of 0.36 pM with a linear calibration range from 0.5 to 10 pM and exhibited excellent sequence selectivity. The above studies proved that RCA is a flexible assay method that can be combined with different platforms for application in different situations to satisfy different demands.

5. Conclusions and Perspective

The quality of the water environment is closely related to human health, food, energy and the economy. As mentioned earlier, RCA-based analysis technology is a reliable alternative for detecting various targets to track environmental pollutants. Due to its simplicity and high selectivity, different types of environmental pollutants can be monitored, and an increasing number of RCA-based analytical methods have been established. As a simple, efficient and temperature-free nucleic acid amplification tool, RCA has now become a powerful tool in the field of environmental monitoring. In particular, when RCAs and functional nucleic acids, including aptamers and DNA enzymes, as well as other assay platforms, PCR, ELISA, microfluidics, surface plasmon resonance (SPR), and nanoparticles, are integrated into ultrasensitive detection of various targets, including nucleic acids, proteins, small molecules, viruses, and cells. From the point of view of materials science, RCA project versatility makes it an exciting tool for the preparation of DNA building blocks and the construction of highly ordered nanostructures and new materials that may have practical applications in biosensing and environmental monitoring.

Furthermore, based on the synthesis of multivalent ligand binding variable RCA, many other multivalent ligand systems are difficult to achieve, including the number of ligands, density, type, and spatial organization, and they represent a new chemical biology tool in environmental monitoring. Despite the many advantages of the RCA system, there are still some challenges to overcome in practical applications. First, preparing a large number of high-purity circular templates may be an unavoidable challenge. For example, in enzymatic ligation methods, in addition to circular DNA, linear multimer byproducts can sometimes be formed. This problem can be partially solved by using low concentrations of DNA in the ligation reaction. Then, the circular DNA product can be purified from the linear byproduct by gel electrophoresis or exonuclease treatment, which only degrades noncircular linear DNA molecules. In addition, it is known that the enzymatic ligation process is effective for relatively large DNA substrates but may not be suitable for making small DNA loops (~30 nt). This may be due to an insufficient number of enzyme binding sites and/or is caused by the induced strain after restriction enzyme digestion. The closed loop of short oligonucleotides. However, this defect can be solved by chemically circularizing DNA oligonucleotides. This method can generate both small (~14 nt) and large circular templates (415 nt) with very good yields (up to 85%) of circular DNA molecules. In addition, other challenges in the practical application of RCA systems include mass production, purification, and storage because RCA products tend to aggregate for a long time due to non-specific intermolecular and intramolecular cross-linking. In addition, due to the large molecular weight of RCA products, nonspecific binding may occur when used in complex water environments such as wastewater. These problems can be minimized by fine-tuning the parameters, including RCA product length, order, composition, and stiffness. For instance, the authors found that the incorporation of polyT (rather than random sequence) spacers between the aptamer domains of RCA products can reduce nonspecific interactions between the multivalent aptamer system and various targets. Finally, computer-aided methods can be used to design RCA sequences and short chains to construct predictable DNA nanostructures to minimize unwanted nonspecific interactions. The analysis method based on RCA is an innovative approach for the simple and rapid detection of various targets in environmental monitoring and can even be applied to on-site detection. This method will open up a new direction for environmental pollution assessment, drug abuse trend assessment, public health assessment, and other fields.

Author Contributions: Conceptualization, K.Z. and K.M.; methodology, K.Z.; software, K.Z.; validation, K.Z., H.Z. and K.M.; formal analysis, K.Z.; investigation, K.Z.; resources, K.Z.; data curation, K.Z., H.C. and Y.J.; writing—original draft preparation, K.Z. and K.M.; writing—review and editing, K.Z., H.Z., H.C., Y.J., K.M., Z.Y.; visualization, K.Z.; supervision, H.Z. and K.M.; project administration, H.Z.; funding acquisition, H.Z. and K.M. All authors have read and agreed to the published version of the manuscript.

Funding: This research was funded by the National Natural Science Foundation of China (42107486), the Science and Technology Program of Guizhou Province (Qiankehe Zhicheng [2020] 4Y190, Qiankehe Zhicheng [2019] 2856), Scientific and Technological Innovation Talent Team of Guizhou Province [2019] 5618, STS of CAS (KFJ-STS-QYZD-185), and the China Postdoctoral Science Foundation (2020M673302).

Data Availability Statement: No new data were created or analyzed in this study. Data sharing is not applicable to this article.

Acknowledgments: The authors acknowledge support from the National Natural Science Foundation of China (42107486), the Science and Technology Program of Guizhou Province (Qiankehe Zhicheng [2020] 4Y190, Qiankehe Zhicheng [2019] 2856), Scientific and Technological Innovation Talent Team of Guizhou Province [2019] 5618, STS of CAS (KFJ-STS-QYZD-185), and the China Postdoctoral Science Foundation (2020M673302); Z.Y. is thankful for a UK NERC Fellowship (NE/R013349/2) and Royal Academy of Engineering Frontier Follow-up Grant (FF\1920\1\36).

Conflicts of Interest: The authors declare no competing financial interest.

References

1. Blair, B.D.; Crago, J.P.; Hedman, C.J.; Klaper, R.D. Pharmaceuticals and personal care products found in the Great Lakes above concentrations of environmental concern. *Chemosphere* **2013**, *93*, 2116–2123. [CrossRef]
2. Eckert, E.M.; Di Cesare, A.; Kettner, M.T.; Arias-Andres, M.; Fontaneto, D.; Grossart, H.P.; Corno, G. Microplastics increase impact of treated wastewater on freshwater microbial community. *Environ. Pollut.* **2018**, *234*, 495–502. [CrossRef]
3. Wang, Z.; Han, S.; Cai, M.; Du, P.; Zhang, Z.; Li, X. Environmental behaviour of methamphetamine and ketamine in aquatic ecosystem: Degradation, bioaccumulation, distribution, and associated shift in toxicity and bacterial community. *Water Res.* **2020**, *174*, 115585. [CrossRef]
4. Yu, Y.Y.; Huang, Q.X.; Wang, Z.F.; Zhang, K.; Tang, C.M.; Cui, J.L.; Feng, J.L.; Peng, X.Z. Occurrence and behaviour of pharmaceuticals, steroid hormones, and endocrine-disrupting personal care products in wastewater and the recipient river water of the Pearl River Delta, South China. *J. Environ. Monitor.* **2011**, *13*, 871–878. [CrossRef] [PubMed]
5. Cheng, D.L.; Ngo, H.H.; Guo, W.S.; Liu, Y.W.; Zhou, J.L.; Chang, S.W.; Nguyen, D.D.; Bui, X.T.; Zhang, X.B. Bioprocessing for elimination antibiotics and hormones from swine wastewater. *Sci. Total Environ.* **2018**, *621*, 1664–1682. [CrossRef] [PubMed]
6. Hamid, H.; Eskicioglu, C. Fate of oestrogenic hormones in wastewater and sludge treatment: A review of properties and analytical detection techniques in sludge matrix. *Water Res.* **2012**, *46*, 5813–5833. [CrossRef]
7. Ganjali, M.R.; Faridbod, F.; Davarkhah, N.; Shahtaheri, S.J.; Norouzi, P. All Solid State Graphene Based Potentiometric Sensors for Monitoring of Mercury Ions in Waste Water Samples. *Int. J. Environ. Res.* **2015**, *9*, 333–340.
8. Gavrilescu, M.; Demnerova, K.; Aamand, J.; Agathoss, S.; Fava, F. Emerging pollutants in the environment: Present and future challenges in biomonitoring, ecological risks and bioremediation. *New Biotechnol.* **2015**, *32*, 147–156. [CrossRef]
9. Henderson, R.K.; Baker, A.; Murphy, K.R.; Hamblya, A.; Stuetz, R.M.; Khan, S.J. Fluorescence as a potential monitoring tool for recycled water systems: A review. *Water Res.* **2009**, *43*, 863–881. [CrossRef] [PubMed]
10. Sousa, J.C.G.; Ribeiro, A.R.; Barbosa, M.O.; Pereira, M.F.R.; Silva, A.M.T. A review on environmental monitoring of water organic pollutants identified by EU guidelines. *J. Hazard. Mater.* **2018**, *344*, 146–162. [CrossRef]
11. Ali, M.M.; Li, F.; Zhang, Z.; Zhang, K.; Kang, D.-K.; Ankrum, J.A.; Le, X.C.; Zhao, W. Rolling circle amplification: A versatile tool for chemical biology, materials science and medicine. *Chem. Soc. Rev.* **2014**, *43*, 3324–3341. [CrossRef]

12. Lizardi, P.M.; Huang, X.H.; Zhu, Z.R.; Bray-Ward, P.; Thomas, D.C.; Ward, D.C. Mutation detection and single-molecule counting using isothermal rolling-circle amplification. *Nat. Genet.* **1998**, *19*, 225–232. [CrossRef] [PubMed]
13. Murakami, T.; Sumaoka, J.; Komiyama, M. Sensitive isothermal detection of nucleic-acid sequence by primer generation-rolling circle amplification. *Nucleic Acids Res.* **2009**, *37*, e19. [CrossRef] [PubMed]
14. Zhao, W.A.; Ali, M.M.; Brook, M.A.; Li, Y.F. Rolling circle amplification: Applications in nanotechnology and biodetection with functional nucleic acids. *Angew. Chem. Int. Ed.* **2008**, *47*, 6330–6337. [CrossRef] [PubMed]
15. Zhao, Y.X.; Chen, F.; Li, Q.; Wang, L.H.; Fan, C.H. Isothermal Amplification of Nucleic Acids. *Chem. Rev.* **2015**, *115*, 12491–12545. [CrossRef] [PubMed]
16. Guo, Y.N.; Wang, Y.; Liu, S.; Yu, J.H.; Wang, H.Z.; Wang, Y.L.; Huang, J.D. Label-free and highly sensitive electrochemical detection of E-coli based on rolling circle amplifications coupled peroxidase-mimicking DNAzyme amplification. *Biosens. Bioelectron.* **2016**, *75*, 315–319. [CrossRef]
17. Zhang, K.Y.; Lv, S.Z.; Lu, M.H.; Tang, D.P. Photoelectrochemical biosensing of disease marker on p-type Cu-doped Zn0.3Cd0.7S based on RCA and exonuclease III amplification. *Biosens. Bioelectron.* **2018**, *117*, 590–596. [CrossRef]
18. Qiu, Z.L.; Shu, J.; He, Y.; Lin, Z.Z.; Zhang, K.Y.; Lv, S.Z.; Tang, D.P. CdTe/CdSe quantum dot-based fluorescent aptasensor with hemin/G-quadruplex DNzyme for sensitive detection of lysozyme using rolling circle amplification and strand hybridization. *Biosens. Bioelectron.* **2017**, *87*, 18–24. [CrossRef]
19. Sun, D.P.; Lu, J.; Luo, Z.F.; Zhang, L.Y.; Liu, P.Q.; Chen, Z.G. Competitive electrochemical platform for ultrasensitive cytosensing of liver cancer cells by using nanotetrahedra structure with rolling circle amplification. *Biosens. Bioelectron.* **2018**, *120*, 8–14. [CrossRef]
20. Zhang, K.Y.; Lv, S.Z.; Lin, Z.Z.; Li, M.J.; Tang, D.P. Bio-bar-code-based photoelectrochemical immunoassay for sensitive detection of prostate-specific antigen using rolling circle amplification and enzymatic biocatalytic precipitation. *Biosens. Bioelectron.* **2018**, *101*, 159–166. [CrossRef]
21. Chen, A.Y.; Ma, S.Y.; Zhuo, Y.; Chai, Y.Q.; Yuan, R. In Situ Electrochemical Generation of Electrochemiluminescent Silver Naonoclusters on Target-Cycling Synchronized Rolling Circle Amplification Platform for MicroRNA Detection. *Anal. Chem.* **2016**, *88*, 3203–3210. [CrossRef] [PubMed]
22. Fan, T.; Du, Y.; Yao, Y.; Wu, J.; Meng, S.; Luo, J.; Zhang, X.; Yang, D.; Wang, C.; Qian, Y.; et al Rolling circle amplification triggered poly adenine-gold nanoparticles production for label-free electrochemical detection of thrombin. *Sens. Actuators B Chem.* **2018**, *266*, 9–18. [CrossRef]
23. He, Y.; Yang, X.; Yuan, R.; Chai, Y.Q. "Off" to "On" Surface-Enhanced Raman Spectroscopy Platform with Padlock Probe-Based Exponential Rolling Circle Amplification for Ultrasensitive Detection of MicroRNA 155. *Anal. Chem.* **2017**, *89*, 2866–2872. [CrossRef] [PubMed]
24. Qiu, Z.L.; Shu, J.; Tang, D.P. Near-Infrared-to-Ultraviolet Light-Mediated Photoelectrochemical Aptasensing Platform for Cancer Biomarker Based on Core Shell NaYF4:Yb,Tm@TiO2 Upconversion Microrods. *Anal. Chem.* **2018**, *90*, 1021–1028. [CrossRef] [PubMed]
25. Kim, T.Y.; Lim, M.C.; Woo, M.A.; Jun, B.H. Radial Flow Assay Using Gold Nanoparticles and Rolling Circle Amplification to Detect Mercuric Ions. *Nanomaterials* **2018**, *8*, 81. [CrossRef] [PubMed]
26. Wen, J.; Li, W.S.; Li, J.Q.; Tao, B.B.; Xu, Y.Q.; Li, H.J.; Lu, A.P.; Sun, S.G. Study on rolling circle amplification of Ebola virus and fluorescence detection based on graphene oxide. *Sens. Actuators B Chem.* **2016**, *227*, 655–659. [CrossRef]
27. Zhao, J.; Lei, Y.-M.; Chai, Y.-Q.; Yuan, R.; Zhuo, Y. Novel electrochemiluminescence of perylene derivative and its application to mercury ion detection based on a dual amplification strategy. *Biosens. Bioelectron.* **2016**, *86*, 720–727. [CrossRef]
28. Osborne, R.J.; Thornton, C.A. Cell-free cloning of highly expanded CTG repeats by amplification of dimerized expanded repeats. *Nucleic Acids Res.* **2008**, *36*, e24. [CrossRef]
29. Wang, F.; Lu, C.H.; Liu, X.Q.; Freage, L.; Willner, I. Amplified and Multiplexed Detection of DNA Using the Dendritic Rolling Circle Amplified Synthesis of DNAzyme Reporter Units. *Anal. Chem.* **2014**, *86*, 1614–1621. [CrossRef]
30. Du, Y.C.; Zhu, Y.J.; Li, X.Y.; Kong, D.M. Amplified detection of genome-containing biological targets using terminal deoxynucleotidyl transferase-assisted rolling circle amplification. *Chem. Commun.* **2018**, *54*, 682–685. [CrossRef]
31. Inoue, J.; Shigemori, Y.; Mikawa, T. Improvements of rolling circle amplification (RCA) efficiency and accuracy using Thermus thermophilus SSB mutant protein. *Nucleic Acids Res.* **2006**, *34*, e69. [CrossRef] [PubMed]
32. Li, J.S.; Deng, T.; Chu, X.; Yang, R.H.; Jiang, J.H.; Shen, G.L.; Yu, R.Q. Rolling Circle Amplification Combined with Gold Nanoparticle Aggregates for Highly Sensitive Identification of Single-Nucleotide Polymorphisms. *Anal. Chem.* **2010**, *82*, 2811–2816. [CrossRef] [PubMed]
33. Qi, X.Q.; Bakht, S.; Devos, K.M.; Gale, M.D.; Osbourn, A. L-RCA (ligation-rolling circle amplification): A general method for genotyping of shingle nucleotide polymorphisms (SNPs). *Nucleic Acids Res.* **2001**, *29*, e116. [CrossRef] [PubMed]
34. Zhou, H.X.; Wang, H.; Liu, C.H.; Wang, H.H.; Duan, X.R.; Li, Z.P. Ultrasensitive genotyping with target-specifically generated circular DNA templates and RNA FRET probes. *Chem. Commun.* **2015**, *51*, 11556–11559. [CrossRef] [PubMed]
35. Ko, O.; Han, S.; Lee, J.B. Selective release of DNA nanostructures from DNA hydrogel. *J. Ind. Eng. Chem.* **2020**, *84*, 46–51. [CrossRef]
36. Liu, M.; Zhang, Q.; Li, Z.P.; Gu, J.; Brennan, J.D.; Li, Y.F. Programming a topologically constrained DNA nanostructure into a sensor. *Nat. Commun.* **2016**, *7*, 12074. [CrossRef]

37. Xu, H.; Zhang, S.X.; Ouyang, C.H.; Wang, Z.M.; Wu, D.; Liu, Y.Y.; Jiang, Y.F.; Wu, Z.S. DNA nanostructures from palindromic rolling circle amplification for the fluorescent detection of cancer-related microRNAs. *Talanta* **2019**, *192*, 175–181. [CrossRef]
38. Zhang, Z.Q.; Zhang, H.Z.; Wang, F.; Zhang, G.D.; Zhou, T.; Wang, X.F.; Liu, S.Z.; Liu, T.T. DNA Block Macromolecules Based on Rolling Circle Amplification Act as Scaffolds to Build Large-Scale Origami Nanostructures. *Macromol. Rapid. Commun.* **2018**, *39*, 1800263. [CrossRef]
39. Mohsen, M.G.; Kool, E.T. The Discovery of Rolling Circle Amplification and Rolling Circle Transcription. *Acc. Chem. Res.* **2016**, *49*, 2540–2550. [CrossRef]
40. Fire, A.; Xu, S.Q. Rolling Replication of Short Dna Circles. *Proc. Natl. Acad. Sci. USA* **1995**, *92*, 4641–4645. [CrossRef]
41. Liu, D.Y.; Daubendiek, S.L.; Zillman, M.A.; Ryan, K.; Kool, E.T. Rolling circle DNA synthesis: Small circular oligonucleotides as efficient templates for DNA polymerases. *J. Am. Chem. Soc.* **1996**, *118*, 1587–1594. [CrossRef]
42. Blanco, L.; Bernad, A.; Lazaro, J.M.; Martin, G.; Garmendia, C.; Salas, M. Highly efficient DNA synthesis by the phage phi 29 DNA polymerase. Symmetrical mode of DNA replication. *J. Biol. Chem.* **1989**, *264*, 8935–8940. [CrossRef]
43. Krzywkowski, T.; Kuhnemund, M.; Wu, D.; Nilsson, M. Limited reverse transcriptase activity of phi29 DNA polymerase. *Nucleic Acids Res.* **2018**, *46*, 3625–3632. [CrossRef] [PubMed]
44. Neumann, F.; Hernandez-Neuta, I.; Grabbe, M.; Madaboosi, N.; Albert, J.; Nilsson, M. Padlock Probe Assay for Detection and Subtyping of Seasonal Influenza. *Clin. Chem.* **2018**, *64*, 1704–1712. [CrossRef]
45. Nilsson, M.; Malmgren, H.; Samiotaki, M.; Kwiatkowski, M.; Chowdhary, B.P.; Landegren, U. Padlock probes: Circularizing oligonucleotides for localized DNA detection. *Science* **1994**, *265*, 2085–2088. [CrossRef] [PubMed]
46. Tang, S.M.; Wei, H.; Hu, T.Y.; Jiang, J.Q.; Chang, J.L.; Guan, Y.F.; Zhao, G.J. Suppression of rolling circle amplification by nucleotide analogues in circular template for three DNA polymerases. *Biosci. Biotech. Bioch.* **2016**, *80*, 1555–1561. [CrossRef]
47. Li, D.X.; Zhang, T.T.; Yang, F.; Yuan, R.; Xiang, Y. Efficient and Exponential Rolling Circle Amplification Molecular Network Leads to Ultrasensitive and Label-Free Detection of MicroRNA. *Anal. Chem.* **2020**, *92*, 2074–2079. [CrossRef]
48. Li, X.Y.; Cui, Y.X.; Du, Y.C.; Tang, A.N.; Kong, D.M. Label-Free Telomerase Detection in Single Cell Using a Five-Base Telomerase Product-Triggered Exponential Rolling Circle Amplification Strategy. *ACS Sens.* **2019**, *4*, 1090–1096. [CrossRef]
49. Xu, H.; Xue, C.; Zhang, R.B.; Chen, Y.R.; Li, F.; Shen, Z.F.; Jia, L.; Wu, Z.S. Exponential rolling circle amplification and its sensing application for highly sensitive DNA detection of tumour suppressor gene. *Sens. Actuators B Chem.* **2017**, *243*, 1240–1247. [CrossRef]
50. Pumford, E.A.; Lu, J.; Spaczai, I.; Prasetyo, M.E.; Zheng, E.M.; Zhang, H.; Kamei, D.T. Developments in integrating nucleic acid isothermal amplification and detection systems for point-of-care diagnostics. *Biosens. Bioelectron.* **2020**, *170*, 112674. [CrossRef]
51. Ulanovsky, L.; Bodner, M.; Trifonov, E.N.; Choder, M. Curved DNA: Design, synthesis, and circularization. *Proc. Natl. Acad. Sci. USA* **1986**, *83*, 862–866. [CrossRef]
52. Jin, G.; Wang, C.; Yang, L.; Li, X.; Guo, L.; Qiu, B.; Lin, Z.; Chen, G. Hyperbranched rolling circle amplification based electro-chemiluminescence aptasensor for ultrasensitive detection of thrombin. *Biosens. Bioelectron.* **2015**, *63*, 166–171. [CrossRef]
53. Wang, X.M.; Teng, D.; Guan, Q.F.; Tian, F.; Wang, J.H. Detection of genetically modified crops using multiplex asymmetric polymerase chain reaction and asymmetric hyperbranched rolling circle amplification coupled with reverse dot blot. *Food Chem.* **2015**, *173*, 1022–1029. [CrossRef]
54. Yang, L.; Tao, Y.; Yue, G.; Li, R.; Qin, B.; Guo, L.; Lin, Z.; Yang, H.-H. Highly Selective and Sensitive Electrochemiluminescence Biosensor for p53 DNA Sequence Based on Nicking Endonuclease Assisted Target Recycling and Hyperbranched Rolling Circle Amplification. *Anal. Chem.* **2016**, *88*, 5097–5103. [CrossRef] [PubMed]
55. Zhang, L.-R.; Zhu, G.; Zhang, C.-Y. Homogeneous and Label-Free Detection of MicroRNAs Using Bifunctional Strand Displacement Amplification-Mediated Hyperbranched Rolling Circle Amplification. *Anal. Chem.* **2014**, *86*, 6703–6709. [CrossRef] [PubMed]
56. Dahl, F.; Baner, J.; Gullberg, M.; Mendel-Hartvig, M.; Landegren, U.; Nilsson, M. Circle-to-circle amplification for precise and sensitive DNA analysis. *Proc. Natl. Acad. Sci. USA* **2004**, *101*, 4548–4553. [CrossRef] [PubMed]
57. Liu, M.; Yin, Q.X.; McConnell, E.M.; Chang, Y.Y.; Brennan, J.D.; Li, Y.F. DNAzyme Feedback Amplification: Relaying Molecular Recognition to Exponential DNA Amplification. *Chem. Eur. J.* **2018**, *24*, 4473–4479. [CrossRef]
58. Zhao, Y.H.; Wang, Y.; Liu, S.; Wang, C.L.; Liang, J.X.; Li, S.S.; Qu, X.N.; Zhang, R.F.; Yu, J.H.; Huang, J.D. Triple-helix molecular-switch-actuated exponential rolling circular amplification for ultrasensitive fluorescence detection of miRNAs. *Analyst* **2019**, *144*, 5245–5253. [CrossRef]
59. Jiang, Y.; Zou, S.; Cao, X. A simple dendrimer-aptamer based microfluidic platform for *E. coli* O157:H7 detection and signal intensification by rolling circle amplification. *Sens. Actuators B Chem.* **2017**, *251*, 976–984. [CrossRef]
60. Peng, X.; Liang, W.-B.; Wen, Z.-B.; Xiong, C.-Y.; Zheng, Y.-N.; Chai, Y.-Q.; Yuan, R. Ultrasensitive Fluorescent Assay Based on a Rolling-Circle-Amplification-Assisted Multisite-Strand-Displacement-Reaction Signal-Amplification Strategy. *Anal. Chem.* **2018**, *90*, 7474–7479. [CrossRef]
61. Wang, J.; Dong, H.-Y.; Zhou, Y.; Han, L.-Y.; Zhang, T.; Lin, M.; Wang, C.; Xu, H.; Wu, Z.-S.; Jia, L. Immunomagnetic antibody plus aptamer pseudo-DNA nanocatenane followed by rolling circle amplication for highly sensitive CTC detection. *Biosens. Bioelectron.* **2018**, *122*, 239–246. [CrossRef] [PubMed]
62. Wang, J.; Mao, S.; Li, H.-F.; Lin, J.-M. Multi-DNAzymes-functionalized gold nanoparticles for ultrasensitive chemiluminescence detection of thrombin on microchip. *Anal. Chim. Acta* **2018**, *1027*, 76–82. [CrossRef] [PubMed]

63. Tang, S.R.; Tong, P.; Li, H.; Tang, J.; Zhang, L. Ultrasensitive electrochemical detection of Pb2+ based on rolling circle amplification and quantum dots tagging. *Biosens. Bioelectron.* **2013**, *42*, 608–611. [CrossRef] [PubMed]

64. Cai, W.; Xie, S.B.; Zhang, J.; Tang, D.Y.; Tang, Y. Immobilized-free miniaturized electrochemical sensing system for Pb2+ detection based on dual Pb2+-DNAzyme assistant feedback amplification strategy. *Biosens. Bioelectron.* **2018**, *117*, 312–318. [CrossRef] [PubMed]

65. Cheng, W.; Zhang, W.; Yan, Y.R.; Shen, B.; Zhu, D.; Lei, P.H.; Ding, S.J. A novel electrochemical biosensor for ultrasensitive and specific detection of DNA based on molecular beacon mediated circular strand displacement and rolling circle amplification. *Biosens. Bioelectron.* **2014**, *62*, 274–279. [CrossRef]

66. Mittal, S.; Thakur, S.; Mantha, A.K.; Kaur, H. Bio-analytical applications of nicking endonucleases assisted signal-amplification strategies for detection of cancer biomarkers -DNA methyl transferase and microRNA. *Biosens. Bioelectron.* **2019**, *124*, 233–243. [CrossRef]

67. Bialy, R.M.; Ali, M.M.; Li, Y.F.; Brennan, J.D. Protein-Mediated Suppression of Rolling Circle Amplification for Biosensing with an Aptamer-Containing DNA Primer. *Chem. Eur. J.* **2020**, *26*, 5085–5092. [CrossRef]

68. Fan, T.T.; Mao, Y.; Liu, F.; Zhang, W.; Lin, J.S.; Yin, J.X.; Tan, Y.; Huang, X.T.; Jiang, Y.Y. Label-free fluorescence detection of circulating microRNAs based on duplex-specific nuclease-assisted target recycling coupled with rolling circle amplification. *Talanta* **2019**, *200*, 480–486. [CrossRef]

69. Duy, J.; Smith, R.L.; Collins, S.D.; Connell, L.B. A field-deployable colorimetric bioassay for the rapid and specific detection of ribosomal RNA. *Biosens. Bioelectron.* **2014**, *52*, 433–437. [CrossRef]

70. Wen, Y.Q.; Xu, Y.; Mao, X.H.; Wei, Y.L.; Song, H.Y.; Chen, N.; Huang, Q.; Fan, C.H.; Li, D. DNAzyme-Based Rolling-Circle Amplification DNA Machine for Ultrasensitive Analysis of MicroRNA in Drosophila Larva. *Anal. Chem.* **2012**, *84*, 7664–7669. [CrossRef]

71. Tang, L.H.; Liu, Y.; Ali, M.M.; Kang, D.K.; Zhao, W.A.; Li, J.H. Colorimetric and Ultrasensitive Bioassay Based on a Dual-Amplification System Using Aptamer and DNAzyme. *Anal. Chem.* **2012**, *84*, 4711–4717. [CrossRef] [PubMed]

72. Du, J.; Xu, Q.F.; Lu, X.Q.; Zhang, C.Y. A Label-Free Bioluminescent Sensor for Real-Time Monitoring Polynucleotide Kinase Activity. *Anal. Chem.* **2014**, *86*, 8481–8488. [CrossRef] [PubMed]

73. Mashimo, Y.; Mie, M.; Suzuki, S.; Kobatake, E. Detection of small RNA molecules by a combination of branched rolling circle amplification and bioluminescent pyrophosphate assay. *Anal. Bioanal. Chem.* **2011**, *401*, 221–227. [CrossRef] [PubMed]

74. Chen, J.; Tong, P.; Lin, Y.; Lu, W.; He, Y.; Lu, M.; Zhang, L.; Chen, G. Highly sensitive fluorescent sensor for mercury based on hyperbranched rolling circle amplification. *Analyst* **2015**, *140*, 907–911. [CrossRef]

75. Xie, S.; Tang, Y.; Tang, D. Highly sensitive electrochemical detection of mercuric ions based on sequential nucleic acid amplification and guanine nanowire formation. *Anal. Methods* **2017**, *9*, 5478–5483. [CrossRef]

76. Zhao, Y.; Liu, H.; Chen, F.; Bai, M.; Zhao, J.; Zhao, Y. Trifunctional molecular beacon-mediated quadratic amplification for highly sensitive and rapid detection of mercury(II) ion with tunable dynamic range. *Biosens. Bioelectron.* **2016**, *86*, 892–898. [CrossRef]

77. Lv, J.; Xie, S.; Cai, W.; Zhang, J.; Tang, D.; Tang, Y. Highly effective target converting strategy for ultrasensitive electrochemical assay of Hg2+. *Analyst* **2017**, *142*, 4708–4714. [CrossRef]

78. Wu, S.; Yu, Q.; He, C.; Duan, N. Colorimetric aptasensor for the detection of mercury based on signal intensification by rolling circle amplification. *Spectrochim. Acta Part A* **2020**, *224*, 117387. [CrossRef]

79. Lim, J.W.; Kim, T.-Y.; Choi, S.-W.; Woo, M.-A. 3D-printed rolling circle amplification chip for on-site colorimetric detection of inorganic mercury in drinking water. *Food Chem.* **2019**, *300*, 125177. [CrossRef]

80. Lu, W.; Lin, C.; Yang, J.; Wang, X.; Yao, B.; Wang, M. A DNAzyme assay coupled with effective magnetic separation and rolling circle amplification for detection of lead cations with a smartphone camera. *Anal. Bioanal. Chem.* **2019**, *411*, 5383–5391. [CrossRef]

81. Tang, D.; Xia, B.; Tang, Y.; Zhang, J.; Zhou, Q. Metal-ion-induced DNAzyme on magnetic beads for detection of lead(II) by using rolling circle amplification, glucose oxidase, and readout of pH changes. *Microchim. Acta* **2019**, *186*, 318. [CrossRef]

82. Cheng, X.; Yu, X.; Chen, L.; Zhang, H.; Wu, Y.; Fu, F. Visual detection of ultra-trace levels of uranyl ions using magnetic bead-based DNAzyme recognition in combination with rolling circle amplification. *Microchim. Acta* **2017**, *184*, 4259–4267. [CrossRef]

83. Li, X.; Song, J.; Xue, Q.-W.; You, F.-H.; Lu, X.; Kong, Y.-C.; Ma, S.-Y.; Jiang, W.; Li, C.-Z. A Label-Free and Sensitive Fluorescent Qualitative Assay for Bisphenol A Based on Rolling Circle Amplification/Exonuclease III-Combined Cascade Amplification. *Nanomaterials* **2016**, *6*, 190. [CrossRef] [PubMed]

84. Ma, Q.; Li, P.; Gao, Z.; Yau Li, S.F. Rapid, sensitive and highly specific label-free fluorescence biosensor for microRNA by branched rolling circle amplification. *Sens. Actuators B Chem.* **2019**, *281*, 424–431. [CrossRef]

85. Xu, Y.; Wang, H.; Luan, C.; Fu, F.; Chen, B.; Liu, H.; Zhao, Y. Porous Hydrogel Encapsulated Photonic Barcodes for Multiplex MicroRNA Quantification. *Adv. Funct. Mater.* **2018**, *28*, 1704458. [CrossRef]

86. Xu, L.-P.; Chen, Y.; Yang, G.; Shi, W.; Dai, B.; Li, G.; Cao, Y.; Wen, Y.; Zhang, X.; Wang, S. Ultratrace DNA Detection Based on the Condensing-Enrichment Effect of Superwettable Microchips. *Adv. Mater.* **2015**, *27*, 6878–6884. [CrossRef]

87. Zhang, J.; Fu, R.; Xie, L.; Li, Q.; Zhou, W.; Wang, R.; Ye, J.; Wang, D.; Xue, N.; Lin, X.; et al. A smart device for label-free and real-time detection of gene point mutations based on the high dark phase contrast of vapour condensation. *Lab Chip* **2015**, *15*, 3891–3896. [CrossRef]

88. He, Z.; Wei, J.; Gan, C.; Liu, W.; Liu, Y. A rolling circle amplification signal-enhanced immunosensor for ultrasensitive microcystin-LR detection based on a magnetic graphene-functionalized electrode. *RSC Adv.* **2017**, *7*, 39906–39913. [CrossRef]

89. Hui, C.Y.; Liu, M.; Li, Y.; Brennan, J.D. A Paper Sensor Printed with Multifunctional Bio/Nano Materials. *Angew. Chem. Int. Ed.* **2018**, *57*, 4549–4553. [CrossRef]

90. Zhang, C.; Chen, G.; Wang, Y.; Zhou, J.; Li, C. Establishment and application of hyperbranched rolling circle amplification coupled with lateral flow dipstick for the sensitive detection of *Karenia mikimotoi*. *Harmful Algae* **2019**, *84*, 151–160. [CrossRef]

91. Zhang, C.; Sun, R.; Wang, Y.; Chen, G.; Guo, C.; Zhou, J. Comparative detection of *Karenia mikimotoi* by exponential rolling circle amplification (E-RCA) and double-ligation E-RCA. *J. Appl. Psychol.* **2019**, *31*, 505–518. [CrossRef]

92. Zhang, C.; Chen, G.; Wang, Y.; Sun, R.; Nie, X.; Zhou, J. MHBMDAA: Membrane-based DNA array with high resolution and sensitivity for toxic microalgae monitoring. *Harmful Algae* **2018**, *80*, 107–116. [CrossRef]

93. Najafzadeh, M.J.; Vicente, V.A.; Feng, P.; Naseri, A.; Sun, J.; Rezaei-Matehkolaei, A.; de Hoog, G.S. Rapid Identification of Seven Waterborne Exophiala Species by RCA DNA Padlock Probes. *Mycopathologia* **2018**, *183*, 669–677. [CrossRef] [PubMed]

94. Yang, X.; Yang, K.; Zhao, X.; Lin, Z.; Liu, Z.; Luo, S.; Zhang, Y.; Wang, Y.; Fu, W. Terahertz spectroscopy for the isothermal detection of bacterial DNA by magnetic bead-based rolling circle amplification. *Analyst* **2017**, *142*, 4661–4669. [CrossRef] [PubMed]

95. Nie, X.; Zhang, C.; Wang, Y.; Guo, C.; Zhou, J.; Chen, G. Application of hyperbranched rolling circle amplification (HRCA) and HRCA-based strip test for the detection of *Chattonella marina*. *Environ. Sci. Pollut. Res.* **2017**, *24*, 15678–15688. [CrossRef]

96. Pearson, V.M.; Caudle, S.B.; Rokyta, D.R. Viral recombination blurs taxonomic lines: Examination of single-stranded DNA viruses in a wastewater treatment plant. *PeerJ* **2016**, *4*, 18. [CrossRef]

97. Chen, G.; Cai, P.; Zhang, C.; Wang, Y.; Zhang, S.; Guo, C.; Lu, D.D. Hyperbranched rolling circle amplification as a novel method for rapid and sensitive detection of *Amphidinium carterae*. *Harmful Algae* **2015**, *47*, 66–74. [CrossRef]

98. Bejhed, R.S.; Zardán Gómez de la Torre, T.; Svedlindh, P.; Strömberg, M. Optomagnetic read-out enables easy, rapid, and cost-efficient qualitative biplex detection of bacterial DNA sequences. *Biotechnol. J.* **2015**, *10*, 469–472. [CrossRef]

99. Ge, C.; Yuan, R.; Yi, L.; Yang, J.; Zhang, H.; Li, L.; Nian, W.; Yi, G. Target-induced aptamer displacement on gold nanoparticles and rolling circle amplification for ultrasensitive live *Salmonella typhimurium* electrochemical biosensing. *J. Electroanal.Chem.* **2018**, *826*, 174–180. [CrossRef]

100. Gao, Z.F.; Liu, R.; Wang, J.; Dai, J.; Huang, W.-H.; Liu, M.; Wang, S.; Xia, F.; Jiang, L. Controlling Droplet Motion on an Organogel Surface by Tuning the Chain Length of DNA and Its Biosensing Application. *Chem* **2018**, *4*, 2929–2943. [CrossRef]

101. Mao, K.; Zhang, H.; Wang, Z.L.; Cao, H.R.; Zhang, K.K.; Li, X.Q.; Yang, Z.G. Nanomaterial-based aptamer sensors for arsenic detection. *Biosens. Bioelectron.* **2020**, *148*, 111785. [CrossRef] [PubMed]

102. Aragay, G.; Pons, J.; Merkoci, A. Recent Trends in Macro-, Micro-, and Nanomaterial-Based Tools and Strategies for Heavy-Metal Detection. *Chem. Rev.* **2011**, *111*, 3433–3458. [CrossRef]

103. Gumpu, M.B.; Sethuraman, S.; Krishnan, U.M.; Rayappan, J.B.B. A review on detection of heavy metal ions in water—An electrochemical approach. *Sens. Actuators B Chem.* **2015**, *213*, 515–533. [CrossRef]

104. Kim, H.N.; Ren, W.X.; Kim, J.S.; Yoon, J. Fluorescent and colorimetric sensors for detection of lead, cadmium, and mercury ions. *Chem. Soc. Rev.* **2012**, *41*, 3210–3244. [CrossRef] [PubMed]

105. Zhan, S.; Wu, Y.; Wang, L.; Zhan, X.; Zhou, P. A mini-review on functional nucleic acids-based heavy metal ion detection. *Biosens. Bioelectron.* **2016**, *86*, 353–368. [CrossRef] [PubMed]

106. Chang, C.; Chen, C.; Yin, R.; Shen, Y.; Mao, K.; Yang, Z.; Feng, X.; Zhang, H. Bioaccumulation of Hg in Rice Leaf Facilitates Selenium Bioaccumulation in Rice (*Oryza sativa* L.) Leaf in the Wanshan Mercury Mine. *Environ. Sci. Technol.* **2020**, *54*, 3228–3236. [CrossRef] [PubMed]

107. Guo, M.; Wang, J.; Du, R.; Liu, Y.; Chi, J.; He, X.; Huang, K.; Luo, Y.; Xu, W. A test strip platform based on a whole-cell microbial biosensor for simultaneous on-site detection of total inorganic mercury pollutants in cosmetics without the need for predigestion. *Biosens. Bioelectron.* **2020**, *150*, 111899. [CrossRef]

108. Ma, X.Y.; Miao, P. Silver nanoparticle@DNA tetrahedron-based colorimetric detection of HIV-related DNA with cascade strand displacement amplification. *J. Mater. Chem. B* **2019**, *7*, 2608–2612. [CrossRef] [PubMed]

109. Wang, K.; Fan, D.Q.; Liu, Y.Q.; Dong, S.J. Cascaded multiple amplification strategy for ultrasensitive detection of HIV/HCV virus DNA. *Biosens. Bioelectron.* **2017**, *87*, 116–121. [CrossRef]

110. Hu, X.L.; Li, C.; Feng, C.; Mao, X.X.; Xiang, Y.; Li, G.X. One-step colorimetric detection of an antibody based on protein-induced unfolding of a G-quadruplex switch. *Chem. Commun.* **2017**, *53*, 4692–4694. [CrossRef]

111. Chong, H.Q.; Ching, C.B. Development of Colorimetric-Based Whole-Cell Biosensor for Organophosphorus Compounds by Engineering Transcription Regulator DmpR. *ACS Synth. Biol.* **2016**, *5*, 1290–1298. [CrossRef]

112. Tao, Y.; Li, M.Q.; Kim, B.; Auguste, D.T. Incorporating gold nanoclusters and target-directed liposomes as a synergistic amplified colorimetric sensor for HER2-positive breast cancer cell detection. *Theranostics* **2017**, *7*, 899–911. [CrossRef]

113. Ye, X.S.; Shi, H.; He, X.X.; Wang, K.M.; He, D.G.; Yan, L.A.; Xu, F.Z.; Lei, Y.L.; Tang, J.L.; Yu, Y.R. Iodide-Responsive Cu-Au Nanoparticle-Based Colorimetric Platform for Ultrasensitive Detection of Target Cancer Cells. *Anal. Chem.* **2015**, *87*, 7141–7147. [CrossRef]

114. Liu, L.; Lin, H.W. Paper-Based Colorimetric Array Test Strip for Selective and Semiquantitative Multi-Ion Analysis: Simultaneous Detection of Hg2+, Ag+, and Cu2+. *Anal. Chem.* **2014**, *86*, 8829–8834. [CrossRef]

115. Ji, R.; Niu, W.; Chen, S.; Xu, W.; Ji, X.; Yuan, L.; Zhao, H.; Geng, M.; Qiu, J.; Li, C. Target-inspired Pb2+-dependent DNAzyme for ultrasensitive electrochemical sensor based on MoS2-AuPt nanocomposites and hemin/G-quadruplex DNAzyme as signal amplifier. *Biosens. Bioelectron.* **2019**, *144*, 111560. [CrossRef] [PubMed]

116. Liu, X.F.; Wang, Y.T.; Hua, X.X.; Huang, Y.Q.; Feng, X.M.; Fan, Q.L.; Huang, W. Rapid Detection of Lead Ion (II) Based on Cationic Conjugated Polymer and Aptamer. *Chin. J. Anal. Chem.* **2016**, *44*, 1092–1098.

117. Tsekenis, G.; Filippidou, M.K.; Chatzipetrou, M.; Tsouti, V.; Zergioti, I.; Chatzandroulis, S. Heavy metal ion detection using a capacitive micromechanical biosensor array for environmental monitoring. *Sens. Actuators B Chem.* **2015**, *208*, 628–635. [CrossRef]

118. Chen, X.; Ba, Y.; Ma, L.J.; Cai, X.; Yin, Y.; Wang, K.H.; Guo, J.G.; Zhang, Y.J.; Chen, J.N.; Guo, X.; et al. Characterization of microRNAs in serum: A novel class of biomarkers for diagnosis of cancer and other diseases. *Cell Res.* **2008**, *18*, 997–1006. [CrossRef] [PubMed]

119. Dong, H.F.; Lei, J.P.; Ding, L.; Wen, Y.Q.; Ju, H.X.; Zhang, X.J. MicroRNA: Function, Detection, and Bioanalysis. *Chem. Rev.* **2013**, *113*, 6207–6233. [CrossRef] [PubMed]

120. Chen, Y.X.; Huang, K.J.; Niu, K.X. Recent advances in signal amplification strategy based on oligonucleotide and nanomaterials for microRNA detection-a review. *Biosens. Bioelectron.* **2018**, *99*, 612–624. [CrossRef]

121. Ma, D.D.; Huang, C.X.; Zheng, J.; Tang, J.R.; Li, J.S.; Yang, J.F.; Yang, R.H. Quantitative detection of exosomal microRNA extracted from human blood based on surface-enhanced Raman scattering. *Biosens. Bioelectron.* **2018**, *101*, 167–173. [CrossRef] [PubMed]

122. Peng, L.C.; Zhang, P.; Chai, Y.Q.; Yuan, R. Bi-directional DNA Walking Machine and Its Application in an Enzyme-Free Electrochemiluminescence Biosensor for Sensitive Detection of MicroRNAs. *Anal. Chem.* **2017**, *89*, 5036–5042. [CrossRef] [PubMed]

123. Zhuang, J.Y.; Han, B.; Liu, W.C.; Zhou, J.F.; Liu, K.W.; Yang, D.P.; Tang, D.P. Liposome-amplified photoelectrochemical immunoassay for highly sensitive monitoring of disease biomarkers based on a split-type strategy. *Biosens. Bioelectron.* **2018**, *99*, 230–236. [CrossRef]

124. Dave, V.P.; Ngo, T.A.; Pernestig, A.-K.; Tilevik, D.; Kant, K.; Nguyen, T.; Wolff, A.; Bang, D.D. MicroRNA amplification and detection technologies: Opportunities and challenges for point of care diagnostics. *Lab. Investig.* **2019**, *99*, 452–469. [CrossRef]

125. Zhou, D.; Zhang, H.; Lin, H.; Jiang, J.; Lu, Z. New High Throughput Method to Analyze the Methylation Pattern of Individual DNA Molecules. *Nanosci. Nanotech. Let.* **2018**, *10*, 1554–1561. [CrossRef]

126. Schwarzenbach, R.P.; Egli, T.; Hofstetter, T.B.; von Gunten, U.; Wehrli, B. Global Water Pollution and Human Health. *Annu. Rev. Environ. Resour.* **2010**, *35*, 109–136. [CrossRef]

127. Hammond, J.L.; Formisano, N.; Estrela, P.; Carrara, S.; Tkac, J. Electrochemical biosensors and nanobiosensors. *Biosens. Technol. Detect. Biomol.* **2016**, *60*, 69–80.

128. Jyoti, A.; Tomar, R.S. Detection of pathogenic bacteria using nanobiosensors. *Environ. Chem. Lett.* **2017**, *15*, 1–6. [CrossRef]

129. Ranjbar, S.; Shahrokhian, S. Design and fabrication of an electrochemical aptasensor using Au nanoparticles/carbon nanoparticles/cellulose nanofibres nanocomposite for rapid and sensitive detection of Staphylococcus aureus. *Bioelectrochemistry* **2018**, *123*, 70–76. [CrossRef] [PubMed]

130. Swaminathan, B.; Feng, P. Rapid Detection of Food-borne Pathogenic Bacteria. *Annu. Rev. Microbiol.* **1994**, *48*, 401–426. [CrossRef] [PubMed]

131. Umesha, S.; Manukumar, H.M. Advanced molecular diagnostic techniques for detection of food-borne pathogens: Current applications and future challenges. *Crit. Rev. Food Sci. Nutr.* **2018**, *58*, 84–104. [CrossRef] [PubMed]

132. Rohde, A.; Hammerl, J.A.; Appel, B.; Dieckmann, R.; Al Dahouk, S. FISHing for bacteria in food—A promising tool for the reliable detection of pathogenic bacteria? *Food Microbiol.* **2015**, *46*, 395–407. [CrossRef]

133. Yang, Z.; Kasprzyk-Hordern, B.; Frost, C.G.; Estrela, P.; Thomas, K.V. Community Sewage Sensors for Monitoring Public Health. *Environ. Sci. Technol.* **2015**, *49*, 5845–5846. [CrossRef]

134. Bickman, S.R.; Campbell, K.; Elliott, C.; Murphy, C.; O'Kennedy, R.; Papst, P.; Lochhead, M.J. An Innovative Portable Biosensor System for the Rapid Detection of Freshwater Cyanobacterial Algal Bloom Toxins. *Environ. Sci. Technol.* **2018**, *52*, 11691–11698. [CrossRef]

135. Preece, E.P.; Hardy, F.J.; Moore, B.C.; Bryan, M. A review of microcystin detections in Estuarine and Marine waters: Environmental implications and human health risk. *Harmful Algae* **2017**, *61*, 31–45. [CrossRef]

136. Zhao, Y.; Shi, L.; Kuang, H.; Xu, C. DNA-Driven Nanoparticle Assemblies for Biosensing and Bioimaging. *Top Curr. Chem.* **2020**, *378*, 18. [CrossRef]

137. Tian, B.; Gao, F.; Fock, J.; Dufva, M.; Hansen, M.F. Homogeneous circle-to-circle amplification for real-time optomagnetic detection of SARS-CoV-2 RdRp coding sequence. *Biosens. Bioelectron.* **2020**, *165*, 112356. [CrossRef] [PubMed]

138. Mao, K.; Zhang, H.; Yang, Z. Can a Paper-Based Device Trace COVID-19 Sources with Wastewater-Based Epidemiology? *Environ. Sci. Technol.* **2020**, *54*, 3733–3735. [CrossRef]

139. Na, W.; Nam, D.; Lee, H.; Shin, S. Rapid molecular diagnosis of infectious viruses in microfluidics using DNA hydrogel formation. *Biosens. Bioelectron.* **2018**, *108*, 9–13. [CrossRef]

140. Lu, C.-H.; Willner, B.; Willner, I. DNA Nanotechnology: From Sensing and DNA Machines to Drug-Delivery Systems. *ACS Nano* **2013**, *7*, 8320–8332. [CrossRef]

141. De Avila, B.E.F.; Angell, C.; Soto, F.; Lopez-Ramirez, M.A.; Baez, D.F.; Xie, S.B.; Wang, J.; Chen, Y. Acoustically Propelled Nanomotors for Intracellular siRNA Delivery. *ACS Nano* **2016**, *10*, 4997–5005. [CrossRef] [PubMed]

142. Shimron, S.; Cecconello, A.; Lu, C.-H.; Willner, I. Metal Nanoparticle-Functionalized DNA Tweezers: From Mechanically Programmed Nanostructures to Switchable Fluorescence Properties. *Nano Lett.* **2013**, *13*, 3791–3795. [CrossRef] [PubMed]

143. Li, Y.; Liu, H.; Huang, H.; Deng, J.; Fang, L.; Luo, J.; Zhang, S.; Huang, J.; Liang, W.; Zheng, J. A sensitive electrochemical strategy via multiple amplification reactions for the detection of *E. coli* O157: H7. *Biosens. Bioelectron.* **2020**, *147*, 111752. [CrossRef]
144. Wang, X.; Liu, Z.; Fan, F.; Hou, Y.; Yang, H.; Meng, X.; Zhang, Y.; Ren, F. Microfluidic chip and its application in autophagy detection. *Trends Anal. Chem.* **2019**, *117*, 300–315. [CrossRef]
145. Heo, H.Y.; Chung, S.; Kim, Y.T.; Kim, D.H.; Seo, T.S. A valveless rotary microfluidic device for multiplex point mutation identification based on ligation-rolling circle amplification. *Biosens. Bioelectron.* **2016**, *78*, 140–146. [CrossRef]
146. Liu, M.; Hui, C.Y.; Zhang, Q.; Gu, J.; Kannan, B.; Jahanshahi-Anbuhi, S.; Filipe, C.D.; Brennan, J.D.; Li, Y. Target-Induced and Equipment-Free DNA Amplification with a Simple Paper Device. *Angew. Chem. Int. Ed. Engl.* **2016**, *55*, 2709–2713. [CrossRef]
147. Huang, S.; Feng, M.; Li, J.; Liu, Y.; Xiao, Q. Voltammetric determination of attomolar levels of a sequence derived from the genom of hepatitis B virus by using molecular beacon mediated circular strand displacement and rolling circle amplification. *Microchim. Acta* **2018**, *185*, 206. [CrossRef]
148. Shen, C.; Liu, S.; Li, X.; Zhao, D.; Yang, M. Immunoelectrochemical detection of thehuman epidermal growth factor receptor 2 (HER2) via gold nanoparticle-based rolling circle amplification. *Microchim. Acta* **2018**, *185*, 547. [CrossRef]
149. Yi, X.; Li, L.; Peng, Y.; Guo, L. A universal electrochemical sensing system for small biomolecules using target-mediated sticky ends-based ligation-rolling circle amplification. *Biosens. Bioelectron.* **2014**, *57*, 103–109. [CrossRef]
150. Jia, Y.; Sun, F.; Na, N.; Ouyang, J. Detection of p53 DNA using commercially available personal glucose meters based on rolling circle amplification coupled with nicking enzyme signal amplification. *Anal. Chim. Acta* **2019**, *1060*, 64–70. [CrossRef]

Article

Rapid Multiplex Strip Test for the Detection of Circulating Tumor DNA Mutations for Liquid Biopsy Applications

Panagiota M. Kalligosfyri [1,†], Sofia Nikou [2,†], Sofia Karteri [3], Haralabos P. Kalofonos [3], Vasiliki Bravou [2,*] and Despina P. Kalogianni [1,*]

[1] Department of Chemistry, University of Patras, Rio, 26504 Patras, Greece; pkalligosfyri@gmail.com
[2] Department of Anatomy-Histology-Embryology, Medical School, University of Patras, Rio, 26504 Patras, Greece; sonikou@upatras.gr
[3] Division of Oncology, Department of Internal Medicine, University Hospital of Patras, Rio, 26504 Patras, Greece; skarteri@gmail.com (S.K.); kalofonos@upatras.gr (H.P.K.)
[*] Correspondence: vibra@upatras.gr (V.B.); kalogian@upatras.gr (D.P.K.)
[†] These authors contributed equally to this work.

Abstract: In the era of personalized medicine, molecular profiling of patient tumors has become the standard practice, especially for patients with advanced disease. Activating point mutations of the KRAS proto-oncogene are clinically relevant for many types of cancer, including colorectal cancer (CRC). While several approaches have been developed for tumor genotyping, liquid biopsy has been gaining much attention in the clinical setting. Analysis of circulating tumor DNA for genetic alterations has been challenging, and many methodologies with both advantages and disadvantages have been developed. We here developed a gold nanoparticle-based rapid strip test that has been applied for the first time for the multiplex detection of KRAS mutations in circulating tumor DNA (ctDNA) of CRC patients. The method involved ctDNA isolation, PCR-amplification of the KRAS gene, multiplex primer extension (PEXT) reaction, and detection with a multiplex strip test. We have optimized the efficiency and specificity of the multiplex strip test in synthetic DNA targets, in colorectal cancer cell lines, in tissue samples, and in blood-derived ctDNA from patients with advanced colorectal cancer. The proposed strip test achieved rapid and easy multiplex detection (normal allele and three major single-point mutations) of the clinically relevant KRAS mutations in ctDNA in blood samples of CRC patients with high specificity and repeatability. This multiplex strip test represents a minimally invasive, rapid, low-cost, and promising diagnostic tool for the detection of clinically relevant mutations in cancer patients.

Keywords: colorectal cancer; KRAS; lateral flow assay; dipstick; biosensor; gold nanoparticles

Citation: Kalligosfyri, P.M.; Nikou, S.; Karteri, S.; Kalofonos, H.P.; Bravou, V.; Kalogianni, D.P. Rapid Multiplex Strip Test for the Detection of Circulating Tumor DNA Mutations for Liquid Biopsy Applications. *Biosensors* **2022**, *12*, 97. https://doi.org/10.3390/bios12020097

Received: 22 December 2021
Accepted: 1 February 2022
Published: 4 February 2022

Publisher's Note: MDPI stays neutral with regard to jurisdictional claims in published maps and institutional affiliations.

1. Introduction

Personalized medicine is based on recommendations according to genomic "drivers" of tumorigenesis [1]. Cancer develops through a multistage process, with the accumulation of somatic mutations and genetic alterations leading to increased cell proliferation and tumorigenesis [2]. Such mutations are known as "drivers" and are crucial for monitoring disease progression and resistance to targeted therapeutic agents [3,4]. Amongst known driver mutations, activating mutations in the KRAS or KRAS2 (Kirsten rat sarcoma virus 2 homolog) oncogene are found in several malignancies including colorectal cancer (CRC) [5–8]. Activating KRAS mutations are point mutations mainly affecting KRAS amino acid residues 12, 13, and 61, which reduce the intrinsic KRAS and GTPase activating protein–promoted GTP hydrolysis [9,10]. KRAS mutations occur early in the development of colorectal cancer (CRC) and are strongly associated with resistance to therapies [10,11]. CRC is the second most common cause of cancer-related mortality in Europe [12]. About 40% of CRC cases are KRAS-mutant-related, meaning that they cannot benefit from EGFR-targeted therapies [13].

In clinical practice, the majority of molecular profiling tests, including KRAS mutation detection, are performed in tissue biopsies. However, excision regions do not sufficiently depict intratumoral heterogeneity, and in some cases, tissue biopsy is not feasible. Liquid biopsy, which mainly includes analysis of cell-free DNA (cfDNA), represents a convenient real-time tool for mutational profiling in a minimally invasive manner. cfDNA is double-stranded, fragmented extracellular DNA released from normal and cancer cells through cell death (apoptosis or necrosis) or by active secretion with extracellular vesicles (exosomes and prostasomes) into the bloodstream. Plasma circulating tumor DNA (ctDNA) that originates specifically from tumors represents a small fragment (<1.0%) of the total cell-free DNA (cfDNA) and is only identified via the detection of cancer-related mutations [14,15]. Several factors, such as the tumor type, stage, and burden, as well as the tumor proliferation rate and turnover, affect the amount of ctDNA in body fluids [16–18]. Moreover, the average size of ctDNA ranges from small fragments of 70–200 bp which are secreted due to cellular apoptosis to longer fragments of 200 bp–21 Kbp generated by necrosis. Thus, the extremely low abundance of ctDNA, especially in early cancer stages, along with its high fragmentation, render ctDNA a challenging analyte [19,20]. The analytical methods reported for ctDNA detection include PCR-based techniques (such as digital PCR, methylation-specific PCR, and real-time PCR), next-generation sequencing (NGS), mass spectrometry, and DNA microarrays. However, many of these techniques have drawbacks in ctDNA analysis or do not meet the requirements of a diagnostic tool applicable in the clinical setting. This is due to attributes including low detectability and specificity, high cost of analysis and expensive instrumentation, long analysis time, and extensive sample pretreatment [21–24]. Given the feasibility of using ctDNA in tracking and monitoring tumor dynamics and resistance to therapy, the use of ctDNA as a marker for detecting driver mutations needs to be further developed. KRAS driver mutations in codon 12 or 13 mutations are a major predictive biomarker of poor response to therapy in patients with CRC [10,25]. Therefore, rapid and precise identification of KRAS mutations in ctDNA is required for improving the response rate and survival in CRC patients [25]. Strip-type rapid tests, and biosensors in general, represent excellent candidates for liquid biopsy applications [26]. Various biosensors have been developed so far for ctDNA detection. However, only a few reports present applications for ctDNA analysis in patients' blood samples.

Herein, we have developed a rapid multiplex strip test that comprises of a gold-nanoparticle-based optical DNA biosensor for KRAS screening in cancer. Based on a previous developed flow strip assay for detecting KRAS mutations [27], the current multiplex strip has been applied, for the first time, for multi-analyte liquid biopsy applications. The wild-type KRAS and three major single-point KRAS mutations (G12D, G12A, G12V) were simultaneously detected in cfDNA/ctDNA with a single strip test. The strip-type DNA biosensor was tested and optimized in synthetic DNA targets, in colorectal cancer cell lines, in tissue samples and was finally applied to blood-derived cfDNA from healthy individuals and ctDNA from patients with advanced colorectal cancer.

2. Materials and Methods

2.1. Reagents and Apparatus

Polymerase chain reactions (PCR) were performed using the HotStarTaq Master Mix Kit (Qiagen, Hilden, Germany). Sections from FFPE tissue samples were cut with Accu-cut SRM 200 Rotary Microtome (The Netherlands, Europe, BV). The Nucleospin DNA FFPE XS kit used for tissue DNA extraction was obtained from Macherey-Nagel (Düren, Germany). All synthetic oligonucleotides were purchased from Eurofins Genomics (Ebersberg, Germany) (Table 1). All other reagents and apparatus used were previously reported [27].

Table 1. Sequences of the PCR and PEXT primers, the synthetic DNA targets, and the anti-tag sequences attached onto the polystyrene microspheres.

Oligonucleotide	Sequence ($5' \rightarrow 3'$)
PCR Primers	
KRAS_Forward	GCCTGCTGAAAATGACTGAATA
KRAS_Reverse	CAAGAGACAGGTTTCTCCATCA
Synthetic Targets	
KRAS-Normal	CTGAATTAGCTGTATCGTCAAGGCACTCTTGCCTACGCCACCAGCTCCAACTACCACAAG
KRAS-G12D	CTGAATTAGCTGTATCGTCAAGGCACTCTTGCCTACGCCATCAGCTCCAACTACCACAAG
KRAS-G12V	CTGAATTAGCTGTATCGTCAAGGCACTCTTGCCTACGCCAACAGCTCCAACTACCACAAG
KRAS-G12A	CTGAATTAGCTGTATCGTCAAGGCACTCTTGCCTACGCCAGCAGCTCCAACTACCACAAG
Anti-Tag Sequences	
KRAS-NORMAL	NH_2-GGATACCGCTGCACCCATCGCCAC
KRAS-G12D	NH_2-CGTTTTAAGTTCGGATGGTGACGT
KRAS-G12V	NH_2-AGCGCACTGGTGGATGCTGGACTG
KRAS-G12A	NH_2-CTTGCTGAACTTCTGACTACGACT
Tag-PEXT Primers	
KRAS-NORMAL	GTGGCGATGGGTGCAGCGGTATCCCCGAATTCTCTCCTTGTGGTAGTTGGAGCTGG
KRAS-G12D	ACGTCACCATCCGAACTTAAAACGCCGAATTCTCTCCTTGTGGTAGTTGGAGCTGA
KRAS-G12V	CAGTCCAGCATCCACCAGTCGGCTCCGAATTCTCTCCTTGTGGTAGTTGGAGCTGT
KRAS-G12A	AGTCGTAGTCAGAAGTTCAGCAAGCCGAATTCTCTCCTTGTGGTAGTTGGAGCTGC

2.2. Cell Lines and Clinical Samples

Human colorectal cancer cell lines, namely Caco2 and LS174T, were used in the present study. The human colorectal cancer cell lines were obtained from the American Type Culture Collection (ATCC, Manassas, Virginia USA). The mutant KRAS cell line LS174T (c.35G>A, p.G12D) and the wild-type KRAS cell line Caco2 were cultured and harvested as previously described [27].

The study included formalin-fixed paraffin-embedded (FFPE) tissue samples from three CRC patients that were retrieved from the archives of the Department of Pathology, University Hospital of Patras, Greece and peripheral blood samples from four healthy individuals and five CRC patients prior to or close to the initiation of chemotherapy treatment. All patients were treated at the Division of Oncology, Department of Medicine, Medical School, University of Patras and had known KRAS mutations detected in FFPE tissue samples by next-generation sequencing (NGS) (Ion Gene Studio S5 Prime System, Thermo Fisher Scientific) using the Ion AmpliSeq NGS Panel (Thermo Fisher Scientific). Relevant information was retrieved from the patients' oncology records. Patient characteristics are summarized in Table S1.

Blood samples were obtained after written informed consent, and the study has been approved by the Ethics and Scientific Committee of the University General Hospital of Patras, Greece (Protocol Number 680/15.10.2019) and by the Institutional Ethics & Research Committee of the University of Patras, Greece (Protocol Number 86436/14.10.2019), in strict compliance with the ethical standards of the institutional and/or national research committee and the 1964 Declaration of Helsinki and its later amendments.

2.3. DNA Extraction from Cell Lines and Tissue Samples

DNA was extracted from cell lines using the NucleoSpin DNA Rapid Lysis kit, according to manufacturer's instructions, after washing of the cell pellet with $1 \times$ PBS, pH 7.4. DNA was also isolated from formalin-fixed paraffin-embedded (FFPE) tissue samples using the NucleoSpin DNA FFPE XS kit. All extractions were performed using 10 µm sections (3 sections from each FFPE block). All sections were initially deparaffinized using xylene (60 °C) and hydrated in descending alcohol series. DNA was then extracted from the FFPE tissue kit, following the manufacturer's instructions. Finally, the purity and concentration of isolated DNA was determined by UV spectrometry at 260/280 nm (Table S2).

2.4. Cell-Free DNA (cfDNA) Extraction from Blood Samples

CfDNA was extracted from the plasma of blood samples. The plasma was carefully separated from the pellets by centrifugation for 10 min at 2500 rpm, at room temperature. The isolation of cfDNA from plasma samples was performed using the NucleoSpin Plasma XS kit, according to the manufacturer's instructions. The cfDNA was obtained by an elution step with 20–30 µL of the provided elution buffer and centrifugation at $11,000\times g$ for 30 s. The purity and the concentration of the isolated cfDNA was measured with the NanoDrop 1000 spectrophotometer (Table S2).

2.5. Preparation of Streptavidin-Conjugated Gold Nanoparticles (SA-AuNPs)

Prior to the conjugation reaction, the pH of the AuNPs solution was adjusted to 6.0 with 10 mM phosphate buffer, pH 6.8. An amount of 1.5 µg of streptavidin was then mixed with 200 µL of gold nanoparticles and the mixture was incubated for 2 h at room temperature, in the dark, with frequent stirring. After the incubation step, BSA was used as blocking agent at a final concentration of 10 g/L. The mixture was incubated for 30 min at room temperature. A 20 min centrifugation step at $3300\times$ g was performed for the collection of the SA-functionalized AuNPs. The SA-AuNPs conjugates were finally reconstituted in 20 µL of proper storage buffer ($1 \times$ PBS pH 7.4, 1% BSA, 0.25% sucrose, 1% Tween-20, 0.05% NaN_3) and stored at 4 °C for further use.

2.6. Synthesis of Functionalized Microspheres

Four different sets of functionalized carboxylated polystyrene microspheres were prepared. The four sets corresponded to the normal allele and the three single-point mutations of the KRAS gene that were examined in the present work. Each set was coupled to a specific oligonucleotide sequence (anti-tag sequence) through an amino group at its 5' end as previously reported [28] but with some modifications. Briefly, a mixture of 12.8 µL of the carboxylated polystyrene microspheres and 125 µL of MES buffer, pH 4.5, was centrifuged for 2 min at $15,700\times g$, and the supernatant was discarded. The microspheres were resuspended in 40 µL MES buffer, pH 4.5, and sonicated for 2 min. An amount of 400 pmol of each anti-tag sequence was separately conjugated to the microspheres, using 1.25 µL of a fresh EDC, 0.4 mg/µL. The coupling reaction was held in the dark and at room temperature for 30 min with occasionally stirring. Then, the addition of the same amount of EDC was repeated. After incubation, the suspension was centrifuged for 2 min at $15,700\times g$, after the addition of 2 µL of 10% Tween-20, to collect the conjugated microspheres. The prepared anti/tag-microspheres were then washed two times with 100 µL of 0.2% Tween-20 in $1 \times$ TE buffer, redispersed in 100 µL of $1 \times$ TE buffer (pH 8.0), and kept at 4 °C until use.

2.7. KRAS Gene Amplification

The KRAS gene at exon 12, that contained the single-point mutations of interest, was amplified by PCR at a final volume of 25 µL. The PCR reaction pool contained $1 \times$ HotStarTaq Master Mix (with 1.5 mM $MgCl_2$), KRAS primers at a concentration of 0.6 µM and 100 ng of isolated DNA. The conditions of the reaction were: a first step of denaturation at 95 °C for 15 min that was followed by 35 cycles of 95 °C for 30 s, 55 °C for 30 s and 72 °C for 30 s, and a final extension step at 72 °C for 10 min. The PCR products were then electrophorized in a 2% agarose gel, using ethidium bromide for DNA staining. The PCR products were finally quantified by densitometric analysis using the open-source ImageJ image processing and gel analyzer tool by National Institutes of Health (NIH) (Bethesda, Maryland, USA) and Laboratory for Optical and Computational Instrumentation, University of Wisconsin (Wisconsin, USA).

2.8. Multiplex Primer Extension Reaction (PEXT) for Single-Point Mutation Discrimination in ctDNA

A multiplex primer extension (PEXT) reaction using four specific primers for the tested polymorphisms of the KRAS gene was conducted. The reaction (20 µL) consisted of $1 \times$

Vent (exo-) buffer with 1.5 mM MgCl2, 2.5 µM each of the dNTPs except from dCTP, 2.5 µM of biotin-dCTP, 0.05 µM of each of the specific primers, 0.5 U of the DNA polymerase Vent (exo-), and 200–400 fmol of each PCR product. The PEXT reaction was performed as follows: denaturation at 95 °C for 3 min, followed by 25 cycles at 95 °C for 15 s, at 58 °C for 10 s, and at 72 °C for 15 s.

2.9. Fabrication of the Multiplex Rapid Strip Test

The strip test consisted of four parts: an immersion pad, a conjugate pad, a nitrocellulose membrane, and an absorbent pad with a total size of 5 mm × 70 mm. The four parts were welded onto a plastic adhesive backing pad with overlapping ends for continuous flow. Biotinylated BSA (b-BSA) (Supplementary Material) was deposited onto the membrane (0.5 µL of 50 ng/µL), in order to construct the control zone of the strip, as an assurance of the functionality of the test. Secondly, a volume of 1 µL of each of the four sets of the microspheres coupled to the four anti-tag sequences were also placed onto the membrane, in order to form four spatially distinct test spots as presented in Figure 1. The spots were then dried for 10 min at room temperature, and the strip was ready for use.

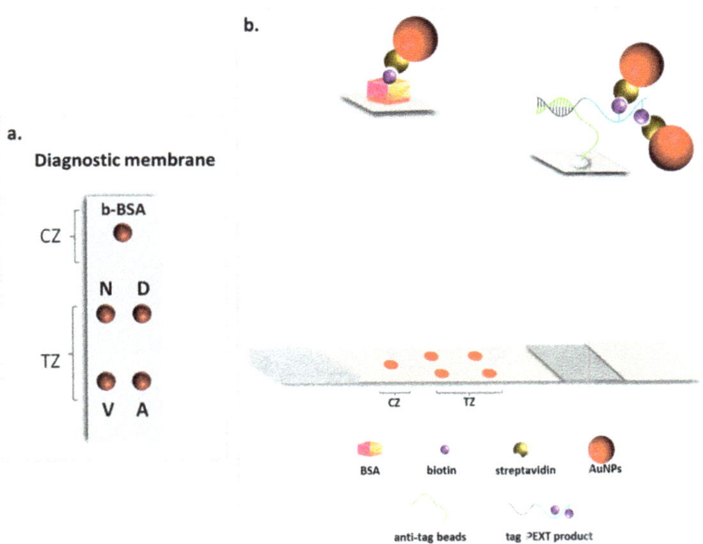

Figure 1. (**a**) Layout of the test and the control spots on the diagnostic membrane of the multiplex strip test. (**b**) The principle of the multiplex strip test. b-BSA: biotinylated BSA, N: normal allele, D: G12D mutant allele, V: G12V mutant allele, A: G12A mutant allele, TZ: test zone, CZ: control zone, AuNPs: gold nanoparticles.

2.10. Multi-Allele Detection by the Rapid Strip Test

The PEXT products were denatured by heating at 95 °C for 3 min and placed immediately on ice for 2 min. A 5 µL aliquot of the denatured PEXT products was applied onto the conjugate pad of the strip just above an 8 µL aliquot of prepared streptavidin-gold nanoparticles (SA-AuNPs) conjugates (Supplementary Material) that were pre-deposited in the same pad. The strip was then immersed into 300 µL of the developing solution that consisted of 4 × SSC pH 7.0, 1.5% glycerol, 1% Tween, 1% BSA and 0.5% SDS. The strip was removed from the solution after 15 min for visual detection of the PEXT products. Finally, images of the strips were acquired by a conventional scanner.

3. Results

We have developed a rapid multiplex strip test based on gold nanoparticles for the detection of gene mutations in circulating tumor DNA for non-invasive liquid biopsy applications. The homo sapiens KRAS oncogene (GTPase (KRAS), transcript variant X2, mRNA—Accession number: XM_011520653), mutations of which guide treatment decisions in colorectal cancer (CRC), was used as a model gene. The proposed strip test was applied, for the first time, for the simultaneous detection of the normal allele and three major single-point mutations of the KRAS gene in cell-free DNA and/or circulating tumor DNA from peripheral blood samples. These mutations are located in exon 12, namely G12D (35G>A), G12V (35G>T), and G12A (35G>C). Optimization parameters involved samples that consisted of (*i*) four single-stranded synthetic DNA targets corresponded to the four KRAS alleles (Table 1) and also DNA isolated from (*ii*) cell lines expressing the wild-type and the mutated (G12D) KRAS gene and (*iii*) FFPE tissue samples from CRC patients carrying known KRAS mutations. After optimization, the method was applied for the genotyping of the KRAS gene in cell-free DNA and circulating tumor DNA in blood samples from healthy individuals and CRC patients with known mutations.

The protocol included: (a) DNA or cfDNA isolation, (b) PCR of the exon 12 of the KRAS gene, (c) multiplex allele-discrimination reaction by a multiplex primer extension (PEXT) reaction, and (d) a multiplex strip test for alleles detection. The size (171 bp) and the amount of the PCR products were estimated by a 2% agarose gel electrophoresis with subsequent ethidium bromide staining (Figure S1). After amplification, each PCR product was subjected to a multiplex PEXT reaction in the presence of four allele-specific PEXT primers that were complementary to the wild-type allele and the three mutations of the KRAS gene exon 12. The primers used in PEXT reaction had the same sequence and differed only in two points: (*i*) they carried a different base at the $3'$ end, complementary to each mutation and the normal allele, respectively; and (*ii*) they carried different 24-bp sequences (tag sequences) at their $5'$ ends, which were complementary to the four anti-tag sequences attached to the four sets of microspheres, in order to capture the PEXT products onto the strip. The PEXT primers were then elongated only if they were fully complementary to the sequence of the KRAS allele by a special DNA polymerase that lacks the $3' \rightarrow 5'$ proofreading exonuclease activity. Biotin moieties were inserted into the PEXT products during the polymerization using biotinylated dCTP. The PEXT products were finally applied onto the strip after being heat denatured for a few minutes.

The diagnostic membrane of the multiplex strip test consisted of a control (CZ) and a test zone (TZ). The TZ reveals the presence of the target in the sample, while the CZ ensures the proper function of the strip test. For the formation of the CZ, biotin was immobilized on the nitrocellulose membrane by deposition of b-BSA. A first red spot was formed at the CZ on the upper site of the membrane, as the excess of SA-AuNPs was bound there through biotin–streptavidin interactions. The TZ of the strip consisted of four test spots of the four different sets of the anti-tag-modified microspheres that were pre-deposited and were spatially discrete on the membrane, in order to capture the PEXT products. The arrangement of the test spots on the membrane was as follows (from the top to the bottom of the membrane): normal and the G12V (35G>T) alleles on the left and the G12D (35G>A) and the G12A (35G>C) alleles on the right side (Figure 1a). Moreover, an 8 μL aliquot of the SA-AuNPs conjugates and a volume of 5 μL of the heat-denatured multiplex PEXT products were deposited on the conjugate pad of the strip, and the strip was then immersed into a microcentrifuge tube containing the developing solution. All the reagents deposited on the conjugated pad were entrained upwards by the developing solution though capillary action. The products from the PEXT reaction were captured at the TZ due to tag/anti-tag hybridization, between the tag tail of the PEXT primers and the anti-tag sequences that were coupled to the immobilized microspheres. The hybridization was finally visually detected by the accumulation of the SA-AuNPs at the same spots through the interaction of streptavidin with the biotin moieties of the PEXT products, forming up to four visual red spots (Figure 1b). The excess of SA-AuNPs was directed to the CZ of the strip, forming another red spot. The visual detection by naked eye was completed after 10–15 min.

The successful preparation of SA-AuNPs conjugates was tested by analyzing the conjugates with two strip tests with immobilized biotinylated and non-biotinylated BSA, respectively. As observed from Figure S2 in the Supplementary Material, the conjugates were specifically captured; thus, a red spot was formed only on the strip on which b-BSA was immobilized due to streptavidin-biotin interaction.

The whole method was first developed and optimized using four 60-bp single-stranded synthetic DNA targets that corresponded to the four analyzed alleles (Table 1). Optimization studies involved the cycling conditions of the PEXT reaction, as well as the performance of the strip test, in order to increase the signal density and eliminate the non-specific interactions at the test zone of the strip. In more detail, the specific parameters studied were: the cfDNA isolation procedure for higher yield and purity, the conditions of the primer extension reactions, and the process of SA-AuNPs preparation that included the concentration of the reagents and the pH of the coupling reaction, as well as the final reconstitution buffer of SA-AuNPs. The composition of the developing solution was also tested, regarding the use of surfactants to minimize the non-specific bindings and the concentration of glycerol to provide the required time for efficient hybridization. Finally, the sample volume applied onto the strip test was optimized. For this purpose, different sample volumes (1–10 µL) that contained only the normal KRAS allele were applied onto the multiplex strip test. As observed from Figure S3 in the Supplementary Material, the optimum sample volume was 5 µL, which gave the strongest signal with high specificity.

The analytical performance of the strip test was also evaluated by assessing the detectability of a biotinylated singe-stranded DNA (ssDNA) (b-dA$_{30}$). The results are presented in Figure S4. As low as 50 amol of ssDNA were detectable with the naked eye with the rapid strip test.

3.1. Synthetic DNA Targets

The synthetic DNA targets were subjected to a multiplex PEXT reaction in the presence of the four PEXT tagged primers. The reaction was conducted for three cycles, using an annealing temperature of 62 °C and 400 fmol of each the synthetic target. All the PEXT products were finally analyzed separately with a single strip test. Each PEXT product was specifically hybridized, and a visible red spot was formed only at the test spot where the complementary anti-tag sequences were immobilized, corresponding to the target allele present in the sample. The images of the strips were scanned using a regular scanner and are presented in Figure 2. We can observe from the images the good specificity of the multi-analyte strip test.

3.2. Cell Lines

DNA isolated from wild-type KRAS and mutant KRAS (G12D) human colon cancer cell lines was also used for optimization experiments. The purified DNA from each cell line was subjected to a single multiplex PEXT reaction containing all four PEXT tagged primers. In comparison to the synthetic targets, a more intense colored signal at the test spots was observed at an annealing temperature of 58 °C and a 15-cycle PEXT reaction, using 400 fmol of the PEXT product. This was attributed to the differences in DNA structures. The synthetic DNA targets are single-stranded oligonucleotides increasing the hybridization efficiency, as well as the non-specific interactions, contrary to the double-stranded genomic DNA. Strip test results from the analysis of the DNA derived from the cell lines are presented in Figure 2. The analysis of the wild-type KRAS cell line (Caco2) resulted in the formation of a red spot only at the TZ of the strip corresponding to the normal allele. As for the cancer cell line LS174T, two red spots were formed both for the normal allele and the G12D KRAS mutation. These results also denoted the high specificity of the multiplex strip test.

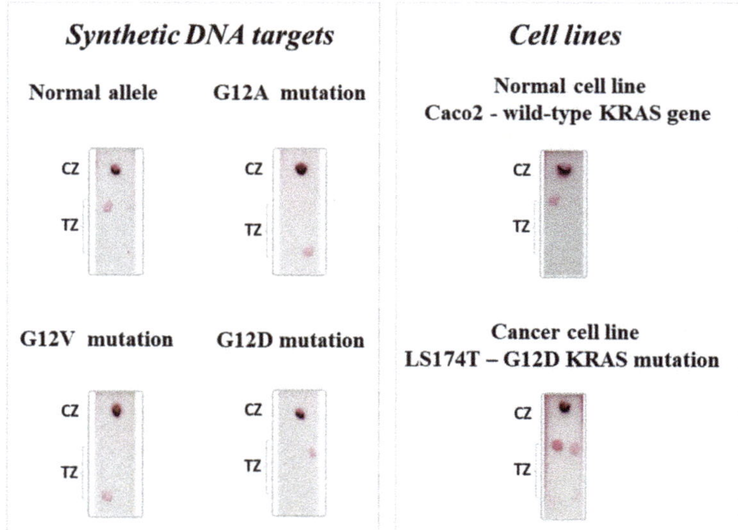

Figure 2. Results of the multiplex strip test for the synthetic DNA targets that correspond to the normal KRAS gene and the three single-point mutations examined. Results of the multiplex strip test for the cell lines that express the wild-type KRAS gene and the G12D KRAS mutant allele. TZ: test zone, CZ: control zone.

3.3. Tissue Samples

The multi-analyte strip-type DNA biosensor was then tested for the detection of KRAS mutations in tissue samples from CRC patients that carried the G12A, G12V, and G12D KRAS gene mutations as previously confirmed by NGS. The optimized parameters required a PEXT reaction of 30 cycles at an annealing temperature of 58 °C, using 200 fmol of the PCR product. The DNA in FPPE tissue samples is partially degraded, leading to a low PCR yield. Thus, only 200 fmol of the PCR product was used for the PEXT reaction, whereas an increase in the cycles of the PEXT reaction was required. The results of the strip test for the tissue samples are shown in Figure 3. Again, only the red spots at the test zones that corresponded to the mutations present in the tissue samples were formed at the membrane of the strips, ensuring the good specificity of the test.

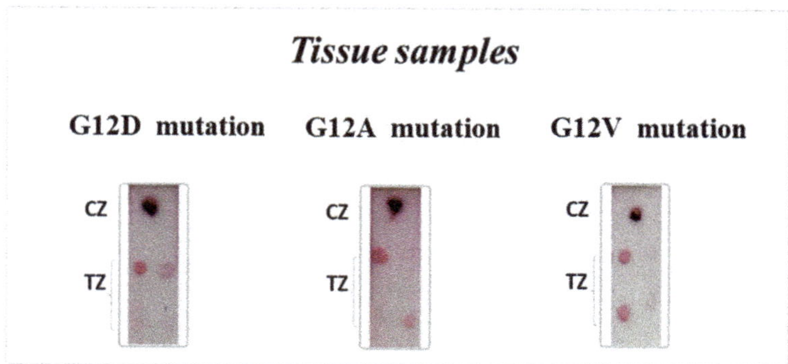

Figure 3. The multiplex strip test results of three tissue samples that have the G12D, G12A, and G12D KRAS mutations. TZ: test zone, CZ: control zone.

3.4. Detectability of the Method

The detectability of the method, in terms of the input DNA prior to the PCR, was determined as follows. Different dilutions (0, 0.1, 0.5, 1, 5, and 100%) of the DNA isolated from the human colorectal cancer cell line LS174T that carries the mutated (G12D) KRAS allele within the context of unmutated DNA isolated from the human cell line Caco2 (normal KRAS) were prepared. In more details, 0.1, 0.5, 1, and 5 ng DNA from the LS174T cell line were mixed with DNA from normal Caco2 cell line to obtain a total amount of 100 ng for all mixtures. A total of 100 ng of the DNA admixtures, as well as 100 ng of the DNA from the KRAS mutant cell line, were then amplified by PCR. Subsequently, 400 fmol of each PCR product was subjected to multiplex PEXT reaction and 5 µL of the PEXT products were analyzed with the multiplex strip test. The results are presented in Figure 4. As observed, 0.1% (0.1 ng) of the DNA form the KRAS mutant cell line was detectable in the background of normal KRAS DNA. However, as LS174T cell line carries both mutated KRAS and normal KRAS allele, meaning that the amount of mutated KRAS is even smaller, we conclude that less than 0.1% or 0.1 ng of the mutated KRAS gene is detectable by the multiplex strip test.

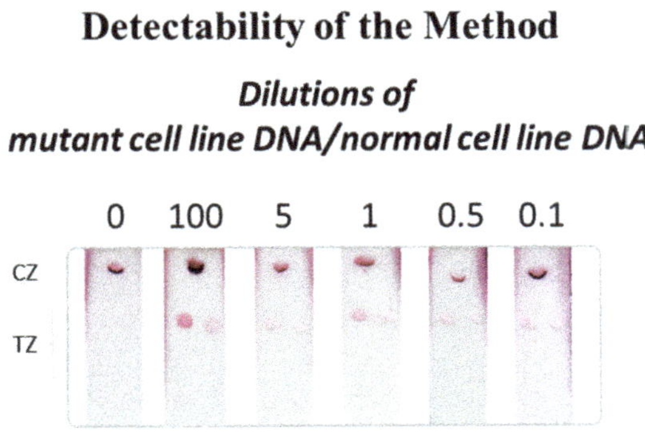

Figure 4. Detectability of the method. Different dilutions of mutant cancer cell line DNA/normal cell line DNA (0.1–100%) were prepared and analyzed by the multiplex strip test after amplification with subsequent multiplex PEXT reaction. TZ: test zone, CZ: control zone.

3.5. Application of the Multiplex Rapid Strip Test to Blood Samples for the Detection of KRAS Mutations in cf/ctDNA

The multiplex strip test was finally evaluated for the detection of KRAS gene mutations in blood samples. For this study, four healthy volunteers and five CRC patients with known KRAS mutations previously confirmed by NGS in tissue samples were used and analyzed by the multiplex strip test. The cf DNA was isolated from the blood samples and underwent the multiplex PEXT reaction in the presence of all four allele-specific PEXT primers. As ctDNA is only a small fraction (<1%) of the cfDNA and KRAS gene is therefore present in extremely low abundancies in blood circulation, a 25-cycle PEXT reaction at 58 °C using 200 fmol of the isolated cfDNA was needed in order to successfully detect normal and mutant KRAS gene in blood samples with high specificity. The PEXT products were directly applied to the multiplex strip test according to the established protocol. For healthy individuals, a red spot appeared at the TZ of the biosensor only for the normal allele. On the other hand, for CRC patients with known KRAS mutations, two red spots were formed at the TZ of the strip test that corresponded to the normal allele and to the corresponding KRAS mutation, as expected (Figure 5) with 100% concordance between our assay in plasma and tissue results by NGS. Moreover, we observed, as shown in Figure 5, that the color intensity of the test spot signal for the gene mutations was lower than the signal of the test spot for the normal allele, and this is due to the fact that ctDNA is present at much lower concentration than the normal cfDNA in blood circulation.

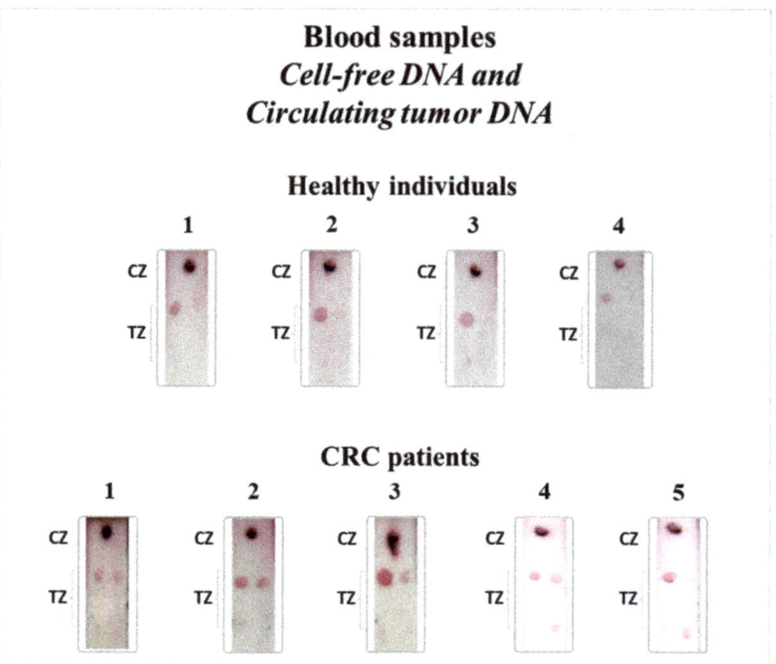

Figure 5. Application of the multiplex strip test for the detection of cell-free DNA and circulating tumor DNA in blood samples of four healthy individuals and five CRC patients. CRC: colorectal cancer, TZ: test zone, CZ: control zone.

3.6. Repeatability of the Multiplex Rapid Strip Test

The repeatability of the developed multiplex strip test was also evaluated and was expressed as % coefficient of variation (%CV) values. For this purpose, isolated DNA (400 fmol) from the wild-type KRAS cell line Caco2 and cfDNA (200 fmol) isolated from blood sample of a CRC patient with a G12D KRAS mutation were subjected to the multiplex PEXT reaction and analyzed in triplicate with the multiplex strip test. The results are presented in Figure 6. The strips were acquired with a conventional scanner and the color density of the red spots at the TZ of the strip was measured in grayscale using the open-source ImageJ image processing software. Finally, the %CVs ($n = 3$) were calculated from the gray values of the test spots of the strip images and were 2.8% for the cell line, 0.5% for the normal allele, and 0.7% for the G12D mutation of the cf/ctDNA from the patient, ensuring the very good repeatability of the multi-analyte strip test.

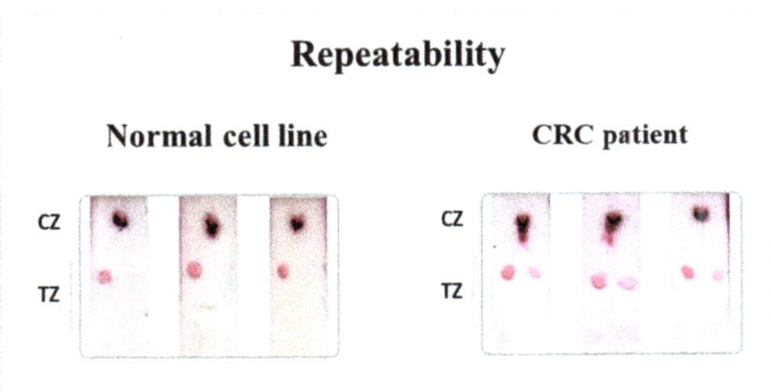

Figure 6. Repeatability of the multiplex strip test for the normal cell line and the blood sample of a CRC patient. TZ: test zone, CZ: control zone.

4. Discussion

Biosensors are the most preferable analytical tools compared to conventional analytical methods for ctDNA analysis or bioanalysis in general, because they offer simplicity, low cost, rapid analysis (in some cases), portability, and multiplicity, along with high detectability, sensitivity, specificity, robustness, and reproducibility. They are also easy to use [29]. Moreover, the use of nanomaterials in such devices emerged as an effort to increase the analytical performance of the biosensors, as these nanomaterials provide signal enhancement without the need for enzyme-assisted amplification [30,31]. Currently, there is still the need for the development of novel biosensors for ctDNA detection and liquid biopsy applications.

Herein, we developed a multiplex strip-type biosensor based on gold nanoparticles for visual simultaneous detection by the naked eye of four alleles, wt and G12D (35G>A), G12V (35G>T), G12A (35G>C) mutations of the KRAS gene, that are frequently observed in CRC patients and guide treatment decisions. The test was optimized using single-stranded synthetic DNA targets and cell lines expressing the wild-type and the mutated (G12D) KRAS gene and also allowed for the detection of three different KRAS mutations in FFPE tissue samples from CRC patients. Importantly, our multiplex strip assay could efficiently and with specificity detect different single point mutations of codon 12 of the KRAS gene in cfDNA of five CRC patients with 100% concordance with tissue results previously obtained by NGS, while only the wt allele was detected in healthy individuals. The protocol includes cfDNA isolation, KRAS gene amplification, allele discrimination by a multiplex PEXT reaction, and simultaneous visual detection of the four alleles by a single strip test. Analysis is completed within 10 min, while the run time is approximately 3.5 h. The strip test shows an LOD $\leq$ 0.1 ng (1:1000 dilution) of mutated KRAS gene as it allowed for the detection of lower than 0.1% or 100 pg mutated KRAS gene in the presence of normal KRAS gene, shows repeatable results, and also has the advantages of portability and universality, as the protocol prior to detection can be easily modified in order to target other gene mutations of interest.

Although many biosensors have been reported in the literature, only a few have been evaluated for ctDNA genotyping in clinical samples, rendering our assay a novel promising tool for liquid biopsy applications. Most of these biosensors are electrochemical due to their high detectability and ease of use. An electrochemical biosensor was developed for the detection of two mutations of the EGFR gene. The detection was based on an amplification refractory mutation system (ARMS) and linear-after-the-exponential PCR to obtain asymmetric biotinylated PCR products. The products were captured to a tetrahedral DNA nanostructure-decorated electrode and detected by avidin-HRP (horseradish peroxidase). This biosensor needs an overnight fabrication and has a >1 h analysis time, while it offers a limit of detection (LOD) of 30 pg, good specificity, and universality potential [32]. An electrochemical biosensor has also been previously developed for the detection of the

PIK3CA E545K mutation in ctDNA. The signal of this biosensor was enhanced through hybridization chain reaction (HCR) that took place on a gold electrode's surface, while the HCR products were biotinylated and detected by streptavidin-alkaline phosphatase through biotin–streptavidin interaction. The sensor's characteristics were >1.5 h of analysis time, about 17 h of fabrication time, quite good specificity, and LOD of 3 pM, but no multiplexing or universality potential [33]. In an effort to enhance the hybridization efficiency and specificity, Das et al., (2019 and 2016) developed an electrochemical sensor that uses combinatorial DNA probes or clutch probes and PNA clamps. The analyzed ctDNA was specifically captured onto the electrode by immobilized capture DNA probes. The electroactive compounds $[Ru(NH_3)_6]^{3+}$ and $[Fe(CN)_6]^{3-}$ were intercalated to the DNA sequences through electrostatic forces, increasing the sensor's response. The authors could simultaneously detect seven different mutations of the KRAS gene and several mutations of the EGFR gene with high specificity. The LOD was 1 fg/µL and the analysis time was about 2 h, while an overnight incubation was needed to construct the sensing electrode [34,35]. Another electrochemical biosensor was reported for the dual detection of mutations and epigenetic methylation of ctDNA. The ctDNA was captured by complementary PNA probes that were coupled to gold nanoparticles. Lead phosphate apoferritin conjugated to anti-5-mC was then bound to the methylation sites of ctDNA, forming a sandwich complex on the electrode. Lead ions were finally released from apoferritin and were detected by square-wave voltammetry. Two mutations of the PIK3CA gene were separately detected by this biosensor with high specificity in >1 h, while the biosensor needs about 30 min to be constructed [36]. A dual-signal amplification system was also reported. This system exploited target recycling through RNase HII action, while DNA dendritic nanostructures were synthesized by terminal transferase, increasing the signal. The electrical signal was generated by the electroactive compound methylene blue. This sensor offered high detectability and specificity for the detection of a single KRAS mutation in ctDNA. The sensor could be universal, while the analysis and fabrication time were about 3 and 4.5 h, respectively [37]. Screen-printed multi-carbon electrodes were also used in a single and multiplex format for ctDNA analysis. Capture DNA probes were immobilized onto the electrode's surface, while the hybridization of the target was performed through ruthenium redox mediator and cyclic voltametric measurements. The fabrication time of the sensor is about 1 h, the analysis time is 3.5 h, while the LOD is 4 copies/ng for the single and 0.58 ng/µL for the multiplex format [37,38]. An urchin-like gold nanocrystal-multiple graphene aerogel was also exploited for ctDNA analysis. In this approach, signal enhancement was accomplished via DNA-induced target recycling. This biosensor had an LOD of 0.033 fM, >1 h analysis time with quite good specificity, and about 4 h fabrication time, but it lacked multiplicity and universality [39,40] RNase HII-aided target recycling was also performed in another SERS assay to detect single-stranded ctDNA in blood samples. T-rich DNA sequences produced by the RNAse were captured on the sensor's surface. Single-walled carbon nanotubes were then attached to the hybrids, while copper nanoparticles were synthesized to the T-rich areas, increasing the SERS signal. Two mutations of the KRAS and PIK3CA were detected within more than 30 min with high specificity. The biosensor needs extensive fabrication (about 54 h) and does not apply for multiplex or universal analysis [41]. Biosensors based on surface-enhanced Raman spectroscopy (SERS) also emerged in the ctDNA analysis. Multiplex detection of ctDNA mutations was also accomplished by a SERS assay. ctDNA was amplified by an allele-specific PCR. The amplified products carried a biotin moiety at one end and contained overhanging sequences at the other end. The products were captured by streptavidin-coated magnetic beads and detected by complementary DNA probes coupled to multicolor SERS nanotags (AuNPs) through hybridization with the overhanging ends. The hybrids were finally analyzed, after magnetic separation, by a portable Raman microscope. With this system, three mutations of the KRAS, BRAF, and NRAS gene were detected within 30 min. The sensor is universal and offers good specificity and sufficient multiplexing potential [42,43]. In another SERS approach, ctDNA was also amplified by allele-specific PCR. The amplified products were detected by positively charged gold/silver nanostars that were electrostatically bound to the negative DNA sequences. The SERS-active compound 2,3,5,6-tetrafluoro-4-mercaptobenzoic acid was used for SERS signaling. SERS spectrum was obtained by a portable Raman microscope. This assay is a mix-and-read method and applied for the detection of one single-point mutation of the BRAF gene. It offers universality, good specificity, and quite good multiplexing potential [44]. The combination of Fe and Au nanoparticles was also used for the detection of ctDNA. Magnetic DNA1-modified amorphous Fe^0 nanoparticles were used to magnetically separate and enrich ctDNA sequence, while the detection is accomplished through DNA2-modified Au nanoparticles and subsequent ICP-MS analysis. The method gave an LOD of 0.1pg/mL [45]. A localized surface plasmon resonance (LSPR) was also developed for ctDNA detection of the KRAS gene. ctDNA was captured on the surface of the sensor by PNA-functionalized gold nanorods. The LOD of this sensor is 2 ng/mL [46]. Table 2 and Table S3 contain a comparison between our work and other biosensors and conventional methods that have been applied for ctDNA analysis in clinical samples. Compared to the previously reported biosensors, our proposed multiplex rapid strip test was efficiently applied for cfDNA analysis of CRC patients' samples and shows great advantages related to specificity and repeatability (%CV < 3%), cost effectiveness, simplicity and ease of use, rapid analysis and fabrication time, portability, universality, and advanced multiplexing potential.

As for lateral flow assays for KRAS gene mutations detection in general, a lateral flow assay based on allele-specific PCR and hybridization to oligonucleotide-decorated AuNPs was developed for the detection of KRAS mutations in genomic DNA isolated from a cancer cell line, but it was not applied for ctDNA analysis [47]. Additionally, a commercially available strip test, the KRAS Strip Assay kit, was used by Kader et al., 2013 for the detection of KRAS mutations in tissue samples from CRC patients. This assay is based on an alkaline phosphatase-catalyzed chromogenic reaction and allele-specific hybridization onto the strip at high temperature (45 °C) to obtain the required specificity. Moreover, the test involves many washing and incubation steps with a total analysis time of about 2 h [48]. Compared to the previously reported biosensors for ctDNA detection in real samples (Table S3) [49–81], the advantages of the proposed multi-plex test are related mostly to the cost, run time, and simplicity, as well as the universality and advanced multiplexing potential.

Albeit the great advantages of the developed test, it should be noted that there are some limitations. Although no false-positive or false-negative results were obtained when the test was applied in cfDNA of healthy donors and CRC patients with 100% concordance with the tissue results previously confirmed by NGS, the number of patient samples used in the study for validation is relatively low. Therefore, further studies with a larger series of CRC patients and comparison with another technique, such as droplet digital PCR or NGS, are required to evaluate the clinical sensitivity and specificity of this newly developed biosensor as a diagnostic tool in liquid biopsy applications. In this context, a receiver operating characteristic (ROC) curve analysis of the results obtained by the two methods would be an effective approach to evaluate the sensitivity and specificity of our assay for diagnostic purposes [82]. Moreover, the multiplex strip test presented herein shows LOD $\leq$ 0.1 ng (1:1000 dilution) of mutated KRAS gene which is comparable with other methods such as ddPCR [82} and most importantly efficiently and specifically detected KRAS wt or mutant alleles in plasma samples with cfDNA concentrations as low as 6 ng/μL. However, considering that ctDNA is a challenging analyte due to the low concentration and high fragmentation and the amount of input DNA is a critical step in developing methods suitable for cfDNA analysis, further optimization in this context would be valuable.

Table 2. Comparison of biosensors for gene mutations detection in ctDNA in real samples.

Gene	Method	Fabrication Time	Analysis Time (after Amplification)	LOD	Precision (%CV)	Multiplicity	Universality	Ref.
EGFR	HPR- and DNA nanostructure-based electrochemical biosensor	overnight	>1 h	30 pg	1.89	2	✓	[31]
PIK3CA	Alkaline-phosphatase and HCR-based electrochemical biosensor	17 h	>1.5 h	3 pM	-	-	-	[32]
KRAS EGFR	Electrochemical sensor	Overnight	>30 min	1 fg/μL	-	array of 40 sensors	-	[33,34]
PIK3CA	Electrochemical platform	>1 h	1 h	10 fM	5.3	-	-	[35]
KRAS	Triple-helix molecular switch-based electrochemical biosensor	>4.5 h	3 h	2.4 aM	5.5–7.4	-	✓	[36]
KRAS	DNA probe-functionalized electrochemical sensor	1 h	3.5 h	4 copies/ng	-	-	-	[37]
KRAS	DNA probe-functionalized electrochemical sensor	1 h	3.5 h	0.58 ng/μL	-	3	-	[38]
KRAS	Urchin-like gold nanocrystal-multiple graphene aerogel	>4 h	>1 h	0.033 fM	-	-	-	[39]
KRAS	RNase assisted SERS platform	58 h	-	0.3 fM	-	-	✓	[40]
BRAF, NRAS	PCR/SERS sensor	>48 h	-	10 copies	8.8	3	✓	[41]
KRAS, BRAF	PCR/ SERS sensor	-	~30 min	-	-	3	-	[42]
BRAF	PCR/SERS sensor	>50 min	-	100 input copies	4	2	-	[43]
KRAS	Fe–Au nanoparticle-coupling/ICP-MS	-	-	0.1 pg/mL	-	7	-	[44]
KRAS	PNA probes on gold nanorods	-	10 min	2 ng/mL	-	-	-	[45]
KRAS	Strip-type biosensor	10 min	10 min	50 amol(10 pM) of ssDNA or <0.1% (100 pg) mutated gene	0.5–2.8	4	✓	This work

HPR: horseradish peroxidise. CP: combinatorial probes. SERS: surface-enhanced Raman spectroscopy. HCR: hybridization chain reaction. ICP-MS: inductively coupled plasma mass spectrometry. PNA: peptide nucleic acid. TMSDR: toehold-mediated strand displacement reaction.

5. Conclusions

In conclusion, we have developed a rapid and low-cost multiplex strip test for genotyping of cell-free/circulating tumor DNA in blood samples. The detection was based on gold nanoparticles. This newly developed strip test was applied for multiplex detection of KRAS mutations in liquid biopsies of CRC patients. Liquid biopsy represents a competing non-invasive tool for tumor genotyping, providing valuable information employed for cancer diagnosis, prognosis, and guidance of treatment decisions. Several methodologies have been used to detect tumor-specific aberrations in ctDNA, many of which are complicated, expensive, and time-consuming. The proposed method is a non-invasive, simple, and low-cost approach for the rapid and accurate analysis of cell-free and circulating tumor DNA. It achieves simultaneous multi-allele detection with a single test with high specificity and repeatability (%CV < 3%), offers an LOD lower than 0.1% (100 pg) mutated KRAS gene, and avoids multiple washing and incubation steps. It is also universal, meaning that it can be applied for the detection of any gene mutation or single nucleotide polymorphism (SNP), since the discrimination of the alleles takes places before the analysis with the strip test. The test was initially developed using DNA from synthetic DNA targets, cancer cell lines, and tissue samples. The proposed strip test was finally applied, for the first time, for the multiplex detection of KRAS mutations in cfDNA and ctDNA isolated from plasma samples from healthy individuals and CRC patients. All four KRAS alleles in ctDNA were simultaneously detected with a single strip test. The visual detection with the multiplex strip test was completed within 15 min, while the whole protocol, including DNA purification, KRAS gene amplification, allele-discrimination reaction, and the multiplex strip test, was completed within 3.5 h. Although further studies are required to validate the clinical sensitivity and specificity of the test, the results obtained so far render this test a promising tool for future liquid biopsy applications.

Supplementary Materials: The following supporting information can be downloaded at: https://www.mdpi.com/article/10.3390/bios12020097/s1, Table S1: Information for the CRC patients. Supplementary Table S2. The quantification results for DNA isolation from the cell lines and ctDNA isolation from the blood samples. Table S3: Conventional methods for the detection of KRAS mutations in ctDNA isolated from blood samples. Figure S1. Electropherograms of the PCR products obtained from DNA and ctDNA from a cell line, a tissue sample and plasma blood samples from CRC patients. M: DNA marker, CRC: colorectal cancer. Figure S2. Evaluation of SA-AuNPs conjugates. b-BSA: biotinylated bovine serum albumin. Figure S3. Optimization study for the sample volume for the analysis with multiplex rapid strip test. Figure S4. Calibration graph for the detection of single-stranded (ssDNA) with the rapid strip test. CZ: control zone, TZ: test zone, b-dA30: biotinylated dA30 oligonucleotide.

Author Contributions: Conceptualization, D.P.K. and V.B.; methodology, P.M.K., S.N., D.P.K. and V.B.; validation, D.P.K.; formal analysis, P.M.K., S.N., and S.K.; data curation, P.M.K., S.N., S.K. and D.P.K., investigation, P.M.K., S.N. and D.P.K.; resources, S.K. and H.P.K.; writing—original draft preparation, S.N. and P.M.K.; writing—review and editing, D.P.K., V.B., S.K. and H.P.K.; visualization, P.M.K.; supervision, D.P.K. and V.B.; funding acquisition, D.P.K., V.B. and H.P.K. All authors have read and agreed to the published version of the manuscript. Figure S3. Optimization study for the sample volume for the analysis with multiplex rapid strip test.

Funding: This research and the APC were funded by Greece and the European Union (European Social Fund—ESF) through the Operational Programme «Human Resources Development, Education and Lifelong Learning 2014–2020» in the context of the project "Gold nanoparticle-based biosensor for rapid liquid biopsy applications" grant number MIS 5047147.

Institutional Review Board Statement: The study was conducted in accordance with the Declaration of Helsinki, and approved by the Ethics and Scientific Committee of the University General Hospital of Patras, Greece (Protocol Number 680/15.10.2019) and by the Institutional Ethics & Research Committee of the University of Patras, Greece (Protocol Number 86436/14.10.2019).

Informed Consent Statement: Informed consent was obtained from all subjects involved in the study.

Acknowledgments: This research is co-financed by Greece and the European Union (European Social Fund—ESF) through the Operational Programme «Human Resources Development, Education and Lifelong Learning 2014–2020» in the context of the project "Gold nanoparticle-based biosensor for rapid liquid biopsy applications" (MIS 5047147). The authors would like to thank Melachrinou M., Director of the Department of Pathology, University of Patras for kindly providing FFPE tissue samples. The authors also acknowledge the valuable contribution of Konstantinos Christopoulos, MSc to shape configurations in Figure 1.

Conflicts of Interest: The authors declared no conflicts of interest.

References

1. El-Deiry, W.S.; Goldberg, R.M.; Lenz, H.-J.; Shields, A.F.; Gibney, G.T.; Tan, A.R.; Brown, J.; Eisenberg, B.; Heath, E.I.; Phuphanich, S.; et al. The current state of molecular testing in the treatment of patients with solid tumors, 2019. *CA Cancer J. Clin.* **2019**, *69*, 305–343. [CrossRef] [PubMed]
2. Hanahan, D.; Weinberg, R.A. The Hallmarks of Cancer. *Cell* **2000**, *100*, 57–70. [CrossRef]
3. Iranzo, J.; Martincorena, I.; Koonin, E.V. Cancer-mutation network and the number and specificity of driver mutations. *Proc. Natl. Acad. Sci. USA* **2018**, *115*, E6010–E6019. [CrossRef] [PubMed]
4. Martincorena, I.; Campbell, P.J. Somatic mutation in cancer and normal cells. *Science* **2015**, *349*, 1483–1489. [CrossRef]
5. Slebos, R.J.C.; Kibbelaar, R.E.; Dalesio, O.; Kooistra, A.; Stam, J.; Meijer, C.J.L.M.; Wagenaar, S.S.; Vanderschueren, R.G.J.R.A.; van Zandwijk, N.; Mooi, W.J.; et al. K-ras Oncogene Activation as a Prognostic Marker in Adenocarcinoma of the Lung. *N. Engl. J. Med.* **1990**, *323*, 561–565. [CrossRef]
6. Bos, J.L. Ras Oncogenes in Human Cancer: A Review. *Cancer Res.* **1989**, *49*, 4682–4689.
7. Cardarella, S.; Ogino, A.; Nishino, M.; Butaney, M.; Shen, J.; Lydon, C.; Yeap, B.Y.; Sholl, L.M.; Johnson, B.E.; Jänne, P.A. Clinical, pathologic, and biologic features associated with BRAF mutations in non–small cell lung cancer. *Clin. Cancer Res.* **2013**, *19*, 4532–4540. [CrossRef]
8. HARVEY, J.J. An Unidentified Virus which causes the Rapid Production of Tumours in Mice. *Nature* **1964**, *204*, 1104–1105. [CrossRef]
9. Rajalingam, K.; Schreck, R.; Rapp, U.R.; Albert, Š. Ras oncogenes and their downstream targets. *Biochim. Biophys. Acta Mol. Cell Res.* **2007**, *1773*, 1177–1195. [CrossRef]
10. Shackelford, R.E.; Whitling, N.A.; McNab, P.; Japa, S.; Coppola, D. KRAS Testing: A Tool for the Implementation of Personalized Medicine. *Genes Cancer* **2012**, *3*, 459–466. [CrossRef]
11. McLellan, E.A.; Owen, R.A.; Stepniewska, K.A.; Sheffield, J.P.; Lemoine, N.R. High frequency of K-ras mutations in sporadic colorectal adenomas. *Gut* **1993**, *34*, 392–396. [CrossRef]
12. Siegel, R.L.; Miller, K.D.; Jemal, A. Cancer statistics, 2019. *CA Cancer J. Clin.* **2019**, *69*, 7–34. [CrossRef]
13. Bettegowda, C.; Sausen, M.; Leary, R.J.; Kinde, I.; Wang, Y.; Agrawal, N.; Bartlett, B.R.; Wang, H.; Luber, B.; Alani, R.M.; et al. Detection of circulating tumor DNA in early- and late-stage human malignancies. *Sci. Transl. Med.* **2014**, *6*, 224ra24. [CrossRef]
14. Elazezy, M.; Joosse, S.A. Techniques of using circulating tumor DNA as a liquid biopsy component in cancer management. *Comput. Struct. Biotechnol. J.* **2018**, *16*, 370–378. [CrossRef]
15. Liebs, S.; Keilholz, U.; Kehler, I.; Schweiger, C.; Haybäck, J.; Nonnenmacher, A. Detection of mutations in circulating cell-free DNA in relation to disease stage in colorectal cancer. *Cancer Med.* **2019**, *8*, 3761–3769. [CrossRef]
16. Kloten, V.; Rüchel, N.; Brüchle, N.O.; Gasthaus, J.; Freudenmacher, N.; Steib, F.; Mijnes, J.; Eschenbruch, J.; Binnebösel, M.; Knüchel, R.; et al. Liquid biopsy in colon cancer: Comparison of different circulating DNA extraction systems following absolute quantification of KRAS mutations using Intplex allele-specific PCR. *Oncotarget* **2017**, *8*, 86253–86263. [CrossRef]
17. Cree, I.A.; Uttley, L.; Buckley Woods, H.; Kikuchi, H.; Reiman, A.; Harnan, S.; Whiteman, B.L.; Philips, S.T.; Messenger, M.; Cox, A.; et al. The evidence base for circulating tumour DNA blood-based biomarkers for the early detection of cancer: A systematic mapping review. *BMC Cancer* **2017**, *17*, 697. [CrossRef]
18. Giannopoulou, L.; Kasimir-Bauer, S.; Lianidou, E.S. Liquid biopsy in ovarian cancer: Recent advances on circulating tumor cells and circulating tumor DNA. *Clin. Chem. Lab. Med.* **2018**, *56*, 186–197. [CrossRef]
19. Chang, Y.; Tolani, B.; Nie, X.; Zhi, X.; Hu, M.; He, B. Review of the clinical applications and technological advances of circulating tumor DNA in cancer monitoring. *Ther. Clin. Risk Manag.* **2017**, *13*, 1363–1374. [CrossRef]
20. Lopez, A.; Harada, K.; Mizrak Kaya, D.; Dong, X.; Song, S.; Ajani, J.A. Liquid biopsies in gastrointestinal malignancies: When is the big day? *Expert Rev. Anticancer Ther.* **2018**, *18*, 19–38. [CrossRef]
21. Han, X.; Wang, J.; Sun, Y. Circulating Tumor DNA as Biomarkers for Cancer Detection. *Genomics Proteom. Bioinform.* **2017**, *15*, 59–72. [CrossRef] [PubMed]
22. Wang, J.; Wuethrich, A.; Sina, A.A.I.; Lane, R.E.; Lin, L.L.; Wang, Y.; Cebon, J.; Behren, A.; Trau, M. Tracking extracellular vesicle phenotypic changes enables treatment monitoring in melanoma. *Sci. Adv.* **2020**, *6*, eaax3223. [CrossRef] [PubMed]
23. Wang, J.; Koo, K.M.; Wang, Y.; Trau, M. Engineering State-of-the-Art Plasmonic Nanomaterials for SERS-Based Clinical Liquid Biopsy Applications. *Adv. Sci.* **2019**, *6*, 1900730. [CrossRef] [PubMed]

24. Das, J.; Kelley, S.O. High-Performance Nucleic Acid Sensors for Liquid Biopsy Applications. *Angew. Chem. Int. Ed.* **2020**, *59*, 2554–2564. [CrossRef]

25. Li, Z.-N.; Zhao, L.; Yu, L.-F.; Wei, M.-J. BRAF and KRAS mutations in metastatic colorectal cancer: Future perspectives for personalized therapy. *Gastroenterol. Rep.* **2020**, *8*, 192–205. [CrossRef]

26. Zheng, W.; Yao, L.; Teng, J.; Yan, C.; Qin, P.; Liu, G.; Chen, W. Lateral Flow Test for Visual Detection of Multiple MicroRNAs. *Sens. Actuators B Chem.* **2018**, *264*, 320–326. [CrossRef]

27. Kalligosfyri, P.; Nikou, S.; Bravou, V.; Kalogianni, D.P. Liquid biopsy genotyping by a simple lateral flow strip assay with visual detection. *Anal. Chim. Acta* **2021**, *1163*, 338470. [CrossRef]

28. Spyrou, E.M.; Kalogianni, D.P.; Tragoulias, S.S.; Ioannou, P.C.; Christopoulos, T.K. Digital camera and smartphone as detectors in paper-based chemiluminometric genotyping of single nucleotide polymorphisms. *Anal. Bioanal. Chem.* **2016**, *408*, 7393–7402. [CrossRef]

29. Li, X.; Ye, M.; Zhang, W.; Tan, D.; Jaffrezic-Renault, N.; Yang, X.; Guo, Z. Liquid biopsy of circulating tumor DNA and biosensor applications. *Biosens. Bioelectron.* **2019**, *126*, 596–607. [CrossRef]

30. Zhu, D.; Liu, W.; Cao, W.; Chao, J.; Su, S.; Wang, L.; Fan, C. Multiple Amplified Electrochemical Detection of MicroRNA-21 Using Hierarchical Flower-like Gold Nanostructures Combined with Gold-enriched Hybridization Chain Reaction. *Electroanalysis* **2018**, *30*, 1349–1356. [CrossRef]

31. Magiati, M.; Sevastou, A.; Kalogianni, D.P. A fluorometric lateral flow assay for visual detection of nucleic acids using a digital camera readout. *Microchim. Acta* **2018**, *185*, 314. [CrossRef]

32. Zhang, H.; Liu, X.; Zhang, C.; Xu, Y.; Su, J.; Lu, X.; Shi, J.; Wang, L.; Landry, M.P.; Zhu, Y.; et al. A DNA tetrahedral structure-mediated ultrasensitive fluorescent microarray platform for nucleic acid test. *Sens. Actuators B Chem.* **2020**, *321*, 128538. [CrossRef]

33. Huang, Y.; Tao, M.; Luo, S.; Zhang, Y.; Situ, B.; Ye, X.; Chen, P.; Jiang, X.; Wang, Q.; Zheng, L. A novel nest hybridization chain reaction based electrochemical assay for sensitive detection of circulating tumor DNA. *Anal. Chim. Acta* **2020**, *1107*, 40–47. [CrossRef]

34. Das, J.; Ivanov, I.; Safaei, T.S.; Sargent, E.H.; Kelley, S.O. Combinatorial Probes for High-Throughput Electrochemical Analysis of Circulating Nucleic Acids in Clinical Samples. *Angew. Chem. Int. Ed.* **2018**, *57*, 3711–3716. [CrossRef]

35. Das, J.; Ivanov, I.; Sargent, E.H.; Kelley, S.O. DNA Clutch Probes for Circulating Tumor DNA Analysis. *J. Am. Chem. Soc.* **2016**, *138*, 11009–11016. [CrossRef]

36. Cai, C.; Guo, Z.; Cao, Y.; Zhang, W.; Chen, Y. A dual biomarker detection platform for quantitating circulating tumor DNA (ctDNA). *Nanotheranostics* **2018**, *2*, 12–20. [CrossRef]

37. Wang, H.-F.; Ma, R.-N.; Sun, F.; Jia, L.-P.; Zhang, W.; Shang, L.; Xue, Q.-W.; Jia, W.-L.; Wang, H.-S. A versatile label-free electrochemical biosensor for circulating tumor DNA based on dual enzyme assisted multiple amplification strategy. *Biosens. Bioelectron.* **2018**, *122*, 224–230. [CrossRef]

38. Attoye, B.; Pou, C.; Blair, E.; Rinaldi, C.; Thomson, F.; Baker, M.J.; Corrigan, D.K. Developing a Low-Cost, Simple-to-Use Electrochemical Sensor for the Detection of Circulating Tumour DNA in Human Fluids. *Biosensors* **2020**, *10*, 156. [CrossRef]

39. Attoye, B.; Baker, M.J.; Thomson, F.; Pou, C.; Corrigan, D.K. Optimisation of an Electrochemical DNA Sensor for Measuring KRAS G12D and G13D Point Mutations in Different Tumour Types. *Biosensors* **2021**, *11*, 42. [CrossRef]

40. Yuanfeng, P.; Ruiyi, L.; Xiulan, S.; Guangli, W.; Zaijun, L. Highly sensitive electrochemical detection of circulating tumor DNA in human blood based on urchin-like gold nanocrystal-multiple graphene aerogel and target DNA-induced recycling double amplification strategy. *Anal. Chim. Acta* **2020**, *1121*, 17–25. [CrossRef]

41. Zhou, Q.; Zheng, J.; Qing, Z.; Zheng, M.; Yang, J.; Yang, S.; Ying, L.; Yang, R. Detection of Circulating Tumor DNA in Human Blood via DNA-Mediated Surface-Enhanced Raman Spectroscopy of Single-Walled Carbon Nanotubes. *Anal. Chem.* **2016**, *88*, 4759–4765. [CrossRef]

42. Wee, E.J.H.; Wang, Y.; Tsao, S.C.-H.; Trau, M. Simple, Sensitive and Accurate Multiplex Detection of Clinically Important Melanoma DNA Mutations in Circulating Tumour DNA with SERS Nanotags. *Theranostics* **2016**, *6*, 1506–1513. [CrossRef]

43. Lyu, N.; Rajendran, V.K.; Diefenbach, R.J.; Charles, K.; Clarke, S.J.; Engel, A.; Sydney 1000 Colorectal Cancer Study Investigators; Rizos, H.; Molloy, M.P.; Wang, Y. Multiplex detection of ctDNA mutations in plasma of colorectal cancer patients by PCR/SERS assay. *Nanotheranostics* **2020**, *4*, 224–232. [CrossRef]

44. Liu, Y.; Lyu, N.; Rajendran, V.K.; Piper, J.; Rodger, A.; Wang, Y. Sensitive and Direct DNA Mutation Detection by Surface-Enhanced Raman Spectroscopy Using Rational Designed and Tunable Plasmonic Nanostructures. *Anal. Chem.* **2020**, *92*, 5708–5716. [CrossRef]

45. Hu, P.; Zhang, S.; Wu, T.; Ni, D.; Fan, W.; Zhu, Y.; Qian, R.; Shi, J. Fe–Au Nanoparticle-Coupling for Ultrasensitive Detections of Circulating Tumor DNA. *Adv. Mater.* **2018**, *30*, 1801690. [CrossRef]

46. Tadimety, A.; Zhang, Y.; Kready, K.M.; Palinski, T.J.; Tsongalis, G.J.; Zhang, J.X.J. Design of peptide nucleic acid probes on plasmonic gold nanorods for detection of circulating tumor DNA point mutations. *Biosens. Bioelectron.* **2019**, *130*, 236–244. [CrossRef]

47. Wang, C.; Chen, X.; Wu, Y.; Li, H.; Wang, Y.; Pan, X.; Tang, T.; Liu, Z.; Li, X. Lateral flow strip for visual detection of K-ras mutations based on allele-specific PCR. *Biotechnol. Lett.* **2016**, *38*, 1709–1714. [CrossRef] [PubMed]

48. Abd El Kader, Y.; Emera, G.; Safwat, E.; Kassem, H.A.; Kassem, N.M. The KRAS StripAssay for detection of KRAS mutation in Egyptian patients with colorectal cancer (CRC): A pilot study. *J. Egypt. Natl. Canc. Inst.* **2013**, *25*, 37–41. [CrossRef] [PubMed]

49. Formica, V.; Lucchetti, J.; Doldo, E.; Riondino, S.; Morelli, C.; Argirò, R.; Renzi, N.; Nitti, D.; Nardecchia, A.; Dell'Aquila, E.; et al. Clinical Utility of Plasma KRAS, NRAS and BRAF Mutational Analysis with Real Time PCR in Metastatic Colorectal Cancer Patients—The Importance of Tissue/Plasma Discordant Cases. *J. Clin. Med.* **2021**, *10*, 87. [CrossRef]

50. Franczak, C.; Witz, A.; Geoffroy, K.; Demange, J.; Rouyer, M.; Husson, M.; Massard, V.; Gavoille, C.; Lambert, A.; Gilson, P.; et al. Evaluation of KRAS, NRAS and BRAF mutations detection in plasma using an automated system for patients with metastatic colorectal cancer. *PLoS ONE* **2020**, *15*, e0227294. [CrossRef] [PubMed]

51. Ono, Y.; Sugitani, A.; Karasaki, H.; Ogata, M.; Nozaki, R.; Sasajima, J.; Yokochi, T.; Asahara, S.; Koizumi, K.; Ando, K.; et al. An improved digital polymerase chain reaction protocol to capture low-copy KRAS mutations in plasma cell-free DNA by resolving 'subsampling' issues. *Mol. Oncol.* **2017**, *11*, 1448–1458. [CrossRef]

52. Milosevic, D.; Mills, J.R.; Campion, M.B.; Vidal-Folch, N.; Voss, J.S.; Halling, K.C.; Highsmith, W.E.; Liu, M.C.; Kipp, B.R.; Grebe, S.K.G. Applying Standard Clinical Chemistry Assay Validation to Droplet Digital PCR Quantitative Liquid Biopsy Testing. *Clin. Chem.* **2018**, *64*, 1732–1742. [CrossRef]

53. Ou, C.-Y.; Vu, T.; Grunwald, J.T.; Toledano, M.; Zimak, J.; Toosky, M.; Shen, B.; Zell, J.A.; Gratton, E.; Abram, T.J.; et al. An ultrasensitive test for profiling circulating tumor DNA using integrated comprehensive droplet digital detection. *Lab Chip* **2019**, *19*, 993–1005. [CrossRef]

54. Pratt, E.D.; Cowan, R.W.; Manning, S.L.; Qiao, E.; Cameron, H.; Schradle, K.; Simeone, D.; Zhen, D.B. Multiplex Enrichment and Detection of Rare KRAS Mutations in Liquid Biopsy Samples using Digital Droplet Pre-Amplification. *Anal. Chem.* **2019**, *91*, 7516–7523. [CrossRef]

55. Alcaide, M.; Cheung, M.; Bushell, K.; Arthur, S.E.; Wong, H.-L.; Karasinska, J.; Renouf, D.; Schaeffer, D.F.; McNamara, S.; du Tertre, M.C.; et al. A Novel Multiplex Droplet Digital PCR Assay to Identify and Quantify KRAS Mutations in Clinical Specimens. *J. Mol. Diagn.* **2019**, *21*, 214–227. [CrossRef]

56. Michaelidou, K.; Koutoulaki, C.; Mavridis, K.; Vorrias, E.; Papadaki, M.A.; Koutsopoulos, A.V.; Mavroudis, D.; Agelaki, S. Detection of KRAS G12/G13 Mutations in Cell Free-DNA by Droplet Digital PCR, Offers Prognostic Information for Patients with Advanced Non-Small Cell Lung Cancer. *Cells* **2020**, *9*, 2514. [CrossRef]

57. Yang, Z.; LaRiviere, M.J.; Ko, J.; Till, J.E.; Christensen, T.; Yee, S.S.; Black, T.A.; Tien, K.; Lin, A.; Shen, H.; et al. A multi-analyte panel consisting of extracellular vesicle miRNAs and mRNAs, cfDNA, and CA19-9 shows utility for diagnosis and staging of pancreatic adenocarcinoma. *Clin. Cancer Res.* **2020**, *26*, 3248–3528. [CrossRef]

58. Decraene, C.; Silveira, A.B.; Bidard, F.-C.; Vallée, A.; Michel, M.; Melaabi, S.; Vincent-Salomon, A.; Saliou, A.; Houy, A.; Milder, M.; et al. Multiple Hotspot Mutations Scanning by Single Droplet Digital PCR. *Clin. Chem.* **2018**, *64*, 317–328. [CrossRef]

59. Zmrzljak, U.P.; Košir, R.; Krivokapić, Z.; Radojković, D.; Nikolić, A. Detection of Somatic Mutations with ddPCR from Liquid Biopsy of Colorectal Cancer Patients. *Genes* **2021**, *12*, 289. [CrossRef]

60. Rowlands, V.; Rutkowski, A.J.; Meuser, E.; Carr, T.H.; Harrington, E.A.; Barrett, J.C. Optimisation of robust singleplex and multiplex droplet digital PCR assays for high confidence mutation detection in circulating tumour DNA. *Sci. Rep.* **2019**, *9*, 12620. [CrossRef]

61. Li, J.; Gan, S.; Blair, A.; Min, K.; Rehage, T.; Hoeppner, C.; Halait, H.; Brophy, V.H. A Highly Verified Assay for KRAS Mutation Detection in Tissue and Plasma of Lung, Colorectal, and Pancreatic Cancer. *Arch. Pathol. Lab. Med.* **2018**, *143*, 183–189. [CrossRef] [PubMed]

62. Li, B.T.; Janku, F.; Jung, B.; Hou, C.; Madwani, K.; Alden, R.; Razavi, P.; Reis-Filho, J.S.; Shen, R.; Isbell, J.M.; et al. Ultra-deep next-generation sequencing of plasma cell-free DNA in patients with advanced lung cancers: Results from the Actionable Genome Consortium. *Ann. Oncol.* **2019**, *30*, 597–603. [CrossRef] [PubMed]

63. Nacchio, M.; Sgariglia, R.; Gristina, V.; Pisapia, P.; Pepe, F.; De Luca, C.; Migliatico, I.; Clery, E.; Greco, L.; Vigliar, E.; et al. KRAS mutations testing in non-small cell lung cancer: The role of Liquid biopsy in the basal setting. *J. Thorac. Dis.* **2020**, *12*, 3836. [CrossRef]

64. Shen, W.; Shan, B.; Liang, S.; Zhang, J.; Yu, Y.; Zhang, Y.; Wang, G.; Bai, Y.; Qian, B.; Lu, J.; et al. Hybrid Capture-based Genomic Profiling of Circulating Tumor DNA From Patients With Advanced Ovarian Cancer. *Pathol. Oncol. Res.* **2021**, *27*, 41. [CrossRef]

65. Takano, S.; Fukasawa, M.; Shindo, H.; Takahashi, E.; Fukasawa, Y.; Kawakami, S.; Hayakawa, H.; Kuratomi, N.; Kadokura, M.; Maekawa, S.; et al. Digital next-generation sequencing of cell-free DNA for pancreatic cancer. *JGH Open* **2021**, *5*, 508–516. [CrossRef]

66. Lee, S.-Y.; Chae, D.-K.; Lee, S.-H.; Lim, Y.; An, J.; Chae, C.H.; Kim, B.C.; Bhak, J.; Bolser, D.; Cho, D.-H. Efficient mutation screening for cervical cancers from circulating tumor DNA in blood. *BMC Cancer* **2020**, *20*, 694. [CrossRef] [PubMed]

67. Tran, L.S.; Pham, H.-A.T.; Tran, V.-U.; Tran, T.-T.; Dang, A.-T.H.; Le, D.-T.; Nguyen, S.-L.; Nguyen, N.-V.; Nguyen, T.-V.; Vo, B.T.; et al. Ultra-deep massively parallel sequencing with unique molecular identifier tagging achieves comparable performance to droplet digital PCR for detection and quantification of circulating tumor DNA from lung cancer patients. *PLoS ONE* **2019**, *14*, e0226193. [CrossRef] [PubMed]

68. Nguyen, H.T.; Tran, D.H.; Ngo, Q.D.; Pham, H.-A.T.; Tran, T.-T.; Tran, V.-U.; Pham, T.-V.N.; Le, T.K.; Le, N.-A.T.; Nguyen, N.M.; et al. Evaluation of a Liquid Biopsy Protocol using Ultra-Deep Massive Parallel Sequencing for Detecting and Quantifying Circulation Tumor DNA in Colorectal Cancer Patients. *Cancer Investig.* **2020**, *38*, 85–93. [CrossRef]

69. Damin, F.; Galbiati, S.; Soriani, N.; Burgio, V.; Ronzoni, M.; Ferrari, M.; Chiari, M. Analysis of KRAS, NRAS and BRAF mutational profile by combination of in-tube hybridization and universal tag-microarray in tumor tissue and plasma of colorectal cancer patients. *PLoS ONE* **2018**, *13*, e0207876. [CrossRef]

70. Wang, J.; Hua, G.; Li, L.; Li, D.; Wang, F.; Wu, J.; Ye, Z.; Zhou, X.; Ye, S.; Yang, J.; et al. Upconversion nanoparticle and gold nanocage satellite assemblies for sensitive ctDNA detection in serum. *Analyst* **2020**, *145*, 5553–5562. [CrossRef]

71. Kerachian, M.A.; Azghandi, M.; Javadmanesh, A.; Ghaffarzadegan, K.; Mozaffari-Jovin, S. Selective capture of plasma cell-free tumor DNA on magnetic beads: A sensitive and versatile tool for liquid biopsy. *Cell. Oncol.* **2020**, *43*, 949–956. [CrossRef]

72. Wang, B.; Wu, S.; Huang, F.; Shen, M.; Jiang, H.; Yu, Y.; Yu, Q.; Yang, Y.; Zhao, Y.; Zhou, Y.; et al. Analytical and clinical validation of a novel amplicon-based NGS assay for evaluation of circulating tumor DNA in metastatic colorectal cancer patients. *Clin. Chem. Lab. Med.* **2019**, *57*, 1501–1510. [CrossRef]

73. Arnold, L.; Alexiadis, V.; Watanaskul, T.; Zarrabi, V.; Poole, J.; Singh, V. Clinical validation of qPCR Target Selector^{TM} assays using highly specific switch-blockers for rare mutation detection. *J. Clin. Pathol.* **2020**, *73*, 648–655. [CrossRef]

74. Giménez-Capitán, A.; Bracht, J.; García, J.J.; Jordana-Ariza, N.; García, B.; Garzón, M.; Mayo-de-las-Casas, C.; Viteri-Ramirez, S.; Martinez-Bueno, A.; Aguilar, A.; et al. Multiplex Detection of Clinically Relevant Mutations in Liquid Biopsies of Cancer Patients Using a Hybridization-Based Platform. *Clin. Chem.* **2021**, *67*, 554–563. [CrossRef]

75. Song, J.; Hegge, J.W.; Mauk, M.G.; Chen, J.; Till, J.E.; Bhagwat, N.; Azink, L.T.; Peng, J.; Sen, M.; Mays, J.; et al. Highly specific enrichment of rare nucleic acid fractions using Thermus thermophilus argonaute with applications in cancer diagnostics. *Nucleic Acids Res.* **2020**, *48*, e19. [CrossRef]

76. Botezatu, I.V.; Kondratova, V.N.; Shelepov, V.P.; Mazurenko, N.N.; Tsyganova, I.V.; Susova, O.Y.; Lichtenstein, A.V. Asymmetric mutant-enriched polymerase chain reaction and quantitative DNA melting analysis of KRAS mutation in colorectal cancer. *Anal. Biochem.* **2020**, *590*, 113517. [CrossRef]

77. Ruiz, C.; Huang, J.; Giardina, S.F.; Feinberg, P.B.; Mirza, A.H.; Bacolod, M.D.; Soper, S.A.; Barany, F. Single-molecule detection of cancer mutations using a novel PCR-LDR-qPCR assay. *Hum. Mutat.* **2020**, *41*, 1051–1068. [CrossRef]

78. Andersen, R.F.; Jakobsen, A. Screening for circulating RAS/RAF mutations by multiplex digital PCR. *Clin. Chim. Acta* **2016**, *458*, 138–143. [CrossRef]

79. Min, S.; Shin, S.; Chung, Y.-J. Detection of KRAS mutations in plasma cell-free DNA of colorectal cancer patients and comparison with cancer panel data for tissue samples of the same cancers. *Genomics Inform.* **2019**, *17*, e42. [CrossRef]

80. Nakagawa, T.; Tanaka, J.; Harada, K.; Shiratori, A.; Shimazaki, Y.; Yokoi, T.; Uematsu, C.; Kohara, Y. 10-Plex Digital Polymerase Chain Reaction with Four-Color Melting Curve Analysis for Simultaneous KRAS and BRAF Genotyping. *Anal. Chem.* **2020**, *92*, 11705–11713. [CrossRef]

81. Olmedillas López, S.; García-Olmo, D.C.; García-Arranz, M.; Guadalajara, H.; Pastor, C.; García-Olmo, D. KRAS G12V Mutation Detection by Droplet Digital PCR in Circulating Cell-Free DNA of Colorectal Cancer Patients. *Int. J. Mol. Sci.* **2016**, *17*, 484. [CrossRef] [PubMed]

82. Frazzi, R.; Bizzarri, V.; Albertazzi, L.; Cusenza, V.Y.; Coppolecchia, L.; Luminari, S.; Ilariucci, F. Droplet digital PCR is a sensitive tool for the detection of TP53 deletions and point mutations in chronic lymphocytic leukaemia. *Br. J. Haematol.* **2020**, *189*, e49–e52. [CrossRef] [PubMed]

 *biosensors*

Article

Development of a Microfluidic Device for CD4⁺ T Cell Isolation and Automated Enumeration from Whole Blood

Robert D. Fennell [1,2], Mazhar Sher [1,2] and Waseem Asghar [1,2,3,*]

1. Asghar-Lab, Micro and Nanotechnology in Medicine, College of Engineering and Computer Science, Boca Raton, FL 33431, USA; rfennel1@my.fau.edu (R.D.F.); msher2015@fau.edu (M.S.)
2. Department of Electrical Engineering and Computer Science, Florida Atlantic University, Boca Raton, FL 33431, USA
3. Department of Biological Sciences (Courtesy Appointment), Florida Atlantic University, Boca Raton, FL 33431, USA
* Correspondence: wasghar@fau.edu

Abstract: The development of point-of-care, cost-effective, and easy-to-use assays for the accurate counting of CD4⁺ T cells remains an important focus for HIV-1 disease management. The CD4⁺ T cell count provides an indication regarding the overall success of HIV-1 treatments. The CD4⁺ T count information is equally important for both resource-constrained regions and areas with extensive resources. Hospitals and other allied facilities may be overwhelmed by epidemics or other disasters. An assay for a physician's office or other home-based setting is becoming increasingly popular. We have developed a technology for the rapid quantification of CD4⁺ T cells. A double antibody selection process, utilizing anti-CD4 and anti-CD3 antibodies, is tested and provides a high specificity. The assay utilizes a microfluidic chip coated with the anti-CD3 antibody, having an improved antibody avidity. As a result of enhanced binding, a higher flow rate can be applied that enables an improved channel washing to reduce non-specific bindings. A wide-field optical imaging system is also developed that provides the rapid quantification of cells. The designed optical setup is portable and low-cost. An ImageJ-based program is developed for the automatic counting of CD4⁺ T cells. We have successfully isolated and counted CD4⁺ T cells with high specificity and efficiency greater than 90%.

Keywords: CD4⁺ T helper cells; microfluidic chip; microbeads; wide-field optical system; ImageJ

Citation: Fennell, R.D.; Sher, M.; Asghar, W. Development of a Microfluidic Device for CD4⁺ T Cell Isolation and Automated Enumeration from Whole Blood. *Biosensors* **2022**, *12*, 12. https://doi.org/10.3390/bios12010012

Received: 12 October 2021
Accepted: 24 December 2021
Published: 28 December 2021

1. Introduction

There is a need to develop accurate cell quantification assays to achieve early-stage disease detection, treatment, and monitoring. Various cell quantification assays have been developed, tested, and validated for a multitude of diseases over the course of time [1]. The Coulter Principle has resulted in the Coulter counters being able to measure cell size and impedance in an electrolyte solution [2,3]. The principle has been extended to interactions with light as well. As a result of these advances, several sophisticated and advanced laboratory-based devices have been tested and approved for accurate cell quantification purposes. Such devices require well-trained personnel in well-equipped laboratories. Resource-limited settings lack these facilities. Hence, there is an unmet need to develop cost-effective, easy-to-use, and rapid disease diagnostic devices at the point-of-care settings, POC. The World Health Organization (WHO) has set these guidelines for future diagnostic equipment using the acronym ASSURED, which stands for affordable, sensitive, specific, user-friendly, rapid and robust, equipment-free, and deliverable. Devices developed based on these guidelines would be equally beneficial for both resource-enabled and resource-limited countries. This lack of adequate resources also applies to physicians' offices, patients' homes, and rapidly growing telemedicine situations. Due to the need for speed and on-site diagnosis, the 'Sample in, Answer-out' type of assays is gaining popularity.

Early disease diagnosis is a critical factor, especially in outbreaks of infectious diseases, such as HIV (human immunodeficiency virus), Ebola, Zika, and SARS-CoV-2 [4–6]. Urgent and timely clinical decisions can help to detect and curtail the spread of infectious diseases. Sending samples and receiving their results from a clinical laboratory often takes days. Higher throughput and rapid assays are the need of the day. Currently, hospitals often create their own testing facilities to reduce the turnaround time for the same-hour diagnosis. The development of POC diagnostic tools would help to make a patient's bedside testing possible. The test results could be obtained within the least amount of time, which enables the physician to make an early clinical decision and explore further options for treatment.

POC devices are supposed to be portable, cost-effective, and environment-friendly [7–9]. It is estimated that the biosensor market will expand further in the coming years. The current developments in cellular communications, smartphone imaging systems, integrated circuit technology, along with disposable microfluidic devices, can be utilized for the future POC devices in resource-limited areas.

The CD4$^+$ T count provides important information about the overall success of HIV treatment. Once HIV is diagnosed, the treatment is evaluated by the CD4$^+$ T lymphocyte cell count and CD4/CD8 ratios. As the disease is treated, several assays are required. Flow cytometry is a reliable and accurate method for the quantification of CD4 cells, but it has high equipment and test costs and requires skilled resources for operation, results analyses, and maintenance. There is a dire need to develop microchip-based assays for the enumeration of CD4$^+$ T cells. The main task is to isolate and quantify CD4$^+$ T cells from a drop of blood. This whole process could lead to a POC assay to be used in medically resource-poor locations that cannot afford expensive diagnostic testing.

The need for microfluidic devices has been explored by a number of researchers [8,10,11]. Earlier, researchers developed a microfluidic device for the label-free CD4$^+$ T cell counting of HIV-infected subjects [11]. They explored the use of a microfluidic module coated with a specific antibody for the isolation of CD4$^+$ T cells and monocytes. A well-controlled fluid flow was used to separate monocytes from CD4$^+$ T lymphocytes. A consistent well-controlled flow is difficult to control in a field location inside the microfluidic channel. Therefore, there is the possibility of an adverse effect on the captured CD4$^+$ T lymphocytes, and, as a result, some of the cells may be forced to detach from the microchip along with the monocytes. In one other study, Moon et.al. have developed a portable microfluidic platform to count the CD4$^+$ T cells of HIV-infected patients [12]. A microchip was coated with anti-CD4 antibody to capture the target CD4$^+$ T cells from a fingerpick volume of whole blood. The captured cells were imaged using the lensless CCD platform. Their setup requires an automated cell counting program developed in MATLAB to identify the CD4$^+$ T lymphocyte cells from monocytes.

Here, we present an alternate approach to identify CD4$^+$ T cell lymphocytes from monocytes. We describe an improved microfluidic chip process that separates a sparse number of CD4$^+$ T lymphocytes from whole blood samples. The developed method utilizes anti-CD4 antibody-conjugated magnetic beads to capture the CD4$^+$ T cells from a drop of whole blood with high specificity. The CD4$^+$ T lymphocytes are separated from monocytes with an anti-CD3 antibody-coated microchip. A washing step is then performed to get rid of the unneeded whole blood components and unattached magnetic beads from the microfluidic chip. The images of the captured CD4$^+$ T cells are then recorded with the help of a custom-built wide-field optical setup. The developed wide-field optical setup helps to cover a large field of view, enabling a sufficient sample size. The images are analyzed using a computer program we developed with Image J, and the cells are counted automatically.

The developed method has several advantages over the existing microfluidic techniques. Here, the avidity improvement is caused by the utilization of an anti-CD3 antibody for coating the bottom glass substrate of the microfluidic chip along with the anti-CD4 antibody coated magnetic beads. This approach has been applied for the first time, to the best of our knowledge, in microfluidic-based CD4$^+$ T cells quantification assays. Along with the enhanced antibody avidity, this approach also enabled much better CD4$^+$ T cell isolations. Initially, we were able to isolate all the CD4$^+$ T cells from the blood samples using the anti-CD4 coated antibody. Then, the monocytes were separated from those initially isolated CD4$^+$ T cells using the anti-CD3 coated to the glass slide inside a microfluidic chip. We have found that the CD4$^+$ T cells separation can be efficiently quantified using this approach with less shear stress on the cell.

2. Materials and Methods

2.1. Fabrication of the Microchip Module

The complete process list, the material source list, and a process datasheet for the microchip fabrication are presented in detail in the Supplementary Materials. The following describes the reagents, microchip design, the process design and development, the optical design and development, followed by the results obtained.

2.2. The Reagents

A complete list of materials and reagents is presented in the Supplementary Materials. Anti-CD3 antibody, UCHT1(biotin), part number ab191112 was obtained from Abcam, and it was used to coat the bottom glass substrate of the microchip. Dynabeads, part number 1145D, were obtained from ThermoFisher (Waltham, MA, USA); they were precoated with a proprietary anti-CD4 antibody.

To avoid bead clumping and unwanted adhesion to the module channel, we used ultra-pure Dulbecco's phosphate-buffered saline solution (DPBS) throughout the washing step. EDTA: ethylenediaminetetraacetic acid was used as an anticoagulant.

PMMA, a polymethylmethacrylate sheet, was utilized as the base material for making microfluidic chips. PMMA can easily be cut with a laser cutter. Various PMMA layers were joined together with the help of double-sided adhesive tape (DSA) to form a composite microfluidic chip.

The bottom layer of a microfluidic device was chemically modified to enable antibody attachment. The in-house processing uses 3-MPS (3-Mercaptopropyl) trimethoxysilane (3-MPS, CN: 175617) and cross-linker GMBS: N-γ-Maleimidobutyryloxy succinimide ester. GMBS is a crosslinker from the amino and sulfhydryl groups. It comes as a solid and is water insoluble. Therefore, it must be mixed with DMSO or DMF [3]. We used DMSO to dissolve the GMBS so that it could be applied to the substrate.

2.3. The Microfluidic Design

A drawing of the microchip module and its photograph are shown in Figure 1. The modules were designed using AutoCAD software. The AutoCAD software was linked to a laser cutter VLS 2.30 laser cutter (VersaLaser, Scottsdale, AZ, USA). PMMA and DSA sheets were cut according to the design requirements as per previously developed method [13–16]. The dimensions of microfluidic channels were selected in such a manner that there were three microchannels in each chip for running two samples and one control.

2.4. Immobilization of Antibody to Glass Substrate

Streptavidin-coated glass slides were prepared in-house. Figure 2 presents the basic process for microchip fabrication and was previously demonstrated [7,10,17]. Briefly, a biotinylated anti-CD3 antibody was applied to the substrate. The anti-CD3, from Abcam, (Catalog No. AB191112) recognizes the CD3 antigen of the TCR/CD3 complex on mature human T cells. The antibody reacts with the epsilon chain of the CD3 complex. The developed assay utilizes the stronger avidity of anti-CD3 antibody coated on a microfluidic chip. More explanation on stronger avidity is given in the Supplement IIC.

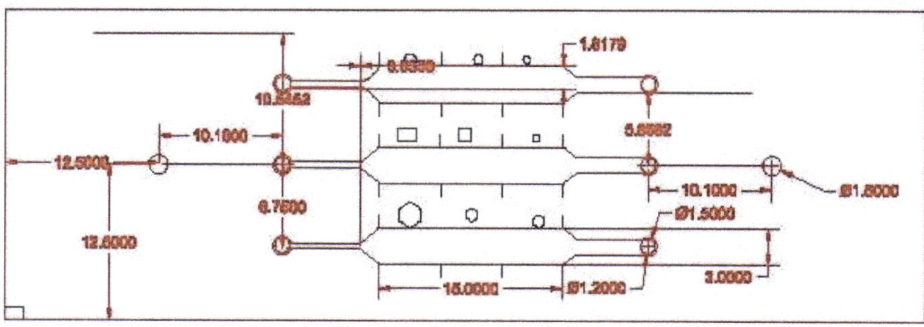

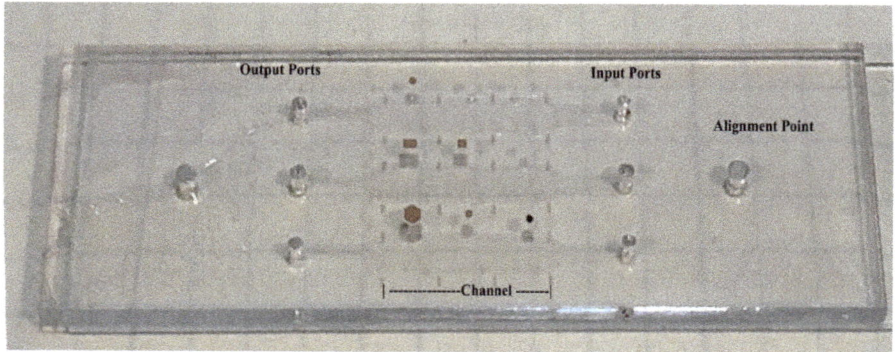

Figure 1. Microchip module with dimensions. The dimensions are in mm. The module is the size of a typical slide, 25 × 75 mm. The channel depth is 0.076 mm.

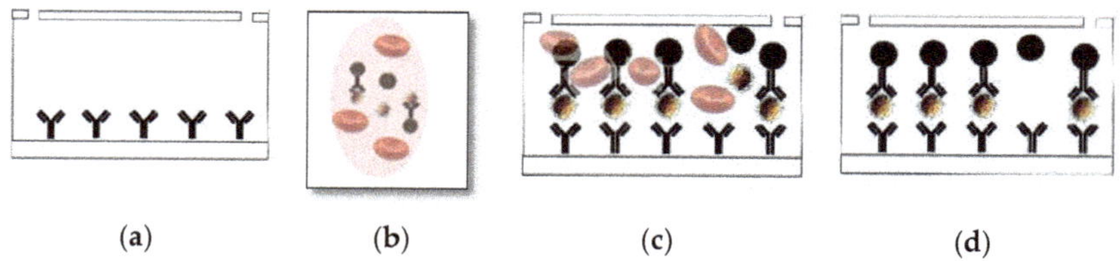

| (a) | (b) | (c) | (d) |

Figure 2. Basic functionalization process: (**a**) prepare substrate with Biotin anti-CD3 antibody; (**b**) mix blood with Dynabeads functionalized with anti-CD4 antibodies; (**c**) wash module with 400 µL DBPS at 50 µL/min; (**d**) observe and count cells.

Our initial process used anti-CD4 antibodies (ab28069) on the substrate and anti-CD3 antibodies on the microspheres. We encountered difficulties in the washing process. In order to clear out the red blood cells, we need at least 50 µL/min flow rate inside microfluidic channels. However, anything greater than 25 µL/min swept out some of the white blood cells also. We reversed the antibodies. This created an improved binding strength to capture CD4⁺ T helper cells to the substrate. We were able to use 50 µL/min cleaning flow with good results. Better adhesion to the prepared substrate was achieved as the Streptavidin layer was already well adhered, enabling a more vigorous wash cycle.

To find out the capture efficiency, the Dynabeads standard magnetic separation method was used to separate out CD4⁺ T cells along with the Dynabeads in a tube. Dynabeads were functionalized by a custom, proprietary anti-CD4 antibody. Initially, Dynabeads were mixed with the blood sample in a tube. Then, the isolation of CD4⁺ T cells was performed by an external magnet; the unbound entities from the blood were removed with the help of a pipette washing. The DPBS was added to the captured beads-cell complexes, and the washing step was repeated again. This second washing step was performed to get rid of any non-desired entity. Hence, CD4$^+$ T cells and beads remained inside the tube. An HC count was then used to count the cells under an inverted microscope (Nikon Eclipse TE2000-S; Nikon, Japan) at 10× [18]. A standard counting method was used. The four corner sections were each counted for the cells of interest, and then the cell count was averaged. The cell count was then multiplied by the dilution factor.

For comparison, the microfluidic chip was also used for capturing CD4$^+$ T cells. The number of CD4$^+$ T cells isolated with the help of Dynabeads was confirmed using an alternate commercially available kit, "EasySep" from Stemcell Technologies [19], which isolated CD4$^+$ T by a negative selection. These cells were then used as a validation check in one of the parallel channels in the microfluidic device to verify the process. As an additional test to confirm white cell identification, NucBlue, a cell permanent counterstain, was used with a Fluorescent Cell Imager (ZOE, Bio-Rad Laboratories, Hercules, CA, USA) with 20× objective lens cell imaging system.

The Dynabeads and CD4$^+$ T cells were injected into the microchip having the anti-CD3 antibody functionalization. Figure 3 portrays a full wide-field view, taken on our lensless equipment, that shows the full view of the microfluidic chip channel.

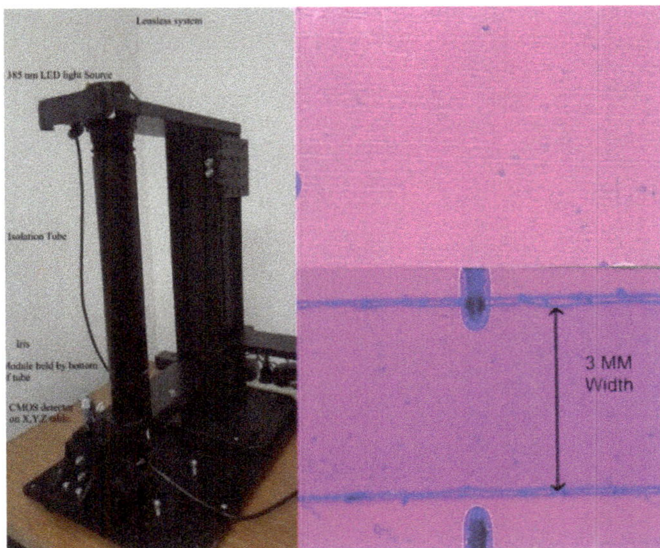

Figure 3. The left image is a lensless imaging system. LED is fitted inside the housing connected to isolation tube followed by 100 microns-sized pinhole. The light passes through the pinhole and shadow patterns of beads/cells are produced on CMOS sensor mounted on the base part. The chip containing captured cells and beads is placed on CMOS sensor before taking the image. The upper right lensless image was enlarged and shows beads and cells in a microfluidic chip module. The bottom right image shows the typical width of field, which is that of the imaging device.

DPBS was used to wash the channels that reduced the clumping of beads and cells. The channel washing was performed at a high flow rate of 50 µL/minute for 8 min. There were only a few beads not attached to CD4$^+$ T cells that were left in the channel.

2.5. Measurement of Capture Efficiency

As previously mentioned, the cell isolation process used the double selection method with anti-CD4 antibody-coated Dynabeads and an anti-CD3 antibody immobilized to the glass slide. The images of the captured cells were taken using the microscope and are called the microscope count, MC. The image areas, microchip channel volume, and dilution amount are taken into account. Multiple images are necessary because the inverted microscope (Nikon Eclipse TE2000-S) trades off magnification for coverage area, and the CD4$^+$ T count can be low per unit area. A wide area imaging technique was developed in earlier work [20] to compensate for this, and the design is described in Section 3 D. Capture efficiency was then the module microscope count (MC) divided by HC, the hemocytometer count. The module count, MC, was done by stitching together 8 photographs down a channel and then counting the CD4$^+$ T cells. In both cases, at this magnification, resolution, and narrow field of view, the low number of cells visible per image requires several different pictures to get a sufficient sample. This is also true for a hemocytometer, HC, where several photos were taken over several sectors. The images were then reviewed and CD4$^+$ T cells were identified and counted per unit volume.

MC: count on inverted microscope (Nikon Eclipse TE2000-S) of the microchip module.

HC: count on inverted microscope (Nikon Eclipse TE2000-S) of the hemocytometer, Dynabeads process.

$$\text{Efficiency} = \text{MC}/\text{HC} \ \% \tag{1}$$

2.6. Measurement of Specificity

To verify our process separation of CD4$^+$ T cells, a cell permeant counterstain was used, NucBlue, (Hoechst 33342), that emits blue fluorescent light at 460 nm when bound to DNA and excited by UV light. A Fluorescent Cell Imager (ZOE, Bio-Rad Laboratories, Hercules, CA, USA) with 20× objective lens was used. Brightfield (BF) and fluorescent images (FC) were created, stored, and merged, as shown in Figure S2 of the Supplement. An ImageJ program was then used to count cells and compare the BF and FC count. Visual confirmation was done with the merged view. Specificity was then the FC divided by BF.

$$\text{Specificity} = \text{FC}/\text{BF} \ \% \tag{2}$$

2.7. Development of a Lensless System to Enable Wide-Field-Imaging

In the development of a lensless imaging system, the key design issue is to generate a near monochromatic light source to ensure a parallel source without scattering. The module needs to be as close as possible to the CCD imaging device. An experimental lab-quality device was built as shown in Figure 3. It has the advantage of an X, Y, Z incremental movement of the microchip module. Multiple channels can easily be observed and brought into view. A portable lensless device is shown in Figure 3 that is suitable for a single channel micromodule. A lensless system is capable of magnification due to the projection of the image over a distance, and that diffraction widens the image with airy disks. The images are not focused due to diffraction and the non-coherence of the wave. Computer enhancement software was used that performs reverse diffraction [21]. The diffraction patterns recorded by the CMOS image sensor were reverse diffracted using the angular spectrum method (ASM). The program takes into account the diffraction distance, wavelength of light, and pixel size. For the lensless case, the resolution was improved from 23 lines per mm to 29 Lp/mm [20]. The imaging setup had a field of view of 6.14 mm × 4.604 mm and had a small pixel size (1.25 μm) that enabled digital magnification of small objects, such as small microbeads or cells.

2.8. Process for Imaging and Counting Cells

The CD4$^+$ T cells isolated from a blood sample can be counted manually. A visual examination is needed for the identification of CD4$^+$ T cells from unattached beads and debris. If the sample can be sufficiently purified to mainly CD4$^+$ T cells and attached Dynabeads, then automatic counting based on the area can be employed to distinguish cells from one another. The ImageJ [22,23] program was customized to automatically count cells that were photographed on an inverted microscope (Nikon Eclipse TE2000-S) at 10×. The Supplement provides details of the process of using ImageJ for various objects and conditions. The parameters were also optimized for images from the Fluorescent Cell

Imager (ZOE, Bio-Rad Laboratories, Hercules, CA, USA) with 20× objective lens and for the images from lensless systems. The parameters are based on the type of object being imaged. We were able to distinguish CD4$^+$ T cells and the beads that were not attached to cells. Dapi or NucBlue stains were used for fluorescence verification.

The Supplement I B. titled "Process for imaging and Counting Cells in ImageJ" and Supplementary Table S1 show the basic ImageJ process parameters for counting the cells.

3. Results

3.1. Capture of CD4$^+$ T Cells and Monocytes Using Dynabeads

Dynabeads functionalized with anti-human CD4 antibody (ThermoFisher, part number 11145D) [24] were utilized for the positive isolation of CD4$^+$ T cells from whole blood samples. EDTA was added to the fresh blood as an anticoagulant. Moreover, ultrapure DPBS was used for washing purposes. The CD4 glycoprotein is expressed on two types of cells, i.e., CD4$^+$ T lymphocytes and CD4$^+$ monocytes. However, CD4$^+$ T lymphocytes also express CD3 glycoprotein, whereas CD4$^+$ monocytes do not express this CD3 antigen. Therefore, we used this biomarker to isolate the pure population of CD4$^+$ T lymphocytes by binding them to the substrate and washing out the monocytes.

3.2. Microfluidic Device Process Verification

The CD4$^+$ T cell isolation process was first verified by the standard method using an inverted microscope (Nikon Eclipse TE2000-S), a hemocytometer, and a Fluorescent Cell Imager (ZOE, Bio-Rad Laboratories, Hercules, CA, USA) microscope. Figure 4 presents the image of the cells obtained with the Fluorescent Cell Imager (ZOE, Bio-Rad Laboratories, Hercules, CA, USA). The lensless system that was developed could then be accurately utilized to the count the CD4$^+$ T cells from the blood samples. The gold standard used for comparison was the standard Dynabeads magnetic separation with a hemocytometer count (HC) on an inverted microscope (Nikon Eclipse TE2000-S). Figure 5 portrays the Dynabeads and Dynabeads with attached CD4$^+$ T cells on a section of the hemocytometer.

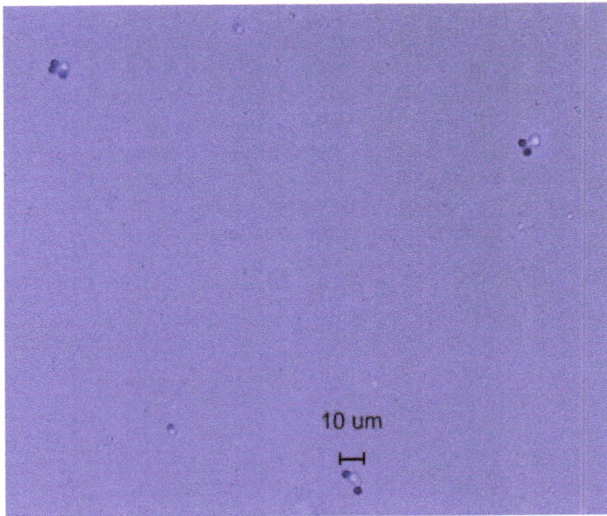

Figure 4. Enlarged merged image from Fluorescent Cell Imager (ZOE, Bio-Rad Laboratories, Hercules, CA, USA) showing beads and white cells.

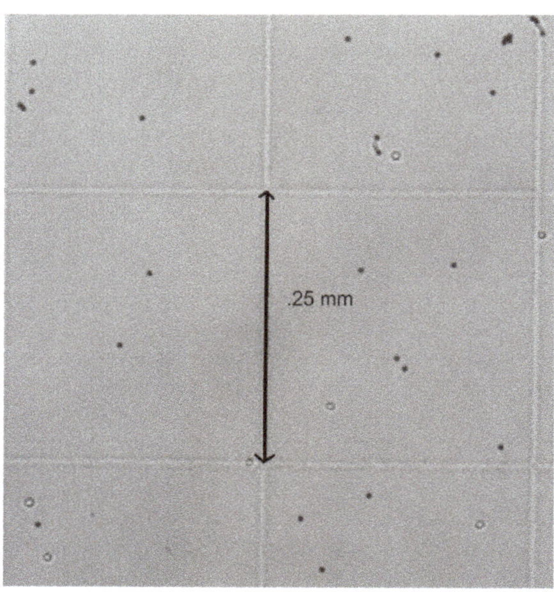

Figure 5. Hemocytometer showing CD4$^+$ T cells and beads from standard Dynabeads process. Note that the normal Dynabeads process does not remove unattached beads since the magnetic force cannot distinguish the difference. Visual identification is required for area calculations.

3.3. Process Verification Results with Blood Samples

The Continental Blood Bank, located locally in Fort Lauderdale, Fl., was the source of several different whole blood samples from which we were able to successfully isolate, identify, and count the CD4$^+$ T cells using a microchip module. A wide-field optical setup and an automatic counting method were used to finalize the count. Three different samples are shown in Table 1. These are the whole blood samples. Sample 3 was first processed with the EasySep CD4$^+$ T isolation kit to isolate the CD4$^+$ T cells. Table 1 exhibits the results of the three different samples for specificity. Figure 4 portrays the CD4$^+$ T cells attached to beads in a merged view of fluorescence and bright-field images that were taken on a Fluorescent Cell Imager (ZOE, Bio-Rad Laboratories, Hercules, CA). The image was enlarged to show the details of various CD4$^+$ T cells containing beads and CD4$^+$ T cells that are also attached to the substrate.

Table 1. Process efficiency of the micromodule based on inverted microscope (Nikon Eclipse TE2000-S) measurements.

Sample	HC Hemocytometer Count	MC Cell Count	Efficiency %
Sample 1 Blood	214	178	86
Sample 2 Blood	380	392	103
Sample 3 Blood (Processed with EasySep kit)	1592	1680	106
Average			98.3 ± 10.8

The microfluidic device process was proven to be effective. Table 1 portrays the efficiency of the process when comparing the microchip-based CD4$^+$ T cell count to the hemocytometer count when using the Nikon as the imaging device. Multiple images are necessary for both the hemocytometer and the microchip module to ensure a sufficient sample. This is a standard procedure for the hemocytometer, and the cells were visually identified and averaged. We were able to take a string of images (eight consecutive images) down the center of the microchip module and form a unified image. Given a known volume and area of the module, the concentration could be calculated. The small field of view offered by a regular optical microscope demonstrated the need to create a wide-field imaging system. We develop these simplified systems as presented in the wide-field imaging section.

Table 2 shows the results of three different blood samples and their specificity. The same image was taken as a fluorescent image and then a bright-field image. The Fluorescent Cell Imager (ZOE, Bio-Rad Laboratories, Hercules, CA, USA) merged view shows a good correlation between the bright field and the blue field. NucBlue was used as the UV dye because it works with both alive and dead cells. It is anticipated that the final field process will be dealing primarily with live cells since the assay time will be short.

Table 2. Fluorescent Cell Imager (ZOE, Bio-Rad Laboratories, Hercules, CA, USA) used for specificity measurement. Cells/image.

Sample	FC Blue Count	BF Bright Field	Specificity %
Sample 4 Blood	345	414	83
Sample 5 Blood	380	449	85
Sample 6 Blood (Processed with EasySep kit)	311	311	100
Average			89.3 ± 9.3

The microchip process has been proven for cell isolation efficiency ($98.3 \pm 10.8\%$) and specificity ($89.3 \pm 9.3\%$), as shown in Tables 1 and 2, respectively. The next step is to show the results when using a lensless or wide-field lens system. The cell isolation efficiency was $88.5 \pm 9.2\%$, as shown in Table 3. The specificity was already proven with the fluorescent imaging system.

Table 3. Lensless cell count efficiency using a lensless or lens optical system.

Sample Blood	Hemocytometer Count (HC)	Lensless Cell Count	Percentage Efficiency
Sample 7 Blood	380	311	82
Sample 8 Blood	480	457	95
Average			88.5 ± 9.2

3.4. Validation of Module Process and Counting Using Wide-Field Imaging

The microfluidic chip modules integrated with a lensless imaging system were designed to have a wide field of view so that the large surface area of the chip can be imaged rapidly. In CD4+ T cell isolation experiments, the number of cells to be counted per unit area is small. It is necessary to have either multiple images or a wide-area imaging system to capture a sufficient sample within the channel of the module.

Our previous work [20] has shown that wide-field lensless and custom-designed lens systems are possible. A low-cost lensless system is also described in that paper. For this work, we used the optical systems we had previously developed. A lensless system has a wide field of view, conforming to the size of the imaging field-of-view, in this case, 6.14×4.6 mm, as shown in Figure 3. The resolution is limited by the size of the pixel (1.25 um in this case) and the image area by the size of the array, 18 Megapixels. Figure 3 also portrays an enlarged section of the microchip module and shows the beads and CD4+

T cells. Table 3 shows that the CD4$^+$ T cell isolation efficiency was at 89 $\pm$ 9.19 for the lensless system.

3.5. Conclusion and Discussion

A microfluidic chip-based assay has been developed that can sort out CD4$^+$ T lymphocytes and count them using a microfluidic device, rather than magnetic separation, with excellent specificity and efficiency. The developed process was proven to be successful and provides 89.3 $\pm$ 9.3% capture specificity and 98.3 $\pm$ 10.8% capture efficiency inside microfluidic channels. The processes have been simplified such that a portable microchip design is possible for point-of-care use.

A wide-field optical system was explored, and its properties and limitations were examined. Lensless equipment was designed, built, and tested to complement the microchip design. This imaging system allows a larger sample area to enable the accurate counting and recognition of biological samples that normally have a low count per unit volume but at a lower cost per assay. The ImageJ-based algorithm was structured and used to automatically count cells. In the future, artificial intelligence could be used to better identify different types of cells.

Simple systems, such as lensless or wide-field lens systems, can be used at the point-of-care locations. This microfluidic device will enable wide-field optical systems to capture a sufficient area to count a low number of cell assays. Overall, the developed method is cost-effective, easy to use, and rapid, and can be used for counting CD4$^+$ T cells at the point-of-care settings.

Supplementary Materials: The following supporting information can be downloaded at: https://www.mdpi.com/article/10.3390/bios12010012/s1, Figure S1. Process Flow Data Sheet. Figure S2. From left to right The Bright field Image on Fluorescent Cell Imager (ZOE, Bio-Rad Laboratories, Hercules, CA, USA). The Blue Field Image and The Merge Image, used to calculate Specification. Note the large im-age but small size of cells. Table S1. Key ImageJ parameters for analyze

Author Contributions: Conceptualization, W.A., R.D.F. and M.S.; methodology, W.A., R.D.F. and M.S.; software, R.D.F.; validation, R.D.F., M.S. and W.A.; formal analysis, W.A., R.D.F. and M.S.; investigation, R.D.F., M.S. and W.A.; resources, W.A.; data curation, R.D.F. and M.S.; writing—original draft preparation R.D.F., M.S. and W.A.; writing—review and editing, R.D.F., M.S. and W.A.; visualization, R.D.F.; supervision, W.A. and M.S.; project administration, W.A. and M.S.; funding acquisition, W.A. All authors have read and agreed to the published version of the manuscript.

Funding: This research was funded by National Institute of Health (NIH) grant number R56AI138659, NSF CAREER Award 1942487, and Humanity in Science Award.

Institutional Review Board Statement: Not applicable.

Informed Consent Statement: Not applicable.

Data Availability Statement: Not applicable.

Acknowledgments: We are thankful to William Rhodes for help on near and far-field theory, Oge Marques for computer analysis of images, Benjamin Coleman for technical discussions, and Charles Perry Wienthal for allowing us to use CEECS lab facilities to fabricate various 3D components. We also thank Diya Jain (Pine Crest School) for helping in proofreading the manuscript.

Conflicts of Interest: The authors declare no conflict of interest.

Abbreviations

Antibody (Clone 13B8.2, CN: COIM0704) was purchased from Fisher Scientific (Fair Lawn, NJ. USA)

APTES slide	APTES, 3-Aminopropyl Trimethoxysilane treated slide
ASSURED	Affordable, sensitive, specific, user-friendly, rapid and robust, equipment-free, and deliverable
BSA	lyophilized bovine serum albumin (BSA, CN a2153). A small stable protein of 607 amino acids, it is used as a blocker to help antibodies find antigens by binding to non-specific binding sites
DMSO	Dimethyl sulfoxide is polar organic solvent that is miscible with water. We mixed the GMBS with DMSO so that it can be further cut and soluble in water. We can then apply to our substrate. So that GMBS can adhere to the substrates, we have a double bond to the surface
DPBS	D phosphate-buffered saline solution
DSA	Double sided adhesive tape. Forms the channel and binds the top and bottom layer in the micro module
EDTA	Ethylenediaminetetraacetic acid; prevents the coagulation of blood cells by absorbing mental ions to prevent binding of beads and blood
GMBS	N-γ-Maleimidobutyryloxy succinimide ester; this is a chemical that is a cross linker from amino and sulfhydryl groups. It comes as a solid and is water insoluble. Therefore, must be mixed with DMSO or DMF [3]. We use DMSO
HC	Hemocytometer count
HIV	Human immunodeficiency virus
Maleimide Activated Neutravidin	This version can bond directly without using GMBS to cross link to slide
MC	Microscope count
MPS	(3-Mercaptopropyl) trimethoxysilane (3-MPS, CN: 175617), b. This compound is an adhesive; it has four hydrogen acceptor molecules and one donator
	It is the first layer to adhere to a glass slide that has donor atoms
Neutravidin protein	This protein is used to attach to the GMBS and the anti-CD4 antibody. Ka = 10^{15} M^{-1}
NucBlue	This stain is best for live and dead cells. It is from Invitrogen and similar to DAPI
PMMA	Polymethymetacrylate sheet. It is an acrylic sheet that can be easily cut with a laser cutter
POC	Point-of-care
WHO	World Health Organization

Other items used were: anti-CD4 beads, absolute alcohol: ethanol 200 proof, OH$^-$, and glass slides

References

1. Kosack, C.S.; Page, A.-L.; Klatser, P.R. A guide to aid the selection of diagnostic tests. *Bull. World Health Organ.* **2017**, *95*, 639. [CrossRef] [PubMed]
2. Hassan, U.; Reddy, B., Jr.; Damhorst, G.; Sonoiki, O.; Ghonge, T.; Yang, C.; Bashir, R. A microfluidic biochip for complete blood cell counts at the point-of-care. *Technology* **2015**, *3*, 201–213. [CrossRef] [PubMed]
3. Hassan, U.; Watkins, N.; Edwards, C.; Bashir, R. Flow metering characterization within an electrical cell counting microfluidic device. *Lab Chip* **2014**, *14*, 1469–1476. [CrossRef] [PubMed]
4. Sher, M.; Coleman, B.; Caputi, M.; Asghar, W. Development of a Point-of-Care Assay for HIV-1 Viral Load Using Higher Refractive Index Antibody-Coated Microbeads. *Sensors* **2021**, *21*, 1819. [CrossRef] [PubMed]
5. Sher, M.; Faheem, A.; Asghar, W.; Cinti, S. Nano-Engineered Screen-Printed Electrodes: A dynamic tool for detection of Viruses. *TrAC Trends Anal. Chem.* **2021**, *143*, 116374. [CrossRef] [PubMed]
6. Kabir, M.A.; Zilouchian, H.; Sher, M.; Asghar, W. Development of a flow-free automated colorimetric detection assay integrated with smartphone for Zika NS1. *Diagnostics* **2020**, *10*, 42. [CrossRef] [PubMed]
7. Shafiee, H.; Asghar, W.; Inci, F.; Yuksekkaya, M.; Jahangir, M.; Zhang, M.H.; Durmus, N.G.; Gurkan, U.A.; Kuritzkes, D.R.; Demirci, U. Paper and flexible substrates as materials for biosensing platforms to detect multiple biotargets. *Sci. Rep.* **2015**, *5*, 8719. [CrossRef] [PubMed]
8. Wasserberg, D.; Zhang, X.; Breukers, C.; Connell, B.J.; Baeten, E.; van den Blink, D.; Benet, E.S.; Bloem, A.C.; Nijhuis, M.; Wensing, A.M.J.; et al. All-printed cell counting chambers with on-chip sample preparation for point-of-care CD4 counting. *Biosens. Bioelectron.* **2018**, *117*, 659–668. [CrossRef] [PubMed]
9. Sher, M.; Zhuang, R.; Demirci, U.; Asghar, W. Paper-based analytical devices for clinical diagnosis: Recent advances in the fabrication techniques and sensing mechanisms. *Expert Rev. Mol. Diagn.* **2017**, *17*, 351–366. [CrossRef] [PubMed]
10. Asghar, W.; Yuksekkaya, M.; Shafiee, H.; Zhang, M.; Ozen, M.O.; Inci, F.; Kocakulak, M.; Demirci, U. Engineering long shelf life multi-layer biologically active surfaces on microfluidic devices for point of care applications. *Sci. Rep.* **2016**, *6*, 21163. [CrossRef] [PubMed]

11. Cheng, X.; Irimia, D.; Dixon, M.; Sekine, K.; Demirci, U.; Zamir, L.; Tompkins, R.G.; Rodriguez, W.; Toner, M. A microfluidic device for practical label-free CD4[+] T cell counting of HIV-infected subjects. *Lab Chip* **2007**, *7*, 170–178. [CrossRef] [PubMed]

12. Moon, S.; Gurkan, U.A.; Blander, J.; Fawzi, W.W.; Aboud, S.; Mugusi, F.; Kuritzkes, D.R.; Demirci, U. Enumeration of CD4[+] T-cells using a portable microchip count platform in Tanzanian HIV-infected patients. *PLoS ONE* **2011**, *6*, e21409. [CrossRef] [PubMed]

13. Asghar, W.; Sher, M.; Khan, N.S.; Vyas, J.M.; Demirci, U. Microfluidic chip for detection of fungal infections. *ACS Omega* **2019**, *4*, 7474–7481. [CrossRef] [PubMed]

14. Ilyas, S.; Sher, M.; Du, E.; Asghar, W. Smartphone-Based sickle cell disease detection and monitoring for point-of-care settings. *Biosens. Bioelectron.* **2020**, *165*, 112417. [CrossRef] [PubMed]

15. Rappa, K.; Samargia, J.; Sher, M.; Pino, J.S.; Rodriguez, H.F.; Asghar, W. Quantitative analysis of sperm rheotaxis using a microfluidic device. *Microfluid. Nanofluid.* **2018**, *22*, 100. [CrossRef]

16. Coarsey, C.; Coleman, B.; Kabir, M.A.; Sher, M.; Asghar, W. Development of a flow-free magnetic actuation platform for an automated microfluidic ELISA. *RSC Adv.* **2019**, *9*, 8159–8168. [CrossRef] [PubMed]

17. Sher, M.; Asghar, W. Development of a multiplex fully automated assay for rapid quantification of CD4[+] T cells from whole blood. *Biosens. Bioelectron.* **2019**, *142*, 111490. [CrossRef] [PubMed]

18. Counting-Cells-Using-A-Hemocytometer. Available online: https://docs.abcam.com/pdf/protocols/counting-cells-using-a-hemocytometer.pdf (accessed on 23 December 2021).

19. EasySep Cell Separation. Available online: https://www.stemcell.com/products/brands/easysep-cell-separation.html (accessed on 23 December 2021).

20. Fennell, R.D.; Sher, M.; Asghar, W. Design, development, and performance comparison of wide field lensless and lens-based optical systems for point-of-care biological applications. *Opt. Lasers Eng.* **2021**, *137*, 106326. [CrossRef] [PubMed]

21. Stork, D.G.; Ozcan, A.; Gill, P.R. Imaging Without Lenses: New imaging systems, microscopes, and sensors rely on computation, rather than traditional lenses, to produce a digital image. *Am. Sci.* **2018**, *106*, 28–34. [CrossRef]

22. Abràmoff, M.D.; Magalhães, P.J.; Ram, S.J. Image processing with ImageJ. *Biophotonics Int.* **2004**, *11*, 36–42.

23. Imagej. Available online: https://imagej.nih.gov/ij/download.html (accessed on 23 December 2021).

24. Dynabeads_Cell_Separation_Guide. Available online: https://assets.fishersci.com/TFS-Assets/LSG/brochures/Dynabeads_cell_separation_guide.pdf (accessed on 23 December 2021).

biosensors

Article

Electrochemical Detection Platform Based on RGO Functionalized with Diazonium Salt for DNA Hybridization

Elena A. Chiticaru [1], Luisa Pilan [2,*] and Mariana Ioniță [1,3,*]

[1] Faculty of Medical Engineering, University Politehnica of Bucharest, Gh Polizu 1-7, 011061 Bucharest, Romania; elenaa.chiticaru@yahoo.com

[2] Department of Inorganic Chemistry, Physical Chemistry and Electrochemistry, University Politehnica of Bucharest, Gh Polizu 1-7, 011061 Bucharest, Romania

[3] Advanced Polymer Materials Group, University Politehnica of Bucharest, Gh Polizu 1-7, 011061 Bucharest, Romania

* Correspondence: luisa.pilan@upb.ro (L.P.); mariana.ionita@polimi.it (M.I.)

Abstract: In this paper, we propose an improved electrochemical platform based on graphene for the detection of DNA hybridization. Commercial screen-printed carbon electrodes (SPCEs) were used for this purpose due to their ease of functionalization and miniaturization opportunities. SPCEs were modified with reduced graphene oxide (RGO), offering a suitable surface for further functionalization. Therefore, aryl-carboxyl groups were integrated onto RGO-modified electrodes by electrochemical reduction of the corresponding diazonium salt to provide enough reaction sites for the covalent immobilization of amino-modified DNA probes. Our final goal was to determine the optimum conditions needed to fabricate a simple, label-free RGO-based electrochemical platform to detect the hybridization between two complementary single-stranded DNA molecules. Each modification step in the fabrication process was monitored by cyclic voltammetry (CV) and electrochemical impedance spectroscopy (EIS) using $[Fe(CN)_6]^{3-/4-}$ as a redox reporter. Although, the diazonium electrografted layer displayed the expected blocking effect of the charge transfer, the next steps in the modification procedure resulted in enhanced electron transfer properties of the electrode interface. We suggest that the improvement in the charge transfer after the DNA hybridization process could be exploited as a prospective sensing feature. The morphological and structural characterization of the modified electrodes performed by scanning electron microscopy (SEM) and Raman spectroscopy, respectively, were used to validate different modification steps in the platform fabrication process.

Keywords: graphene; reduced graphene oxide; electrochemistry; electrochemical impedance spectroscopy; DNA biosensor; diazonium chemistry; screen-printed electrodes; DNA hybridization

Citation: Chiticaru, E.A.; Pilan, L.; Ioniță, M. Electrochemical Detection Platform Based on RGO Functionalized with Diazonium Salt for DNA Hybridization. *Biosensors* **2022**, *12*, 39. https://doi.org/10.3390/bios12010039

Received: 7 December 2021
Accepted: 10 January 2022
Published: 13 January 2022

Publisher's Note: MDPI stays neutral with regard to jurisdictional claims in published maps and institutional affiliations.

1. Introduction

Deoxyribonucleic acid (DNA) is a biopolymer that can self-assemble from two single-strands (ss) in a unique way that conforms to the Watson−Crick base pairing rules [1]. This process of specific self-assembly between two complementary polynucleotides is called hybridization and it is centered around the hydrogen bonds formed between the nucleobase base pairs, producing a double-helix structure [2]. The complementarity of DNA bases has been highly exploited in nanotechnology [3], in medicine [4] and in sensing applications [5,6]. The high specificity of the self-pairing between single-stranded DNA molecules can be used to detect specific sequences or biomarkers, which are related to bacterial [7] or viral [8] infections, for example. Moreover, the hybridization event can also be employed in the detection of DNA or messenger ribonucleic acid (mRNA) oligonucleotides, which can be extremely valuable in point-of-care applications [9], as well as for the diagnosis of various genetic mutations and diseases [10,11].

Electrochemical DNA biosensors are highly advantageous for the detection of particular ssDNA molecules due to their fast response time, inexpensive instrumentation and miniaturization potential [12]. Due to these advantages, DNA biosensors have been used so far in various applications, such as clinical diagnostics [13], drug interactions [14] and detection [15], and environmental monitoring [16]. The development of simple and effective electrochemical methods is of high interest in nucleic acid detection [17]. In this context, label-free approaches have attracted widespread interest as they can simplify the manufacturing processes and are free from the possible unfavorable effects of the labels, which proves their effectiveness for target sensing [18,19]. Many of these label-free detection strategies exploit the variation in the electrochemical signal of some charged freely diffusing redox species as a result of the change in their accessibility to the electrode surface driven by the target analyte binding [20]. One of the most employed redox species is the negatively charged $[Fe(CN)_6]^{3-/4-}$, and the hybridization events are detected by several methods, such as cyclic voltammetry (CV), differential pulse voltammetry (DPV), square wave voltammetry (SWV), and electrochemical impedance spectroscopy (EIS).

The discovery of graphene and its related materials offers great opportunities for the design of superior electrochemical DNA biosensors due to their unique properties, such as high surface area, which can increase the immobilization capacity, increase the electron transfer rate, and can contribute to excellent biosensor sensitivity [21]. The functional groups present on the surface of graphene oxide (GO) can often be considered a disadvantage in sensing applications because they limit the electrical conductivity of the material. However, there are some strategies that exploit these functionalities by either removing them through different approaches (chemical, electrochemical, thermal, etc.) [22] or by increasing their reactivity through carbodiimide chemistry [23] in order to be further covalently linked to amino-terminated molecules. As pointed out in the recent literature, reduced graphene oxide (RGO) material offers a multitude of advantages compared to other biosensing transducer materials by providing a balance between the beneficial properties of pristine graphene and GO, and through its feasibility for various functionalization schemes suitable for the biorecognition elements immobilization [24,25]. It was confirmed by different studies [26–28] that graphene functionalization is substantially enhanced when the reaction is performed under electrochemical control. The Fermi level of graphene can be shifted by adjusting the applied electrochemical potential, which is a simpler way to increase the reactivity of the material than using aggressive chemicals to disrupt the covalent sp^2 bonds [29]. Among numerous functionalization strategies, the electrochemical reduction of diazonium salts [30–34] has been proposed as a versatile method that affords the grafting of a large variety of molecules at conductive substrates [35,36]. So far it has been applied for different types of carbon-based substrates such as glassy carbon [37], screen-printed carbon [38], carbon paste [39], graphite pencil [40] or carbon nanomaterials electrodes [41,42], that function as transducers in a large number of sensors for bioanalytes, such as proteins [43], DNA [44], RNA or cells [45]. Although great effort has been dedicated to this subject, there is still no refined method available to mass-produce highly stable, efficient biosensors with real applicability in biomedicine. Therefore, our aim was to design an improved DNA detection platform in terms of stability, sensitivity and reproducibility, by modulating the electrochemical parameters.

In a previous study [25], we have developed an electrochemical platform for DNA detection by physical adsorption of DNA probe (DNAp) molecules onto RGO-modified electrodes. Moreover, we have introduced a new protocol for the pretreatment of screen-printed carbon electrodes (SPCEs) that were used in our work, which improved their electrochemical properties. Considering that the immobilization of oligonucleotides through physical adsorption does not permit a high level of precision and control for DNA attachment on the substrate, we are proposing a new approach that consists in the covalent immobilization of amino-modified DNA probes onto a carboxylic functionalized RGO surface. This strategy provides stronger and highly specific binding between the bioreceptor and the working electrode by eliminating the possibility of DNA desorption [12], as opposed to the non-covalent method [46,47]. The amino-modified ssDNA probes are covalently attached at the RGO substrate through stable amide link via carbodiimide coupling reaction. Unlike other literature studies that report the same coupling chemistry for the biorecognition element at the carboxylic functions of the GO [48], this work also exploits the significant advantages of diazonium chemistry to provide the stable and controlled functionalization of the RGO electrode with aryl carboxylic groups (Ar-COOH) [46], with the implied benefits for ssDNA probe immobilization. A reproducible platform for DNA hybridization detection was envisaged by putting special emphasis on the electrochemically controlled fabrication steps for both the preparation of the RGO substrate electrode and the covalent binding of the biorecognition element. In addition to the portability and cost-effectiveness of SPCEs, the reproducible modification of these substrates with RGO and the controlled covalent binding of the biorecognition element can enable the fabrication of competitive, stable, reproducible and novel platforms for DNA hybridization detection. The effect of the successive steps in the sensor fabrication process was established by the electrochemical properties of the modified electrodes quantified by the charge transfer resistance (Rct) and the electron transfer rate constant (k^0) parameters. The electrochemical results were in agreement with morphological and structural characterization studies performed by scanning electron microscopy (SEM) and Raman spectroscopy, respectively. The improvement in the charge transfer after the DNA hybridization process could be exploited as a prospective sensing feature.

2. Materials and Methods

2.1. Reagents and Materials

Graphene oxide dispersion in water (2 mg/mL), KCl, HCl, H_2NaO_4P, HNa_2O_4P, N-(3-Dimethylaminopropyl)-N'-ethylcarbodiimide hydrochloride (EDC), N-Hydroxysulfosucci nimide sodium salt (NHS), $NaNO_2$ and 4-aminobenzoic acid (4-ABA) were supplied by Sigma-Aldrich (St. Louis, MO, USA). Potassium ferricyanide ($K_3[Fe(CN)_6]$) and potassium ferrocyanide ($K_4[Fe(CN)_6] \times 3H_2O$) were purchased from Merck Co., (Darmstadt, Germany). Single-stranded DNA probe modified with amino groups (5'-AmC6-TTT CAA CAT CAG TCT GAT AAG CTA TCT CCC-3'), single-stranded DNA target (DNAt, 5'-GGG AGA TAG CTT ATC AGA CTG ATG TTG AAA-3') and 10 mM Tris, 0.1 mM EDTA (IDTE) buffer were acquired from Integrated DNA Technologies, Inc (Coralville, IA, USA). The SPCEs were thoroughly rinsed with ultrapure water (Adrona Crystal EX water purification system, 18.2 M$\Omega \times$ cm resistivity) after each modification. Screen-printed carbon electrodes (SPCE-DRP 110) were purchased from Metrohm DropSens, Spain.

2.2. Procedures

2.2.1. Electrochemical Measurements

The SPCEs consisted of a 4 mm diameter working electrode (WE), a silver pseudo-reference electrode (RE), and a carbon counter electrode (CE). Each SPCE modification was investigated by cyclic voltammetry and electrochemical impedance spectroscopy, which were performed at room temperature using a potentiostat/galvanostat Autolab PGSTAT 204 (Metrohm Autolab, Netherlands) equipped with NOVA 2.1 software. The SPCE was attached to a connector (DSC) device from Metrohm Dropsens, functioning as an interface between the SPCE and the potentiostat. The electrochemical measurements were recorded by adding 100 µL 0.1 M KCl electrolyte solution containing 1 mM $[Fe(CN)_6]^{3-/4-}$ redox probe on the SPCE. CV curves were recorded by scanning the potential between -0.2 V and $+0.6$ V, at a sweep of 0.05 V/s (unless stated otherwise). The impedance spectra were recorded in the frequency range of 0.01–10^5 Hz, with 10 mV AC amplitude at an applied potential of 0.16 V, which is the formal potential of the $[Fe(CN)_6]^{4-/3-}$ redox probe vs. the Ag pseudo reference electrode.

2.2.2. Morphological and Structural Characterization

A FEI high-resolution focused ion beam scanning electron microscopy (FIB-SEM) system model Versa 3D DualBeam (FEI Company, Hillsboro, OR, USA) was used to characterize the selected sample series. The plane view (0° tilt) samples' surface morphology were investigated by detecting the secondary electrons (SE) signals in High-Vacuum operation mode (6.1×10^{-4} Pa) at a working distance of 10 mm, using 10 kV as the accelerating voltage and a spot size of 4.5. Moreover, the SmartSCAN scanning strategy and DCFI drift suppression features of Versa 3D DualBeam tool were involved to fully ensure the imaging stability.

Raman spectroscopy was performed in order to investigate the structural changes on the surface of the electrodes. Raman spectra were obtained with a Renishaw inVia Raman confocal spectrometer (Renishaw, Brno-Černovic, Czech Republic), using a laser excitation wavelength of 473 nm and the $100\times$ objective.

2.2.3. RGO-Modified Electrode Preparation Procedure

The electrodes coated with electrochemically reduced graphene oxide (further denoted as RGO/SCPE) were prepared following the procedure described in [25]. Before any modification, the SPCEs were pretreated by 5 voltametric cycles performed in 0.1 M HCl, from +0.5 to −1.5 V, at the scan rate of 0.05 V/s, followed by 5 cycles in 0.1 M phosphate buffer solution (PBS), pH 7, from 0 to +2 V, 0.05 V/s. This activation treatment ensures a higher electron transfer rate for the redox probe at SPCE, illustrated by a smaller peak to peak separation in the CV signal [25]. Subsequently, prior to the graphene modification step, 3 μL PBS was deposited on the WE surface, and then the electrodes were washed with ultrapure water and dried at 60 °C. As mentioned previously [25], this step is highly important because it changes the surface properties of the carbon surface and it guarantees a reproducible GO modification. Once the SPCEs cooled down, 3 μL of GO dispersion (0.3 mg/mL) was cast on their surface, then dried at 60 °C for 2 h and kept at room temperature overnight. Finally, the GO/SPCEs were reduced electrochemically by 10 CV cycles applying a potential between 0 and −1.5 V, at a scan rate of 0.1 V/s, in 0.5 M KCl. The measurements were performed in triplicate in order to ensure the reproducibility of the procedure.

2.2.4. RGO/SPCE Functionalization by Diazonium Chemistry

The functionalization of RGO/SPCEs with carboxyphenyl layer was conducted with in situ generated aryldiazonium cations that involved the electrochemical reduction of the corresponding aniline (4-aminobenzoic acid) in acidic media. Specifically, 2 mM 4-aminobenzoic acid was solubilized in 0.5 M aqueous HCl, to which 2 mM of sodium nitrite was added to generate the aryl diazonium salt. After 5 min stirring at room temperature, 100 μL of mixture was immediately added onto the electrode surface, so as to cover all connections. The electrochemical procedure employed for grafting was based on the protocol described by Eissa et al. [49] and consisted in potential cycling for one cycle between +0.3 V and −0.6 V, with a scan rate of 0.1 V/s. At this stage, an irreversible reduction peak should be observed as a preliminary validation of the functionalization process. After surface derivatization, the electrodes were rinsed with high amounts of ultrapure water. The blocking properties of the modified electrodes evaluated by CV and EIS in the presence of soluble electroactive $[Fe(CN)_6]^{3-/4-}$ species also confirmed this grafting step.

2.2.5. Fabrication of DNA Biosensor

The DNA biosensor was fabricated by covalently immobilizing the amino ssDNA probe onto RGO/SPCE using carbodiimide chemistry. For this purpose, 0.2 M EDC and 0.05 M NHS in 0.1 M PBS buffer solution of pH 6, were used as activation agents for the carboxyl groups confined at the electrode surface. After 2 h activation at room temperature, the Ar-COOH/RGO/SPCEs were immersed overnight in the ssDNA probe solutions at room temperature to obtain the detecting electrodes. Fresh DNA probes were prepared from the stock solution of 100 μM to 10, 5, 1, and 0.5 μM using IDTE buffer to dilute. The hybridization testing of the developed platform for the complementary target DNA sequence was achieved by dropping an appropriate concentration of target DNA solution (in IDTE buffer) onto the SPCE modified with the recognition layer and incubating at room temperature for 2 h. For longer times (i.e., 4 h and 6 h), no significant change in the electrochemical studies occurred, indicating that the ssDNA probes immobilized onto SPCE were completely hybridized by the target DNA after 2 h.

3. Results and Discussions

3.1. Morphological Characterization

The morphology of the modified electrodes was investigated by SEM, which is also a useful technique to determine if the surface of the working electrodes is uniformly covered with graphenic material. Several images were acquired from different spots on the working electrodes and the most representative are presented in Figure 1. First of all, the bare SPCE (Figure 1A) was studied and it revealed a rough porous surface similar to the one previously obtained by us [25]. The image corresponding to GO/SPCE (Figure 1B) shows homogenous coverage of the electrode surface with thin, well-dispersed graphene oxide sheets, while RGO deposited on SPCE (Figure 1C) displays the typical graphene morphology with wrinkled layers uniformly covering the whole working electrode. Finally, Ar-COOH/RGO/SPCE (Figure 1D) reveals a morphology similar to the other electrodes coated with graphenic species, showing thin graphene layers with creased margins. All in all, the SEM images emphasize the flexibility of the graphenic sheets and confirm the homogenous coverage of the electrodes with a thin layer of GO dispersion, while the porous morphology of the carbon electrode can still be observed underneath.

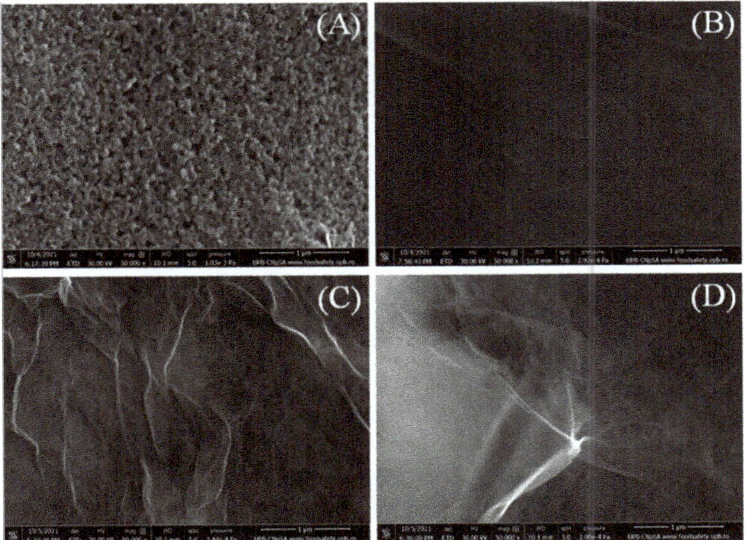

Figure 1. SEM images of bare SPCE (**A**), GO/SPCE (**B**), RGO/SCPE (**C**), and Ar-COOH/RGO/SPCE (**D**). Images recorded at 50 kX magnification.

3.2. Structural Characterization

Raman spectra (Figure 2) were recorded after each step of electrode modification to investigate the changes in the graphenic structure. I_D/I_G ratio is higher for RGO compared to GO due to the many structural defects that appeared as a consequence of removing functional groups from the GO surface. Once Ar-COOH groups were grafted on the electrode surface, more sp^2 carbon bonds were formed and the G peak increased more than the D band. The I_D/I_G ratio increased again after activation in EDC-NHS and decreased after electrode incubation in the ssDNA probe, confirming the immobilization of oligonucleotides on the activated Ar-COOH/RGO/SPCE platform. Moreover, the position of D and G vibrational bands was at approximately 1360 cm^{-1} and 1590 cm^{-1}, respectively, which is characteristic of graphene and its derivatives [50]. These two peaks change their position after each electrode modification, recording the biggest shift after ssDNA probe immobilization, which strongly suggests that the bioreceptor is indeed bound to the electrode surface. The positions of the D and G bands along with the I_D/I_G ratios are presented in Table 1.

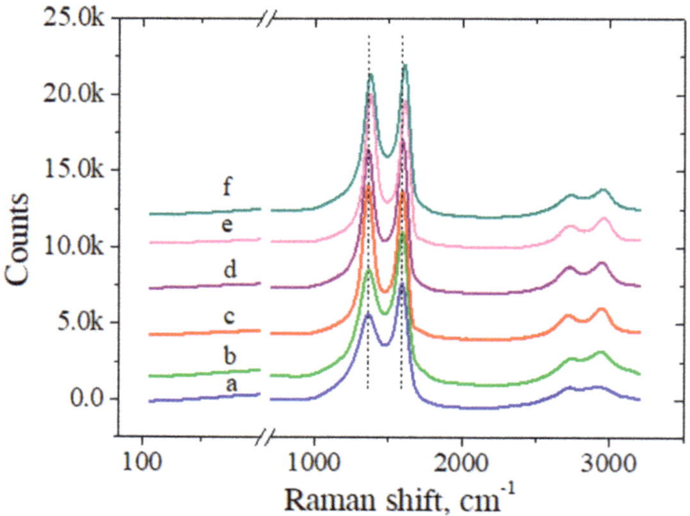

Figure 2. Raman spectra of bare SPCE (**a**), GO/SPCE (**b**), RGO/SCPE (**c**), Ar–COOH/RGO/SPCE (**d**), Ar–COOH/RGO/SPCE activated in EDC–NHS (**e**), and ssDNAp/Ar–COOH/RGO/SPCE (**f**).

Table 1. Positions of D and G bands and their I_D/I_G ratio from Raman spectra.

Electrode Modification	D Band [cm^{-1}]	G Band [cm^{-1}]	I_D/I_G Ratio
SPCE	1362.9	1589.4	0.7167
GO/SPCE	1363	1591.4	0.7739
RGO/SPCE	1359.9	1591.7	1.0228
Ar–COOH/RGO/SPCE	1360.9	1593.6	0.9647
Ar–COOH/RGO/SPCE activated	1375.3	1608.3	1.0232
ssDNAp/Ar–COOH/RGO/SPCE	1373.6	1604	0.9713

3.3. Electrochemical Characterization

3.3.1. Carboxyphenyl Electrografted RGO Electrodes

The modification of commercial electrodes with RGO by applying our former optimized protocol [25] with minor changes has been described in Section 2. As presented in Figure 3, the modification of SPCEs with GO caused a decrease in the current peaks that correlated with a high charge transfer resistance, characteristic to the poor conductive properties of the oxygenated graphene. Nevertheless, the conductivity was improved with the electrochemical reduction of GO, as can be seen by a substantial increase in the redox peaks and a decrease in R_{ct}. The subsequent diazonium electrografting of the electrode surface with 4-carboxyphenyl moieties induced a significant decrease in peak currents attributed to the electrostatic repulsion of negatively charged COOH groups (at neutral pH) for the $[Fe(CN)_6]^{3-/4-}$ ions (Figure 3A). The Nyquist plot of the carboxylic-grafted RGO displayed a semicircle for the whole frequency domain, indicating that the interfacial charge transfer dominates over mass transport effects (Figure 3B). The modification of SPCE with GO dispersion was also confirmed by SEM analysis, which displayed a change in the WE morphology (see Section 3.1), while the changes in the electrochemical properties observed after each SPCE modification were well correlated with the variations in the I_D/I_G ratios from the Raman spectra reported in Section 3.2.

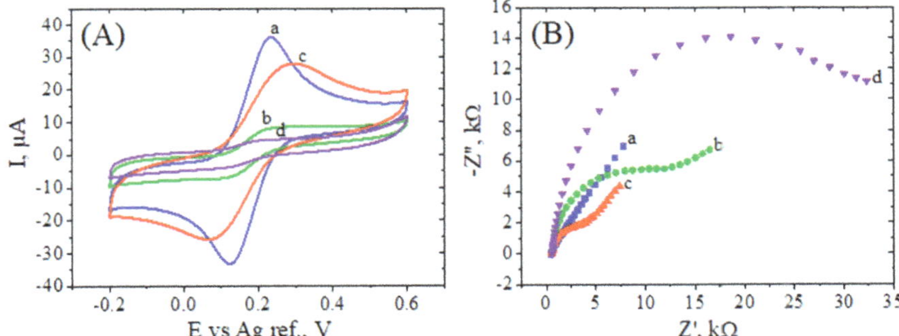

Figure 3. Electrochemical characterization for bare SPCE (**a**), GO/SPCE (**b**), RGO/SPCE (**c**), and Ar–COOH/RGO/SPCE (**d**) modified electrodes: CVs, 0.05 V/s (**A**) and EIS Nyquist plots at 0.16 V vs. Ag ref. (**B**) recorded in 1 mM $[Fe(CN)_6]^{3-/4-}$, 0.1 M KCl.

The typical mechanism for graphene covalent functionalization with aryl diazonium salts consists in the transport of a delocalized electron from the graphene lattice to the aryl diazonium cation that releases a nitrogen molecule and leads to a highly reactive aryl radical [51]. These radicals then form covalent bonds with the carbon atoms of the graphene structure, causing the conversion of the sp^2-carbon in the graphene sheet to sp^3 hybridization [52]. Although there are many challenges related to the precise control of the structure of the aryl layers, the use of diazonium chemistry offers advantages in terms of simplicity and long-term stability of the biorecognition layer [53]. Moreover, modulating the degree of functionalization by using the electrochemical route for the reduction of aryl diazonium salts represents an efficient alternative to traditional methods for graphene functionalization [54].

The functionalization of the RGO surface was performed by electrochemical reduction of in situ generated carboxyphenyl diazonium salt in acidic aqueous solution by CV, from 0.3 to −0.6 V vs. Ag ref. with a scan rate of 0.1 V/s. Upon first scanning, an irreversible reduction wave was observed at −0.5 V vs. Ag ref., which can be attributed to the formation of the 4-carboxyphenyl radicals with subsequent covalent bond formation with the graphene surface. In a second performed cycle, the cathodic peak disappears, indicating the coverage of the RGO surface with the carboxyphenyl groups that block further electron transfer between the diazonium cations in the solution and the modified electrode (Figure 4). This behavior agrees well with previously reported studies on the reduction of diazonium cations onto other carbon materials in various forms [33,36]. These results demonstrated that one reduction cycle is sufficient to saturate the electrode surface with functional carboxyphenyl groups, and in the biosensor fabrication process we employed one CV cycle at 0.1 V/s, which ensured a reproducible electrode. Moreover, extended grafting was avoided in order to prevent the formation of poorly conducting multilayers, which can create a significant barrier to electron transfer between the electrode and the redox species in the solution [55,56].

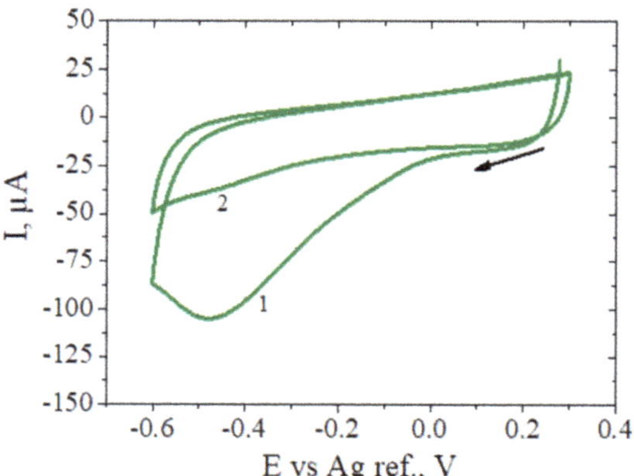

Figure 4. CVs corresponding to the 1st and 2nd cycles of the reduction of 4–carboxyphenyl diazonium salt (generated in situ) at the RGO/SPCE (potential sweep from +0.3 V to −0.6 V at 0.1 V/s).

3.3.2. Amino-Modified ssDNA Probe Immobilization

The amine-modified ssDNA probe was covalently attached to the carboxyl groups of the aryl layer through carbodiimide coupling, in the presence of EDC and NHS as activating agents. In order to determine a saturation level, different concentrations for the ssDNA solution in IDTE buffer (0.5, 1, 5, and 10 µM) were tested for the coupling reaction. As depicted in Figure 5A, the CV tests for the ssDNA-modified electrodes in the presence of $[Fe(CN)_6]^{3-/4-}$ redox probes revealed consecutive losses in the peak currents' intensity with an increasing concentration up to 5–10 µM, at which values, the response signal stagnated. Based on these results, we considered that at 10 µM the electrode surface was fully covered with specifically linked oligonucleotides, and this concentration of ssDNA probe was chosen to further check the hybridization process. Interestingly, this decrease in the peak currents was accompanied by a decrease in the peaks' potential separation (ΔE_p). The ΔE_p trend in the CV response was sustained by the results obtained by faradaic impedance measurements (Figure 5B), with a decrease in R_{ct} upon ssDNA probe immobilization, which suggests an improvement in the kinetics of the $[Fe(CN)_6]^{3-/4-}$ redox species at the ssDNA-modified electrodes. This decrease in the R_{ct}, despite the presence of the negatively-charged DNA backbone, which is prone to repulse the ferri/ferrocyanide ion, can be explained by an improvement in the hydrophilicity [57] of the interface in comparison with the electrode simply modified with the aryl grafted layer.

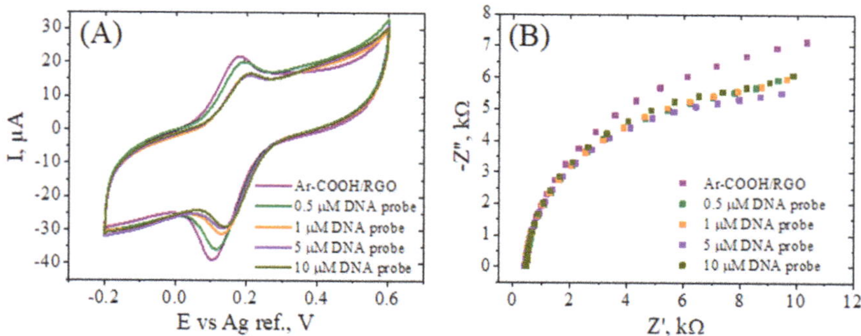

Figure 5. CV (**A**) and EIS Nyquist plot (**B**) recorded in 1 mM $[Fe(CN)_6]^{3-/4-}$, 0.1 M KCl, for Ar–COOH/RGO/SPCE activated in EDC–NHS, and incubated with 0.5, 1, 5, and 10 µM amino–modified ssDNA probe.

3.3.3. The Sensor Response for DNA Target Molecule

The hybridization between the probe and target DNA was monitored by using the same ferri/ferrocyanide mixture as a redox indicator, while the EIS technique was employed to evaluate the recognition event (Figure 6). Table 2 presents the charge transfer resistance values for the different steps in the electrode modification, obtained by fitting the Nyquist spectra using a Randles equivalent circuit (inset Figure 6). Modification of the RGO electrode with Ar-COOH groups induced a large increase in the R_{ct} value due to the electrostatic repulsion between the redox probe and the negatively charged carboxylic groups. Still, as a result of the activation reaction with EDC/NHS and the formation of the NHS ester, a significant decrease in its value was then observed. Although single-stranded DNA probe molecules are also negatively charged and prone to repulse the ferri/ferrocyanide ions, the further slight decrease in R_{ct} might be explained by an increase in the hydrophilic character of DNA functionalized electrodes compared with Ar-COOH/RGO/SPCE, which aids the access of the negative redox couple to the electrode surface [44].

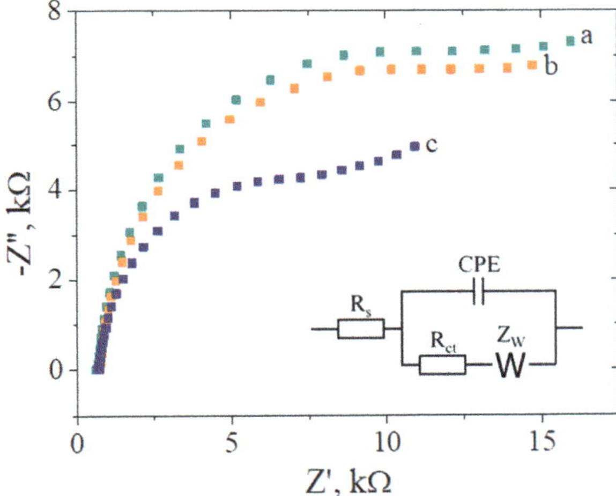

Figure 6. EIS Nyquist plot recorded in 1 mM [Fe(CN)$_6$]$^{3-/4-}$, 0.1 M KCl, for the Ar–COOH/RGO–modified electrodes after activation of the carboxylic groups (**a**), after carbodiimide coupling with 10 µM NH$_2$–DNA probe (**b**) and after hybridization with 200 nM DNA target (**c**). Inset: Randles equivalent circuit used for fitting.

Table 2. Charge transfer resistance values obtained from EIS measurements for the different stages in the electrode modification, along with the quality of fitting χ^2 and standard deviation (SD).

Electrode Modification	Rct	χ^2	SD
GO	10.5	0.0149	0.2121
RGO	1.7	0.0143	0.3111
Ar–COOH/RGO	36.3	0.0295	9.3338
Ar–COOH/RGO activated	13.5	0.0367	0.2828
ssDNAp/Ar–COOH/RGO	11.5	0.0236	0.9192
DNAt/ssDNAp/Ar–COOH/RGO	7.4	0.0178	0.7212

The spectrum (c) in Figure 6 shows a significant decrease in R_{ct} upon DNA target binding, which constitutes the basis of the detection process. This significant decrease in charge-transfer resistance during hybridization can be associated with the morphological changes resulting from the formation of the "rigid rod" double stranded (ds) DNA. Similar behavior has been observed by Gooding et al. [58] for the free electrochemical detection of DNA hybridization at SAM-modified Au electrodes. According to their studies, the hybridization opens the interface, allowing redox species in the solution to access the surface and the electrochemical reaction to occur. This atypical decrease in R_{ct} can be justified considering that ssDNA probes behave as flexible molecules that lie down across the modified electrode, thus contributing with direct electrostatic repulsion to the negatively charged redox probe. Conversely, the more rigid dsDNA structure generated after the hybridization process provides increased access of the $[Fe(CN)_6]^{3-/4-}$ redox probe to the electrode surface, and thus, improved electron transfer. This significant change in R_{ct} following DNA hybridization suggests that this electrochemical platform has great potential as a high sensitivity DNA biosensor.

3.3.4. Assessment of the Electron Transfer Kinetics at the RGO-Modified Electrodes

In order to compare the charge transfer properties of the modified electrodes, we further evaluated heterogeneous electron transfer rate constants for the $[Fe(CN)_6]^{3-/4-}$ redox couple. Scan dependence CV studies performed at RGO/SPCE, Ar-COOH/RGO/SPCE, ssDNAp/Ar-COOH/RGO/SPCE, and DNAt/ssDNAp/Ar-COOH/RGO/SPCE (Figure 7) displayed an increase in the peak-to-peak separation potential for all electrodes except Ar-COOH/RGO/SPCE (not shown), for which passivation occurs due to grafted carboxylic layers that suppress further electrochemistry of $Fe(CN)_6^{3-/4-}$ species. In all other cases, anodic and cathodic peak currents increase with increasing potential scan rate and are linearly related to the square root of the scan rates in the range of 0.005–0.1 V/s (see the insets in Figure 7), suggesting that the oxido-reduction process at the electrode surface is controlled by diffusion. The Nicholson methodology [59] was applied to estimate k^0 for the quasi-reversible process of $[Fe(CN)_6]^{3-/4-}$ at the modified electrodes. The values of the diffusion coefficients of the oxidized and reduced forms of $[Fe(CN)_6]^{3-/4-}$ $D_O = 7.6 \times 10^{-6}$ cm^2 s^{-1}, $D_R = 6.5 \times 10^{-6}$ cm^2 s^{-1} were taken from the literature [60], and the dimensionless kinetic parameter Ψ was calculated by using the equation given by Lavagnini et al. [61]. The calculated k^0 values for the Ar-COOH/RGO/SPCE-modified electrodes after the ssDNA probe immobilization, and the hybridization with DNA target are 1.65×10^{-3} cm s^{-1}, and 2.06×10^{-3} cm s^{-1}, respectively. This trend is in agreement with the EIS results reported before, showing an improvement in the charge transfer after DNA hybridization process.

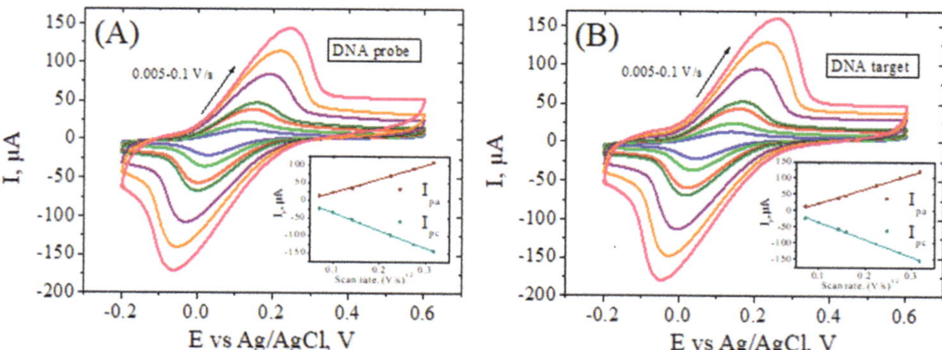

Figure 7. CV curves of ssDNAp/Ar–COOH/RGO/SPCE (**A**), and ssDNAp/Ar–COOH/RGO/SPCE after hybridization with DNA target (**B**) recorded at different scan rates in 1 mM $[Fe(CN)_6]^{3-/4-}$, 0.1 M KCl solution. Inset: plot of anodic and cathodic peaks vs. square root of scan rate.

The same pattern in the sensor response was observed throughout the concentration range of 1–200 nM, i.e., a smaller peak separation by DNA target addition, suggesting an increase in the charge transfer rate with target concentration (Figure 8). These evident changes in heterogeneous electron transfer as a result of the hybridization process should support the further use of this detection platform for the advancement of DNA biosensors.

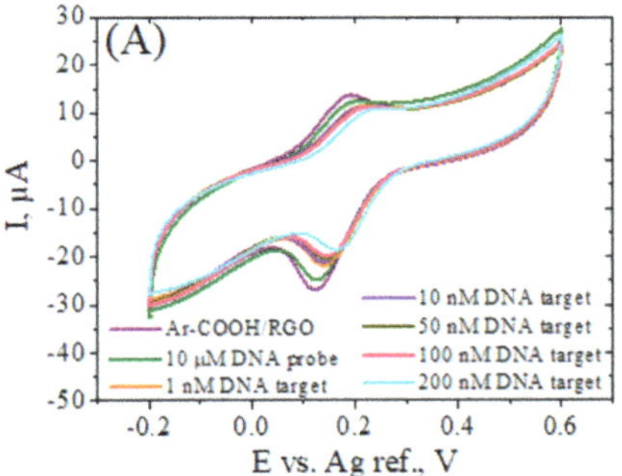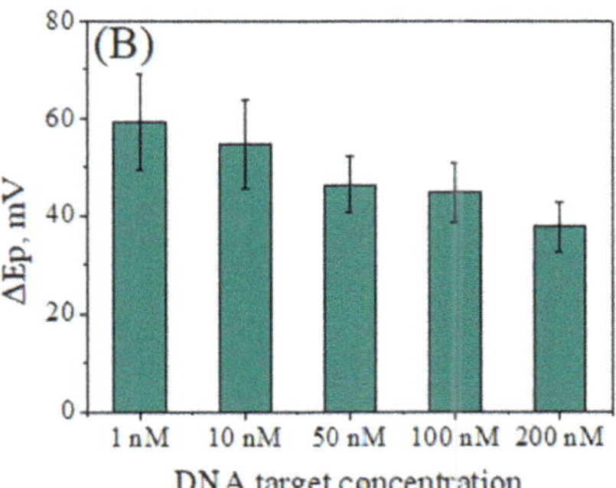

Figure 8. CV curves (**A**) and the corresponding peak separation values, ΔEp (**B**) of ssDNAp/Ar–COOH/RGO/SPCE after hybridization with different concentration of DNA target, recorded at a scan rate of 0.05 Vs^{-1} in 1 mM [Fe(CN)$_6$]$^{3-/4-}$, 0.1 M KCl solution.

4. Conclusions

In summary, we have developed an RGO-based stable and reproducible biosensing platform capable of achieving label-free and sensitive detection of DNA hybridization. To the best of our knowledge, no attempts have been presented to date in the literature to develop electrochemical label-free DNA detection platforms at graphene electrodes by ssDNA probe immobilization through diazonium chemistry. The diazonium-grafted layers allow for the controlled immobilization of oligonucleotide probes at the RGO substrate, which results in stability and hybridization efficiency. Our label-free diazonium-based strategy relies on the changes in the accessibility of a redox probe in the solution to the RGO electrode surface, as a consequence of the changes in the flexibility of the DNA oligonucleotides upon hybridization. This conclusion was supported by the kinetic study performed at the RGO-modified electrodes with ssDNA probes and in the presence of target, which demonstrated a faster redox reaction after the hybridization process as the resulting rigid dsDNA structure facilitates the redox probe access to the electrode surface. In addition, the modification of SPCE with GO dispersion was evidenced by SEM analysis, while the electrochemical features observed by CV and EIS after each electrode modification were in agreement with the variations in the I_D/I_G ratio from the Raman spectra. This new, simple, reproducible protocol proposed by us has great potential to be employed for the future manufacturing of miniaturized, sensitive and stable DNA biosensors.

Author Contributions: Conceptualization, M.I. and E.A.C.; methodology, E.A.C. and L.P.; writing—original draft preparation, E.A.C.; writing—review and editing, M.I. and L.P.; supervision, M.I.; project administration, M.I.; funding acquisition, M.I. All authors have read and agreed to the published version of the manuscript.

Funding: This work was supported by a grant from the Ministry of Research and Innovation, Operational Program Competitiveness Axis 1—Section E, Program co-financed by the European Regional Development Fund under the project number 154/25.11.2016, P_37_221/2015, "A novel graphene biosensor testing osteogenic potency; capturing best stem cell performance for regenerative medicine" (GRABTOP).

Conflicts of Interest: The authors declare no conflict of interest.

References

1. Cisse:, I.I.; Kim, H.; Ha, T. A rule of seven in Watson-Crick base-pairing of mismatched sequences. *Nat. Struct. Mol. Biol.* **2012**, *19*, 623–627. [CrossRef] [PubMed]
2. Ouldridge, T.E.; Sulc, P.; Romano, F.; Doye, J.P.K.; Louis, A.A. DNA hybridization kinetics: Zippering, internal displacement and sequence dependence. *Nucleic Acids Res.* **2013**, *41*, 8886–8895. [CrossRef] [PubMed]
3. DeLuca, M.; Shi, Z.; Castro, C.E.; Arya, G. Dynamic DNA nanotechnology: Toward functional nanoscale devices. *Nanoscale Horiz.* **2020**, *5*, 182–201. [CrossRef]

4.   Kimna, C.; Lieleg, O. Molecular micromanagement: DNA nanotechnology establishes spatio-temporal control for precision medicine. *Biophys. Rev.* **2020**, *1*, 011305. [CrossRef]

5.   Trotter, M.; Borst, N.; Thewes, R.; von Stetten, F. Review: Electrochemical DNA sensing—Principles, commercial systems, and applications. *Biosens. Bioelectron.* **2020**, *154*, 112069. [CrossRef]

6.   Wu, X.; Mu, F.; Wang, Y.; Zhao, H. Graphene and graphene-based nanomaterials for DNA detection: A review. *Molecules* **2018**, *23*, 2050. [CrossRef]

7.   Pei, Q.; Song, X.; Liu, S.; Wang, J.; Leng, X.; Cui, X.; Yu, J.; Wang, Y.; Huang, J. A facile signal-on electrochemical DNA sensing platform for ultrasensitive detection of pathogenic bacteria based on Exo III-assisted autonomous multiple-cycle amplification. *Analyst* **2019**, *144*, 3023–3029. [CrossRef] [PubMed]

8.   Rasouli, E.; Shahnavaz, Z.; Basirun, W.J.; Rezayi, M.; Avan, A.; Ghayour-Mobarhan, M.; Khandanlou, R.; Johan, M.R. Advancements in electrochemical DNA sensor for detection of human papilloma virus-A review. *Anal. Biochem.* **2018**, *556*, 136–144. [CrossRef]

9.   Mohanraj, J.; Durgalakshmi, D.; Rakkesh, R.A.; Balakumar, S.; Rajendran, S.; Karimi-Maleh, H. Facile synthesis of paper based graphene electrodes for point of care devices: A double stranded DNA (dsDNA) biosensor. *J. Colloid Interface Sci.* **2020**, *566*, 463–472. [CrossRef] [PubMed]

10.   Zeng, N.; Xiang, J. Detection of KRAS G12D point mutation level by anchor-like DNA electrochemical biosensor. *Talanta* **2019**, *198*, 111–117. [CrossRef] [PubMed]

11.   Huang, Y.; Xu, J.; Liu, J.; Wang, X.; Chen, B. Disease-related detection with electrochemical biosensors: A review. *Sensors* **2017**, *17*, 2375. [CrossRef]

12.   Rashid, J.I.A.; Yusof, N.A. The strategies of DNA immobilization and hybridization detection mechanism in the construction of electrochemical DNA sensor: A review. *Sens. Bio-Sens. Res.* **2017**, *16*, 19–31. [CrossRef]

13.   Mahmoodi, P.; Rezayi, M.; Rasouli, E.; Avan, A.; Gholami, M.; Mobarhan, M.G.; Karimi, E.; Alias, Y. Early-stage cervical cancer diagnosis based on an ultra-sensitive electrochemical DNA nanobiosensor for HPV-18 detection in real samples. *J. Nanobiotechnology* **2020**, *18*, 1–12. [CrossRef] [PubMed]

14.   Bolat, G. Investigation of poly (CTAB-MWCNTs) composite based electrochemical DNA biosensor and interaction study with anticancer drug Irinotecan. *Microchem. J.* **2020**, *159*, 105426. [CrossRef]

15.   Javar, H.A.; Garkani-Nejad, Z.; Dehghannoudeh, G.; Mahmoudi-Moghaddam, H. Development of a new electrochemical DNA biosensor based on Eu3+—Doped NiO for determination of amsacrine as an anti-cancer drug: Electrochemical, spectroscopic and docking studies. *Anal. Chim. Acta* **2020**, *1133*, 48–57. [CrossRef] [PubMed]

16.   Kumar, V.; Guleria, P. Application of DNA-Nanosensor for Environmental Monitoring: Recent Advances and Perspectives. *Curr. Pollut. Rep.* **2020**, 1–21. [CrossRef]

17.   Chen, Z.; Liu, X.; Liu, D.; Li, F.; Wang, L.; Liu, S. Ultrasensitive electrochemical DNA biosensor fabrication by coupling an integral multifunctional zirconia-reduced graphene oxide-thionine nanocomposite and exonuclease I-assisted cleavage. *Front. Chem.* **2020**, *8*, 521. [CrossRef]

18.   Zhu, D.; Liu, W.; Zhao, D.; Hao, Q.; Li, J.; Huang, J.; Shi, J.; Chao, J.; Su, S.; Wang, L. Label-free electrochemical sensing platform for microRNA-21 detection using thionine and gold nanoparticles co-functionalized MoS2 nanosheet. *ACS Appl. Mater. Interfaces* **2017**, *9*, 35597–35603. [CrossRef]

19.   Kavita, V. DNA biosensors-a review. *J. Bioeng. Biomed. Sci.* **2017**, *7*, 222.

20.   Ferapontova, E.E. DNA electrochemistry and electrochemical sensors for nucleic acids. *Annu. Rev. Anal. Chem.* **2018**, *11*, 197–218. [CrossRef] [PubMed]

21.   Cho, I.-H.; Kim, D.H.; Park, S. Electrochemical biosensors: Perspective on functional nanomaterials for on-site analysis. *Biomater. Res.* **2020**, *24*, 6. [CrossRef] [PubMed]

22.   Alazmi, A.; Rasul, S.; Patole, S.P.; Costa, P.M. Comparative study of synthesis and reduction methods for graphene oxide. *Polyhedron* **2016**, *116*, 153–161. [CrossRef]

23.   Benvidi, A.; Rajabzadeh, N.; Mazloum-Ardakani, M.; Heidari, M.M.; Mulchandani, A. Simple and label-free electrochemical impedance Amelogenin gene hybridization biosensing based on reduced graphene oxide. *Biosens. Bioelectron.* **2014**, *58*, 145–152. [CrossRef] [PubMed]

24.   Gosai, A.; Khondakar, K.R.; Ma, X.; Ali, M. Application of Functionalized Graphene Oxide Based Biosensors for Health Monitoring: Simple Graphene Derivatives to 3D Printed Platforms. *Biosensors* **2021**, *11*, 384. [CrossRef]

25.   Chiticaru, E.A.; Pilan, L.; Damian, C.-M.; Vasile, E.; Burns, J.S.; Ioniţă, M. Influence of Graphene Oxide Concentration when Fabricating an Electrochemical Biosensor for DNA Detection. *Biosensors* **2019**, *9*, 113. [CrossRef]

26.   Sarkar, S.; Bekyarova, E.; Haddon, R.C. Reversible Grafting of α-Naphthylmethyl Radicals to Epitaxial Graphene. *Angew. Chem. Int. Ed.* **2012**, *51*, 4901–4904. [CrossRef] [PubMed]

27.   Xia, Z.Y.; Giambastiani, G.; Christodoulou, C.; Nardi, M.V.; Koch, N.; Treossi, E.; Bellani, V.; Pezzini, S.; Corticelli, F.; Morandi, V. Synergic exfoliation of graphene with organic molecules and inorganic ions for the electrochemical production of flexible electrodes. *Chem. Plus. Chem.* **2014**, *79*, 439–446. [CrossRef] [PubMed]

28.   Xia, Z.; Leonardi, F.; Gobbi, M.; Liu, Y.; Bellani, V.; Liscio, A.; Kovtun, A.; Li, R.; Feng, X.; Orgiu, E. Electrochemical functionalization of graphene at the nanoscale with self-assembling diazonium salts. *ACS Nano* **2016**, *10*, 7125–7134. [CrossRef] [PubMed]

29. Shih, C.-J.; Wang, Q.H.; Jin, Z.; Paulus, G.L.; Blankschtein, D.; Jarillo-Herrero, P.; Strano, M.S. Disorder imposed limits of mono-and bilayer graphene electronic modification using covalent chemistry. *Nano Lett.* **2013**, *13*, 809–817. [CrossRef]

30. Paulus, G.L.; Wang, Q.H.; Strano, M.S. Covalent electron transfer chemistry of graphene with diazonium salts. *Acc. Chem. Res.* **2013**, *46*, 160–170. [CrossRef]

31. Kongsfelt, M.S.; Ceccato, M.; Nilsson, L.; Jørgensen, B.; Hornekær, L.; Pedersen, S.U.; Daasbjerg, K. Chemical modifications of graphene using diazonium chemistry. In Proceedings of the Annual World Conference on Carbon 2010, Clemson, SC, USA, 11–16 July 2010.

32. Raicopol, M.; Vlsceanu, I.; Lupulescu, I.; Brezoiu, A.M.; Pilan, L. Amperometric glucose biosensors based on functionalized electrochemically reduced graphene oxide. *UPB Sci. Bull. Ser. B* **2016**, *78*, 131–142.

33. Ott, C.; Raicopol, M.D.; Andronescu, C.; Vasile, E.; Hanganu, A.; Pruna, A.; Pilan, L. Functionalized polypyrrole/sulfonated graphene nanocomposites: Improved biosensing platforms through aryl diazonium electrochemistry. *Synth. Met.* **2018**, *235*, 20–28. [CrossRef]

34. Ge, L.; Wang, W.; Li, F. Electro-grafted electrode with graphene-oxide-like DNA affinity for ratiometric homogeneous electrochemical biosensing of microRNA. *Anal. Chem.* **2017**, *89*, 11560–11567. [CrossRef] [PubMed]

35. Allongue, P.; Delamar, M.; Desbat, B.; Fagebaume, O.; Hitmi, R.; Pinson, J.; Saveant, J.-M. Covalent modification of carbon surfaces by aryl radicals generated from the electrochemical reduction of diazonium salts. *J. Am. Chem. Soc.* **1997**, *119*, 201–207. [CrossRef]

36. Gan, L.; Zhang, D.; Guo, X. Electrochemistry: An efficient way to chemically modify individual monolayers of graphene. *Small* **2012**, *8*, 1326–1330. [CrossRef] [PubMed]

37. Tavakkoli, Z.; Goljani, H.; Sepehrmansourie, H.; Nematollahi, D.; Zolfigol, M.A. New insight into the electrochemical reduction of different aryldiazonium salts in aqueous solutions. *RSC Adv.* **2021**, *11*, 25811–25815. [CrossRef]

38. Yáñez-Sedeño, P.; Campuzano, S.; Pingarrón, J.M. Integrated Affinity Biosensing Platforms on Screen-Printed Electrodes Electrografted with Diazonium Salts. *Sensors* **2018**, *18*, 675. [CrossRef]

39. Yuliandari, P.; Wibowo, R.; Nurani, D.A. Para-carboxyphenyl diazonium-modified carbon paste electrode for analysis Cu (II) in water. In *AIP Conference Proceedings*; AIP Publishing LLC: Melville, NY, USA, 2021; p. 020002.

40. Gökçe, G.; Ben Aissa, S.; Nemčeková, K.; Catanante, G.; Raouafi, N.; Marty, J.-L. Aptamer-modified pencil graphite electrodes for the impedimetric determination of ochratoxin A. *Food Control* **2020**, *115*, 107271. [CrossRef]

41. Raicopol, M.; Necula, L.; Ionita, M.; Pilan, L. Electrochemical reduction of aryl diazonium salts: A versatile way for carbon nanotubes functionalisation. *Surf. Interface Anal.* **2012**, *44*, 1081–1085. [CrossRef]

42. Randriamahazaka, H.; Ghilane, J. Electrografting and Controlled Surface Functionalization of Carbon Based Surfaces for Electroanalysis. *Electroanalysis* **2016**, *28*, 13–26. [CrossRef]

43. Gillan, L.; Teerinen, T.; Johansson, L.-S.; Smolander, M. Controlled diazonium electrodeposition towards a biosensor for C-reactive protein. *Sens. Int.* **2021**, *2*, 100060. [CrossRef]

44. Mousavisani, S.Z.; Raoof, J.-B.; Turner, A.P.F.; Ojani, R.; Mak, W.C. Label-free DNA sensor based on diazonium immobilisation for detection of DNA damage in breast cancer 1 gene. *Sens. Actuators B Chem.* **2018**, *264*, 59–66. [CrossRef]

45. Polsky, R.; Harper, J.C.; Wheeler, D.R.; Arango, D.C.; Brozik, S.M. Electrically addressable cell immobilization using phenylboronic acid diazonium salts. *Angew. Chem. Int. Ed.* **2008**, *120*, 2671–2674. [CrossRef]

46. Hu, Y.; Li, F.; Han, D.; Wu, T.; Zhang, Q.; Niu, L.; Bao, Y. Simple and label-free electrochemical assay for signal-on DNA hybridization directly at undecorated graphene oxide. *Anal. Chim. Acta* **2012**, *753*, 82–89. [CrossRef] [PubMed]

47. Giovanni, M.; Bonanni, A.; Pumera, M. Detection of DNA hybridization on chemically modified graphene platforms. *Analyst* **2012**, *137*, 580–583. [CrossRef] [PubMed]

48. Gong, Q.; Yang, H.; Dong, Y.; Zhang, W. A sensitive impedimetric DNA biosensor for the determination of the HIV gene based on electrochemically reduced graphene oxide. *Anal. Methods* **2015**, *7*, 2554–2562. [CrossRef]

49. Eissa, S.; Jimenez, G.C.; Mahvash, F.; Guermoune, A.; Tlili, C.; Szkopek, T.; Zourob, M.; Siaj, M. Functionalized CVD monolayer graphene for label-free impedimetric biosensing. *Nano Res.* **2015**, *8*, 1698–1709. [CrossRef]

50. Wu, J.-B.; Lin, M.-L.; Cong, X.; Liu, H.-N.; Tan, P.-H. Raman spectroscopy of graphene-based materials and its applications in related devices. *Chem. Soc. Rev.* **2018**, *47*, 1822–1873. [CrossRef] [PubMed]

51. Sun, Z.; Guo, D.; Wang, S.; Wang, C.; Yu, Y.; Ma, D.; Zheng, R.; Yan, P. Efficient covalent modification of graphene by diazo chemistry. *RSC Adv.* **2016**, *6*, 65422–65425. [CrossRef]

52. Jiang, D.-e.; Sumpter, B.G.; Dai, S. How do aryl groups attach to a graphene sheet? *J. Phys. Chem. B* **2006**, *110*, 23628–23632. [CrossRef]

53. Pilan, L. Tailoring the performance of electrochemical biosensors based on carbon nanomaterials via aryldiazonium electrografting. *Bioelectrochemistry* **2021**, *138*, 107697. [CrossRef]

54. Ambrosio, G.; Brown, A.; Daukiya, L.; Drera, G.; Di Santo, G.; Petaccia, L.; De Feyter, S.; Sangaletti, L.; Pagliara, S. Impact of covalent functionalization by diazonium chemistry on the electronic properties of graphene on SiC. *Nanoscale* **2020**, *12*, 9032–9037. [CrossRef]

55. Lee, L.; Ma, H.; Brooksby, P.A.; Brown, S.A.; Leroux, Y.R.; Hapiot, P.; Downard, A.J. Covalently anchored carboxyphenyl monolayer via aryldiazonium ion grafting: A well-defined reactive tether layer for on-surface chemistry. *Langmuir* **2014**, *30*, 7104–7111. [CrossRef] [PubMed]

56. Phal, S.; Shimizu, K.; Mwanza, D.; Mashazi, P.; Shchukarev, A.; Tesfalidet, S. Electrografting of 4-carboxybenzenediazonium on glassy carbon electrode: The effect of concentration on the formation of mono and multilayers. *Molecules* **2020**, *25*, 4575. [CrossRef]
57. Wallen, R.; Gokarn, N.; Bercea, P.; Grzincic, E.; Bandyopadhyay, K. Mediated electron transfer at vertically aligned single-walled carbon nanotube electrodes during detection of DNA hybridization. *Nanoscale Res. Lett.* **2015**, *10*, 1–11. [CrossRef] [PubMed]
58. Gooding, J.J.; Chou, A.; Mearns, F.J.; Wong, E.; Jericho, K.L. The ion gating effect: Using a change in flexibility to allow label free electrochemical detection of DNA hybridisation. *Chem. Commun.* **2003**, *15*, 1938–1939. [CrossRef]
59. Nicholson, R.S. Theory and application of cyclic voltammetry for measurement of electrode reaction kinetics. *Anal. Chem.* **1965**, *37*, 1351–1355. [CrossRef]
60. Ojeda, I.; Barrejón, M.; Arellano, L.M.; González-Cortés, A.; Yáñez-Sedeño, P.; Langa, F.; Pingarrón, J.M. Grafted-double walled carbon nanotubes as electrochemical platforms for immobilization of antibodies using a metallic-complex chelating polymer: Application to the determination of adiponectin cytokine in serum. *Biosens. Bioelectron.* **2015**, *74*, 24–29. [CrossRef]
61. Lavagnini, I.; Antiochia, R.; Magno, F. An extended method for the practical evaluation of the standard rate constant from cyclic voltammetric data. *Electroynalysis* **2004**, *16*, 505–506. [CrossRef]

biosensors

Article

Plastic Antibody of Polypyrrole/Multiwall Carbon Nanotubes on Screen-Printed Electrodes for Cystatin C Detection

Rui S. Gomes [1,2,3], Blanca Azucena Gomez-Rodríguez [4], Ruben Fernandes [5,6,7], M. Goreti F. Sales [1,2,3], Felismina T. C. Moreira [1,3,*] and Rosa F. Dutra [4,*]

[1] BioMark@ISEP, School of Engineering, Polytechnic Institute of Porto, 4249-015 Porto, Portugal; deb12008@fe.up.pt (R.S.G.); goreti.sales@eq.uc.pt (M.G.F.S.)
[2] BioMark@UC, Department of Chemical Engineering, Faculty of Science and Technology, University Coimbra, 3030-790 Coimbra, Portugal
[3] CEB, Centre of Biological Engineering, University of Minho, 4715-000 Braga, Portugal
[4] Biomedical Engineering Laboratory, Federal University of Pernambuco, Recife 50670-901, PE, Brazil; blanca.gr@teziutlan.tecnm.mx
[5] LaBMI–PORTIC, Laboratory of Medical and Industrial Biotechnology–Porto Research, Technology & Innovation Centre, Polytechnic Institute of Porto, 4200-472 Porto, Portugal; rfernandes@ess.ipp.pt
[6] Escola Superior de Saúde, Polytechnic Institute of Porto, 4200-072 Porto, Portugal
[7] i3S-Institute of Health and Research Innovation, University of Porto, 4200-135 Porto, Portugal
* Correspondence: ftm@isep.ipp.pt (F.T.C.M.); rosa.dutra@ufpe.br (R.F.D.)

Abstract: This work reports the design of a novel plastic antibody for cystatin C (Cys-C), an acute kidney injury biomarker, and its application in point-of-care (PoC) testing. The synthetic antibody was obtained by tailoring a molecularly imprinted polymer (MIP) on a carbon screen-printed electrode (SPE). The MIP was obtained by electropolymerizing pyrrole (Py) with carboxylated Py (Py-COOH) in the presence of Cys-C and multiwall carbon nanotubes (MWCNTs). Cys-C was removed from the molecularly imprinted poly(Py) matrix (MPPy) by urea treatment. As a control, a non-imprinted poly(Py) matrix (NPPy) was obtained by the same procedure, but without Cys-C. The assembly of the MIP material was evaluated in situ by Raman spectroscopy and the binding ability of Cys-C was evaluated by the cyclic voltammetry (CV) and differential pulse voltammetry (DPV) electrochemical techniques. The MIP sensor responses were measured by the DPV anodic peaks obtained in the presence of ferro/ferricyanide. The peak currents decreased linearly from 0.5 to 20.0 ng/mL of Cys-C at each 20 min successive incubation and a limit of detection below 0.5 ng/mL was obtained at pH 6.0. The MPPy/SPE was used to analyze Cys-C in spiked serum samples, showing recoveries <3%. This device showed promising features in terms of simplicity, cost and sensitivity for acute kidney injury diagnosis at the point of care.

Keywords: cystatin C; molecularly imprinted polymer; electrochemical biosensor; polypyrene; multiwall carbon nanotubes; acute kidney injury

Citation: Gomes, R.S.; Gomez-Rodríguez, B.A.; Fernandes, R.; Sales, M.G.F.; Moreira, F.T.C.; Dutra, R.F. Plastic Antibody of Polypyrrole/Multiwall Carbon Nanotubes on Screen-Printed Electrodes for Cystatin C Detection. *Biosensors* 2021, 11, 175. https://doi.org/10.3390/bios11060175

Received: 18 April 2021
Accepted: 26 May 2021
Published: 31 May 2021

1. Introduction

Worldwide, chronic kidney disease (CKD) constitutes a great economic impact with high rates of morbidity and mortality [1]. Renal substitutive therapy is a burden, especially for low-income countries, since their socio-economic conditions restrain effective programs of non-preventable cardiovascular diseases and diabetes, drug therapy, tobacco control, promotion of physical activity and the reduction of salt intake through legislation and food [2]. Patients undergoing chronic renal replacement therapy have an increased chance of acute kidney injury (AKI) and consequent death [3]. AKI is characterized by a rapid decline in the glomerular filtration rate, thus, biomarkers should be continuously monitored in chronic renal patients to increase their survival rates [3,4]. Cystatin C (Cys-C) is a single non-glycosylated polypeptide chain consisting of 120 amino acid residues with a molecular mass of 13 kDa, and is more specific as a renal biomarker for glomerular filtration than creatinine, because it does not depend on gender, age, diet and muscle mass [4–8].

Thus, development of new methods for monitoring Cys-C that are user-friendly and practical in point-of-care (PoC) settings could represent a strategy to follow patients with AKI. Due to the major advances in nanotechnology [9], several biosensors for Cys-C have been developed for this purpose. Most of these are immunosensors and employ naturally derived antibodies. They include impedimetric systems with an interdigitated electrode [10] or a three-electrode system [11], spectroelectrochemical systems using TiO_2 nanotube arrays [12] or gold nanopillar substrates [13], electrochemiluminescence systems using Au/Pd/Pt nanoflowers modified with MoS2 nanosheets [14] or a graphene sheet modified with a layer of rubrene [15] or reflectometric interference spectroscopy systems employing glass substrates [16]. As an alternative to immunosensors, enzymatic-based sensors have also been developed [17]. Overall, all these sensing materials are of natural origin, making the devices less stable, less reproducible and more expensive.

Alternative synthetic recognition materials include molecularly imprinted polymers (MIPs), working as biorecognition elements that ensure that electrochemical responses come from Cys-C. MIPs are known as plastic antibodies for mimicking the behavior of antibodies obtained from biological sources. These polymers can selectively recognize a given target molecule to which they are designed. If applied as recognition units of biosensors, these receptors provide very high selectivity, so the use of MIPs as recognition units in biochemical sensors is gaining increasing interest. Overall, the presence of a biorecognition element is essential, or else the response of the biosensor to real samples would be linked to all compounds present in the sample and not only to Cys-C, as the working electrode would absorb/adsorb any protein/biomolecule present. Thus far, there is no MIP material for Cys-C that we know of. In general, MIP materials offer long-term stability, are inexpensive and can be tailor-made on demand, for almost any intended target compound [18,19]. Moreira et al. (2013) are among the pioneers of this field, proving that the integration of plastic antibodies for detecting protein biomarkers in simple biosensing technology is possible [20]. MIP materials may be tuned to display conductive features, fulfilling electron transfer requirements and benefiting the subsequent electrochemical responses.

MIP materials may be composed of conducting polymers, such as polypyrrole (PPy), a widely known conducting polymer that could be employed as a polymeric network of MIP material, offering easy electropolymerization [21]. In combination with carbon nanotubes (CNTs), the nanocomposite PPy/CNT exhibits a high doping and de-doping rate and a high capacity of charge storage, making it a supercapacitor [22]. When a PPy/CNT composite is obtained by electrosynthesis, it achieves a long cycle life, derived from strong π–π bonds between the PPy conjugated structure and the CNT [10].

Thus, this work reports, for the first time, MIP material for Cys-C, produced by electropolymerizing in situ Py-based monomers in the presence of the target protein. The contribution of additional conductive materials based on CNTs was also evaluated. The MIP synthesis was optimized, characterized and suitable and applied to check application feasibility of the final biosensor.

2. Experimental Section

2.1. Apparatus

The electrochemical measurements were conducted with a potentiostat/galvanostat from Metrohm Autolab and a PGSTAT302N. Carbon SPEs (C-SPEs) were from Dropsens (DRP-C110, Oviedo, Spain). The working electrode had a diameter of 4 mm and working/counter electrodes were made of carbon and the reference electrode was made of silver. C-SPEs were interfaced with the potentiostat via a specially designed switch box (BioTID, Porto, Portugal). Raman spectroscopy data were collected by Thermo Scientific DXR equipment, with a confocal microscope and a 532 nm laser. The Raman spectrometer was operated with 2 mW laser power and a 50 μm slit aperture.

2.2. Reagents and Solutions

De-ionized laboratory grade water was employed, and all chemicals were of analytical grade. Potassium hexacyanoferrate III ($K_3[Fe(CN)_6]$), potassium hexacyanoferrate II ($K_4[Fe(CN)_6]$) trihydrate, L-ascorbic acid and sodium acetate were obtained from Riedel-de Häen. Cys-C, MWCNTs and urea were obtained from Fluka. Py, sodium chloride (KCl) and Py-COOH (Py-3-carboxylic acid, 99%) were obtained from Merck. Creatine kinase iso-enzyme was obtained from European Reference Materials. Creatinine and bovine serum albumin were obtained from Amresco.

Electrochemical readings were carried out in 5.0×10^{-3} mol/L $K_3[Fe(CN)_6]$ and $K_4[Fe(CN)_6]$ redox standard solution, prepared in 0.1 mol/L KCl. Selectivity studies used synthetic serum spiked with other compounds that could act as interfering species (creatine kinase-MB 0.2 g/L, ascorbic acid 0.15 g/L, creatinine 1g/L and bovine serum albumin 12 g/L). Spiked serum samples were diluted 10 times in human serum obtained as Cormay®HN.

2.3. Electrochemical Procedures

Electrochemical data were obtained by cyclic voltammetry (CV) and differential pulse voltammetry (DPV). For the control of the surface immobilization, CV was carried out from -0.6 to $+0.6$ V with a scan rate of 20 mV/s and DPV was carried out from -0.3 to $+0.3$ V. Electrochemical readings were obtained for MPPy and NPPy materials using a minimum of three replicate readings ($n < 3$). Calibration curves used DPV data using the NOVA software program.

The calibration curve was made by incubating increasing concentrations of Cys-C standard solutions for 20 min. After each concentration, the electrochemical response of the standard probe $[Fe(CN)_6]^{3-}/^{4-}$ was collected, obtaining in this stage the electrical features. The Cys-C concentrations ranged from 0.5 to 40 ng/mL, prepared in acetate buffer pH 6.0.

Selectivity data were collected by incubating Cys-C standard solutions prepared with diluted spiked Cormay® serum.

2.4. Production of the Plastic Antibody on the C-SPE

C-SPEs were first pre-treated by applying +1.7 V for 200 s to a 0.1M KCl solution. The MPPy film was obtained as described in Figure 1. Electropolymerization of both MPPy and NPPy was achieved by 10 CV scans, with a start potential of -0.5 V, a lower vertex potential of -0.8 V and an upper vertex potential $+0.8$ V, with a scan rate of 20 mV/s. The MPPy material was obtained in a pH 6 acetate buffer solution containing MWCNTs (40 %), Py (80.0 mol/L), 10% Py-COOH (4.0 mmol/L) and 1% Cys-C (0.050 µg/mL). The NPPy material was obtained by the same procedure, but without Cys-C in the solution.

The use of pH 6 and 1% Cys-C followed previous preliminary experiments involving other protein imprinting assemblies. The overall composition was selected according to the experience of the research groups and optimization steps. In addition, the use of a small amount of Py-COOH compared to Py followed the same principle as that of the preparation of a material denoted as SPAM in the literature [23]. In it, the protein is surrounded by a monomer that is different from the overall polymeric matrix, aiming to enhance the capacity of recognizing this protein (it has a higher affinity to the binding site).

Cys-C from was extracted from the polymeric network by incubating the working electrode area of the MPPy/C-SPEs in 0.1M urea for 3 h. The electrode surface was then washed several times in acetate buffer to remove any contaminants on the surface and rinsed with water.

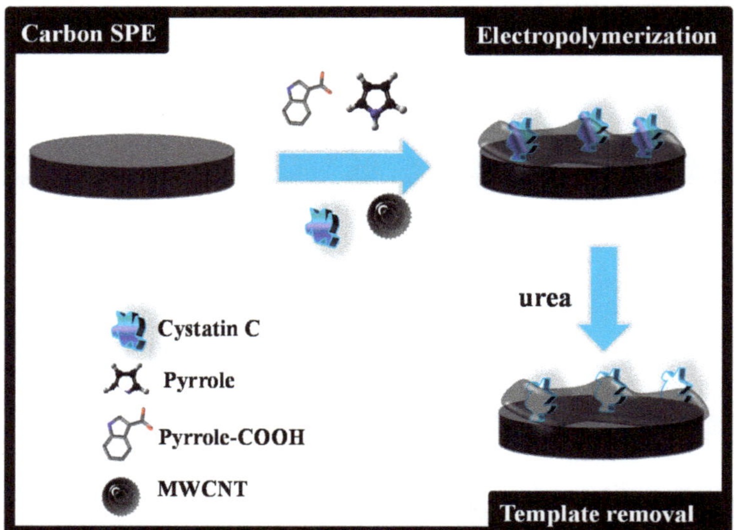

Figure 1. Schematic representation of the assembly process of the imprinted material.

3. Results and Discussion

3.1. Follow-Up of the Surface Modification

The Raman spectra of carbon working electrodes on the C-SPE and their subsequent modification with MPPy and NPPy materials are shown in Figure 2. As expected from the literature, the G- and D-bands are the more relevant peaks [24,25]. The G-band shows the presence of sp^2 hybrid orbitals in rings and chains, while the D-band reveals hexagonal lattice defects of carbon-based materials, including sp^3–carbon hybridization. The G- and D-bands of the pre-treated C-SPE were at 1577.6 cm^{-1} and 1314.2 cm^{-1} Raman shifts [25], and those of the MPPy and NPPy materials moved to higher Raman shift values, thereby confirming the modification made to the substrate. It may be that the modification consisted of the formation of a thin film of PPy with CNTs and with carboxylated moieties from the Py-COOH, because the spectra obtained had no evidence of specific PPy peaks.

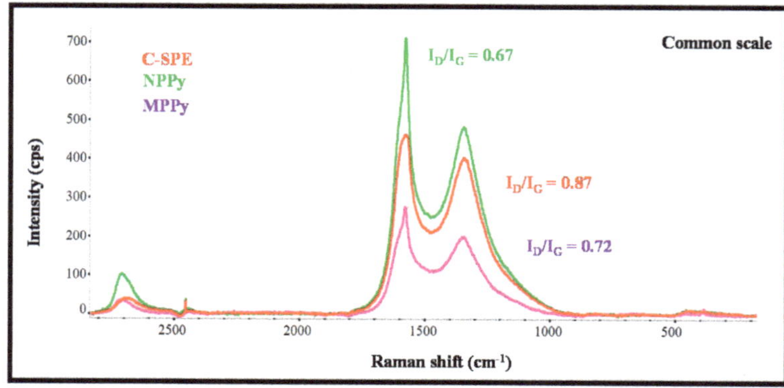

Figure 2. Raman spectra of the carbon working electrode on the SPE (C-SPE) and its subsequent modification with MPPy and NPPy materials.

Moreover, the changes in the ratio of the intensity of G- and D-bands was analyzed, as they reflect changes in the organization of the carbon materials and may help to confirm the chemical modifications made to the carbon electrode. In simple terms, an increasing ID/IG ratio reveals increasing disorder degrees within the materials [26]. The ID/IG ratio of the working electrode of non-modified C-SPE was 0.87, while the ratios of MPPy and NPPy were 0.72 and 0.67, respectively. These decreasing ratios on the polymer-modified electrodes confirmed the presence of a polymeric film because PPy is a conducting polymer with conjugated C=C bonds and CNTs were introduced in the polymeric network. The fact that NPPy had a lower ratio than MPPy was related to two effects: (1) the non-imprinted film does not contain Cys-C molecules entrapped within the network, because they were absent in the synthesis process; (2) the film may have been formed to a greater extent, because the presence of a protein at the time of polymer growth hinders polymer formation.

Overall, Raman analysis confirmed the presence of the polymeric materials on both MPPy and NPPy devices and the differences between imprinted and non-imprinted devices seem to confirm a different composition of the sensing materials.

3.2. Polypyrrole-Based Sensing Element

A PPy-based material was employed as the basis for creating a sensing element of Cys-C, due to its excellent conductivity features that may enhance sensitivity. This was done by electrochemical polymerization, in which the oxidation of the monomers under a suitable anodic potential or current produces a radical cation. These radical species are highly unstable, reacting with each other to create a radical dimer, which in turn is transformed into a trimmer, leading, in time, a to longer polymer chain. Apart from being able to produce a 3D network capable of recognizing Cys-C, the resulting polymer should lead to a stable electrochemical signal. This is fundamental to ensuring that the electrochemical signal changes are related only to Cys-C and not to random background changes or signal drifting. This depends mostly on the conditions selected to obtain the polymeric network. Thus, different NPPy devices were prepared first by CV, for optimizing the production of this polymeric film.

The effect of the potential range of the CV scan was evaluated first, as it contributes to the way the polymer grows, keeping in mind that the oxidation peak of the monomer Py is included within this range. Three different potential ranges were tested: (i) -0.80 to $+0.65$ V, (ii) -0.80 to $+0.80$ V and (iii) -0.80 to $+1.40$ V. In general, the increasing potentials increased the number of radical species formed and thereby the extent of polymer formation, as more energy was being introduced into the electrochemical system. While the higher potential values had the possibility of leading to the overoxidation of the polymeric film, the lower values displayed poorer reproducibility. This poor reproducibility reflected the fact that the voltage required to reach the maximum Py oxidation (maximum current) was far from being achieved. Thus, the selected range of potential was within -0.80 to $+0.80$V, which ensured the production of a reproducible and stable polymeric layer.

The stability of the NPPy film formed is also intrinsically linked to the potential range selected, according to our experience. A total of 10 CV scans was used in each condition, with a scan rate of 20 mV/s. In general, the number of selected CV scans is based on the stability of the sensor after successive readings with a redox probe and washing steps with the buffer. Using 10 CV scans, the sensor displayed reproducible and stable electrochemical data.

The behavior of the resulting films was evaluated by CV and DPV readings in the presence of the standard redox probe solution. The data obtained are shown in Figure 3. When the films were obtained in the potential range -0.80 to $+0.65$ V, the peak separation was lower than 0.15V, with the redox probe showing a quasi-reversible behavior. However, the successive incubations of the acetate buffer solution yielded unstable readings. This could the attributed to the incomplete polymerization of the monomer species, which in this condition would undergo chemical changes when subject to the electrochemical reading conditions of the probe. The preparation of films within the potential range -0.80 to $+0.80$ V led to increased currents that revealed the presence of an increased electroactive area, associated with a higher conductivity. In contrast, for the potential range -0.80 to

+1.4 V, the electrochemical measurements of the redox probe after successive readings were very stable, but the current signal was much reduced by the formation of a highly insulating film. This result reflected the overoxidation of Py, which was close to +1.1 V [27].

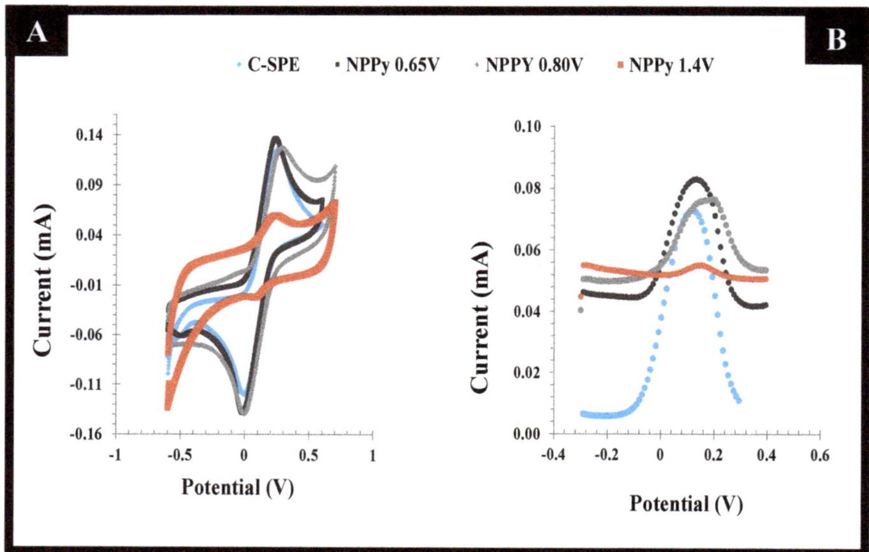

Figure 3. CV (**left**) and DPV (**right**) voltammograms of a 5.0 mmol/L [Fe (CN)6]3−/4− solution in KCl 0.1 mol/L on PPy films prepared with C-SPE electrodes with different potential ranges, starting at −0.80 V and finishing at +0.65, +0.80 or +1.40 V.

Overall, according to both CV and DPV studies of the different films, the best compromise between electron transfer and sensor surface stability was obtained for PPy films produced within −0.80 to +0.80 V. This experimental condition was used in further studies.

3.3. MWCNT Effect

In general, MWCNTs improve electron transfer and superficial area in electrochemical systems, thereby improving the electrical properties of electrochemical biosensors. Thus, to produce a conductive MPPy of improved electrical features, MWCNTs were added to the reaction mixture, forming a nanocomposite material of PPy/CNT. This nanocomposite is also known to act as a supercapacitor due to its good properties of charge delivery and energy/electron storage. Overall, the inclusion of CNTs within the PPy network formed a 3D structure of increased conductivity, derived from the cross-talk between the conjugated π–π bonds of the PPy structure and of the CNT sidewalls [28].

The effect of the addition of MWCNTs to the PPy network was evaluated by CV scanning of the standard redox probe. The data obtained was shown in Figure 4. A reduction of 30% in the peak-to-peak potential separation (ΔE) was observed with the addition of MWCNTs. The typical ΔE of PPy decreased from 0.382 to 0.264 V by adding MWCNTs to the polymerizing mixture, reflecting an improvement in the conductivity of the material. In addition, the faradaic current was also increased by the presence of the MWCNTs, just as expected [29].

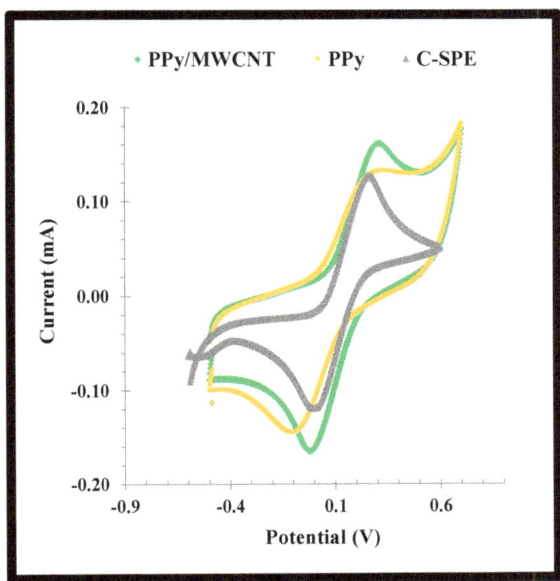

Figure 4. CV voltammograms of a solution of a 5.0 mmol/L [Fe (CN)6]$^{3-}$/$^{4-}$ solution in KCl 0.1 mol/L casted on the NPPy material assembled from -0.80 V to $+0.80$V, with or without MWCNTs, on substrates of C-SPEs.

The effect of the PY was also evaluated by measuring the specific capacitance (Cs). It was calculated by means of Equation (1), in which i is current, V potential window, scan rate and m mass-specific capacitance (F/g). The results indicated that the presence of PPy on the cleaned electrode surface increased the Cs about 12.6%.

$$Cs = (\int i.dV)/(v.m.\Delta V) \tag{1}$$

3.4. Synthesis of the Imprinted Material

The synthesis of the MPPy material was carried out by following the previously selected conditions, adapted to the target compound. Specifically, human Cys-C exists in two isoforms with the isoelectric points 9.2 and 7.8 [30], meaning that it is positively charged in serum, or in pH 6, the pH selected for this work [31]. Thus, the addition of negatively charged monomer species could intensify the binding of Cys-C to the final polymeric network [32]. This would enhance the sensitivity of the final device.

Thus, the MPPy film was obtained by adding Py-COOH to Cys-C and allowing (over 30 min) self-arrangement between these compounds, by means of ionic interactions. In brief, the negatively charged carboxylic groups of the Py-COOH, at pH 6, were expected to bind to the positively charged amino groups from the protein, by means of ionic interactions. This self-arranged structure was then added to the Py/MWCNT solution, to undergo electropolymerization on the working electrode of the C-SPE. Electropolymerization was conducted by CV, under the selected optimum conditions. Cys-C was then extracted from the polymeric network by treatment with a urea solution (5 µL of 0.1 M of urea dissolved in water were cast on the working electrode area), to leave vacant binding positions with a complementary charge and shape to the target protein.

The C-SPE surface modification was followed by CV and DPV procedures. The CV profiles of the iron redox probe obtained after formation of the polymeric film (MPPy and NPPy) clearly confirmed the presence of this polymer on the C-SPE (Figure 5). This was confirmed by the increasing of the overall area of the CV voltammograms after formation of the polymeric layer. This behavior also evidenced the presence of a more capacitive surface.

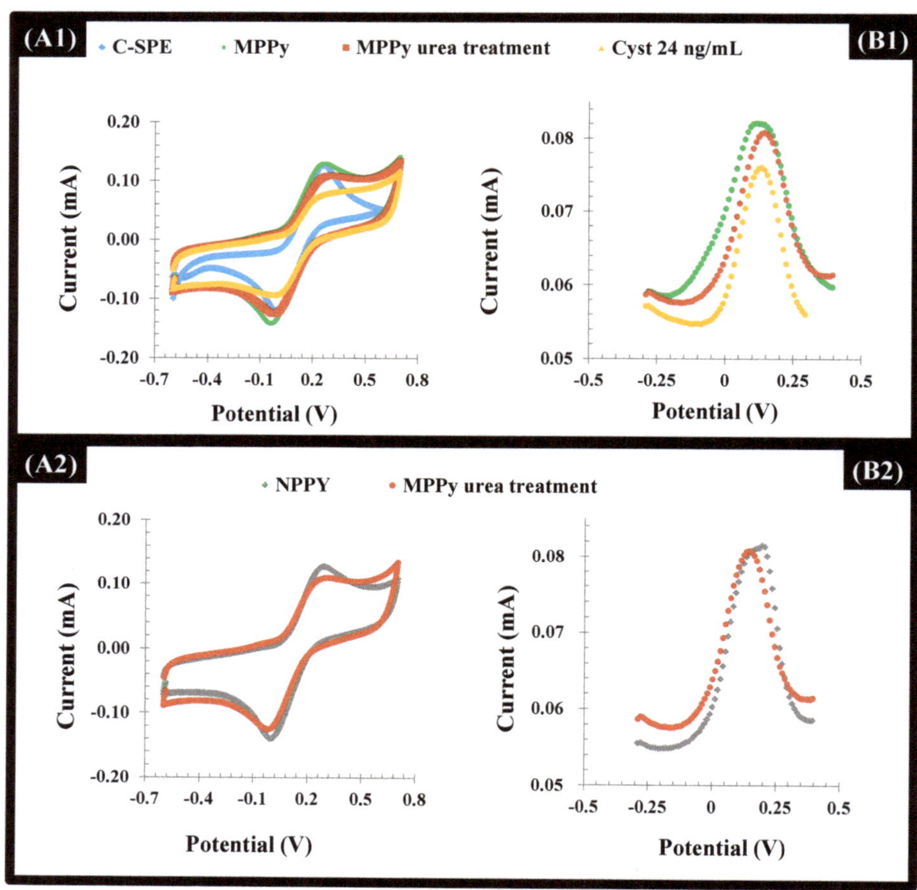

Figure 5. CV (**A**) and DPV (**B**) voltammograms of a 5.0 mmol/L [Fe (CN)6]$^{3-}$/$^{4-}$ solution in KCl 0.1 mol/L corresponding to the assembly of the MPPy and NPPy devices, in the several stages of this process (the blank C-SPE, the electropolymerization on top of it and the subsequent urea treatment), along with the incubation of Cys-C solution (24 ng/mL) on the MPPy film. (**A1** and **A2**)-CV measurements; (**B1** and **B2**):-DPV measurements.

After the template removal, the overall net current decreased in both the MPPy and NPPy (Figure 5). Although the removal of the Cys-C occurred only in the MPPy, the impact of urea treatment upon electrochemical features of the polymer was more prominent, resulting in insignificant changes for both. This behavior can be attributed to the anionic charge on the Py/PyCOOH that was strongly affected by acid treatment. However, biorecognizing cavities were preserved since only MPPy yielded decreasing net current after the Cys-C incubation. In addition to this, it became clear that after 3 h of incubation with the extraction solution, washing and incubation with the target protein, it was possible to observe a higher binding capacity of the MIP material, when compared with the NIP material. Longer periods of incubation were not tried for technical reasons.

3.5. Analytical Performance of the Sensor

The analytical performance of the biosensor was evaluated in acetate buffer, pH 6.0. This was carried out by incubating each standard solution for 20 min, washing it out and reading after the signal of the standard redox probe by squared-wave voltammetry (SWV). The time given for incubation was typically set from 15 to 30 min. In general, longer times may improve sensitivity as there is more time to allow Cys-C adsorption, while shorter times may improve selectivity as there is little time for non-specific adsorption [33,34]. Therefore, as a compromise, without further experiments for this selection, the incubation time was set to an intermediate value of 20 min.

The different Cys-C standard solutions were incubated consecutively in increasing concentrations, up to 30 ng/mL, and the voltammograms obtained are shown in Figure 6 (left). The peaks of the redox probe were centered at 0.15 V and showed decreasing currents (I) for increasing Cys-C concentrations.

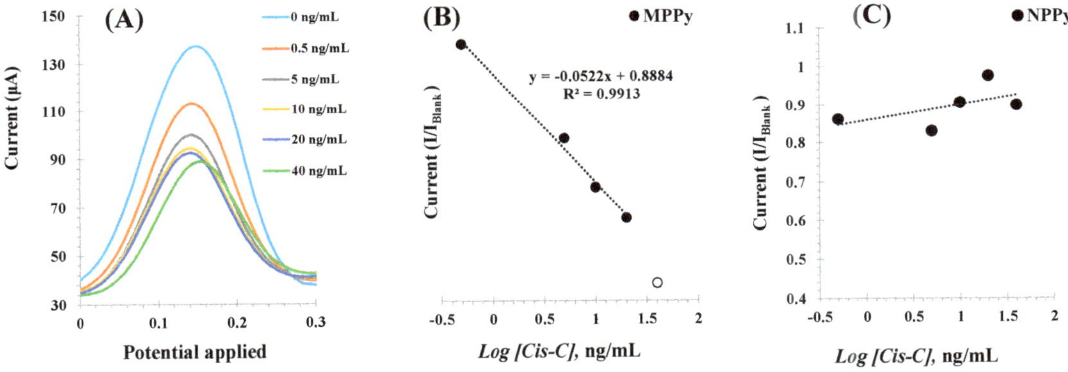

Figure 6. SWV voltammograms (**A**) corresponding to the incubation of increasing concentrations of Cys-C (in ng/mL) and the electrochemical signal obtained with a 5.0 mmol/L [Fe(CN)$_6$]$^{3-/4-}$ solution prepared in KCl 0.1 mol/L, and the corresponding calibration curves of the MPPy (**B**) and NPPy (**C**) devices. Assays performed in triplicate.

The calibrations plotted current (I) responses as a function of the logarithm of Cys-C concentrations, as shown in Figure 6B Under optimum conditions, the MPPy sensor displayed a dynamic response range between 0.5 and 30.0 ng/mL, with a limit of detection (LOD) of 0.5 ng/mL. The typical linear equation was current (mA) = −0.0021 × [Log (Cys-C, ng/mL)] + 0.0024, with a squared correlation coefficient of 0.972 and a standard deviation of repeated assays < 5%.

Identical calibration procedures were performed with the NPPy control films, to assess the dimension of non-specific responses Figure 6C. As these films were prepared without the target protein, there were no binding sites available and any interaction with Cys-C would reveal the presence of a non-specific interaction with the polymer. The results obtained are shown in Figure 6Aand showed a random behavior of the NPPy film in the same range of protein concentration. Thus, this behavior demonstrated that within the concentration range studied, the response of the MPPy was being controlled by the interaction of Cys-C with the imprinted binding sites and a non-specific response was not observed. The reproducibility and repeatability were less than 10%.

3.6. Selectivity and Application

Selectivity was assessed by checking the analytical response of the MPPy devices on a background of diluted spiked Cormay® HN serum, which corresponds to human serum from normal individuals, spiked with specific interfering compounds. The interfering compounds added were creatine kinase-MB 0.2g/L, ascorbic acid 0.15g/L, creatinine 1g/L and bovine serum albumin 12g/L. In general, negligible changes were found when the MIP sensor was incubated with serum samples when compared with blank signal from the buffer. This diluted serum was also spiked with Cys-C to check the ability of the system to respond with accuracy. The electrochemical data were obtained in single sample analysis and the resulting concentration values were extracted from the calibration curve. The spiked samples were prepared in two different Cys-C concentrations, equal to 2.0 and 5.0 ng/mL (Table 1). This analysis was performed in triplicate. The obtained data confirmed a good correlation between added and found amounts of Cys-C, with recovery values ranging from 88 to 99% and standard errors < 5%. This confirmed the accuracy and the reproducibility of the analytical responses.

Table 1. Analytical data obtained with the MPPy biosensor with diluted serum spiked with Cys-C standard solutions.

Sample	Added (ng/mL)	Found (ng/mL)	Recovery (%)	RSD (%)
1	2.0	1.76	87.8	2.20
2	5.0	4.93	98.6	1.42

4. Conclusions

This work produced a selective and stable MIP-based biosensor for the detection of the kidney biomarker Cys-C at PoC. The MPPy was assembled with a quick and simple procedure, yielding good reproducibility, accuracy, and linearity. The use of Py monomers ensured good electrocatalytic properties of the electrode, which were expected to enhance the sensitivity of the electrochemical system. Moreover, the introduction of MWCNTs in the PPy matrix was strategically explored to form a more porous and larger surface, allowing the formation of MPPy 3D structures with good sensitivity. The rebinding ability of this biosensor device was unique when compared to natural biomolecules because it offers high stability and low cost with similar performances. Moreover, this is a disposable device, not tested for regeneration, with suitable features for PoC use.

Overall, this method surpasses previously used methods for monitoring Cys-C in serum, offering faster execution and lower cost. The LODs achieved are of interest for clinical applications, allowing the PoC detection of cystatin C. Moreover, this approach may also be translated to other protein biomarkers.

Author Contributions: Conceptualization, F.T.C.M. and R.F.D.; methodology, R.S.G. and B.A.G.-R.; validation, R.S.G.; formal analysis, M.G.F.S.; investigation, R.S.G. and R.F.; resources, F.T.C.M., M.G.F.S. and R.F.D.; writing—original draft preparation, R.S.G. and F.T.C.M.; writing—review and editing, R.S.G., F.T.C.M., M.G.F.S., R.F.D. and R.F.; supervision, F.T.C.M. and R.F.D.; project administration R.F.D.; funding acquisition, R.F.D., F.T.C.M. and M.G.F.S. All authors have read and agreed to the published version of the manuscript.

Funding: This research was funded by POCTEP/FEDER, grant number 0624_2IQBIONEURO_6_E", by CNPQ (National Council for Scientific and Technological Development, Brazil), grant number 440605/2016-4 and by FCT (Fundação para a Ciência e Tecnologia, Portugal) with the Grant number SAICT-POL/24325/2016.

Institutional Review Board Statement: Not applicable. The study was conducted without the involvement of human or animals subjects. Also, no human or animal cells or tissues were used in the present work.

Informed Consent Statement: Not applicable.

Data Availability Statement: Not applicable.

Conflicts of Interest: The authors declare no conflict of interest.

References

1. Luyckx, V.A.; Tonelli, M.; Stanifer, J.W. The global burden of kidney disease and the sustainable development goals. *Bull. World Health Organ.* **2018**, *96*, 414. [CrossRef] [PubMed]
2. Onopiuk, A.; Tokarzewicz, A.; Gorodkiewicz, E. Cystatin C: A Kidney Function Biomarker. *Adv. Clin. Chem.* **2015**, *68*, 57–69.
3. Nejat, M.; Pickering, J.W.; Walker, R.J.; Westhuyzen, J.; Shaw, G.M.; Frampton, C.M.; Endre, Z.H. Urinary cystatin C is diagnostic of acute kidney injury and sepsis, and predicts mortality in the intensive care unit. *Crit. Care* **2010**, *14*, 1–13. [CrossRef] [PubMed]
4. Jha, V.; Garcia-Garcia, G.; Iseki, K.; Li, Z.; Naicker, S.; Plattner, B.; Saran, R.; Wang, A.Y.M.; Yang, C.W. Chronic kidney disease: Global dimension and perspectives. *Lancet* **2013**, *382*, 260–272. [CrossRef]
5. Fox, J.A.; Dudley, A.G.; Bates, C.; Cannon, G.M. Cystatin C as a Marker of Early Renal Insufficiency in Children with Congenital Neuropathic Bladder. *J. Urol.* **2014**, *191*, 1602–1607. [CrossRef]
6. Ravn, B.; Prowle, J.R.; Martensson, J.; Martling, C.R.; Bell, M. Superiority of Serum Cystatin C Over Creatinine in Prediction of Long-Term Prognosis at Discharge From ICU. *Crit. Care Med.* **2017**, *45*, E932–E940. [CrossRef] [PubMed]
7. Shlipak, M.G.; Matsushita, K.; Arnlov, J.; Inker, L.A.; Katz, R.; Polkinghorne, K.R.; Rothenbacher, D.; Sarnak, M.J.; Astor, B.C.; Coresh, J.; et al. Cystatin C versus Creatinine in Determining Risk Based on Kidney Function. *N. Engl. J. Med.* **2013**, *369*, 932–943. [CrossRef]
8. Shlipak, M.G.; Mattes, M.D.; Peralta, C.A. Update on Cystatin C: Incorporation Into Clinical Practice. *Am. J. Kidney Dis.* **2013**, *62*, 595–603. [CrossRef] [PubMed]
9. Trojanowicz, M. Impact of nanotechnology on design of advanced screen-printed electrodes for different analytical applications. *Trends Anal. Chem.* **2016**, *84*, 22–47. [CrossRef]
10. Ferreira, P.A.B.; Araujo, M.C.M.; Prado, C.M.; de Lima, R.A.; Rodriguez, B.A.G.; Dutra, R.F. An ultrasensitive Cystatin C renal failure immunosensor based on a PPy/CNT electrochemical capacitor grafted on interdigitated electrode. *Colloids Surf. B Biointerfaces* **2020**, *189*, 1–8. [CrossRef] [PubMed]
11. Devi, K.S.S.; Krishnan, U.M. Microfluidic electrochemical immunosensor for the determination of cystatin C in human serum. *Microchim. Acta* **2020**, *187*, 1–12. [CrossRef] [PubMed]
12. Mi, L.; Wang, P.Y.; Yan, J.R.; Qian, J.; Lu, J.S.; Yu, J.C.; Wang, Y.Z.; Liu, H.; Zhu, M.; Wan, Y.K.; et al. A novel photoelectrochemical immunosensor by integration of nanobody and TiO2 nanotubes for sensitive detection of serum cystatin C. *Anal. Chim. Acta* **2016**, *902*, 107–114. [CrossRef]
13. Hassanain, W.A.; Izake, E.L.; Ayoko, G.A. Spectroelectrochemical Nanosensor for the Determination of Cystatin C in Human Blood. *Anal. Chem.* **2018**, *90*, 10843–10850. [CrossRef]
14. Li, Y.J.; Wang, Y.H.; Bai, L.J.; Lv, H.Y.; Huang, W.; Liu, S.C.; Ding, S.J.; Zhao, M. Ultrasensitive electrochemiluminescent immunosensing based on trimetallic Au-Pd-Pt/MoS2 nanosheet as coreaction accelerator and self-enhanced ABEI-centric complex. *Anal. Chim. Acta* **2020**, *1125*, 86–93. [CrossRef] [PubMed]
15. Zhao, M.; Bai, L.J.; Cheng, W.; Duan, X.L.; Wu, H.P.; Ding, S.J. Monolayer rubrene functionalized graphene-based eletrochemiluminescence biosensor for serum cystatin C detection with immunorecognition-induced 3D DNA machine. *Biosens. Bioelectron.* **2019**, *127*, 126–134. [CrossRef]
16. Bleher, O.; Ehni, M.; Gauglitz, G. Label-free quantification of cystatin C as an improved marker for renal failure. *Anal. Bioanal. Chem.* **2012**, *402*, 349–356. [CrossRef] [PubMed]
17. Desai, D.; Kumar, A.; Bose, D.; Datta, M. Ultrasensitive sensor for detection of early stage chronic kidney disease in human. *Biosens. Bioelectron.* **2018**, *105*, 90–94. [CrossRef] [PubMed]
18. Frasco, M.F.; Truta, L.; Sales, M.G.F.; Moreira, F.T.C. Imprinting Technology in Electrochemical Biomimetic Sensors. *Sensors* **2017**, *17*, 523. [CrossRef]
19. Irshad, M.; Iqbal, N.; Mujahid, A.; Afzal, A.; Hussain, T.; Sharif, A.; Ahmad, E.; Athar, M.M. Molecularly Imprinted Nanomaterials for Sensor Applications. *Nanomaterials* **2013**, *3*, 615–637. [CrossRef]
20. Moreira, F.T.C.; Dutra, R.A.F.; Noronha, J.P.C.; Cunha, A.L.; Sales, M.G.F. Artificial antibodies for troponin T by its imprinting on the surface of multiwalled carbon nanotubes: Its use as sensory surfaces. *Biosens. Bioelectron.* **2011**, *28*, 243–250. [CrossRef]
21. Hoefer, M.; Bandaru, P.R. Determination and enhancement of the capacitance contributions in carbon nanotube based electrode systems. *Appl. Phys. Lett.* **2009**, *95*, 1–4. [CrossRef]
22. Canobre, S.C.; Xavier, F.F.S.; Fagundes, W.S.; de Freitas, A.C.; Amaral, F.A. Performance of the Chemical and Electrochemical Composites of PPy/CNT as Electrodes in Type I Supercapacitors. *J. Nanomater.* **2015**, *2015*, 1–14. [CrossRef]
23. Moreira, F.T.C.; Sharma, S.; Dutra, R.A.F.; Noronha, J.P.C.; Cass, A.E.G.; Sales, M.G.F. Smart plastic antibody material (SPAM) tailored on disposable screen printed electrodes for protein recognition: Application to myoglobin detection. *Biosens. Bioelectron.* **2013**, *45*, 237–244. [CrossRef] [PubMed]
24. Fan, X.; Yang, Z.W.; He, N. Hierarchical nanostructured polypyrrole/graphene composites as supercapacitor electrode. *RSC Adv.* **2015**, *5*, 15096–15102. [CrossRef]
25. Moreira, F.T.C.; Truta, L.; Sales, M.G.F. Biomimetic materials assembled on a photovoltaic cell as a novel biosensing approach to cancer biomarker detection. *Sci. Rep.* **2018**, *8*, 1–11. [CrossRef] [PubMed]
26. Wang, J.; Zhang, S.; Zhou, J.; Liu, R.; Du, R.; Xu, H.; Liu, Z.; Zhang, J.; Liu, Z. Identifying sp-sp(2) carbon materials by Raman and infrared spectroscopies. *Phys. Chem. Chem. Phys.* **2014**, *16*, 11303–11309. [CrossRef]

27. Debiemme-Chouvy, C.; Tran, T.T.M. An insight into the overoxidation of polypyrrole materials. *Electrochem. Commun.* **2008**, *10*, 947–950. [CrossRef]

28. Fang, Y.; Xu, L.; Wang, M.D. High-Throughput Preparation of Silk Fibroin Nanofibers by Modified Bubble-Electrospinning. *Nanomaterials* **2018**, *8*, 471. [CrossRef]

29. Afzal, A.; Abuilaiwi, F.A.; Habib, A.; Awais, M.; Waje, S.B.; Atieh, M.A. Polypyrrole/carbon nanotube supercapacitors: Technological advances and challenges. *J. Power Sources* **2017**, *352*, 174–186. [CrossRef]

30. Popovic, T.; Brzin, J.; Ritonja, A.; Turk, V. Different forms of human cystatin-c. *Biol. Chem. Hoppe-Seyler* **1990**, *371*, 575–580. [CrossRef]

31. Çuhadar, S. Serum Cystatin C as a Biomarker. In *Biomark. Kidney Disease*; Biomarkers in Disease: Methods, Discoveries and Applications; Patel, V., Preedy, V., Eds.; Springer: Dordrech, The Netherlands, 2016; pp. 445–461.

32. Silva, B.V.M.; Rodriguez, B.A.G.; Sales, G.F.; Sotomayor, M.D.T.; Dutra, R.F. An ultrasensitive human cardiac troponin T graphene screen-printed electrode based on electropolymerized-molecularly imprinted conducting polymer. *Biosens. Bioelectron.* **2016**, *77*, 978–985. [CrossRef] [PubMed]

33. Gomes, R.; Moreira, F.T.C.; Fernandes, R.; Sales, M.G.F. Sensing CA 15-3 in point-of-care by electropolymerizing O-phenylenediamine (oPDA) on Au-screen printed electrodes. *PLoS ONE* **2018**, *13*, e0196656. [CrossRef] [PubMed]

34. Ribeiro, J.A.; Pereira, C.M.; Silva, A.F.; Goreti, M.; Sales, F. Disposable electrochemical detection of breast cancer tumour marker CA 15-3 using poly(Toluidine Blue) as imprinted polymer receptor. *Biosens. Bioelectron.* **2018**, *109*, 246–254. [CrossRef] [PubMed]

biosensors

Article

Display of Microbial Glucose Dehydrogenase and Cholesterol Oxidase on the Yeast Cell Surface for the Detection of Blood Biochemical Parameters

Shiyao Zhao [1,†], Dong Guo [2,†], Quanchao Zhu [1], Weiwang Dou [1] and Wenjun Guan [1,*]

1   Institute of Pharmaceutical Biotechnology and the Children's Hospital, Zhejiang University School of Medicine, Hangzhou 310012, China; zhaoshiyao@zju.edu.cn (S.Z.); 11818296@zju.edu.cn (Q.Z.); douww@zju.edu.cn (W.D.)

2   College of Pharmaceutical Sciences, Zhejiang University, Hangzhou 310012, China; Y215190015@zju.edu.cn

*   Correspondence: guanwj@zju.edu.cn; Tel.: +86-0571-88206477

†   These authors contributed equally to this work.

Abstract: High levels of blood glucose are always associated with numerous complications including cholesterol abnormalities. Therefore, it is important to simultaneously monitor blood glucose and cholesterol levels in patients with diabetes during the management of chronic diseases. In this study, a glucose dehydrogenase from *Aspergillus oryzae* TI and a cholesterol oxidase from *Chromobacterium* sp. DS-1 were displayed on the surface of *Saccharomyces cerevisiae*, respectively, using the yeast surface display system at a high copy number. In addition, two whole-cell biosensors were constructed through the immobilization of the above yeast cells on electrodes, for electrochemical detection of glucose and cholesterol. The assay time was 8.5 s for the glucose biosensors and 30 s for the cholesterol biosensors. Under optimal conditions, the cholesterol biosensor exhibited a linear range from 2 to 6 mmol·L^{-1}. The glucose biosensor responded efficiently to the presence of glucose at a concentration range of 20–600 mg·dL^{-1} (1.4–33.3 mmol·L^{-1}) and showed excellent anti-xylose interference properties. Both biosensors exhibited good performance at room temperature and remained stable over a three-week storage period.

Keywords: whole-cell biosensor; yeast surface display; cholesterol oxidase; glucose dehydrogenase; electrochemical detection

Citation: Zhao, S.; Guo, D.; Zhu, Q.; Dou, W.; Guan, W. Display of Microbial Glucose Dehydrogenase and Cholesterol Oxidase on the Yeast Cell Surface for the Detection of Blood Biochemical Parameters. *Biosensors* **2021**, *11*, 13. https://doi.org/10.3390/bios11010013

Received: 20 November 2020
Accepted: 28 December 2020
Published: 30 December 2020

Publisher's Note: MDPI stays neutral with regard to jurisdictional clai-ms in published maps and institutio-nal affiliations.

1. Introduction

Diabetes is currently a global epidemic with over 400 million cases and the prevalence of which is rapidly increasing [1,2]. The disease is associated with numerous complications, especially cardiovascular diseases. Indices of blood cholesterol are often used as the thresholds for risk assessment and guides to therapy [3–5]. Therefore, effective monitoring of blood glucose and cholesterol plays a crucial role in the management of diabetes. Currently, large biochemical analyzers are the mainstream methods of detecting diabetes related blood parameters in clinical practice [6–8]. However, the approach requires specialized instruments and complicated experimental procedures [9] although it is highly reliable and gives high precision. With increasing demand for point-of-care testing (POCT), biosensors have been proposed as an attractive alternative as they are quick, convenient and economically feasible [10]. Among them, the optical and electrochemical based POCT biosensors gradually become the leader choices [11]. Because significant turbidity of most real samples and strong light from environment always bring errors, the detection accuracy of optical biosensors is easily interfered [12]. Hence, electrochemical biosensors are more favorable in practical applications [13]. At present, the most well-known brands of glucose POCT biosensors based on electrochemical technology on the market are Roche, Johnson & Johnson, Bayer, Abbott, et al.

259

Enzyme-based biosensors represent the most popular group of electrochemical biosensors currently available [14]. Notably, glucose oxidase or glucose dehydrogenase is the most widely used enzyme for glucose detection [15,16]. Glucose dehydrogenase forms a complex with cofactors such as flavin adenine dinucleotide (FAD), nicotinamide adenine dinucleotide (NAD), or pyrroloquinoline quinone (PQQ) [17]. Compared with glucose oxidase and other kinds of glucose dehydrogenases, the FAD-dependent glucose dehydrogenase, predominantly found in *Aspergillus*, has obvious sensing advantages due to its favorable substrate specificity and insensitivity to oxygen [18]. On the other hand, cholesterol oxidase, which is a FAD-dependent oxidoreductase derived from bacteria, is most widely utilized for the detection of cholesterol [19]. Nonetheless, the high cost of purification and poor stability of the free enzymes still present a challenge for the large-scale production and practical application of these enzyme-based biosensors [20]. Therefore, electrochemical whole-cell biosensors based on the surface display technology have been the focus of several studies since they may provide an effective way of overcoming the challenges associated with their enzyme-based counterparts [21].

Surface display is a powerful technology that allows the presentation of multiple proteins on the surface of microbes such as bacteria and yeast [22]. For instance, previous studies displayed a *Bacillus subtilis* derived NAD-dependent glucose dehydrogenase and its mutants on the surface of *Escherichia coli* using the ice nucleation protein (INP) as an anchoring motif [23,24]. However, no report currently exists on microbial cell-surface display of cholesterol oxidase. The yeast surface display (YSD) system is more advantageous in displaying complex eukaryotic proteins which require post-translational modification or have high molecular mass [25]. The most common yeast display system, pioneered by Boder and Wittrup, employs the Aga1 and Aga2 subunits of a-agglutinin to anchor the target enzymes onto the cell wall of *Saccharomyces cerevisiae* [26]. The anchoring of enzymes on the yeast cell surface allows for direct enzymatic reaction with substrates without the need for purification of enzymes and this can significantly reduce the cost of preparation as well as application of enzyme-based biosensors [27]. Moreover, the surface of yeast cells provides a biocompatible microenvironment that helps in maintaining the stability of enzymes [28]. The above advantages indicate that displaying glucose dehydrogenase and cholesterol oxidase on the surface of yeast cells may potentially be useful in the development of electrochemical biosensing platforms for the detection of blood glucose and cholesterol.

In this study, a FAD-dependent glucose dehydrogenase gene from *Aspergillus oryzae* TI [29] and a cholesterol oxidase gene from *Chromobacterium* sp. DS-1 [30] were cloned and the corresponding enzymes were displayed on the surface of *S. cerevisiae* through the Aga1-Aga2 system, respectively. Afterwards, the enzyme-displayed yeast cells were immobilized onto electrodes to construct electrochemical biosensors for the detection of glucose and cholesterol. The catalytic activity of the surface-displayed enzymes was measured carefully and the performance of the two related biosensors was evaluated subsequently. The results demonstrated that this strategy had the advantages of simplicity, economic feasibility, and stability in the application of glucose and cholesterol biosensors.

2. Materials and Methods

2.1. Strains, Media and Reagents

The strains used in this study are described in Table S1. *E. coli* TG1 [31] was used for recombinant DNA manipulations and was cultured in Luria-Bertani (LB) medium at 37 °C. *S. cerevisiae* EBY100 [26] was obtained from Ziyun Biotech (Hangzhou, Zhejiang, China) and used for yeast cell surface display. The *S. cerevisiae* EBY100 was first grown in a seed medium containing 0.67% yeast nitrogen base, 0.5% casamino acid and 2% glucose before being transferred into the induction medium for surface display of the enzymes. The composition of induction medium was the same as that of the seed medium only that glucose was replaced with galactose. The protein expression was induced at 20 °C with continuous shaking at 220 rpm. Restriction enzymes were purchased from Takara Bio

(Shiga, Japan) while all the other biochemical reagents were of at least analytical grade and purchased from Merck (Darmstadt, Germany), Sangon Biotech (Shanghai, China), Aladdin (Shanghai, China) or Sinopharm Chemical Reagent (Shanghai, China).

2.2. Construction of Vectors

All the vectors used in this study are listed in Table S1. The cholesterol oxidase gene derived from *Chromobacterium* sp. DS-1 (named *CHO1*, Sequence ID: AB456533.1) and glucose dehydrogenase gene derived from *Aspergillus oryzae* TI (named *GDH1*, Sequence ID: XM_002372558.1) were codon optimized and synthesized by Generay Biotech (Shanghai, China). To construct the vectors pYD1-CHO1 and pYD1-GDH1, the synthetic genes were digested with *Bam*HI and *Eco*RI before being ligated into the multiple cloning site of vector pYD1 [32]. In addition, the extended linkers were prepared by annealing synthetic complementary single-stranded DNAs of the target sequences followed by PCR. The linkers added one or two proline-alanine-serine (PAS) sequences (ASPAAPAPASPAAPAPSAPA) to the original GS linker (GGGGSGGGGSGGGGS) in pYD1-CHO1. Thereafter, the amplification products were cloned into the *Hind*III and *Bam*HI sites of vector pYD1-CHO1 to generate pYD1-CHO1-PASx1 and pYD1-CHO1-PASx2. All the resulting vectors were verified through PCR and sequencing.

2.3. Freeze-Drying of Yeast Cells

After induction, the enzyme-displayed yeast cells harboring vector pYD1-CHO1 or pYD1-GDH1 were harvested through centrifugation, washed with 1 × phosphate buffered saline (PBS, pH 7.4) and resuspended in the cryoprotectant buffer which consisted of 1 × PBS and 5% glycerol. Afterwards, the cell suspensions were freeze-dried using a freeze dryer (LGJ-10, Henan Brother Equipment Co., Ltd., Zhengzhou, China). The freeze-dried cell samples were then stored at 4 °C for later use.

2.4. Enzyme Activity Assays

Glucose dehydrogenase activity of the Gdh1-displayed yeast cells was assessed using 2,6-dichlorophenol-indophenol (DCPIP) and phenazine methosulfate (PMS) according to a previously published protocol [33]. Briefly, the working solution was configured, and the final concentration of glucose was 201 mmol·L^{-1}. 1.5 mL of the working solution was equilibrated at 37 °C for about 5 min, and then the 20 OD$_{600}$ freeze-dried cell samples were resuspended in this solution to initiate the reaction. During a 5-min reaction, a decrease in the optical density of the supernatant was measured at 600 nm using a spectrophotometer, with water as the reference. Finally, change in absorbance per minute was used to calculate enzyme activity.

Cholesterol oxidase activity of the Cho1-displayed yeast cells was determined through the oxidative coupling of 4-aminoantipyrine and phenol as previously described [34]. Configuration of the working solution was performed according to this method and the final concentration of cholesterol in the solution was 0.89 mmol·L^{-1}. After incubating the 1.5 mL working solution at 37 °C for 5 min, the 20 OD$_{600}$ freeze-dried cell samples were resuspended with this solution to start the reaction. The amount of H$_2$O$_2$ catalyzed by cholesterol oxidase was then calculated by measuring the increase in OD$_{500}$ of the supernatant per minute, to define the value of enzyme activity.

2.5. Fabrication of the Whole-Cell Biosensors

The biosensor used in this study consisted of a two-electrode system made from carbon paste or gold. The carbon paste-based two-electrode strips were formed by successively printing silver ink, carbon ink and insulating ink on a polyethylene terephthalate (PET) material through the screen-printing technique. The gold two-electrode strips were purchased from Jinhong Technology (Beijing, China). The size of the reaction chamber for each two-electrode strip was 5 mm long, 1.95 mm wide and 0.125 mm high. The 20 OD$_{600}$ freeze-dried YSD cells were resuspended in 50 μL of the respective buffers before adding

0.05 g of FAD to mix. The YSD cell solution and electrochemical solution which contained microcrystalline cellulose, polyvinylpyrrolidone, octyl polyethylene glycol phenyl ether, phenazine ethosulfate, trehalose and the electron mediator (potassium ferricyanide or hexaammineruthenium (III) chloride), were mixed in a ratio of 1:4. The reaction chamber of each two-electrode strip was spotted with 1 μL of the above mixed solution and dried at 30 °C for 20 min. In theory, each resulting biosensor contained approximately 0.08 OD_{600} YSD cells. The mechanism of detection is described in Figure S1, and a schematic illustration of the reaction chamber is shown in Figure S2. Performance of the screen-printed carbon electrodes was evaluated before the formal testing (Figure S3).

2.6. Preparation of Whole Blood Samples

Whole blood samples were collected into sodium heparin tubes to prevent hemolysis. The initial hematocrit (HCT) was determined by testing the volume of plasma and red blood cells after centrifugation before adjusting it to 42% by adding or removing plasma. The initial concentration of glucose in whole blood samples was measured using a YSI glucose analyzer (YSI2300, YSI Life Sciences, Yellow Springs, OH, USA). Glucose solutions were supplemented to obtain the desired concentrations in whole blood samples.

2.7. Electrochemical Measurements

All electrochemical measurements were performed using an electrochemical workstation (CHI660E, CH Instruments, Shanghai, China) or a portable electrochemical monitor (305A, Jiangsu Yuyue Medical Instruments Co., Ltd., Jiangsu, China). Amperometric method, in which a constant potential was applied to the working electrode and the current was measured after a certain reaction time, was used in this work. For the glucose biosensor, a DC voltage of 0.3 V was applied on the working electrode for 8.5 s and a final current value at 8.5 s was read for data analysis. In addition, for the cholesterol biosensor, a DC voltage of 0.3 V was applied on the working electrode for 30 s and a final current value at 30 s was used for data analysis.

3. Results

3.1. Surface Display of Glucose Dehydrogenase

Previous reports suggested that the glucose dehydrogenase derived from *Aspergillus oryzae* TI is a FAD-dependent oxidoreductase, which has the characteristics of high substrate specificity against glucose, excellent thermal stability and is not affected by dissolved oxygen [29]. This enzyme is composed of 593 amino acids, including a signal peptide ranging from 1–22 amino acids at the N-terminus [35]. To display Gdh1 on the surface of yeast, the signal peptide truncated gene sequence (1713 bp) of Gdh1 was synthesized according to the codon preference in *S. cerevisiae* and introduced into the multiple cloning site of vector pYD1. The resulting vector pYD1-GDH1 carrying the *AGA2-GDH1* fusion gene (Figure 1a) was then transformed into *S. cerevisiae* EBY100, which harbored Aga1 as a cell wall anchoring motif, to generate the G1 strain (Figure 1b). The expression of *AGA1* as well as *AGA2* was driven by the galactose-inducible *GAL1* promoter. The G1 strain was cultured in a 2% glucose medium to the mid-logarithmic growth phase, and then was transferred to a 2% galactose medium to induce the persistent expression of Gdh1. Afterwards, the G1 yeast cells were collected at the end of fermentation and freeze dried.

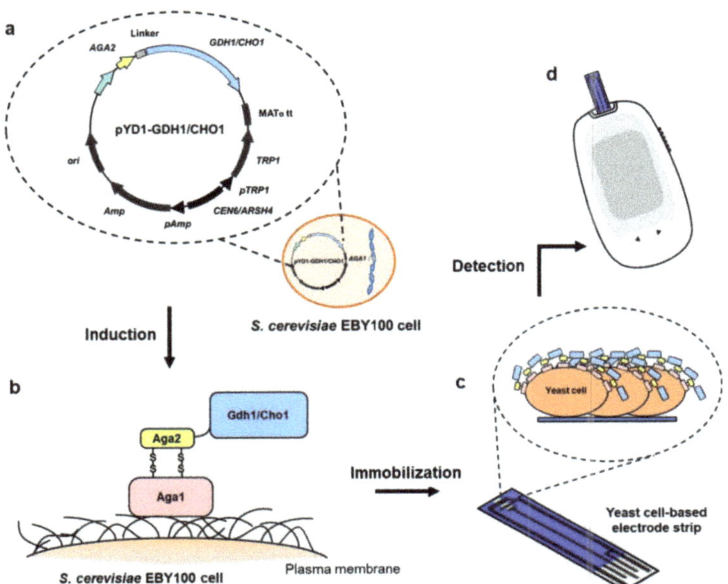

Figure 1. The strategy for constructing the surface-displayed glucose and cholesterol biosensors. (**a**) A map of the pYD1-GDH1 or pYD1-CHO1 vector. The enzyme Gdh1 or Cho1 (blue) was expressed as a C-terminal fusion to the Aga2 subunit of a-agglutinin (yellow) and connected by a linker (grey). (**b**) Displaying Gdh1 or Cho1 on the surface of yeast. The inducing Aga2 subunit associates with the a-agglutinin Aga1 subunit through two disulfide bonds. The Aga2-enzyme fusion protein was subsequently secreted to the extracellular space where Aga1 could be anchored to the cell wall through β-1, 6-glucan covalent linkage. (**c**) Fabrication of the whole-cell biosensors by immobilizing the YSD cells. (**d**) Electrochemical detection of glucose or cholesterol using a portable electrochemical monitor.

Maximum catalytic activity (0.7 U per OD_{600} cells) of the surface-displayed Gdh1 was observed after 48 h of galactose induction (Figure 2a). Therefore, 48 h was selected as the optimal induction time for the G1 strain and used for subsequent experiments. The effect of temperature, pH and storage time on the activity of Gdh1 was also evaluated. The results showed that Gdh1 was active from 10 to 60 °C and the activity reached a peak at 30 °C. In addition, the enzyme retained over 95% of its activity from 20 °C to 40 °C, indicating that it had good thermal stability and could be applied over a broad range of temperature (Figure 2b). The optimum pH for Gdh1 activity was observed to be 7.0. More than 75% of Gdh1 activity was maintained within the pH range of 6.5–7.5 (Figure 2c). However, the storage stability of the surface-displayed Gdh1 appeared to be limited. Activity of the enzyme began to decrease following storage at 4 °C for one week and retained just 20% of the initial activity after 3 weeks of storage (Figure 2d). This implied that the subsequent immobilization process should be performed as soon as possible after obtaining the freeze-dried G1 strain cells to avoid the loss of enzyme activity.

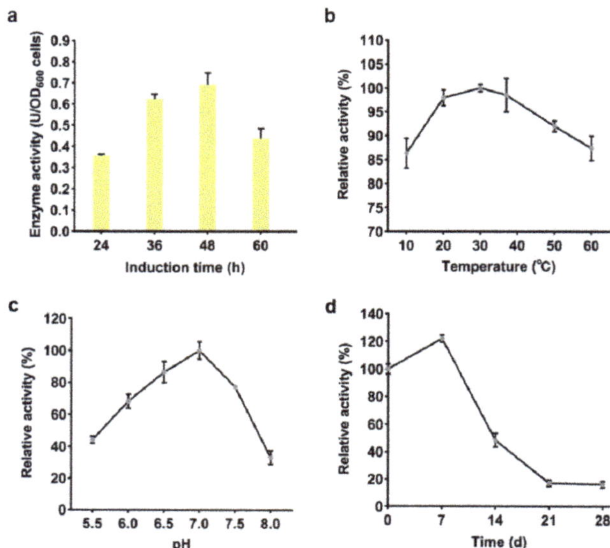

Figure 2. Determination of the catalytic activity of surface-displayed Gdh1. (**a**) Activity of the surface-displayed Gdh1 at different time points after galactose induction. (**b–d**) Effect of (**b**) temperature, (**c**) pH, and (**d**) storage time on the activity of Gdh1. Error bars indicate the SD of samples tested in triplicate.

3.2. Development of an Electrochemical Glucose Biosensor Based on the Surface-Displayed Gdh1

To develop a whole-cell glucose biosensor, the freeze-dried G1 yeast cells (0.08 OD_{600} cells per strip) with different artificial electron mediators and buffer solutions, were immobilized on the surface of screen-printed carbon electrodes. Electrochemical detection was then performed. Although the glucose biosensors exhibited a similar slope of the concentration-response current curve when different electron mediators were used, the glucose biosensor using hexaammineruthenium (III) chloride as electron mediator presented lower background current than that of potassium ferricyanide (Figure 3a). In addition, the impact of phosphate buffer, heppso-malic acid buffer and fumaric acid buffer was evaluated to examine which one of them was better suited for the glucose biosensor. The results showed that the biosensor based on the 0.1 $mol \cdot L^{-1}$ phosphate buffer (pH 7.0) system had a greater response current at a higher glucose concentration (Figure 3b). Consequently, under optimized electrode conditions, the biosensor was employed for the detection of glucose with successive additions. The linear detection range of the glucose biosensor was 20–600 $mg \cdot dL^{-1}$ (1.4–33.3 $mmol \cdot L^{-1}$) and the testing sensitivity was shown to be 25 $mg \cdot dL^{-1}$ at the concentration range of 20 to 600 $mg \cdot dL^{-1}$ (Figure 3c,d).

Xylose is the most common interferent to biosensors using glucose dehydrogenase. Therefore, the study examined the selectivity of the optimized glucose biosensor by adding extra xylose into the glucose substrate solution. Addition of xylose had almost no impact on the biosensor developed by this study. However, the biosensor immobilized with commercial FAD-dependent glucose dehydrogenase (581 $U \cdot mg^{-1}$, Code: GLD-361, Toyobo Co., Ltd., Osaka, Japan) was affected substantially. 189% and 29% of additional interference current was generated in 70 $mg \cdot dL^{-1}$ and 300 $mg \cdot dL^{-1}$ of glucose substrate solution after adding 90 $mg \cdot dL^{-1}$ of xylose respectively (Figure 3e). This indicated that the optimized glucose biosensor had a good ability to get rid of xylose interference and great potential for practical application.

The biosensors were stored at room temperature for 21 days and changes in current responses towards 280 $mg \cdot dL^{-1}$ of glucose were measured. The findings revealed that the biosensors retained 79% of the initial response current after storage for 21 days (Figure 3f).

Furthermore, the biosensor was used to determine the concentration of glucose in whole blood samples to explore the possibility of clinical application. The results showed that the linear response range of the biosensor to glucose in whole blood samples was 20–600 mg·dL^{-1} (Figure 3g). Although at present the accuracy of the glucose biosensor did not completely meet the requirements of the standard [36], it exhibited the potential of detecting real samples (Supplementary Figure S4, Table S3).

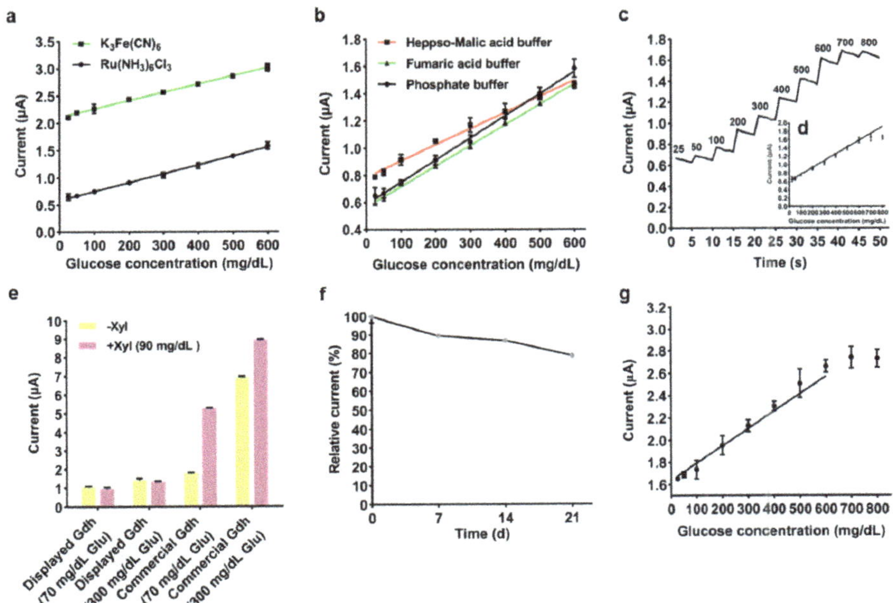

Figure 3. Characterization of the whole-cell glucose biosensor. Effect of (**a**) electron mediator and (**b**) buffer on the glucose biosensor. (**c**) Amperometric response for successive addition of 25, 50 and 100 mg·dL^{-1} of glucose. (**d**) The plot of linear regression. The (**e**) anti-xylose performance and (**f**) storage stability of the optimized glucose biosensor. (**g**) The response current of the optimized biosensor to glucose in whole blood samples. Electrochemical detection parameters: voltage 0.3 V, acquisition time 8.5 s. Error bars indicate the SD of samples tested in triplicate.

3.3. Surface Display of Cholesterol Oxidase

The strategy for developing a glucose biosensor based on YSD cells proved to be feasible and showed potential for practical application in monitoring glucose levels in whole blood samples. Therefore, the study used a similar approach for the detection of cholesterol. Previous studies reported that the cholesterol oxidase derived from *Chromobacterium* sp. DS-1 had favorable protein structural characteristics [37], excellent stability [30] and high enzyme activity [38]. This enzyme consists of 584 amino acids with the first 44 being a signal peptide [38]. In the present study, the codon optimized gene sequence (1620 bp) encoding Cho1 without the signal peptide was synthesized and the resulting vector pYD1-CHO1 was constructed (Figure 1a). Thereafter, pYD1-CHO1 was transformed into *S. cerevisiae* EBY100 to generate the C1 strain, which successfully displayed Cho1 on the surface of yeast (Figure 1b). It was observed that the C1 strain had a significantly lower growth rate compared with the G1 strain or the P1 strain containing the empty vector pYD1. This implied that displaying Cho1 on the surface might impose a growth burden to the cells (Figure S5).

The optimal induction time for the C1 strain was shown to be 36 h, at which enzyme activity reached the maximum value of 3.4×10^{-3} U per 20 OD$_{600}$ freeze-dried cells in vitro

(Figure 4a). The optimum reaction temperature for the surface-displayed Cho1 was also found to be 30 °C. More than 60% of its activity occurred between 20 °C and 40 °C, while about 18% activity was retained at 60 °C (Figure 4b). Cho1 had an optimum pH of 6.5 and over 50% of its activity occurred in the pH range of 5.5 to 7.5 (Figure 4c). Afterwards, the freeze-dried C1 strain cells were stored at 4 °C and their residual activity determined intermittently within a one-month period, to investigate the stability of Cho1 in vitro. No significant decrease in activity was observed and approximately 95% of original enzyme activity could still be detected after the 4-week storage period. This suggested that the C1 strain cells had good storage stability (Figure 4d).

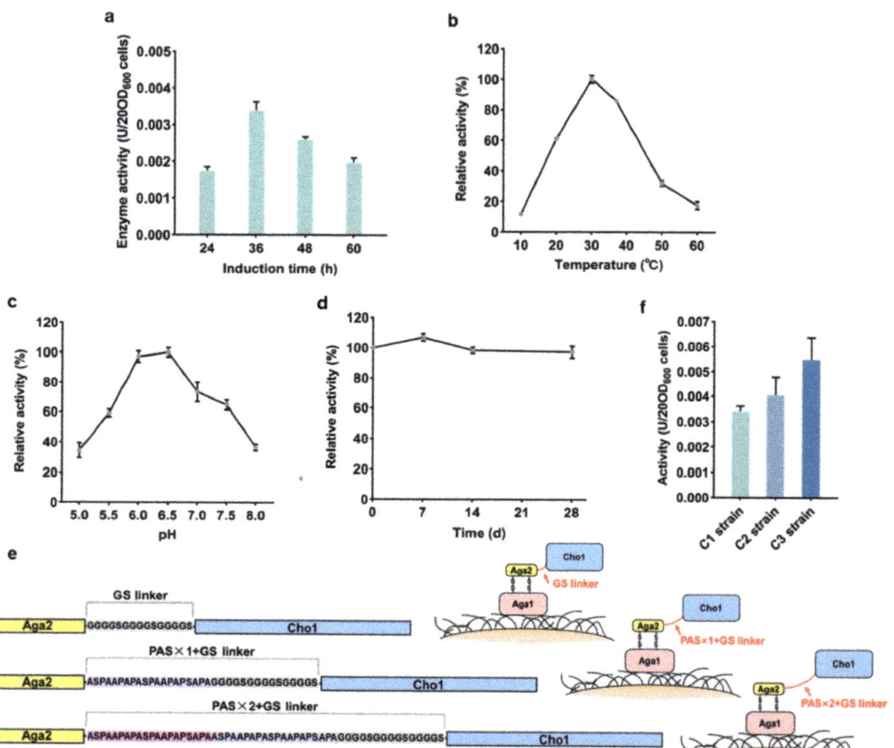

Figure 4. Determination of the catalytic activity of surface-displayed Cho1. (**a**) Activity of the displayed Cho1 at different time points after galactose induction. (**b–d**) Effect of (**b**) temperature, (**c**) pH and (**d**) storage time on the activity of Cho1. (**e**) Schematic representation of the linkers connecting Aga2 and Cho1. The original GS linker (GGGGSGGGGSGGGGS) was lengthened with one or two PAS sequences (ASPAAPAPASPAAPAPSAPA) to form the PAS × 1 + GS and PAS × 2 + GS linkers, respectively. (**f**) Activity of Cho1 with different linkers. Error bars indicate the SD of samples tested in triplicate.

The maximum activity of Cho1 was determined as 3.4×10^{-3} U per 20 OD_{600} freeze-dried yeast cells and this was almost two orders of magnitude lower than that of Gdh1. This implied that the activity of surface-displayed Cho1 had a limited potential to meet the requirements for practical application. Therefore, an attempt was made to improve its activity by increasing the length of linker sequence, which was located between *AGA2* and *CHO1* in the pYD1-CHO1 vector. Based on previous reports, the original GS linker was modified by adding one or two PAS linkers, which consisted of 20 amino acids including proline, alanine and serine, with the advantages of being hydrophilic, uncharged and structureless [39,40] (Figure 4e). The resulting vector pYD1-CHO1-PASx1 or pYD1-CHO1-PASx2 was then transformed into *S. cerevisiae* EBY100 to generate the C2 or C3 strains, respectively. The results showed that the displayed Cho1 with a PAS $\times$ 1 + GS linker had a 19% increase in activity, while the one with a PAS $\times$ 2 + GS linker exhibited a 62% increase, which was 5.5×10^{-3} U per 20 OD_{600} freeze-dried cells (Figure 4f). Owing to the longer linkers, the displayed enzyme obtained a greater distance to leave the cell surface, hence had enough conformational space, resulting in a decrease in the loss of enzymatic activity.

3.4. Development of an Electrochemical Cholesterol Biosensor Based on Surface-Displayed Cho1

The freeze-dried C3 yeast cells (0.08 OD_{600} cells per strip) with different artificial electron mediators and buffer solutions, were immobilized onto the surface of screen-printed carbon or gold electrodes, respectively, to develop a whole-cell cholesterol biosensor. Like that of the glucose biosensor, hexaammineruthenium (III) chloride had more advantages as an artificial electron mediator in this cholesterol biosensor, exhibiting a lower background current (Figure 5a). Moreover, the study compared the current response of the cholesterol biosensor in phosphate buffer, hepes buffer and TES buffer. As a result, the biosensor based on 0.1 $mol \cdot L^{-1}$ phosphate buffer (pH 6.5) exhibited a smaller background current and a larger difference in response current from 2 to 6 $mol \cdot L^{-1}$ concentration of cholesterol (Figure 5b). In addition, using a gold electrode instead of a carbon paste one not only slightly improved the detection sensitivity of the cholesterol biosensor, but also contributed to the larger background current (Figure 5c).

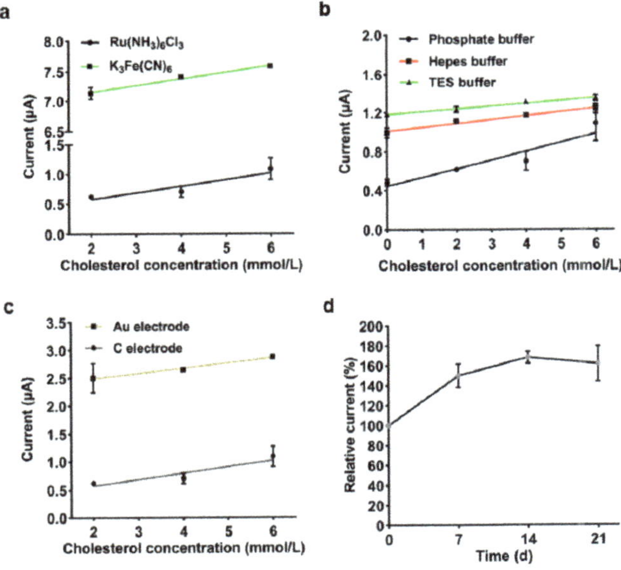

Figure 5. Characterization of the whole-cell cholesterol biosensor. (**a**–**c**) Effect of (**a**) electron mediator, (**b**) buffer and (**c**) electrode material on the cholesterol biosensor. (**d**) Storage stability of the optimized cholesterol biosensor. Electrochemical detection parameters: voltage 0.3 V, acquisition time 30 s. Error bars indicate the SD of samples tested in triplicate.

The cholesterol biosensors were stored at room temperature and the changes in current responses towards 4 $mol \cdot L^{-1}$ cholesterol were measured weekly to explore their storage stability. After 21 days of storage, the response currents of the biosensors began to gradually decrease, and the loss was less than 20% of the optimal response current (Figure 5d). It was also shown that the response currents of the biosensors increased during the first seven days of storage.

4. Discussion

The advantage of a surface-displayed enzyme is that the costly protein purification step can be skipped, and stability is effectively improved. In this study, two whole-cell biosensors based on YSD were developed for the detection of glucose and cholesterol. The results showed that the optimized glucose biosensor had a broad detection linear range of 20–600 mg·dL^{-1} while that of the cholesterol biosensor was 2–6 mmol·L^{-1}. This indicated that the biosensors had a great potential for clinical application in the future. Additionally, the strategy presented here may provide insights on the development of other biosensors based on the cell surface display technology.

Although the cholesterol oxidase (Cho1) used in this study was reported to have outstanding activity (Supplementary Table S2), the surface-displayed Cho1 did not show the anticipated enzyme activity. This resulted to a relatively poor detection capability of the biosensor when compared with Gdh1, manifesting as a narrow linear range. Screening the highly active Cho1 mutants by directed evolution offers an available approach to problem solving [41]. It is still unclear why displaying Cho1 on the surface of yeast resulted to slow growth of the C1 strain although this may have been responsible for its low activity (Supplementary Figure S5). Moreover, given that Cho1 was obtained from a bacterium (*Chromobacterium* sp. DS-1) but displayed on the surface of a fungus (*S. cerevisiae*), the mismatched protein folding system might have resulted to partial misfolding of Cho1 and subsequently low activity. Therefore, the identification of a highly active cholesterol oxidase from a fungus should be the focus of future studies.

Since yeast cells occupy more space, the amount of surface-displayed enzyme immobilized on the electrode surface was limited, leading to a lower response current compared to the purified enzymes (Figure 3e). Consequently, crushing the cells to collect pieces of the cell wall and directly immobilize them onto the surface of the electrode might solve this problem and will be the focus of future research.

Supplementary Materials: The following are available online at https://www.mdpi.com/2079-637 4/11/1/13/s1, Table S1: Strains and vectors used in this study, Table S2: Comparison of different cholesterol oxidases reported previously, Table S3: Accuracy evaluation of the glucose biosensor, Figure S1: The detection mechanism of glucose or cholesterol biosensor, Figure S2: Schematic diagram of the reaction chamber of the electrode strips, Figure S3: Performance evaluation of the screen-printed carbon electrodes, Figure S4: Accuracy evaluation of the glucose biosensor, Figure S5: Growth curve of the P1, G1 and C1 strains in the induction medium containing 2% galactose.

Author Contributions: Conceptualization, W.G.; methodology, S.Z. and D.G.; validation, S.Z. and D.G.; formal analysis, S.Z. and D.G.; investigation, S.Z., D.G., Q.Z., and W.D.; data curation, S.Z. and D.G.; writing—original draft preparation, W.G. and S.Z.; writing—review and editing, W.G. and S.Z.; supervision, W.G. All authors have read and agreed to the published version of the manuscript.

Funding: This research was funded by National Key Research and Development Program of China (2019YFA0905400).

Institutional Review Board Statement: Not applicable.

Informed Consent Statement: Not applicable.

Data Availability Statement: Not applicable.

Conflicts of Interest: The authors declare no conflict of interest.

References

1. World Health Organization (WHO). *Global Report on Diabetes*; WHO: Geneva, Switzerland, 2016.
2. Cho, N.; Shaw, J.; Karuranga, S.; Huang, Y.; Fernandes, J.D.R.; Ohlrogge, A.; Malanda, B. IDF Diabetes Atlas: Global estimates of diabetes prevalence for 2017 and projections for 2045. *Diabetes Res. Clin. Pract.* **2018**, *138*, 271–281.
3. Levelt, E.; Rodgers, C.T.; Clarke, W.T.; Mahmod, M.; Ariga, R.; Francis, J.M.; Liu, A.; Wijesurendra, R.S.; Dass, S.; Sabharwal, N.; et al. Cardiac energetics, oxygenation, and perfusion during increased workload in patients with type 2 diabetes mellitus. *Eur. Heart J.* **2016**, *37*, 3461–3469.

4. Anderson, T.J.; Grégoire, J.; Pearson, G.J.; Barry, A.R.; Couture, P.; Dawes, M.; Francis, G.A.; Genest, J.; Grover, S.; Gupta, M.; et al. 2016 Canadian Cardiovascular Society guidelines for the management of dyslipidemia for the prevention of cardiovascular disease in the adult. *Can. J. Cardiol.* **2016**, *32*, 1263–1282.

5. Grundy, S.M.; Stone, N.J.; Bailey, A.L.; Beam, C.; Birtcher, K.K.; Blumenthal, R.S.; Braun, L.T.; de Ferranti, S.; Faiella-Tommasino, J.; Forman, D.E.; et al. 2018 AHA/ACC/AACVPR/AAPA/ABC/ACPM/ADA/AGS/APhA/ASPC/ NLA/PCNA guideline on the management of blood cholesterol: Executive summary. *J. Am. Coll. Cardiol.* **2019**, *73*, 3168.

6. Chen, Y.; Liu, Q.; Yong, S.; Lee, T.K. High accuracy analysis of glucose in human serum by isotope dilution liquid chromatography-tandem mass spectrometry. *Clin. Chim. Acta* **2012**, *413*, 808–813. [PubMed]

7. Qureshi, R.N.; Kaal, E.; Janssen, H.-G.; Schoenmakers, P.J.; Kok, W.T. Determination of cholesterol and triglycerides in serum lipoproteins using flow field-flow fractionation coupled to gas chromatography-mass spectrometry. *Anal. Chim. Acta* **2011**, *706*, 361–366. [PubMed]

8. Reinicke, M.; Schröter, J.; Müller-Klieser, D.; Helmschrodt, C.; Ceglarek, U. Free oxysterols and bile acids including conjugates-Simultaneous quantification in human plasma and cerebrospinal fluid by liquid chromatography-tandem mass spectrometry. *Anal. Chim. Acta* **2018**, *1037*, 245–255. [PubMed]

9. Li, L.; Dutkiewicz, E.; Huang, Y.; Zhou, H.; Hsu, C. Analytical methods for cholesterol quantification. *J. Food Drug Anal.* **2019**, *27*, 375–386. [PubMed]

10. Luppa, P.B.; Bietenbeck, A.; Beaudoin, C.; Giannetti, A. Clinically relevant analytical techniques, organizational concepts for application and future perspectives of point-of-care testing. *Biotechnol. Adv.* **2016**, *34*, 139–160.

11. Gauglitz, G. Point-of-care platforms. *Annu. Rev. Anal. Chem.* **2014**, *7*, 297–315.

12. Derina, K.; Korotkova, E.; Barek, J. Non-enzymatic electrochemical approaches to cholesterol determination. *J. Pharm. Biomed. Anal.* **2020**, *191*, 113538. [PubMed]

13. Putzbach, W.; Ronkainen, N.J. Immobilization techniques in the fabrication of nanomaterial-based electrochemical biosensors: A review. *Sensors* **2013**, *13*, 4811–4840. [PubMed]

14. Kucherenko, I.; Soldatkin, O.; Dzyadevych, S.; Soldatkin, A. Electrochemical biosensors based on multienzyme systems: Main groups, advantages and limitations—A review. *Anal. Chim. Acta* **2020**, *1111*, 114–131. [PubMed]

15. Bruen, D.; Delaney, C.; Florea, L.; Diamond, D. Glucose sensing for diabetes monitoring: Recent developments. *Sensors* **2017**, *17*, 1866.

16. Stolarczyk, K.; Rogalski, J.; Bilewicz, R. NAD(P)-dependent glucose dehydrogenase: Applications for biosensors, bioelectrodes, and biofuel cells. *Bioelectrochemistry* **2020**, *135*, 107574.

17. Narla, S.; Jones, M.; Hermayer, K.; Zhu, Y. Critical care glucose point-of-care testing. In *Advances in Clinical Chemistry*; Elsevier: Amsterdam, The Netherlands, 2016; Volume 76, pp. 97–121.

18. Okuda-Shimazaki, J.; Yoshida, H.; Sode, K. FAD dependent glucose dehydrogenases-Discovery and engineering of representative glucose sensing enzymes. *Bioelectrochemistry* **2020**, *132*, 107414.

19. Narwal, V.; Deswal, R.; Batra, B.; Kalra, V.; Hooda, R.; Sharma, M.; Rana, J.S. Cholesterol biosensors: A review. *Steroids* **2019**, *143*, 6–17.

20. Kirk, O.; Borchert, T.V.; Fuglsang, C.C. Industrial enzyme applications. *Curr. Opin. Biotechnol.* **2002**, *13*, 345–351.

21. Park, M. Surface display technology for biosensor applications: A review. *Sensors* **2020**, *20*, 2775.

22. Pham, M.-L.; Polakovič, M. Microbial cell surface display of oxidoreductases: Concepts and applications. *Int. J. Biol. Macromol.* **2020**, *165*, 835–841.

23. Liang, B.; Li, L.; Tang, X.; Lang, Q.; Wang, H.; Li, F.; Shi, J.; Shen, W.; Palchetti, I.; Mascini, M. Microbial surface display of glucose dehydrogenase for amperometric glucose biosensor. *Biosens. Bioelectron.* **2013**, *45*, 19–24. [CrossRef] [PubMed]

24. Liang, B.; Lang, Q.; Tang, X.; Liu, A. Simultaneously improving stability and specificity of cell surface displayed glucose dehydrogenase mutants to construct whole-cell biocatalyst for glucose biosensor application. *Bioresour. Technol.* **2013**, *147*, 492–498. [CrossRef] [PubMed]

25. Ueda, M. *Yeast Cell Surface Engineering: Biological Mechanisms and Practical Applications*; Springer: Singapore, 2019.

26. Boder, E.T.; Wittrup, K.D. Yeast surface display for screening combinatorial polypeptide libraries. *Nat. Biotechnol.* **1997**, *15*, 553–557. [CrossRef] [PubMed]

27. Shibasaki, S.; Maeda, H.; Ueda, M. Molecular display technology using yeast-Arming technology. *Anal. Sci.* **2009**, *25*, 41–49. [CrossRef] [PubMed]

28. Fan, S.; Liang, B.; Xiao, X.; Bai, L.; Tang, X.; Lojou, E.; Cosnier, S.; Liu, A. Controllable display of sequential enzymes on yeast surface with enhanced biocatalytic activity toward efficient enzymatic biofuel cells. *J. Am. Chem. Soc.* **2020**, *142*, 3222–3230. [CrossRef] [PubMed]

29. Omura, H.; Sanada, H.; Yada, T.; Morita, T.; Kuyama, M.; Ikeda, T.; Kano, K.; Tsujimura, S. Coenzyme-Binding Glucose Dehydrogenase. U.S. Patent 7,514,250, 7 April 2009.

30. Doukyu, N.; Shibata, K.; Ogino, H.; Sagermann, M. Purification and characterization of *Chromobacterium* sp. DS-1 cholesterol oxidase with thermal, organic solvent, and detergent tolerance. *Appl. Microbiol. Biotechnol.* **2008**, *80*, 59–70. [CrossRef]

31. Gibson, T. *Studies on the Epstein-Barr Virus Genome*; University of Cambridge: Cambridge, UK, 1984.

32. Kieke, M.C.; Cho, B.K.; Boder, E.T.; Kranz, D.M.; Wittrup, K.D. Isolation of anti-T cell receptor scFv mutants by yeast surface display. *Protein Eng.* **1997**, *10*, 1303–1310. [CrossRef]

33. Kirmair, L.; Skerra, A. Biochemical analysis of recombinant AlkJ from *Pseudomonas putida* reveals a membrane-associated, flavin adenine dinucleotide-dependent dehydrogenase suitable for the biosynthetic production of aliphatic aldehydes. *Appl. Environ. Microbiol.* **2014**, *80*, 2468–2477. [CrossRef]

34. Allain, C.C.; Poon, L.S.; Chan, C.S.; Richmond, W.; Fu, P.C. Enzymatic determination of total serum cholesterol. *Clin. Chem.* **1974**, *20*, 470–475. [CrossRef]

35. Kitabayashi, M.; Tsuji, Y.; Kawaminami, H.; Kishimoto, T.; Nishiya, Y. Method for Highly Expressing Recombinant Glucose Dehydrogenase Derived from Filamentous Fungi. U.S. Patent 7,741,100, 22 June 2010.

36. International Organization for Standardization. *In Vitro Diagnostic Test Systems-Requirements for Blood-Glucose Monitoring Systems for Self-Testing in Managing Diabetes Mellitus*; ISO 15197; ISO: Geneva, Switzerland, 2013.

37. Sagermann, M.; Ohtaki, A.; Newton, K.; Doukyu, N. Structural characterization of the organic solvent-stable cholesterol oxidase from *Chromobacterium* sp. DS-1. *J. Struct. Biol.* **2010**, *170*, 32–40. [CrossRef]

38. Doukyu, N.; Shibata, K.; Ogino, H.; Sagermann, M. Cloning, sequence analysis, and expression of a gene encoding *Chromobacterium* sp. DS-1 cholesterol oxidase. *Appl. Microbiol. Biotechnol.* **2009**, *82*, 479. [CrossRef] [PubMed]

39. Schlapschy, M.; Binder, U.; Börger, C.; Theobald, I.; Wachinger, K.; Kisling, S.; Haller, D.; Skerra, A. PASylation: A biological alternative to PEGylation for extending the plasma half-life of pharmaceutically active proteins. *Protein Eng. Des. Sel.* **2013**, *26*, 489–501. [CrossRef] [PubMed]

40. Stern, L.A.; Schrack, I.A.; Johnson, S.M.; Deshpande, A.; Bennett, N.R.; Harasymiw, L.A.; Gardner, M.K.; Hackel, B.J. Geometry and expression enhance enrichment of functional yeast-displayed ligands via cell panning. *Biotechnol. Bioeng.* **2016**, *113*, 2328–2341. [CrossRef] [PubMed]

41. Qu, G.; Li, A.; Acevedo-Rocha, C.G.; Sun, Z.; Reetz, M.T. The crucial role of methodology development in directed evolution of selective enzymes. *Angew. Chem. Int. Ed.* **2019**. [CrossRef]

 *biosensors*

Article

Interdigitated Sensor Optimization for Blood Sample Analysis

Julien Claudel, Thanh-Tuan Ngo, Djilali Kourtiche and Mustapha Nadi *

Institut Jean Lamour, Lorraine University (CNRS—UMR 7198), 54011 Nancy, France; julien.claudel@univ-lorraine.fr (J.C.); ngothanhtuan@humg.edu.vn (T.-T.N.); djilali.kourtiche@univ-lorraine.fr (D.K.)
* Correspondence: mustapha.nadi@univ-lorraine.fr; Tel.: +33-372-7426-41

Received: 17 November 2020; Accepted: 14 December 2020; Published: 16 December 2020

Abstract: Interdigitated (ITD) sensors are specially adapted for the bioimpedance analysis (BIA) of low-volume (microliter scale) biological samples. Impedance spectroscopy is a fast method involving simple and easy biological sample preparation. The geometry of an ITD sensor makes it easier to deposit a sample at the microscopic scale of the electrodes. At this scale, the electrode size induces an increase in the double-layer effect, which may completely limit interesting bandwidths in the impedance measurements. This work focuses on ITD sensor frequency band optimization via an original study of the impact of the metalization ratio α. An electrical sensor model was studied to determine the best α ratio. A ratio of 0.6 was able to improve the low-frequency cutoff by a factor of up to 2.5. This theoretical approach was confirmed by measurements of blood samples with three sensors. The optimized sensor was able to extract the intrinsic electrical properties of blood in the frequency band of interest.

Keywords: biosensors; interdigitated electrodes; impedance spectroscopy; blood analysis

1. Introduction

Many improvements in the sensitivity and selectivity of biosensors have been made in the last two decades [1–5]. They are mainly due to the sensors (lab-on-a-chip) designed in micro-nanotechnology facilities. For biological applications [6], charge transfer sensors [7], impedance-based sensors [8–10], and capacitance-based sensors are often used. Impedance spectroscopy is a well-known and powerful technique for biological characterizations on both the macroscopic and microscopic scales [11]. The electrodes apply an electric field to the sample being tested and measure electrical signals. They also can provide information on the relative permittivity and electrical conductivity of biosamples that correspond to intrinsic parameters [12]. Impedance-based sensors are divided into four main categories: cell trap sensors, cytometric sensors, matrix sensors and interdigitated sensors. The sensors in the first three categories have the best sensitivities because only one cell is analyzed at a time. Trap sensors often use micro-hole or microcavity [13] systems to isolate cells from each other. Cytometric sensors use microchannels to focus on one cell at a time [14], and cells are dynamically characterized during their passage into a measurement area situated inside the channel. Matrix electrode systems use multiple electrodes to perform numerous measurements at a time [15]. Despite the fact that these three techniques have better sensitivities, they are also the most difficult to implement. Thus, interdigitated (ITD) sensors [16], composed of just one layer of coplanar electrodes, remain competitive for low-volume and low-concentration biological samples. For example, they are perfectly suitable for cell surface cultures [17] and low-volume biological sampling, as in DNA analyses [18]. Beyond the simple interface, the geometrical properties of the electrodes can have a significant impact on the efficiency of the biosensor [19,20]. They can be optimized a priori during the sensors' design step according to the targeted application and the nature of the cells to be analyzed [14]. In this work, we propose a method for optimizing the ITD sensor frequency band involving the metalization ratio.

First, we propose a complete electrical equivalent model that takes into account all the sensor's parameters, such as the electrode length, gap, width, and electrical properties of the medium. Interface capacitive effects, also known as the double-layer effects, are also modeled to assess the impact on the global impedance spectrum.

In the second part, analytical simulations are performed to analyze the effects of geometrical parameters on bioimpedance measurements. It is important to maintain a sufficiently wide bandwidth in order to be able to characterize a biosample over many decades. To achieve optimization, the impact of the metalization ratio α on the bandwidth was studied. This ratio is defined as $\alpha = W/(S + W)$, where G is the gap (m) and W is the width (m) of the IDT sensor digits (see Section 2.1 below).

The last section focuses on the experimental validation of the two previous sections. Three sensor designs with different degrees of optimization corresponding to different values of the ratio α were fabricated using a standard microfabrication process. Characterizations were performed in calibrated electrolytic solutions to validate our model. Finally, the capability of our optimized sensor in terms of characterizing a blood sample was tested.

2. Theoretical Considerations

2.1. Sensor Structure and Cell Factor

An IDT sensor is composed of two comb-like electrodes deposited on an insulated substrate. Each electrode's digits have a width W and a length L, and there is a gap S between digits [21,22], as presented in Figure 1a. This specific electrode configuration produces an elliptic current displacement inside the sample, as shown in Figure 1b. Changes in W and S allow the electrical penetration depth in the sample to be adjusted. Ninety-five percent of the electric excitation signal power is concentrated within $2(S + W)$. This makes IDT sensors perfectly suitable for low-surface sample characterizations of, for instance, cell cultures or small-volume samples. An electrical equivalent model for an IDT sensor loaded with an electrolyte is presented in Figure 1c. The electrolyte is generally used as a simple reference medium for impedance-based sensor characterization.

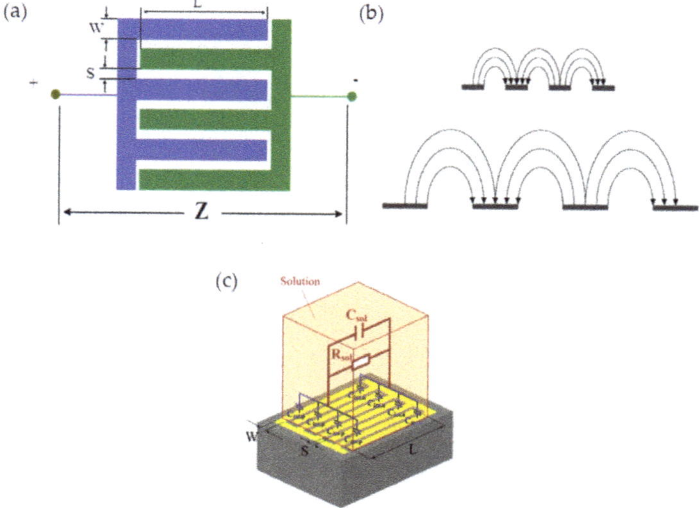

Figure 1. (a) Geometric parameters of interdigitated sensors; (b) Electric current displacement between electrodes; (c) Electrical model of an interdigitated sensor and sample (an ionic solution).

According to Olthuis et al. [23], R_{sol} and C_{sol}, the resistance and capacitance of the ionic solution, can be linked via Equation (1) to electrolyte conductivity and permittivity using a cell factor K_{cell}. This factor can be calculated as a function of IDT geometries using Equation (2).

$$R_{sol} = \frac{K_{cell}}{\sigma_{sol}}; \quad C_{sol} = \frac{\varepsilon_0 \varepsilon_{r,sol}}{K_{cell}},$$

(1)

$$K_{cell} = \frac{2}{L(N-1)} \frac{K(k)}{K\left(\sqrt{1-k^2}\right)},$$

(2)

where σ_{sol} is the electrolyte conductivity (S·m^{-1}), ε_{sol} is the electrolyte relative permittivity, N is the number of digits, L is the digit length (m), W is the digit width (m), S is the space between two digits (m), and K_{cell} is the cell factor (m^{-1}). The function $K(k)$ is the incomplete integral of the first module k, and is calculated with Equation (3). The metalization ratio is defined by $\alpha = W/(S + W)$.

$$K(k) = \int_0^1 \frac{1}{\sqrt{\left(1-t^2\right)\left(1-k^2 t^2\right)}} dt \ \text{with}\ k = \cos\left(\frac{\pi}{2}\frac{W}{S+W}\right).$$

(3)

where σ_{sol} is the electrolyte conductivity (S·m^{-1}), ε_{sol} is the electrolyte relative permittivity, N is the number of digits, L is the digit length (m), W is the digit width (m), S is the space between two digits (m), and K_{cell} is the cell factor (m^{-1}). The function $K(k)$ is the incomplete integral of the first module k, and is calculated with Equation (3). This function is used to formulate the elliptic electric field distribution, as described above. The metalization ratio is defined by $\alpha = W/(S + W)$.

2.2. Double Layer Impedance

The double-layer impedance represents the interface effects that occur when the polarized electrodes are in contact with an electrolyte. The double layer corresponds to two parallel layers of charge on a thin section (nanometric scale) of the electrode surface. These effects act as a barrier for low-frequency measurements and need to be taken into account in global modeling. Generally, it is necessary to limit these effects to increase the bandwidth of interest. A typical model for interface effects is composed of three elements making up an equivalent circuit, as presented in Figure 2. The charge transfer resistance R_{CT} represents the resistive effect, and the double-layer capacitance C_{DL} represents the capacitive effect. The Warburg impedance Z_W with a constant phase ($-\pi/2$) is used to model the non-linear decrease in impedance induced by the diffusion of ionic species. These elements can be very difficult to model for complex electrolytes, such as blood plasma. Moreover, when the frequency is high enough, the decrease in C_{DL}, the impedance short-circuit R_{CT} and Z_w, and the interface impedance become negligible compared to the impedance of the sample. Hence, double-layer impedance is generally only modeled by C_{DL} when only the sample impedance is studied. This capacitance effect is proportional to the electrode surface and captured as the surface capacitance C_0, with typical values ranging from 10 to 50 µF/cm^2. For interdigitated sensors, global induced capacitance $C_{interface}$ can be calculated with the unitary digit capacitances $C_{int,p}$ and $C_{int,n}$ using Equations (4)–(6).

$$C_{int,p} = C_{int,n} = LWC_0,$$

(4)

$$C_{interface} = \frac{N}{4}LWC_0,$$

(5)

$$Z_{interface} = \frac{1}{j\omega C_{interface}},$$

(6)

where $C_{int,p}$ and $C_{int,n}$ are the capacitances at each electrode digits (F), $C_{interface}$ is the global capacitance at the sensor (F), $Z_{interface}$ is the induced impedance in series with the sample impedance (Ω), and C_0 is the surface capacitance at the interface electrode/electrolyte (F/m^{-1}).

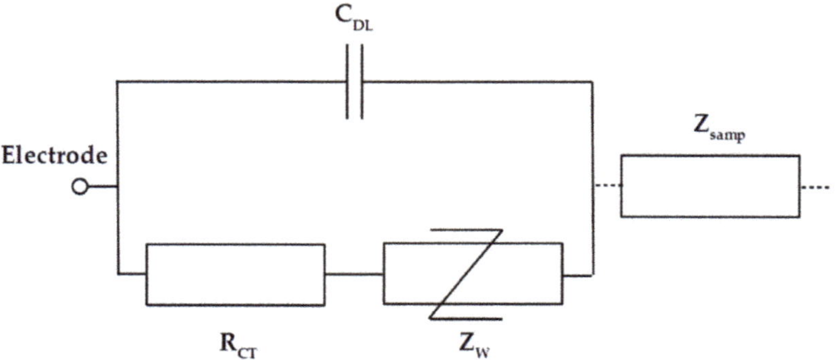

Figure 2. Electrical equivalent circuit for the double-layer impedance of one electrode.

2.3. Sample Impedance

In a linear, homogeneous, and isotropic samples, measured impedance Z_{samp} (Ω) and admittance Y_{samp} (S) depend on the sample's intrinsic properties (conductivity σ_{samp} and permittivity ε_{samp}) and the sensor factor K_{cell}, as seen in Equation (7) [24]. Z_{samp} is the sample's impedance and does not take into account other effects, such as double-layer capacitance.

$$Z_{samp} = \frac{K_{cell}}{\sigma_{samp} + j\omega\varepsilon_0\varepsilon_{r,samp}} \rightarrow Y_{samp} = \frac{\sigma_{samp} + j\omega\varepsilon_0\varepsilon_{r,samp}}{K_{cell}} \rightarrow \begin{cases} G_{samp} = \frac{\sigma_{samp}}{K_{cell}} \\ C_{samp} = \frac{\varepsilon_0\varepsilon_{r,samp}}{K_{cell}} \end{cases} . \tag{7}$$

For a simple sample, such as that of an electrolyte, σ_{samp} and $\varepsilon_{r,samp}$ can be considered constants. For more complex biological samples, such as blood, these values depend on frequency and are referred to as complex conductivity $\sigma_{samp}(\omega)$ and complex relative permittivity $\varepsilon_{r,samp}(\omega)$, respectively. Figure 3 shows the typical values for the complex conductivity and relative permittivity of blood [12]. The two levels of relative permittivity represent the effects of cell membrane capacitance and water permittivity, respectively. The two levels of conductivity represent the effects of extracellular medium conductivity and the combination of extracellular/cytoplasm conductivity, respectively. The passage from the first to second level is known as the β dispersion and occurs when the impedance of the capacitive cell membrane becomes negligible compared to the cytoplasmic impedance. The second increasing of conductivity is called the γ dispersion and appears at microwave frequencies (GHz). It corresponds to water molecular excitation and is not studied here. The blood conductivity diagram shows the necessity of performing the conductivity extraction correctly before and after β dispersion when characterizing a biological sample. In the case of a blood sample, this dispersion begins at approximately 250 kHz.

After adding the interface effect in series with the samples, the global measured impedance becomes

$$\begin{cases} G_{Tot}(\omega) = \dfrac{\omega^2 G_{samp} C_{interface}^2}{G_{samp}^2 + \omega^2\left(C_{samp} + C_{interface}\right)^2} \\ C_{Tot}(\omega) = \dfrac{C_{interface} G_{samp}^2 + \omega^2 C_{samp} C_{interface}\left(C_{samp} + C_{interface}\right)}{G_{samp}^2 + \omega^2\left(C_{samp} + C_{interface}\right)^2} \end{cases} . \tag{8}$$

$$\lim_{\omega \to 0} C_{Tot} = C_{interface} \tag{9}$$

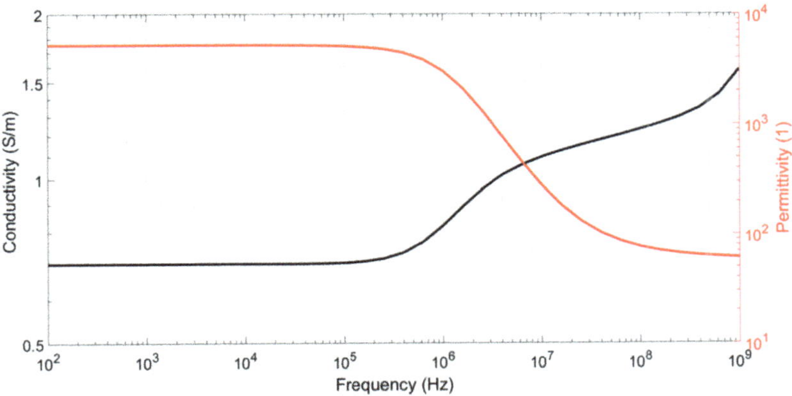

Figure 3. Typical complex conductivity and relative permittivity values for a blood sample.

G_{Tot} and C_{Tot} correspond to global measured conductance and capacitance induced by the sample and interface impedances, respectively. At low frequencies, the global capacitance tends to $C_{interface}$, as presented in Equation (9). When the frequency increased, $Z_{interface}$ became negligible compared to Z_{samp}, and total impedance can be considered equal to Z_{samp}. The typical curve profiles for a biological sample and an electrolytic sample, showing the double-layer capacitance effect, are shown in Figure 4. Two plateaus are clearly visible and correspond to blood conductivity. If the lower cutoff frequency is too high, the first plateau can be completely hidden by double-layer impedance and interfere with the measurement. The low cutoff frequency is defined in Equation (10) below via an analogy with low and high passive first-order filters using the conductivity of the medium.

$$f_{low} = \frac{1}{2\pi R_{samp} C_{interface}}$$

$$\text{when } f = f_{low} \text{, } |Y_{samp}| = \frac{G_{samp}}{\sqrt{2}} \text{ and } |Z_{samp}| = \sqrt{2}\, R_{samp} \tag{10}$$

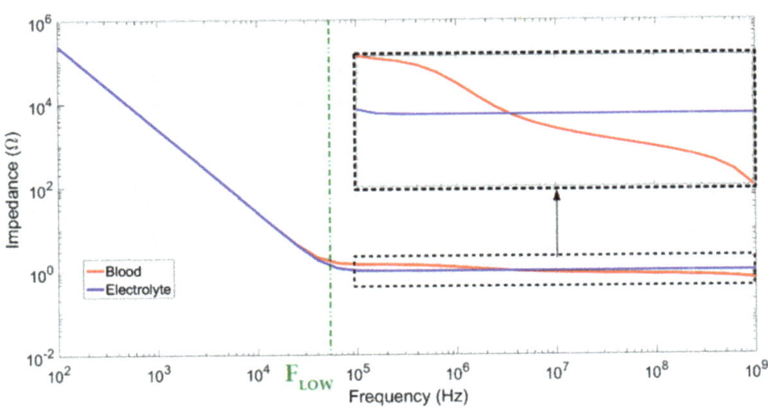

Figure 4. Typical curve profiles for biological and electrolyte samples.

Since $Z_{interface}$ is not purely capacitive, the first relation in Equation (10) may prove difficult to use. Hence, the second equation is usually preferred. In this work, this second equation was used to determine f_{low} from the measured impedance spectrum. R_{samp} and G_{samp} can be extracted from the real part of the impedance or admittance when a plateau is visible in the spectrum (predominance of resistive/conductive effects). These parameters are extracted from the real part of the impedance measured in the center of each plateau. Furthermore, σ_{samp} can be calculated using the extracted R_{samp} or G_{samp} and Equation (7).

3. Sensor Optimization

As mentioned above, the interface impedance acts as a barrier at low frequencies. Optimizing the frequency band consists of reducing the f_{low} value. Since f_{low} depends on both the sample properties and sensor geometry, it is possible to optimize the frequency band by optimizing the sensor design. Equation (10) was developed by using Equations (1), (2) and (5) to check which geometric parameter was the most suitable one to study. Hence, Equation (11) is obtained. As σ_{sol} and C_0 depend on sample properties, decreasing f_{low} is equivalent to decrease the term given in expression (12). It can be seen that this term depends on N, W, and S. In addition, the term $(N - 1)/N$ will be negligible for large N.

$$f_{low} = \frac{1}{\pi} \frac{\sigma_{samp}}{C_0} \frac{(N-1)}{N} \frac{1}{W} \frac{K\left(\sqrt{1-k^2}\right)}{K(k)} \tag{11}$$

$$\frac{(N-1)}{N} \frac{1}{W} \frac{K\left(\sqrt{1-k^2}\right)}{K(k)} \tag{12}$$

Only W and S seem to have a significant impact on f_{low} through the K_{cell} and "$1/W$" terms. It is relevant to study the optimization of α. Clearly, f_{low} could be reduced by increasing W, but doing so implies an increase in sensor size as well as the measurement volume: the smaller the cell constant (large electrodes), the lower the cut-off frequency, but the faster the measurement volume will increase. Here, we are interested in optimizing the sensor for a given volume/measurement surface on a microscopic scale. ($W + S$) and N were set to constant values to maintain the same volume for the investigation. Analytical simulations were performed by varying the α ratio from 0.1 to 1 and holding the other parameters constant. L was set to 2 mm, and ($W + S$) was set to 50 μm. Finally, σ_{samp} was set to 0.6 S/m (conductivity of blood at low frequency), and C_0 was set to 40 μF/cm^2. Three values of N (20, 40 and 50) were tested. The results are shown in Figure 5. These results demonstrate that the value of N does not have a significant impact on the cutoff frequency. In contrast, the value of f_{low} mainly depends on α and is a minimum for $\alpha = 0.6$. These results show that it is possible to decrease the low cutoff frequency by a factor up to approximately 2.5 by optimizing the metalization ratio α.

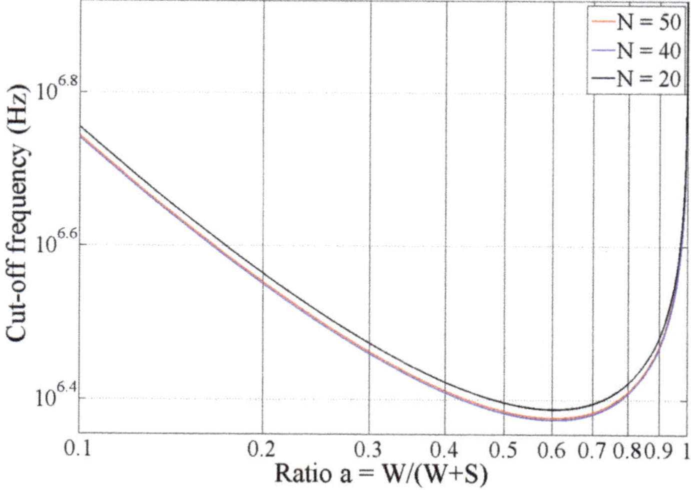

Figure 5. Low cutoff frequency as a function of α.

4. Sensor Realization

4.1. Sensor Manufacturing

To prove our assumptions, three sensors were realized in a clean room using standard microfabrication techniques. Platinum electrodes were structured via sputtering deposition and optical lithography on a glass substrate. Wells were created by using negative SU_8 resin with a thickness of approximately 400 μm. Sensor 1 is illustrated in Figure 6. All sensors presented the same periodicity, electrode number and electrode length. Different α ratios were tested to prove our assumptions. The sensor geometries are listed in Table 1. All sensors presented the same pitch $(W + S)$ and the same surface of 2×2 mm^2 for ease of comparison. The pitch was set to 50 μm to provide a penetration depth of only several cell sizes (from 6 μm to 15 μm for red blood cells and white blood cells, respectively). The smallest W and S values were of the same size order as blood cells. These dimensions allowed increased sensitivity by performing characterizations of a few cell layers in depth.

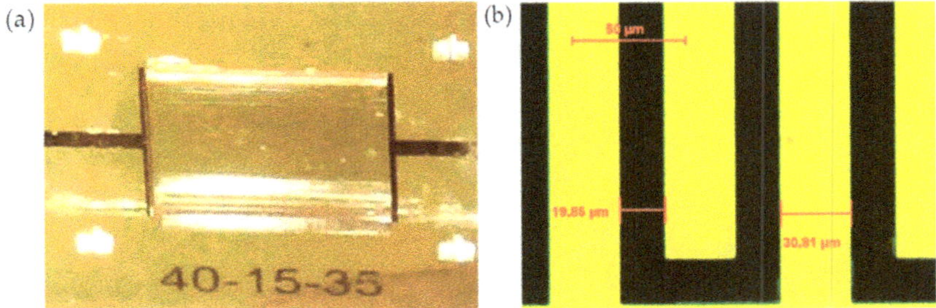

Figure 6. (a) Photograph of Sensors 1 with SU_8 well and blood sample, and (b) optical microscope image of different Sensor 1 digits.

Table 1. Sensor geometries.

Sensor Number	N	L (μm)	W (μm)	S (μm)	Theoretical K_{cell} (m^{-1})	α Ratio
1	40	2000	30	20	21.8	0.6
2	40	2000	15	35	35.8	0.3
3	40	2000	40	10	15.1	0.8

4.2. Reference Measurements

To verify both the validity of our model and the integrity of the sensors, measurements were performed for three different calibrated solutions of 84 μS/cm, 100 μS/cm, and 1413 μS/cm. All measurements were performed by depositing a 2 μL drop into the well with a micropipette. Examples of the resulting impedance diagrams are shown in Figure 7 for Sensor 1. All results are reported in Table 2. The K_{cell} factor was calculated for R_{sol} on the curves' plateaus using Equation (1). The calculated values of K_{cell} were in accordance with the simulated ones. Sensor 3 had a higher error in K_{cell} determination compared to the others, which can be explained by the fact this sensor has the smallest useful gap size ($S = 10$ μm). Using optical lithography, the resolution is approximately 0.5 μm for each edge, or 1 μm for a digit, representing a possible error of 10% for a 10 μm digit, which is close to the K_{cell} error for Sensor 3. Note that Sensor 1, with $\alpha = 0.6$, presents the smallest cutoff frequencies, as predicted in the "Sensor optimization" section. The capacitance effect observed in the higher frequencies can be attributed to the experimental setup limitations, including the capacitance induced by the interfacing Printed Circuit Board (PCB), the connections, and the measurement devices.

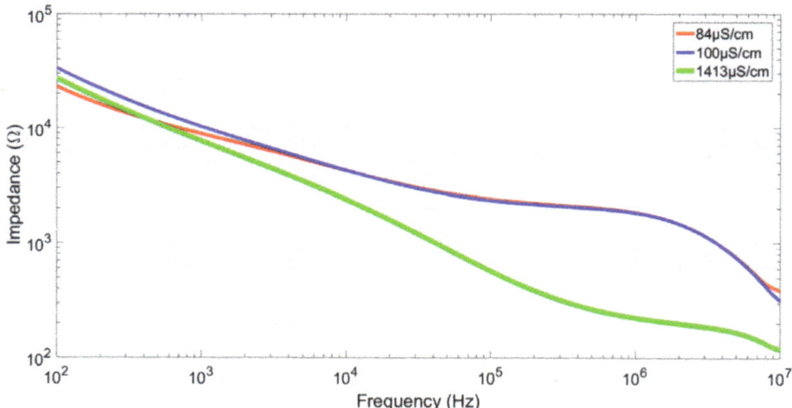

Figure 7. Bode diagrams of the impedances for three different calibrated solutions measured with Sensor 1.

Table 2. Sensor results for the calibrated solutions.

Sensor Number	Theoretical K_{cell} (m^{-1})	α Ratio	84 µS/cm			100 µS/cm			1413 µS/cm		
			fc,L (kHz)	K_{cell}	K_{cell} Error (%)	fc,L (kHz)	K_{cell}	K_{cell} Error (%)	fc,L (kHz)	K_{cell}	K_{cell} Error (%)
1	21.8	0.6	75.6	22.3	2.3	85	22.01	0.96	433	22.65	3.9
2	35.8	0.3	170	35.49	0.87	486	34.99	2.26	1960	35.55	0.7
3	15.1	0.8	135	16.83	11.48	305	16.08	6.49	2400	17	12.57

5. Blood Characterization

5.1. Experimental Setup

Figure 8 represents a general view of the instrumentation setup developed in this work for the electrical characterization of biological media. This instrumentation setup consists of the following elements:

(a) Biofluid samples placed directly on the sensor (Figure 9).

(b) A microscope to observe the position of the volume of liquid.

(c) A thermometer to measure the ambient temperature.

(d) A micropipette (Socorex Micropipette Acura 825)

(e) HF2TA current amplifier (manufactured by Zurich Instrument),

(f) HF2IS impedance spectroscope for the frequency range from 0.7 µHz to 50 MHz.

(g) A computer for observing and processing measurement data using LabVIEW® application.

This application allows us to enter the measurement parameters described above and save the values of the impedance spectra as text files.

The optical microscope allows us to correctly place the drop of the biofluid on the sensor. The interface card containing the sensors (Figure 9) is connected to the HF2IS impedance analyzer and to the HF2TA current amplifier via short SMA cables in order to reduce the length of the measurement loop as much as possible. The impedance spectroscope is connected via a USB cable to a computer to control it and retrieve the data.

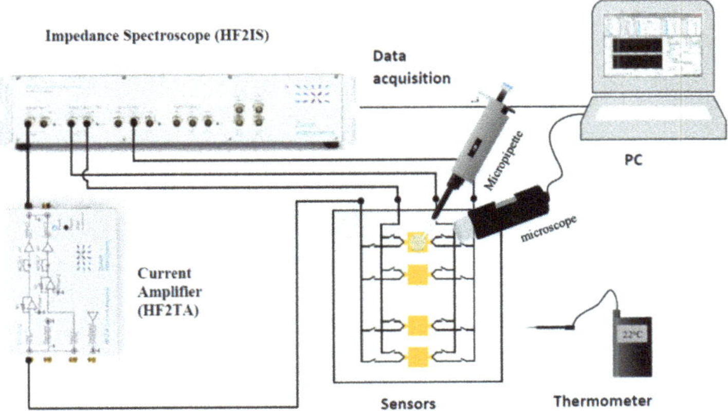

Figure 8. Bio impedance measurement setup.

Figure 9. Interdigital electrode sensors connected to the PCB circuit.

5.2. Sample Preparation

Blood sampling was performed under medical supervision (University Hospital, CHRU Nancy, France) using the same donor. The samples were placed into three 6 mL tubes with heparin. Measurements were taken within 10 min of sampling.

To ensure good repeatability for the measurements of all sensors, each sample was deposited on each sensor using the following procedure:

- The tube was shaken slightly for 1 min before sampling with a micropipette.
- A 2 µL sample was obtained with an adjustable micropipette and deposited into the well (Figure 9). This volume was chosen to ensure that all the sensor cavities were full. As described in Section 2.1, only the first 100 µm of thickness of the sample in contact with the IDT was really due to the small penetration depth of the IDT sensors, ensuring that measurements will not be impaired by round-shaped droplets, the sample surface in contact with the air or cavity walls, or the sample rising above the cavity.
- The impedance spectrum acquisition started 10 s after sample deposition. The impedance measurement was then performed within several seconds at 1 V sinus amplitude.
- The room temperature was maintained at 25 ± 1 °C during the measurement campaign.

5.3. Results

Measurements were performed using the experimental setup described in Section 5.1 and the sampling procedure described above for each sensor. The results are shown in Figure 10 in the form of Bode diagrams for the modules and phases. f_{low} represents the measured low-frequency cut-off using the method proposed in Section 2.3. As in the measurements of the calibrated solution, Sensor 1 has the smallest low-frequency cut-off and the widest bandwidth. Unlike the two other sensors, two plateaus are visible from and after the cutoff frequency for sensor 1. This result proves that only sensor 1 is able to characterize the complete spectrum for blood impedance. For the other sensors, the double-layer impedance remains non-negligible until the cutoff frequency and interferes with sample impedance measurement.

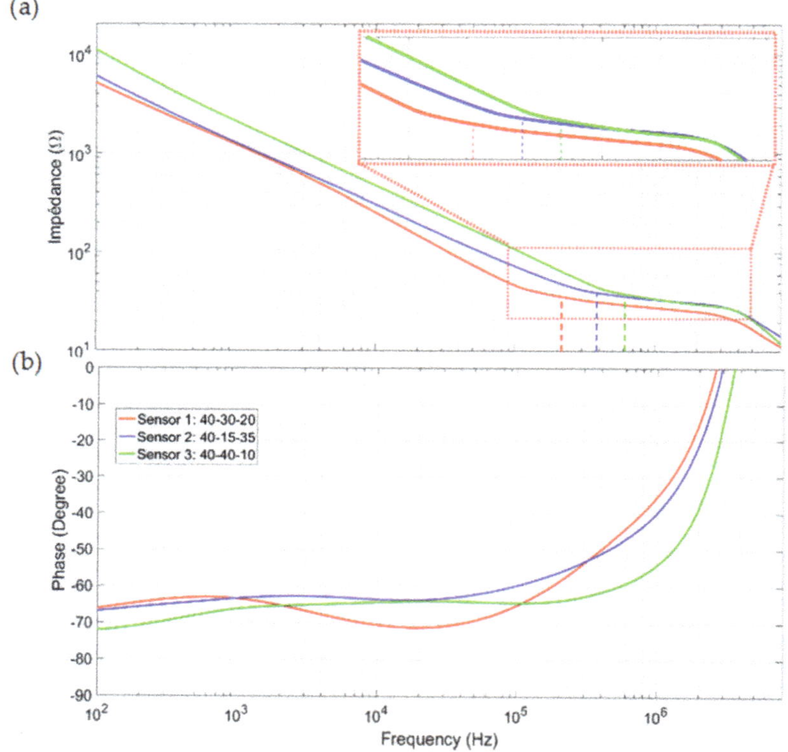

Figure 10. Bode diagrams for blood impedance characterizations (**a**) in modules, and (**b**) in phases.

Low-frequency blood conductivity was calculated using the R_{sol} values measured on the plateaus just after the f_{low} values and Equation (1). The results extracted from the impedance spectrum are summarized in Table 3. The values 0.69, 0.89, and 0.43 S·m^{-1} were obtained for sensors 1, 2, and 3, respectively. These conductivities are of the same order as the 0.7 S/m value obtained by Gabriel [12]. The difference between sensor 1 and the other two sensors can be explained by its lower immunity to the interface effect that disturbs the R_{sol} measurement and conductivity extraction. For sensor 1, f_{low} is slightly lower than at the beginning of the β dispersion band studied in the theoretical section (215 kHz, down from 250 kHz) and allows the first plateau of conductivity to be measured. For the other sensors, the measurements were performed away from this plateau and are incorrect.

Table 3. Sensor results for the calibrated solution.

Sensor Number	N	L (µm)	W (µm)	S (µm)	f_{low} (kHz)	Measured K_{cell} (m^{-1})	α	R_{sol} (Ω)	σ_{blood} (S/m)
1	40	2000	30	20	215	22.32	0.6	32.35	0.69
2	40	2000	15	35	358	35.834	0.3	40.26	0.89
3	40	2000	40	10	614	16.64	0.8	38.70	0.43

6. Conclusions

An analytical model for an IDT sensor was proposed. Using this model, the effect of geometric parameters on global impedance was studied in order to propose an optimized sensor for blood characterization. It appears that the number of electrodes N and digit length L do not contribute significantly to improving sensor bandwidth. To the contrary, the metalization ratio α, which depends only on the digit width W and gap S, is a relevant parameter for optimizing the sensor. The analytical simulations showed that the best optimization is obtained for $\alpha = 0.6$. This α value permits the low cut-off frequency to be reduced by a factor of up to 2.5. To validate our theoretical results, measurements were performed on three different sensors using the same active surface area (2 mm × 2 mm) but different ratios α. The results obtained with the calibrated solutions proved the validity of our model and the possibility improving sensor bandwidth. Finally, blood sample characterizations were performed using the sensors. The optimized sensor was able to characterize the blood sample and extract its intrinsic property (electrical conductivity), achieving good concordance with the reference blood conductivity provided in the literature.

Author Contributions: Conceptualization—T.-T.N., J.C., D.K. and M.N.; methodology—T.-T.N., J.C., D.K. and M.N.; validation—T.-T.N., J.C. and M.N.; data curation—T.-T.N., J.C.; writing, original draft preparation—J.C.; writing, review and editing—J.C., D.K. and M.N.; project administration—M.N., D.K.; funding acquisition—J.C., D.K. and M.N. All authors have read and agreed to the published version of the manuscript.

Funding: This research was done with the support of the "Région Grand Est—France" and the European Regional Development Fund (FEDER). It was also supported by the 80' PRIME program of the Conseil National de la Recherche Scientifique (CNRS).

Acknowledgments: The authors are grateful to the MINALOR Skill Center of the Institut Jean Lamour (IJL) at the Université de Lorraine for the technical support.

Conflicts of Interest: The authors declare no conflict of interest.

References

1. Vikas, A.; Pundir, C.S. Biosensors: Future analytical tools. *Sens. Transducers J.* **2007**, *76*, 935–936.
2. Yoon, J.; Shin, M.; Lee, T.; Choi, J.-W. Highly sensitive biosensors based on biomolecules and functional nanomaterials depending on the types of nanomaterials: A perspective review. *Materials* **2020**, *13*, 299. [CrossRef] [PubMed]
3. Turner, A.P. Biosensors: Sense and sensibility. *Chem. Soc. Rev.* **2013**, *21*, 3184–3196. [CrossRef] [PubMed]
4. Sang, S.; Zhang, W.; Zhao, Y. *Review on the Design Art of Biosensors. State of the Art in Biosensors—General Aspects*; Rinken, T., Ed.; IntechOpen: London, UK, 2013. Available online: https://www.intechopen.com/books/state-of-the-art-in-biosensors-general-aspects/review-on-the-design-art-of-biosensors (accessed on 15 December 2020). [CrossRef]
5. Ahmed, S.; Shaikh, N.; Pathak, N.; Sonawane, A.; Pandey, V.; Maratkar, S. An overview of sensitivity and selectivity of biosensors for environmental applications. In *Tools, Techniques and Protocols for Monitoring Environmental Contaminants*; Elsevier: Amsterdam, The Netherlands, 2019; Chapter 3; pp. 53–73.
6. Chuanrui, C.; Yue, G.; Peining, C.; Huisheng, P. Recent advances of tissue-interfaced chemical biosensors. *J. Mater. Chem. B* **2020**, *8*, 3371–3381.
7. RIgreja, R.; Dias, C.J. Analytical evaluation of the interdigital electrodes capacitance for a multi-layered structure. *Sens. Actuators A Phys.* **2004**, *112*, 291–301. [CrossRef]

8.   Moraes, M.L.; Maki, R.M.; Paulovich, F.V.; Rodrigues Filho, U.P.; De Oliveira MC, F.; Riul, A.; De Souza, N.C.; Ferreira, M.; Gomes, H.L.; Oliveira, O.N. Strategies to optimize biosensors based on impedance spectroscopy to detect phytic acid using layer-by-layer films. *Anal. Chem.* **2010**, *82*, 3239–3246. [CrossRef]

9.   Radhakrishnan, R.; Suni, I.; Bever, C.S.; Hammock, B.D. Impedance biosensors: Applications to sustainability and remaining technical challenges. *ACS Sustain. Chem. Eng.* **2014**, *2*, 1649–1655. [CrossRef]

10.   Song, S.; Xu, H.; Fan, C. Potential diagnostic applications of biosensors: Current and future directions. *Int. J. Nanomed.* **2006**, *1*, 433–440. [CrossRef]

11.   Katz, E.; Willner, I. Probing biomolecular interactions at conductive and semiconductive surfaces by impedance spectroscopy: Routes to impedimetric immunosensors, DNA-sensors, and enzyme biosensors. *Electroanalysis* **2003**, *15*, 913–947. [CrossRef]

12.   Gabriel, C.; Gabriel, S.; Corthout, E. The dielectric properties of biological tissues: I. Literature survey. *Phys. Med. Biol.* **1996**, *41*, 2231–2249. [CrossRef]

13.   Park, Y.; Cha, J.J.; Seo, S.; Yun, J.; Woo Kim, H.; Park, C.; Lee, J.H. Ex vivo characterization of age-associated impedance changes of single vascular endothelial cells using micro electrical impedance spectroscopy with a cell trap. *Biomicrofluidics* **2016**, *10*, 014114. [CrossRef] [PubMed]

14.   Claudel, J.; Alves De Araujo, A.L.; Nadi, M.; Kourtiche, D. Lab-on-a-chip device for yeast cell characterization in low-conductivity media combining cytometry and bio-impedance. *Sensors* **2019**, *19*, 3366. [CrossRef] [PubMed]

15.   Alves de Araujo, A.L.; Claudel, J.; Kourtiche, D.; Nadi, M. Use of an insulation layer on the connection tracks of a biosensor with coplanar electrodes to increase the normalized impedance variation. *Biosensors* **2019**, *9*, 108. [CrossRef]

16.   Ertl, P.; Heer, R. Interdigitated impedance sensors for analysis of biological cells in microfluidic biochips. *EI Elektrotechnik Inf.* **2009**, *126*, 47–50. [CrossRef]

17.   Posseckardt, J.; Schirmer, C.; Kick, A.; Rebatschek, K.; Lamz, T.; Mertig, M. Monitoring of saccharomyces cerevisiae viability by non-Faradaic impedance spectroscopy using interdigitated screen-printed platinum electrodes. *Sens. Actuators B Chem.* **2018**, *255*, 3417–3424. [CrossRef]

18.   Wang, L.; Veselinovic, M.; Yang, L.; Geiss, B.J.; Dandy, D.S.; Chen, T. A sensitive DNA capacitive biosensor using interdigitated electrodes. *Biosens. Bioelectron.* **2017**, *87*, 646–653. [CrossRef]

19.   Ibrahim, M.; Claudel, J.; Kourtiche, D.; Nadi, M. Geometric parameters optimization of planar interdigitated electrodes for bioimpedance spectroscopy. *J. Electr. Bioimpedance* **2013**, *4*, 13–22. [CrossRef]

20.   Ngo, T.T.; Bourjilat, A.; Claudel, J.; Kourtiche, D.; Nadi, M. Design and realization of a planar interdigital microsensor for biological medium characterization. In *Next Generation Sensors and Systems*; Springer: Cham, Switzerland, 2016; pp. 23–54.

21.   Starzyk, F. Parametrisation of interdigit comb capacitor for dielectric impedance spectroscopy. *Arch. Mater. Sci. Eng.* **2008**, *34*, 31–34.

22.   Kidner, N.J.; Homrighaus, Z.J.; Mason, T.O.; Garboczi, E.J. Modeling interdigital electrode structures for the dielectric characterization of electroceramic thin films. *Thin Solid Film.* **2006**, *496*, 539–545. [CrossRef]

23.   Olthuis, W.; Streekstra, W.; Bergveld, P. Theoretical and experimental determination of cell constants of planar-interdigitated electrolyte conductivity sensors. *Sens. Actuators B Chem.* **1995**, *25*, 252–256. [CrossRef]

24.   Steendijk, P.; Mur, G.; Van Der Velde, E.T.; Baan, J. The four-electrode resistivity technique in anisotropic media: Theoretical analysis and application on myocardial tissue in vivo. *IEEE Trans. Biomed. Eng.* **1993**, *40*, 1138–1148. [CrossRef] [PubMed]

Publisher's Note: MDPI stays neutral with regard to jurisdictional claims in published maps and institutional affiliations.

biosensors

Article

RF Remote Blood Glucose Sensor and a Microfluidic Vascular Phantom for Sensor Validation

Muhammad Farhan Affendi Mohamad Yunos [1], Rémi Manczak [2], Cyril Guines [2], Ahmad Fairuzabadi Mohd Mansor [1], Wing Cheung Mak [3], Sheroz Khan [4], Noor Amalina Ramli [1], Arnaud Pothier [2] and Anis Nurashikin Nordin [1,*]

[1] Department of Electrical and Computer Engineering, Kulliyyah of Engineering, International Islamic University Malaysia, Kuala Lumpur 53100, Malaysia; farhanfendi93@gmail.com (M.F.A.M.Y.); fairuzabadimansor@gmail.com (A.F.M.M.); amalina_bella@yahoo.com (N.A.R.)

[2] XLIM-UMR 7252, University of Limoges/CNRS, 87060 Limoges, France; Remi.MANCZAK@chu-limoges.fr (R.M.); Cyril.guines@unilim.fr (C.G.); arnaud.pothier@xlim.fr (A.P.)

[3] Biosensors and Bioelectronics Centre, Department of Physics, Chemistry and Biology (IFM), Linköping University, 58183 Linkoping, Sweden; wing.cheung.mak@liu.se

[4] Manager Department of Electrical Electronics and Renewable Engineering, Onaizah Colleges of Engineering, P.O. Box 2053, Unayzah 56453, Saudi Arabia; cnar32.sheroz@gmail.com

* Correspondence: anisnn@iium.edu.my

Abstract: Diabetes has become a major health problem in society. Invasive glucometers, although precise, only provide discrete measurements at specific times and are unsuitable for long-term monitoring due to the injuries caused on skin and the prohibitive cost of disposables. Remote, continuous, self-monitoring of blood sugar levels allows for active and better management of diabetics. In this work, we present a radio frequency (RF) sensor based on a stepped impedance resonator for remote blood glucose monitoring. When placed on top of a human hand, this RF interdigital sensor allows detection of variation in blood sugar levels by monitoring the changes in the dielectric constant of the material underneath. The designed stepped impedance resonator operates at 3.528 GHz with a Q factor of 1455. A microfluidic device structure that imitates the blood veins in the human hand was fabricated in PDMS to validate that the sensor can measure changes in glucose concentrations. To test the RF sensor, glucose solutions with concentrations ranging from 0 to 240 mg/dL were injected into the fluidic channels and placed underneath the RF sensor. The shifts in the resonance frequencies of the RF sensor were measured using a network analyzer via its S_{11} parameters. Based on the change in resonance frequencies, the sensitivity of the biosensor was found to be 264.2 kHz/mg·dL^{-1} and its LOD was calculated to be 29.89 mg/dL.

Keywords: blood glucose monitoring; diabetes; non-invasive sensor; PDMS; vascular phantom; glucometer; RF sensor

Citation: Yunos, M.F.A.M.; Manczak, R.; Guines, C.; Mansor, A.F.M.; Mak, W.C.; Khan, S.; Ramli, N.A.; Pothier, A.; Nordin, A.N. RF Remote Blood Glucose Sensor and a Microfluidic Vascular Phantom for Sensor Validation. *Biosensors* **2021**, *11*, 494. https://doi.org/10.3390/bios11120494

Received: 14 October 2021
Accepted: 30 November 2021
Published: 3 December 2021

Publisher's Note: MDPI stays neutral with regard to jurisdictional claims in published maps and institutional affiliations.

1. Introduction

Diabetes is a chronic disease that is on the rise across the globe, especially amongst developed nations. To date, there is no known way to prevent or cure diabetes; however, the patient's quality of life can be improved by having periodic monitoring and quantification of glucose levels. Continuous self-monitoring of blood glucose levels allows patients to better control their diet and medication intake, leading to more stable blood sugar levels and fewer incidents of hypo and hyperglycemia [1]. The early generations of glucose sensors were chemical-based, a manual operation mostly involving blood sampling [2]. With the advance of technology and complexity of devices, the manual sampling of blood glucose has become outdated and should be improved for better monitoring of patients. Microwave sensing using small, printed high-frequency sensors is one of the promising techniques to implement non-invasive glucose monitoring. The general principle is to utilize electromagnetic waves to characterize the material under test (MUT), based on its dielectric properties and measuring its resonant frequencies. Previous studies have found that these resonant frequencies correlate to dielectric permittivity of the material and shift with varying concentrations of glucose [3]. Permittivity values are inversely proportional to glucose concentration, where higher concentrations result in lower permittivity values. Different designs of radio frequency (RF) resonator sensors can be used to detect permittivity changes in a material such as circular [4], interdigital transducers (IDT) [5], near-field [6] and microstrip [7]. IDT is the most common design for a wide range of applications such as chemical sensor [8], food inspection [9] and humidity sensor [10,11]. IDT structure has also been used as a biosensor where it was used for direct, rapid quantification and detection of prostate-specific antigens [12]. To enhance the performance of the IDT sensor, stepped impedance structures can be added to the IDT design to improve the quality factor and impedance values of the resonator. These designs, known as stepped impedance resonators (SIR), are compact, have small footprints and allow independent control of the characteristic impedance of the resonator [13]. The RF sensors are advantageous because, unlike electrochemical sensors, they do not require direct contact with the biological samples, allowing them to be reusable, reliable and not vulnerable to performance degradation with service time [14].

Validation of RF sensors can be done using either human tissue phantoms or direct testing using human subjects. Due to the difficulties of obtaining real blood samples, having additional variables such as movements, respiration, temperature variations, for preliminary testing, simulations and testing using phantom models of either animal or liquid phantoms, are preferred [15]. Animal models involving hamsters have been used in previous studies to study correlations between blood glucose and permittivity at frequencies of 10 kHz [16]. In that study, glucose solutions or various concentrations were fed to the hamster and sensor measurements were obtained at the hamster's tail. Nowadays, phantom models are preferred as they are a more humane, cruelty-free method in which synthetic materials are used to mimic human tissues or blood vessels instead of using animals. Human tissue phantoms can also be made by mixing a combination of solutions and gels that produces the same electrical properties as human tissue [17]. The recipe for this combination of deionized water, gelatin, salt, oil, and detergent can be varied and tuned such that this mixture has the same permittivity and conductivity of human tissue at the desired frequency. Cespedes et al., in his research, tuned his mixture for applications to 5.8 GHz [17], while Yilmaz et al. designed a mixture for broadband applications of 0.3 to 20 GHz [18]. More accurate patient-specific phantom models can also be developed using synthetic anatomical models with vascular phantoms that are either 3D printed [19] or fabricated using soft lithography. In this work, we have fabricated microfluidic vascular phantoms in Polydimethylsiloxane (PDMS), which are based on angiogram images of the hand artery to validate the functionality of our sensor with different glucose concentrations.

In this work, we have presented the use of an RF sensor to monitor the different levels of blood glucose concentration. Compared to sensors with different mechanisms, RF sensors are advantageous in terms of their fast response. RF-based sensors enable sensitive differentiation of dielectric properties of the material under test that is seen as a difference in magnitude and frequency shift of S-parameters at resonance frequencies. This paper presents the design, fabrication, and testing of a stepped impedance resonator sensor for continuous in vivo blood glucose monitoring. Section 2 details the design and fabrication of the RF sensor which was fabricated using printed circuit fabrication techniques. Next, the design and fabrication of the vascular phantom using PDMS are described. The vascular phantom consists of two layers, one thin layer to mimic the permittivity of the skin layer and the second layer which contains microfluidic channels that have the design and dimensions of a hand artery. To validate the functionality of the sensor, this vascular phantom is placed underneath the RF sensor, and different glucose concentrations are injected into the channels. The scattering parameters of the sensor are recorded using a network analyzer. Section 3 presents the experimental results obtained, together with the interpretation and discussion of the results. Finally, the main conclusions inferred from this study are gathered in Section 4.

2. Design and Method

2.1. Stepped Impedance Resonator Sensor

The IDT structure is a coplanar capacitor equivalent to the parallel plate capacitor where the electrodes are placed on the same plane to provide a one-sided sensing area for any MUT. These coplanar capacitors can detect changes in permittivity of the material underneath the electrodes. If liquid is placed underneath these electrodes, any changes in the permittivity of the liquid will result in a shift of the resonant frequency of the IDT sensor. In this way, the RF sensor can detect different liquids with different values of permittivity by monitoring the changes in capacitance. The IDT sensor variables are width (w), length (l), and gap (d) between the electrode's digits. Changes in width and gap of the IDT structure affect the capacitance of the sensor and allow the electrical penetration depth of the sensor to be adjusted [20] in accordance with the thickness of the material being tested.

In this work, the IDT sensor was designed to perform detection of glucose concentrations by applying frequencies in the range of 1 to 3 GHz, which are within the range of frequencies that are sensitive to changes in permittivity values of blood plasma [3]. The IDT sensor design, simulation was done using CST Microwave Studio and optimization details have been elaborated in [11]. The finalized design has the following parameters: $w = 0.7$ mm, $l = 14$ mm, $d = 0.5$ mm and number of electrodes = 20.

An important parameter that affects the sensitivity of the sensor is its Quality (Q) factor. The Q factor describes the relation between stored energy and energy usage rate and is used to describe the efficiency of the device. The basic Q-factor equation is dependent on the energy loss of the components in the device such as inductor, capacitor, or resistor. Q can also be calculated using the following equation from the S_{11} frequency response [21]

$$Q = \frac{f}{\Delta f_{3dB}} = 2\pi \frac{Energy\ stored}{Energy\ loss} = \frac{1}{R}\sqrt{\frac{L}{C}} \tag{1}$$

where f is the frequency at minimum loss, and Δf_{3dB} is the difference in frequencies at 3 dB drop from the maximum magnitude. This implies that, as the energy losses increase, or if there are losses in the amplitude of the signal, the R of the device is increased, resulting in a reduced Q factor.

The IDT sensor produced an initial Q of 200 from the simulations. To improve the Q factor, a Stepped Impedance Resonator (SIR) was added to the IDT sensor design. SIR structures are compact and have been used in filters, oscillators and mixers to improve their Q factors [22]. For our device, the SIR consists of two transmission lines of different lengths (L_1 and L_2) with characteristic impedances that are low impedance (Z_1) and high impedance (Z_2), respectively. The design parameters of both the IDT sensor and SIR are shown in Figure 1a. The addition of these two transmission lines (L_1 and L_2) allows us to control the capacitance of the sensor and tune the sensor to produce high Q via better impedance matching. The capacitive Equation (2) was simplified by Chomtong, P. et al. [18], as follows:

$$C_i = \frac{(\varepsilon_r + 1)}{L_T} L_2 (\varepsilon_r + 1)[0.1(n - 3) + 0.11] \tag{2}$$

where C_i is the capacitive value of the IDT, L_T is the total length of the IDT, L_2 is the length of SIR, n is the number of fingers and ε_r is relative permittivity. The design parameter of SIR is controlled by both length and impedance ratio.

To optimize the Q factor, simulations were done using CST Microwave Studio in which the SIR length was varied, and the sensor's frequency response was obtained. The design dimension unit was set in millimeters (mm). The simulations were performed in a frequency range of 1 GHz to 5 GHz. Based on the above equation, L_1 has no significant effect on the performance of SIR and was kept constant for the simulations, while L_2 varied between 21.2 and 21.7 with a step of 0.1. The Q factor of each L_2 value was calculated based on the RF sensor's simulated frequency response, and the plot of Q versus L_2 is shown in Figure 1b. The highest value of Q factor obtained was 1455 when $L_2 = 21.45$ mm. Q factor values of near or over 1000 at frequencies between 100 MHz and 1 GHz are suitable for sensing applications and can adequately reduce high-frequency losses [23].

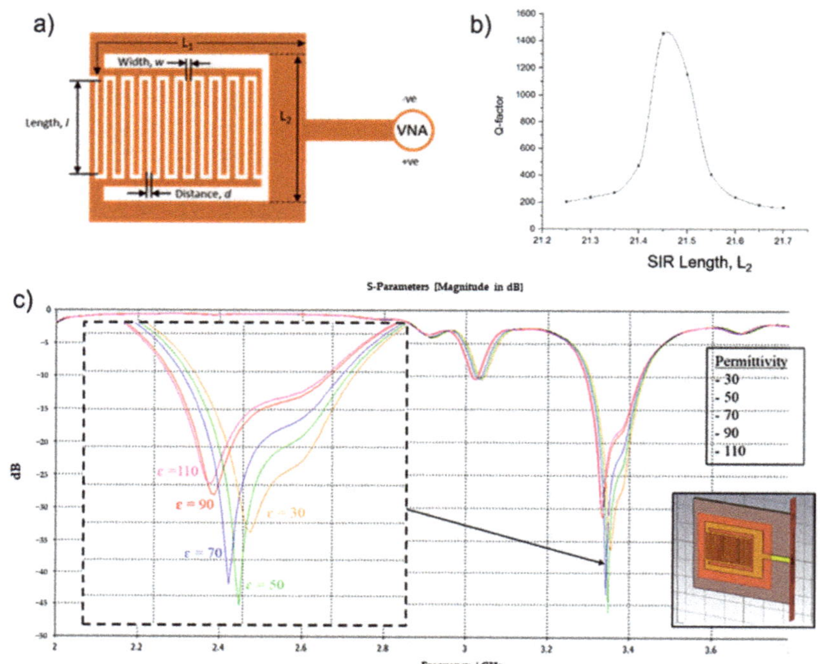

Figure 1. (**a**) RF Sensor comprised of interdigital and Stepped Impedance Resonator (SIR) structures and its design variables. (**b**) Quality (Q) factor values with varying SIR (L_2) lengths. IDT with SIR design and parameter variable (**c**) Simulated frequency response of RF sensor with varying (30 to 110) blood glucose permittivity. Inset: RF sensor model in CST Microwave Studio.

To evaluate the performance of the device for sensing applications, another simulation was done in which the dielectric values near to the sensor's surface were changed to have blood permittivity values. A new material block was placed underneath the sensor design (Figure 1c (inset)) and its permittivity values were varied from 30 to 110 with increments of 20. This method allows us to observe the change in frequency response of the sensor due to the varying of blood glucose levels. Details on the boundary conditions of the sensor can be found in [11]. The sensor's S_{11} frequency response was plotted and is shown in Figure 1c. All values used for blood in the CST simulations are listed in Table 1. From the simulations, higher permittivity values result in a decrease in resonant frequencies. In terms of glucose concentrations, lower glucose concentrations also correspond to higher permittivity values.

Table 1. Important blood parameter values used in the simulation.

Parameter	Values
Electric Conductivity (S/m)	3.05
Density (kg/m^3)	1050
Heat Capacity (J/kg/°C)	3617
Thermal Cond. (W/m/°C)	0.52
Heat Transfer Rate (mL/min/kg)	10,000

Based on the simulation results, three different sensor designs were selected for fabrication: Sensor A (SIR width = 21.40 mm), Sensor B (SIR width = 21.45 mm), Sensor C (SIR width = 21.50 mm). The sensors were fabricated on a Flame-Retardant 4 (FR4) substrate, a common substrate for printed circuits. The FR4 printed circuit board has a thickness of 1.57 mm, with relative permittivity of 4.7, and loss tangent of 0.014. The electrodes are from copper due to its high electrical conductivity. The sensor design was first drawn in Adobe Illustrator and printed on a transparency to form a positive mask. The design is then fabricated onto the FR4 substrate using the same techniques as the printed circuit process which involves exposure, development, etching, and finally stripping. Once the circuit board is ready, it is tested with a multimeter to ensure there is no short and a female SMA connector is attached as the input and output port of the sensor.

2.2. Microfluidic Vascular Phantom

Patient-specific vascular models have been used for surgical training to simulate complex procedures such as endovascular aneurysm repair and coil embolization [19]. In this work, vascular model of the ulnar artery of humans was translated into a microfluidic device using angiogram images of the hand artery. Creating a microfluidic phantom specifically to test the RF sensor instead of human subjects allows us better control and also simplifies the experiments. Using this method, multiple glucose concentrations are injected into the channel repeatedly without harming the subjects. Other environmental factors such as temperature, and flow rate of the liquid being tested, can also be made constant during experimentation.

Dimensions of the ulnar artery were obtained from Fazan et al. [24], who studied 46 male, embalmed human cadavers. In this study, the mean diameter of the ulnar artery on the right hand was found to be approximately 2.5 ± 0.2 mm. The design of the channels was drawn in Adobe Illustrator, as shown in Figure 2a, to closely mimic the angiogram images of the hand artery obtained from [25]. Angiography typically involves injection of a radio-opaque contrast agent into the blood vessels and, using the X-ray imaging, the blood vessel shapes can be captured. The design was then fabricated into a chromium on glass mask, as shown in Figure 2b. Soft lithography methods were used to produce the design mold, as shown in Figure 2c. Fabrication steps are detailed in Section 2.3. The fabricated microfluidic device in PDMS is shown in Figure 2d. The size of the microfluidic device corresponds to the size of the RF sensor, which is 40×50 mm, and alignment markers were included in the mask design to provide constant placement of the channel on the sensor when taking the experimental measurements.

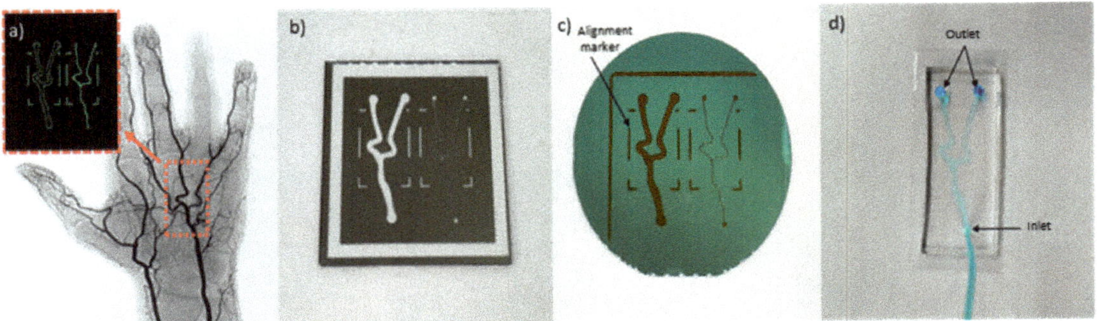

Figure 2. (**a**) Angiogram image of a hand artery from [19]. Inset: Replica image of hand artery redrawn in Adobe Illustrator. (**b**) Design fabricated on chromium on glass mask. (**c**) Microfluidic pattern on dry photoresistant film fabricated using soft lithography. (**d**) Microfluidic device fabricated using PDMS.

When the sensor is placed on a human hand, the capacitive model of the layers can be represented as a 3-layer stack of skin, fat and blood, as shown in Figure 3a. The interdigital sensor is represented as capacitive parallel plates. To accurately model the dielectrics that are present in the hand, a thin layer of PDMS is placed in between the sensor and the microfluidic channel, as shown in Figure 3b. The dielectric value of this thin PDMS layer should match the dielectric values of skin and fat, while the dielectric value of blood is the unknown. The total capacitance of skin and fat can be expressed as follows:

$$\frac{1}{C_T} = \frac{1}{C_{skin}} + \frac{1}{C_{fat}} \tag{3}$$

$$C_T = \varepsilon_0 A \frac{\varepsilon_{skin}\varepsilon_{fat}}{\varepsilon_{fat}d + \varepsilon_{skin}d_{fat}} \tag{4}$$

where C_T is total capacitance, C_{skin} and C_{fat} is the capacitance, ε_{skin} and ε_{fat} is the dielectric constant, and d_{skin} and d_{fat} is the average thickness for skin and fat respectively. The expression (4) was derived from (3). The total capacitance, C_T, can then be used to find the thickness of the thin-layer PDMS, as follows:

$$d_{\text{PDMS}} = \frac{\varepsilon_{\text{PDMS}}\left(\varepsilon_{fat}d_{skin} + \varepsilon_{skin}d_{fat}\right)}{\varepsilon_{skin}\varepsilon_{fat}} \tag{5}$$

where $\varepsilon_{\text{PDMS}}$ is the dielectric constant of PDMS. Substituting the parameter values in Table 2 [26] into Equation (5), the thickness of the bottom layer PDMS was theoretically calculated to be 900 µm. The planar scheme of the RF sensor on a human hand is shown in Figure 3c (left), and the cross-section of the microfluidic vascular phantom is shown in Figure 3c (right).

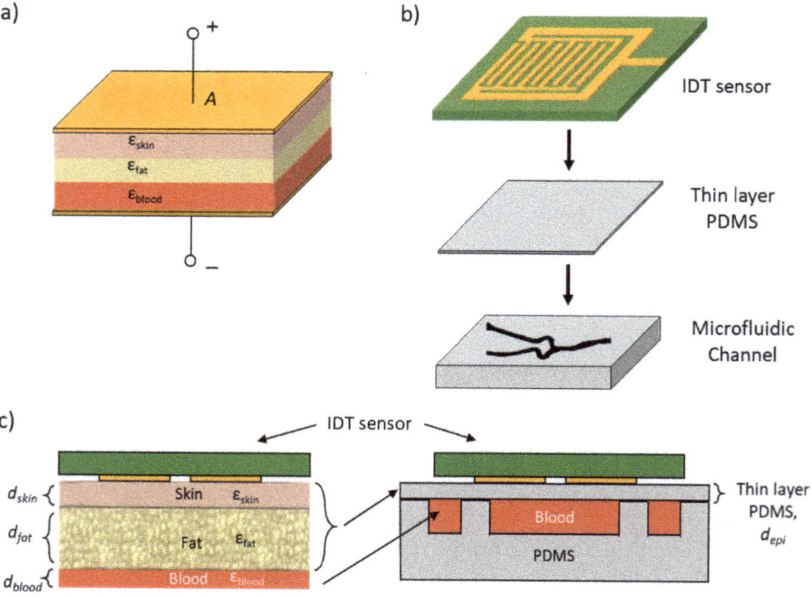

Figure 3. (**a**) Dielectric stack model of skin, fat and blood between two capacitive plates. (**b**) Layer stack of RF sensor on thin PDMS and microfluidic channel used for testing. (**c**) Left: Layer stack of RF sensor on human hand comprising of three layers: skin, fat and blood. Right: Microfluidic vascular phantom layer stack that was used for testing where the thickness of the thin-layer PDMS was designed such that it has the same permittivity values of the three layers (skin, fat and blood).

Table 2. Specific parameter values for skin and fat of human body on ulnar artery.

Parameter	Dielectric Constant, ε	Thickness [mm]
Skin	37.50	1.5
Fat	10.70	3.0
Blood vessel	41.90	2.5
PDMS	2.68	0.9

2.3. Microfluidic Fabrication

The microfluidic blood vessel pattern was developed into a chromium photomask (Delta Mask B.V., The Netherlands) for soft lithography. The final product is shown in Figure 2b. For soft lithography, a silicon wafer is laminated with dry photoresist film at 100 °C using a laminator (RSL-382S) and covered with an adhesion promoter. The microfluidic pattern from the mask was then transferred onto the dry photoresist film using a Suss MicroTec mask aligner MJB4 [27] via photolithography to form the microfluidic mold. After exposure, the sample is heated on a standard hotplate at 80 °C for 2 h. Next, the sample was immersed in a developer solution (cyclohexanone) which dissolves areas of the dry film that were exposed to light. The sample is then baked in an oven or hot plate at temperatures between 100 and 120 °C. This is needed to drive off liquids that may have been absorbed on the substrate and to crosslink the remaining photoresist layer of the mold. Once hardened, this mold can be peeled off from the silicon substrate. These steps are illustrated in Figure 4a–e.

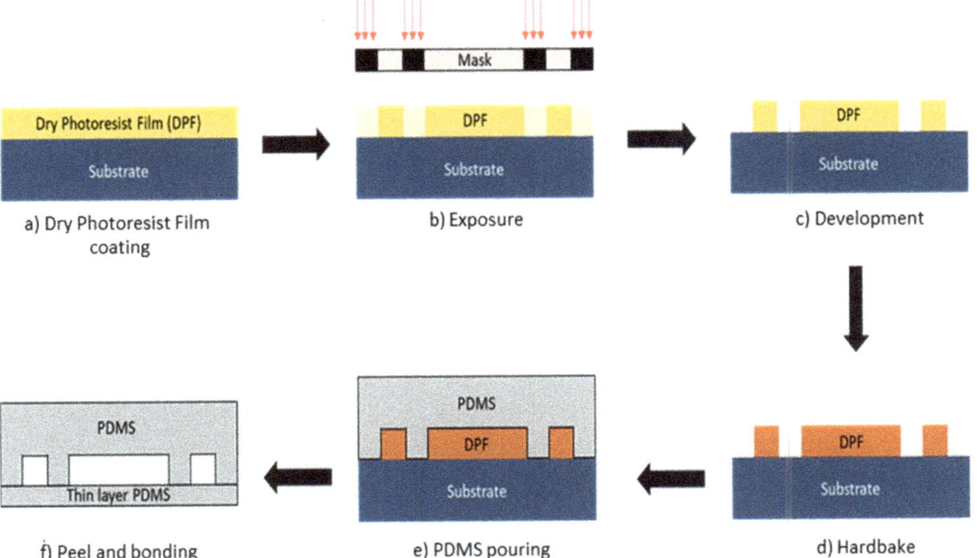

Figure 4. Microfluidic soft lithography fabrication steps. (**a**) Silicon wafer substrate is laminated with dry photoresist film (DPF). (**b**) Microfluidic pattern from mask was exposed with UV light and transferred to DPF. (**c**) Substrate with DPF was immersed in developer solution to remove the layer that is exposed to the UV light. (**d**) The sample was baked to harden the DPF. (**e**) PDMS was poured on to the design mold. (**f**) PDMS was peeled off from the mold and bonded with the thin layer PDMS.

As shown in Figure 4f, there are two PDMS layers, one thin to represent the human skin and fat, and another thicker layer with microfluidic channels that represents the blood vessels. To prepare the PDMS layers, Sylgard 184 silicone elastomer with elastomer curing agent (a crosslinking agent) is mixed at a 10:1 ratio in the plastic cup. Air bubbles were removed from the PDMS mixture by degassing it for 1 h. This mixture can then be used to form both layers: thin-layer PDMS and microfluidic channels. To produce the thin layer, the liquid PDMS was spin coated in a spin coater (Karl SUSS RC8) at a speed of 2500 rpm for 30 s to obtain a layer thickness of around 900 μm. Once the layer is formed, it is baked in an oven at 100 °C for 30 min to solidify the PDMS mixture. The solid PDMS layer was then peeled off and cut into the desired shape using a razor blade.

For the fabrication of microfluidic channels, the liquid PDMS is poured onto the microstructure mold and heated in an oven at 85 °C for 15 min to obtain a hardened, elastomeric replica of the mold. The hardened PDMS layer was then peeled off and cut using a razor blade into a rectangular shape. Two circular holes were punched into the PDMS layer near the two ends of the capillary channel to make the holes for the inlet and outlet media reservoirs. Both layers of PDMS (thin layer and microfluidic layer) were treated with oxygen plasma before assembly. Oxygen plasma treatment promotes adhesion between the two layers and allows very strong bonding that avoids any leakage. The two PDMS layers were then assembled, and a polyethylene tube (length = 900 mm and inner diameter = 840 μm) was connected at the inlet of the microfluidic channel. The tube can be connected later to a disposable syringe and a syringe pump for testing.

Figure 5 shows the fabricated sensor (a), the testing scheme using the microfluidic vascular phantom in lieu of a human hand (b). The microfluidic device was connected to a syringe pump. Measurements of glucose levels using the RF sensor is done without direct contact with blood, allowing the sensor to be reusable as there is no contamination. The fabricated sensor is connected to a network analyzer and the frequency response at each concentration can be obtained. The usage of a microfluidic vascular phantom eases the testing methods and reduces the need of using human subjects at the prototype level.

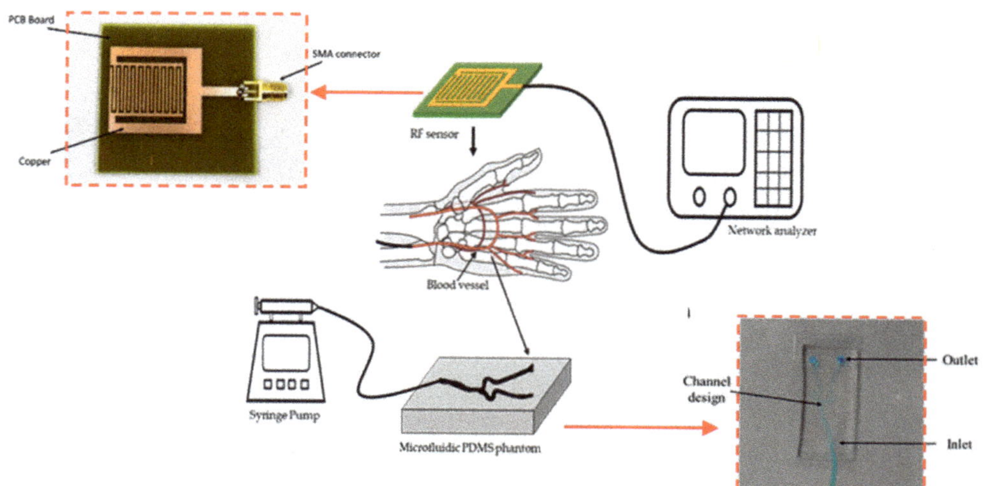

Figure 5. Experimental setup and placement of RF sensor on top of the vascular phantom. A syringe pump was used to inject fluids with different glucose concentrations to mimic the flow of blood with varied sugar levels. (Inset: Fabricated RF Sensor and fabricated microfluidic vascular phantom).

2.4. Glucose Sample Preparation

D(+)-Glucose anhydrous [$C_6H_{12}O_6$] was used to prepare the glucose concentration. The sample glucose (180.16 g/mol) was diluted with DI water at normal room temperature. Glucose solution with concentrations ranging from 30 mg/dL to 240 mg/dL with an increment of 30 mg/dL was prepared to match with the different blood sugar levels (hyperglycemic, normal, hypoglycemic) that may exist in diabetic patients, as shown in Table 3 [28].

Table 3. Range of blood glucose concentrations for different sugar levels in the blood [28].

Blood Sugar Levels	Blood Glucose Concentration (mg/dL)
Hyperglycemia	>200
Normal Glycemia	72–108
Hypoglycemia	<60

2.5. Experimental Setup

Measurement of RF sensor frequency response was made using Keysight Fieldfox RF Analyzer (RFA). Figure 6 shows the placement of the microfluidic device on top of the RF sensor. A male SMA cable was connected to the sensor and the RFA to measure the S_{11} parameters of the sensor. A disposable syringe that contains the appropriate glucose concentration was placed in the syringe pump and the tube was connected to the inlet of the microfluidic device.

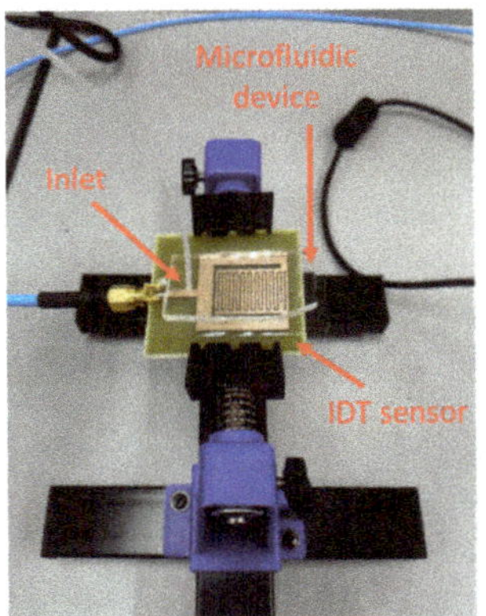

Figure 6. Experimental setup used for RF sensor measurements. Left: The microfluidic device is placed on top of the RF sensor. The inlet of the microfluidic channel is connected to a syringe pump. Right: The RF sensor is connected to a Keysight Fieldfox RF Analyzer.

The RFA was calibrated with a Short, Open, Load, Through (SOLT) SMA calibration kit prior to measurements. The calibration was carried out by measuring a SOLT termination at the point where the sensor will be measured, which is directly at the port. The calibration and measurement frequencies of the RFA were set between 1 GHz and 5 GHz. All three different sensor designs were measured with two different microfluidic devices. The two microfluidic devices only differ in terms of the materials used as their substrate namely: thin-layer PDMS and glass slide. Experimental measurements were done in triplicate where measurement of S_{11} values for each glucose concentration was repeated three times ($n = 3$). The RF sensor was aligned with the fluidic device using the alignment markers to ensure constant placement for each experiment to ensure precision and repeatability of the measurements. The measurements were performed in triplicate for each sensor with each type of microfluidic device to ensure the conformities of the readings.

To imitate the blood fluid properties in human arteries, the flow rate inside the channel must be uniform and constant throughout the measurement. To achieve this requirement, a syringe pump was used separately at the inlet of the microfluidic device. The flow of the pump was initially set at 0.5 mL/min and slowly increased until the speed reached an optimal rate. Air bubbles were formed and trapped in the channel at the junction of the artery model at a flow rate of 0.5 mL/min. This indicates that the pressure of the liquid injection is quite low, thus prone to bubble formation inside the channel. Bubbles fully disappeared once the rate was increased to 3.2 mL/min. This rate also falls inside the range of normal blood flow rate in human arteries, which is 3.0–26 mL/min, as discussed in [29], and was used throughout the experiments. Further increase in the flow rate could result in leakage within the channels.

3. Results and Discussion

Three different sensor designs were measured and produced different quality factors, namely: Sensor A (SIR width = 21.40 mm, Q factor = 980), Sensor B (SIR width = 21.45 mm, Q factor = 1455), Sensor C (SIR width = 21.50 mm, Q factor = 1154). Next, measurements with the microfluidic devices were conducted as shown in Figure 6. RF sensors are advantageous as the response time of the RF sensors only depends on the sweep period of the vector network analyzer used for taking measurements [14]. Vector network analyzers have been reported to have short detection times: 0.85 s in [14] and less than 2 s in [13].

Only measurements of Sensor B with the thin PDMS as its substrate are shown as it produced the highest quality factor. The S_{11} frequency response for RF sensor B was plotted as shown in Figure 7 and the values of frequency (f = 3.2548 GHz), magnitude (S_{11} = −34.873 dB) and phase angle (θ = −155.487) were used to find the capacitance value of thin-layer PDMS. The load impedance formula shown in (6) was used to convert S_{11} to capacitance value.

$$Z_L = Z_0 \left(\frac{1 + S_{11}}{1 - S_{11}} \right) \tag{6}$$

where Z_L is load impedance and Z_0 = 50 ohms. From the results in Z_L, the imaginary part is the series reactance and negative values indicate that it is capacitive. From calculations, Z_L = 48.38−0.723i and the reactance value was converted to find the capacitance of the thin-layer PDMS. The capacitance value of thin-layer PDMS (C_{PDMS} = 6.756 × 10^{-11} F) was used to find the relative permittivity of the PDMS, ε_{PDMS}. The measured ε_{PDMS} was found to be 0.389.

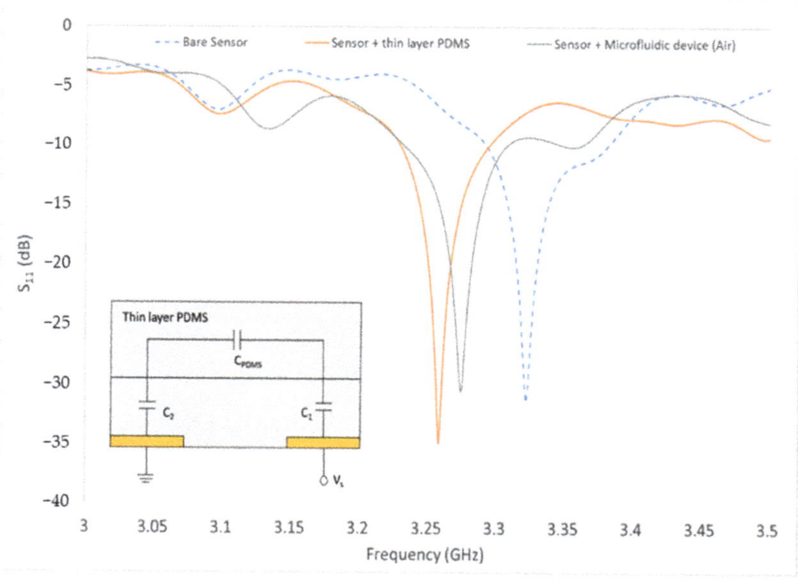

Figure 7. Measurement of the bare sensor, sensor with thin-layer PDMS and sensor with microfluidic device. (Inset: Equivalent circuit of RF sensor with thin-layer PDMS).

Measurement of the microwave reflection coefficient S_{11} was done on a series of aqueous glucose concentrations flowing inside the microfluidic channel. The glucose concentrations ranged from 0 mg/dL (DI water) to 240 mg/dL, which is in accordance with the human blood glucose levels. Figure 8a shows the pattern of resonance frequency shifting towards the different glucose concentrations (0, 120, 240 mg/dL). As can be seen, this graph shows the relationship between the frequency change response against the sample glucose concentration. As the glucose concentrations are increased, the frequency is also increased, shifting the resonance frequency of the S_{11} measurement to the right side after each concentration changed. The RF sensor can detect different liquids with different values of permittivity. The difference in permittivity value will cause changes in the resonance frequency shift.

A linear regression fitted line was plotted on the graph and the correlation coefficient R-square = 0.9325. The regression line indicates that the change of frequency is directly proportional to the concentration changes of the glucose. This suggested that different glucose concentrations have different dielectric permittivity properties, which can be detected by measuring the change in capacitance, and therefore, change in resonance frequency (S_{11}) of the RF sensor. This is in accordance with the Cole–Cole model theory [30] where the variation of reflection coefficient S_{11} leads to the frequency shift over the glucose concentration. Increasing the glucose concentrations in a liquid would generally decrease the permittivity of the liquid over the frequency; thus, in turn shifting the resonance frequency to a higher value.

The graph shows a gradually increasing pattern as the number of concentrations is increased and the trendline from the curve was added to find the sensitivity (slope) and the limit of detection (LOD). Based on the linear regression equation shown in Figure 8b, the detection sensitivity of the biosensor is 264.2 KHz/mg·dL^{-1}. The LOD of the sensor was calculated to be at 29.89 mg/dL using the formula of $3\sigma/m$, where σ is the standard deviation of blank solution and m is the slope of the graph. This implies that the sensor can be used to detect glucose changes greater than approximately 30 mg/dL. The sensor has successfully detected the three human glucose levels: below 60 mg/dL (hypoglycemic), 72–108 mg/dL (normal) and above 200 mg/dL (hyperglycemic). To achieve the optimal sensing performance of the sensor, the regression coefficient needs to be improved to the value of 0.99. This can be done by fixing the location of the cables when setting up the experiment such that it remains constant throughout all the measurements. This will result in improvements in the sensitivity and LOD of the sensor. As a comparison with existing research, Table 4 shows the performance of other types of remote sensors in terms of their LOD, frequency of operation and range of glucose level detection.

Table 4. The value of specific parameters of skin and fat of human body on ulnar artery.

Reference	Sensor	Phantom	Frequency	LOD	Range Glucose Level
[13]	Dual needle	Hamster tail	10 kHz	-	20–500 mg/dL
[14]	Dielectric probe	Oil, gelatin, salt, deionized water, detergent	4–7 GHz	100 mg/dL	0–400 mg/dL
[15]	Dielectric probe	Oil, gelatin, salt, deionized water, detergent	0.3–20 GHz	72 mg/dL	72–216 mg/dL
This work	Stepped impedance resonator	Dual-layer PDMS	1–5 GHz	29.89 mg/dL	0–240 mg/dL

The advantage of this work is that other than the high-Q sensor design, measurements were conducted with a vascular phantom, which is a closer mimic to blood vessels compared to tissue phantoms which were used in other works. The skin and fat dielectric permittivity values were also represented as a single, thin PDMS layer to improve similarity to the human model. In addition, the RF sensor has a measured LOD of 29.89 mg/dL, which is lower compared to the other works. These results show that this biosensor has the potential of being applied as a low-cost non-invasive glucose sensing.

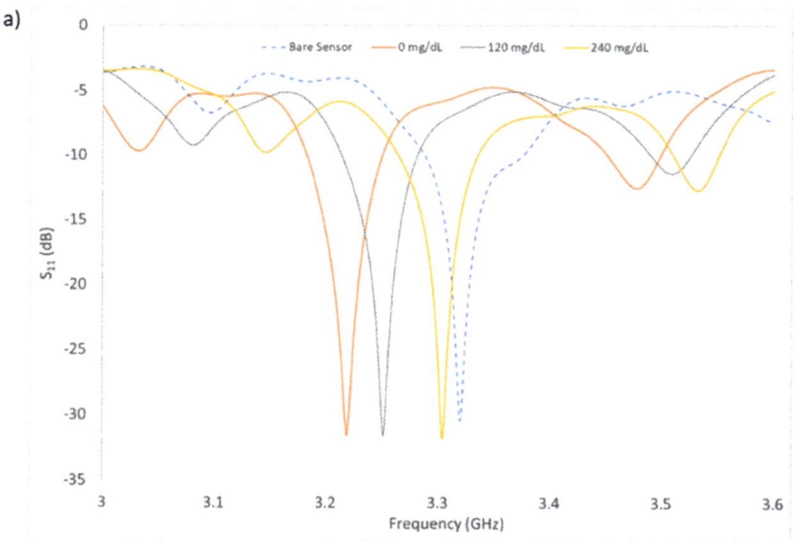

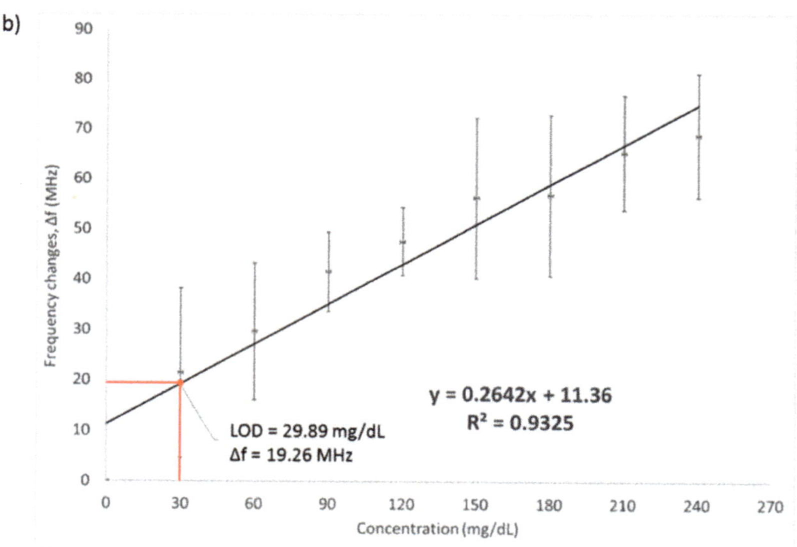

Figure 8. (**a**) S_{11} measurement results of the sensor with varying glucose concentration (0, 120, 240 mg/dL). (**b**) Frequency changes with increasing value of glucose concentration. Limit of detection was found to be 29.89 mg/dL. Measurements were made in triplicates. Error bars are also shown in the graph.

4. Conclusions

In this paper, an interdigital electrode with SIR structures was used for remote sensing of different glucose concentrations. The sensor was fabricated on a double-sided PCB using a conventional PCB manufacturing process. A microfluidic device was also fabricated to imitate the real blood vessel structure of the human hand and form a microfluidic vascular phantom to ease testing. The optimal flow rate to flow glucose inside the channel was found to be at 3.2 mL/min where all bubbles completely disappeared from the channel. A wide range of glucose concentrations (0 mg/dL to 240 mg/dL) was used to test the sensor's performance. The reflection coefficient, S_{11} and the resonance frequency are sensitive to the glucose changes inside the microfluidic hand vascular model. The sensor showed a linearly proportional relationship between resonance frequency changes and varying glucose concentrations. The sensitivity of the sensor was found to be 264.2 kHz/mg·dL^{-1} with LOD of 29.89 mg/dL. This sensor has the potential of being used as a device for continuous, remote monitoring of blood sugar levels for diabetic and prediabetic patients. Non-invasive, continuous monitoring of sugar levels for vulnerable individuals allows early interventions and results in better quality of life.

Author Contributions: M.F.A.M.Y.: writing—original draft preparation, methodology, formal analysis, visualization. R.M.: supervision. C.G.: supervision, resources. A.F.M.M.: methodology. W.C.M.: writing—review and editing, validation. S.K.: conceptualization, resources. N.A.R.: methodology. A.P.: supervision, resources. A.N.N.: conceptualization, supervision, writing—review and editing, validation. All authors have read and agreed to the published version of the manuscript.

Funding: This work is supported by the Malaysia Ministry of Higher Education (MOHE) under the Fundamental Research Grant Scheme (FRGS) FRGS17-030-0596. This work benefited from French government support managed by the National Research Agency under the Investments for the Future program with the reference ANR-10-LABX-0074-01 Sigma-LIM.

Institutional Review Board Statement: Not applicable.

Informed Consent Statement: Not applicable.

Data Availability Statement: The study did not report any data.

Acknowledgments: The authors would like to thank Sumayyah from UniTEN for lending the portable RFA. The authors would also like to thank Afiq Asri for providing guidance in preparing glucose samples and data analysis.

Conflicts of Interest: The authors declare no conflict of interest.

References

1. Coster, S.; Gulliford, M.; Seed, P.; Powrie, J.; Swaminathan, R. Monitoring blood glucose control in diabetes mellitus: A systematic review. *Health Technol. Assess.* **2000**, *4*, 1–94. [CrossRef]
2. Abellán-Llobregat, A.; Jeerapan, I.; Bandodkar, A.; Vidal, L.; Canals, A.; Wang, J.; Morallon, E. A stretchable and screen-printed electrochemical sensor for glucose determination in human perspiration. *Biosens. Bioelectron.* **2017**, *91*, 885–891. [CrossRef] [PubMed]
3. Topsakal, E.; Karacolak, T.; Moreland, E.C. Glucose-dependent dielectric properties of blood plasma. In Proceedings of the 2011 XXXth URSI General Assembly and Scientific Symposium, Istanbul, Turkey, 13 August 2011; pp. 1–4.
4. Choi, H.; Naylon, J.; Luzio, S.; Beutler, J.; Birchall, J.; Martin, C.; Porch, A. Design and in vitro interference test of microwave noninvasive blood glucose monitoring sensor. *IEEE Trans. Microw. Theory Tech.* **2015**, *63*, 3016–3025. [CrossRef]
5. Turgul, V.; Kale, I. Characterization of the complex permittivity of glucose/water solutions for noninvasive RF/Microwave blood glucose sensing. In Proceedings of the 2016 IEEE International Instrumentation and Measurement Technology Conference Proceedings, Taipei, Taiwan, 23 May 2016; pp. 1–5.
6. Gorst, A.; Zavyalova, K.; Mironchev, A. Non-invasive determination of glucose concentration using a near-field sensor. *Biosensors* **2021**, *11*, 62. [CrossRef] [PubMed]
7. Juan, C.G.; Bronchalo, E.; Potelon, B.; Quendo, C.; Sabater-Navarro, J.M. Glucose concentration measurement in human blood plasma solutions with microwave sensors. *Sensors* **2019**, *19*, 3779. [CrossRef]
8. Kitsara, M.; Goustouridis, D.; Chatzandroulis, S.; Chatzichristidi, M.; Raptis, I.; Ganetsos, T.; Igreja, R.; Dias, C. Single chip interdigitated electrode capacitive chemical sensor arrays. *Sens. Actuators B Chem.* **2007**, *127*, 186–192. [CrossRef]
9. Rahman, M.S.B.A.; Mukhopadhyay, S.C.; Yu, P.L. Novel sensors for food inspections. *Sens. Transducers* **2010**, *114*, 1–40.
10. Lei, S.; Deng, C.; Chen, Y.; Li, Y. A novel serial high frequency surface acoustic wave humidity sensor. *Sens. Actuators A Phys.* **2011**, *167*, 231–236. [CrossRef]
11. Bin Yunos, M.F.A.; Nordin, A.N.; Zainuddin, A.; Khan, S. Modeling and development of radio frequency planar interdigital electrode sensors. *Bull. Electr. Eng. Inform.* **2019**, *8*, 978–984.

12. Mishra, S.; Kim, E.-S.; Sharma, P.K.; Wang, Z.-J.; Yang, S.-H.; Kaushik, A.K.; Wang, C.; Li, Y.; Kim, N.-Y. Tailored biofunctionalized biosensor for the label-free sensing of prostate-specific antigen. *ACS Appl. Bio Mater.* **2020**, *3*, 7821–7830. [CrossRef]

13. Adhikari, K.K.; Kim, N.-Y. Ultrahigh-sensitivity mediator-free biosensor based on a microfabricated microwave resonator for the detection of micromolar glucose concentrations. *IEEE Trans. Microw. Theory Tech.* **2015**, *64*, 319–327. [CrossRef]

14. Li, Y.; Yao, Z.; Yue, W.; Zhang, C.; Gao, S.; Wang, C. Reusable, non-invasive, and ultrafast radio frequency biosensor based on optimized integrated passive device fabrication process for quantitative detection of glucose levels. *Sensors* **2020**, *20*, 1565. [CrossRef]

15. Malena, L.; Fiser, O.; Stauffer, P.R.; Drizdal, T.; Vrba, J.; Vrba, D. Feasibility evaluation of metamaterial microwave sensors for non-invasive blood glucose monitoring. *Sensors* **2021**, *21*, 6871. [CrossRef] [PubMed]

16. Park, J.-H.; Kim, C.-S.; Choi, B.-C.; Ham, K.-Y. The correlation of the complex dielectric constant and blood glucose at low frequency. *Biosens. Bioelectron.* **2003**, *19*, 321–324. [CrossRef]

17. Cespedes, F.A. RF Sensing System for Continuous Blood Glucose Monitoring. Ph.D. Thesis, University of South Florida, Tampa Bay, FL, USA, 2017.

18. Yilmaz, T.; Foster, R.; Hao, Y. Broadband tissue mimicking phantoms and a patch resonator for evaluating noninvasive monitoring of blood glucose levels. *IEEE Trans. Antennas Propag.* **2014**, *62*, 3064–3075. [CrossRef]

19. Coles-Black, J.; Bolton, D.; Chuen, J. Accessing 3D printed vascular phantoms for procedural simulation. *Front. Surg.* **2021**, *7*, 158. [CrossRef] [PubMed]

20. Claudel, J.; Ngo, T.-T.; Kourtiche, D.; Nadi, M. Interdigitated sensor optimization for blood sample analysis. *Biosensors* **2020**, *10*, 208. [CrossRef]

21. Zainuddin, A.A.; Nordin, A.N.; Rahim, R.A.; Ralib, A.A.M.; Khan, S.; Guines, C.; Chatras, M.; Pothier, A. Verification of quartz crystal microbalance array using vector network analyzer and OpenQCM. *Indones. J. Electr. Eng. Comput. Sci* **2018**, *10*, 84–93. [CrossRef]

22. Chomtong, P.; Akkaraekthalin, P. A quad-band bandpass filter using stepped impedance resonators with interdigital capacitors. *IEEJ Trans. Electr. Electron. Eng.* **2018**, *13*, 1080–1086. [CrossRef]

23. Amsüss, R.; Saito, S.; Munro, W. Hybridization of quantum systems: Coupling nitrogen–vacancy (NV) centers in diamond to superconducting circuits. In *Quantum Information Processing with Diamond*; Elsevier: Amsterdam, The Netherlands, 2014; pp. 264–290.

24. Fazan, V.P.S.; Borges, C.T.; Da Silva, J.H.; Caetano, A.G.; Filho, O.A.R. Superficial palmar arch: An arterial diameter study. *J. Anat.* **2004**, *204*, 307–311. [CrossRef] [PubMed]

25. Loring, L.A.; Hallisey, M.J. Arteriography and interventional therapy for diseases of the hand. *Radiographics* **1995**, *15*, 1299–1310. [CrossRef] [PubMed]

26. Gabriel, C. Compilation of the Dielectric Properties of Body Tissues at RF and microwave Frequencies. *Physics* **1996**. [CrossRef]

27. Courson, R.; Cargou, S.; Conédéra, V.; Fouet, M.; Blatché, M.-C.; Serpentini, C.L.; Gué, A.M. Low-cost multilevel microchannel lab on chip: DF-1000 series dry film photoresist as a promising enabler. *RSC Adv.* **2014**, *4*, 54847–54853. [CrossRef]

28. Vrba, J.; Vrba, D. A Microwave metamaterial inspired sensor for non-invasive blood glucose monitoring. *Radioengineering* **2015**, *24*, 877–884. [CrossRef]

29. Klarhöfer, M.; Csapo, B.; Balassy, C.; Szeles, J.; Moser, E. High-resolution blood flow velocity measurements in the human finger. *Magn. Reson. Med. Off. J. Int. Soc. Magn. Reson. Med.* **2001**, *45*, 716–719. [CrossRef] [PubMed]

30. Yilmaz, T.; Foster, R.; Hao, Y. Radio-frequency and microwave techniques for non-invasive measurement of blood glucose levels. *Diagnostics* **2019**, *9*, 6. [CrossRef] [PubMed]

 *biosensors*

Communication

Electrochemical Fingerprint Biosensor for Natural Indigo Dye Yielding Plants Analysis

Boyuan Fan [1,†], Qiong Wang [2,3,†], Weihong Wu [1], Qinwei Zhou [1], Dongling Li [2,3], Zenglai Xu [2,3], Li Fu [1,*], Jiangwei Zhu [4], Hassan Karimi-Maleh [5,6,7] and Cheng-Te Lin [8]

1   Key Laboratory of Novel Materials for Sensor of Zhejiang Province, College of Materials and Environmental Engineering, Hangzhou Dianzi University, Hangzhou 310018, China; 191200023@hdu.edu.cn (B.F.); whwu@hdu.edu.cn (W.W.); zhouqw@hdu.edu.cn (Q.Z.)

2   Institute of Botany, Jiangsu Province & Chinese Academy of Sciences (Nanjing Botanical Garden Mem. Sun Yat-Sen), Nanjing 210014, China; wangqiong@cnbg.net (Q.W.); lidongling@cnbg.net (D.L.); xuzenglai@cnbg.net (Z.X.)

3   The Jiangsu Provincial Platform for Conservation and Utilization of Agricultural Germplasm, Nanjing 210014, China

4   Co-Innovation Center for Sustainable Forestry in Southern China, Nanjing Forestry University, Nanjing 210037, China; jwzhu@njfu.edu.cn

5   School of Resources and Environment, University of Electronic Science and Technology of China, Xiyuan Ave, Chengdu 611731, China; hassan@uestc.edu.cn

6   Department of Chemical Engineering, Quchan University of Technology, Quchan 9477177870, Iran

7   Department of Chemical Sciences, Doornfontein Campus, University of Johannesburg, P.O. Box 17011, Johannesburg 2028, South Africa

8   Key Laboratory of Marine Materials and Related Technologies, Zhejiang Key Laboratory of Marine Materials and Protective Technologies, Ningbo Institute of Materials Technology and Engineering (NIMTE), Chinese Academy of Sciences, Ningbo 315201, China; linzhengde@nimte.ac.cn

*   Correspondence: fuli@hdu.edu.cn

†   These authors contribute equally to the work.

Citation: Fan, B.; Wang, Q.; Wu, W.; Zhou, Q.; Li, D.; Xu, Z.; Fu, L.; Zhu, J.; Karimi-Maleh, H.; Lin, C.-T. Electrochemical Fingerprint Biosensor for Natural Indigo Dye Yielding Plants Analysis. *Biosensors* **2021**, *11*, 155. https://doi.org/10.3390/bios11050155

Received: 6 April 2021
Accepted: 12 May 2021
Published: 14 May 2021

Abstract: Indigo is a plant dye that has been used as an important dye by various ancient civilizations throughout history. Today, due to environmental and health concerns, plant indigo is re-entering the market. *Strobilanthes cusia* (Nees) Kuntze is the most widely used species in China for indigo preparation. However, other species under *Strobilanthes* have a similar feature. In this work, 12 *Strobilanthes* spp. were analyzed using electrochemical fingerprinting technology. Depending on their electrochemically active molecules, they can be quickly identified by fingerprinting. In addition, the fingerprint obtained under different conditions can be used to produce scattered patter and heatmap. These patterns make plant identification more convenient. Since the electrochemically active components in plants reflect the differences at the gene level to some extent, the obtained electrochemical fingerprints are further used for the discussion of phylogenetics.

Keywords: electroanalysis; indigo dyes; fast identification; fingerprints; differential pulse voltammetry

1. Introduction

Plant indigo was the most widely used and important dye in the world until the invention of synthetic organic dye aniline violet in 1856. India is widely believed to be the oldest centre of indigo dyeing, and has been Europe's most important importer of the dye since Greco-Roman times [1–4]. Tombs in Egypt have unearthed linen fabrics from around 2400 BC, some of them with delicate indigo lace. In ancient Israel and Palestine, indigo was mixed with green and black dyes for dyeing. Historians in the Egyptian town of Forstadt, south of Cairo, where caravans used to stop during the Middle Ages, have unearthed large quantities of chintz fragments from India [5–8]. Some of them were printed with plant indigo. China, along with Egypt, Peru and India, is the ancient country that applies indigo in the world. Due to the humid climate in Asia, natural fibers are easily damaged. Therefore, ancient fabrics are difficult to preserve in historical relics. However, there are still many discoveries that have been made through archaeological work. Indigo was used to dye silk fabrics unearthed from the Western Han Dynasty tomb in Mawangdui, Changsha. Silk and cotton printing and dyeing products of the Tang Dynasty are preserved in Shosakura in Xinjiang and Japan, including batik cotton fabric with blue background and white flower [9–12].

Indigo is still widely used in traditional textiles, even though other natural plant dyes are rarely used. *Strobilanthes cusia* (Nees) Kuntze is the most widely used species in China for indigo preparation [13,14]. *Strobilanthes* spp. is the second largest genus of the family Acanthaceae distributed in tropical and subtropical regions of Asia. Estimates of the number of species in *Strobilanthes* Blume range from 250 to 450. Many of these species are also used to make indigo. Since many of these species are morphologically similar, identifying plants is not an easy task among non-botanists. With the development of digital image processing and recognition technology, digital images of plants are often used for species identification [15,16]. This approach can be used effectively in species with large morphological differences. However, it is unable to distinguish between species with very similar morphology, and it is especially difficult to do so in the same genus [17]. In this case, spectral analysis and chemical signal analysis are able to overcome this difficulty. For example, Fourier transform infrared (FTIR) spectroscopy allows for the classification of plants based on their different phytochemical compositions [18,19]. However, FTIR spectra mainly reflect certain groups or bonds in the molecule, such as methyl, methylene, carbonyl, cyano, hydroxyl and amine groups. The differences in the spectra in fact reflect differences in the composition of the functional groups and, therefore, do not fully respond to the differences in chemical composition [20]. Conversely, electrochemical fingerprinting can reflect the variability in electrochemically active molecules in the detection system. Previous studies have shown that electrochemical fingerprinting can be successfully applied in plant identification and phylogenetic studies [21,22]. This technology has the potential to be developed for the identification of different commercial plants and the monitoring of corresponding products.

In this study, we selected 12 species from *Strobilanthes* and 2 exotaxa for analysis. Electrochemical taxonomy is a new recently developed technology and is used as an alternative method for plant phylogenetics analysis [23–28]. The electrochemical fingerprints of these species were recorded under different conditions. The patterns of these species were generated for identification. Then, the phylogenetic position of these species was studied.

2. Materials and Methods

Strobilanthes hossei, Strobilanthes japonica, Strobilanthes dimorphotricha, Strobilanthes cusia, Hemigraphis cumingiana, Strobilanthes oliganthus, Strobilanthes hamiltoniana, Strobilanthes austrosinensis, Strobilanthes henryi, Strobilanthes tonkinensis, Strobilanthes schomburgkii, Strobilanthes dyeriana, Strobilanthes hamiltoniana, Strobilanthes biocullata and *Peristrophe japonica* were supplied by Nanjing Botanic Garden. All chemicals were analytical grade and used without purification. All electrochemical fingerprint recordings were conducted using a CHI760 electrochemical workstation. A commercial glassy carbon electrode (GCE), an Ag/AgCl electrode and a Pt electrode were used as the working electrode, reference electrode and counter electrode, respectively.

Ethanol and water were used as solvents for plant leaf extraction. A small amount of leaf (0.01 g) was carefully mixed with 2 mL of solvent. Then, the slurry was sonicated for 5 min for extraction. Then, 2 μL of plant tissue dispersion was drop coated on the working electrode surface and dried naturally. In this study, the electrochemical fingerprinting of two conditions was recorded. The samples extracted with water were recorded under PBS. Samples extracted with ethanol were recorded under ABS. The voltammetric profile (fingerprints) of plant leaf were recorded using differential pulse voltammetry (DPV) in the range −0.1 to 1.5 V in either PBS (pH 7.0, 0.1 M) or ABS (pH 4.5, 0.1 M). Except for the reproducibility test, the fingerprints of herbal tissue were recorded repeated three times in each condition.

All raw data were first treated with a normalization process, where the ratios between the current and the maximum peak current were obtained at different potentials (Scampicchio et al., 2005). The normalized voltammetric data have been used for pattern generation. PCA analysis and cluster analysis were carried out using Origin 2021. The ward linkage method was applied during the cluster analysis.

3. Results and Discussion

Figure 1 shows the voltammetric profiles of *S. hossei, S. japonica, S. dimorphotricha, S. cusia, S. biocullata, S. oliganthus, S. hamiltoniana, S. austrosinensis, S. henryi, S. tonkinensis, S. schomburgkii, S. dyeriana, S. hamiltoniana, H. cumingiana* and *P. japonica* recorded after water extraction in PBS. It can be seen that each DPV curve has electrochemical oxidation signal, which represents that each species contains electrochemical active molecules [29–31]. According to phytochemical studies [32–34], these electrochemically active molecules are phenolic acids, alkaloids, pigments, flavonols and procyanidins. Although we have no way to distinguish each of these molecules in the electrochemical fingerprinting, since many of them have similar chemical structures and are able to oxidize at similar potentials. However, the height and area of these electrochemical oxidation peaks have a positive correlation with the type and number of oxidized molecules. Therefore, comparing the differences in the electrochemical fingerprints of plants can distinguish the differences in electrochemically active molecules in plants. No curves of two samples showed exact same profile, representing the difference of the molecules involved in the electrochemical reaction in each species. This difference is due to differences in the composition and number of electrochemical molecules in different species. The differences also reflect differences at the genetic level [35–37]. Although the composition of a plant is influenced by its environment, is primarily determined by its genes. Three samples have been tested for each species and found to be well reproducible. Therefore, differences in these DPV profiles can be used to identify different species based on the peak locations and peak intensity. However, we see some similarities in the fingerprints of some species, such as *S. hossei* and *S. biocullata*. These two species are not only very similar in leaf morphology, but their flowers are also very similar in morphology and color [38,39].

In order to increase the accuracy of recognition, ethanol was used to extract plant tissues, and fingerprint was collected in ABS. As shown in Figure 2, each species also exhibits electrochemical oxidation behavior under these conditions. Studying Figure 1, it can be seen that the electrochemical oxidation behavior of each plant is not consistent. This is for two reasons. The first is that the species produce different electrochemically active molecules in the extraction of different solvents [40]. Another reason is that in, different pH and buffer solution environments, the oxidation potential of the molecules involved in electrochemical oxidation is not the same [41]. This allows the two species with similar electrochemical behavior, such as *S. hossei* and *S. biocullata*, to exhibit different behaviors here. Therefore, although the two species may be morphologically very similar, their electrochemically active molecules will remain somewhat different in type and amount. Combined with electrochemical fingerprinting, taken under different conditions, this difference can be amplified and provide the opportunity to perform identification.

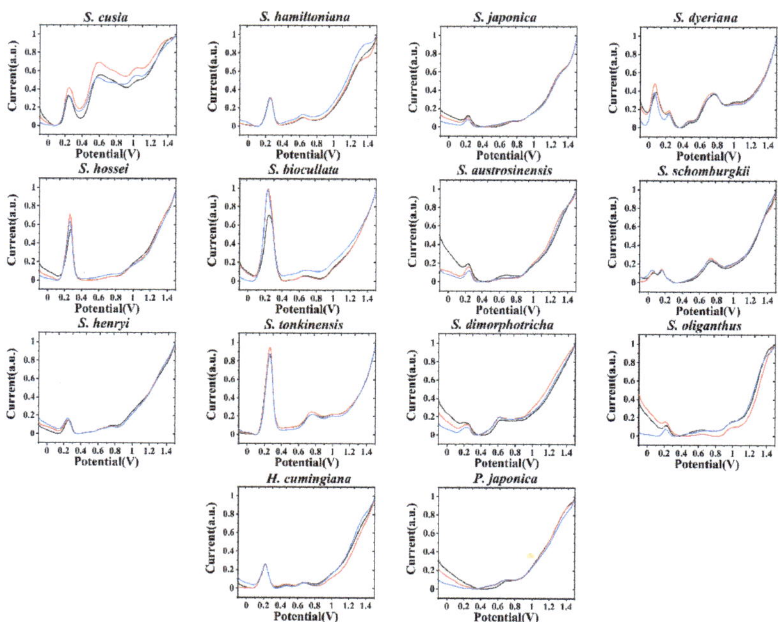

Figure 1. Electrochemical fingerprint of *S. hossei, S. japonica, S. dimorphotricha, S. cusia, S. biocullata, S. oliganthus, S. hamiltoniana, S. austrosinensis, S. henryi, S. tonkinensis, S. schomburgkii, S. dyeriana, S. hamiltoniana, H. cumingiana* and *P. japonica* recorded after water extraction in PBS.

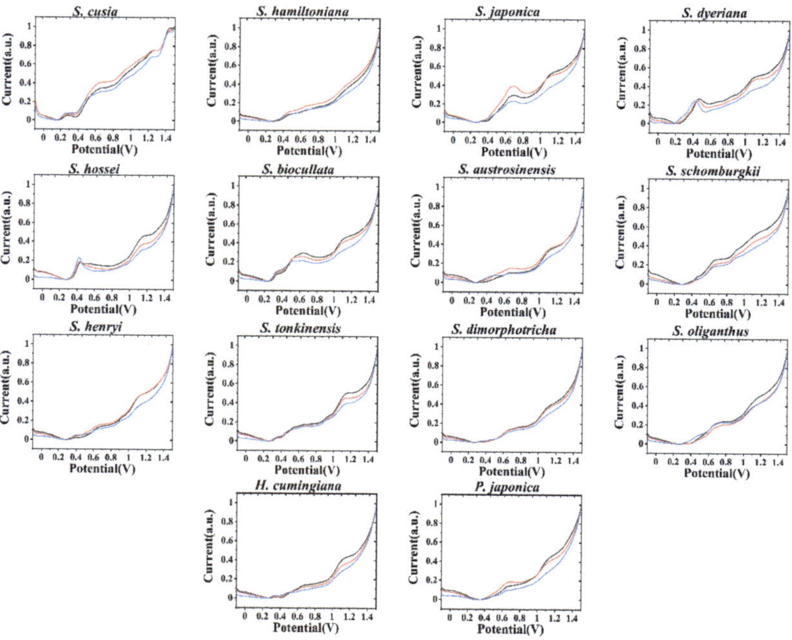

Figure 2. Electrochemical fingerprint of *S. hossei, S. japonica, S. dimorphotricha, S. cusia, S. biocullata, S. oliganthus, S. hamiltoniana, S. austrosinensis, S. henryi, S. tonkinensis, S. schomburgkii, S. dyeriana, S. hamiltoniana, H. cumingiana* and *P. japonica* recorded after ethanol extraction in ABS.

However, it is difficult to directly identify species using DPV profiles, especially when there are many samples. Thus, we combined the data from the two figures to produce a scatter plot pattern for each species. Because the potential information in the two sets of data is exactly the same, we delete their weights. In this case, the scatter plot's X-and Y-axis data are given equal weight, and we can combine the species' electrochemical fingerprints collected in both conditions in a single pattern. Figure 3 shows the scatter patterns of *S. hossei, S. japonica, S. dimorphotricha, S. cusia, S. biocullata, S. oliganthus, S. hamiltoniana, S. austrosinensis, S. henryi, S. tonkinensis, S. schomburgkii, S. dyeriana, S. hamiltoniana, H. cumingiana* and *P. japonica*. As can be seen from the figure, each species has its own unique pattern. By dividing the whole area into several quadrants, and then counting the data points in different quadrants, the unknown sample can be compared with the database. Multivariate variance analysis shows that there is no significant difference between patterns of the same species. However, there are significant differences between the scatter patterns of any two species. Therefore, the scatter pattern is a better recognition pattern than the DPV profile. We conducted the investigation of the reproducibility of scatter pattern. Although all patterns of one species showed a similar shape, a small difference can be observed when overlapping them together. These variations are inevitable for fingerprinting the chemical and metabolic profile of a biological sample. However, the slight differences between individual recordings from the same species cannot affect the identification results. Based on the significant pattern difference between the species, the unknown sample can be compared with the database for identification.

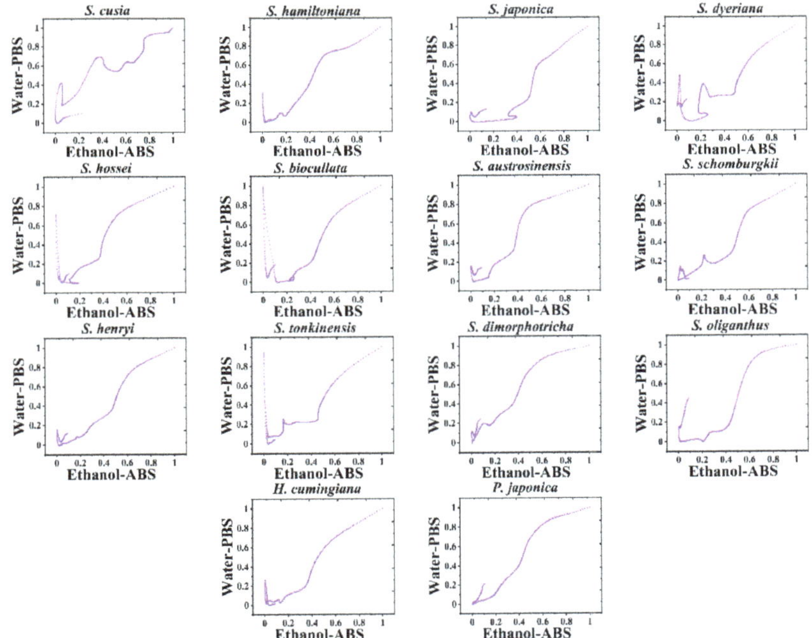

Figure 3. Scatter patterns of *S. hossei, S. japonica, S. dimorphotricha, S. cusia, S. biocullata, S. oliganthus, S. hamiltoniana, S. austrosinensis, S. henryi, S. tonkinensis, S. schomburgkii, S. dyeriana, S. hamiltoniana, H. cumingiana* and *P. japonica*.

We further propose a more intuitive pattern recognition method. In this model, two sets of fingerprints collected in different environments can be used to make a heatmap of the species. As shown in Figure 4, we not only combined the two groups of data, but also displayed a pattern similar to a scatter pattern. In addition, we put a value on the density of the data. The more data points in an area, the darker hot spots will appear. In this pattern recognition mode, we no longer need logarithmic data points for statistics, but only need to locate the range of the hot area. Species can be identified if the hot zones of a species are in a composite database of unknown samples.

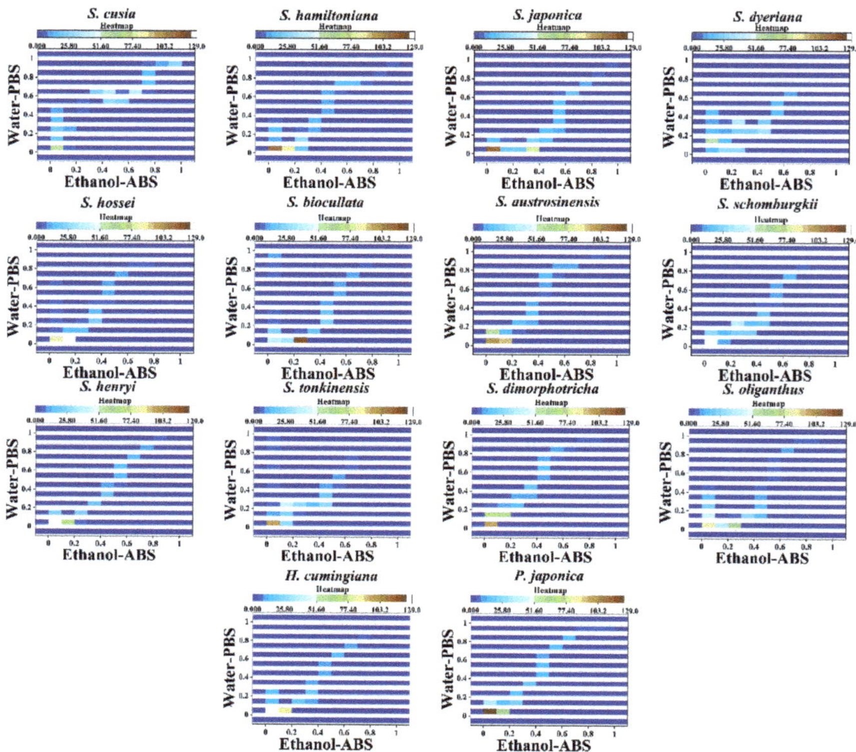

Figure 4. Heatmap of *S. hossei, S. japonica, S. dimorphotricha, S. cusia, S. biocullata, S. oliganthus, S. hamiltoniana, S. austrosinensis, S. henryi, S. tonkinensis, S. schomburgkii, S. dyeriana, S. hamiltoniana, H. cumingiana* and *P. japonica.*

Principal component analysis (PCA) is a common statistical technique used to analyze differences between data groups and between data groups. In this work, we performed a PCA analysis of the homogenized current values collected in both environments for each species.

As shown in Figure 5, *S. Japonica, S. Dimorphotricha,* and *S. Schomburgkii* were grouped together. Meanwhile, *S. austrosinensis, S. oliganthus,* and *H. cumingiana* were grouped into one cluster. The proximity of their data is due to the similarity of electrochemically active molecules in their tissues. This also reflects their genetic similarity.

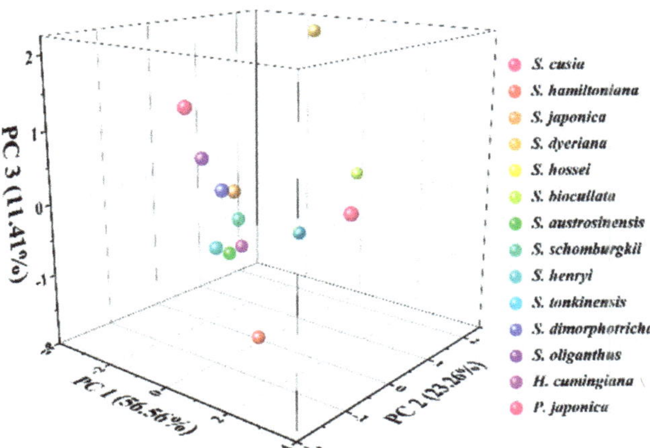

Figure 5. PCA analysis of *S. hossei, S. japonica, S. dimorphotricha, S. cusia, S. biocullata, S. oliganthus, S. hamiltoniana, S. austrosinensis, S. henryi, S. tonkinensis, S. schomburgkii, S. dyeriana, S. hamiltoniana, H. cumingiana* and *P. japonica*.

We further attempted to study the infrageneric relationships of these species, and hierarchical clustering analysis was carried out using electrochemical profiles. As shown in Figure 6, the first group consisted of the species *S. dyeriana, S. hossei, S. tonkinensis* and *S. biocullata*. The second group contains two clades. The clade included *S. austrosinensis, S. oliganthus* and *H. cumingiana*. Another clade included *S. hamiltoniana, S. japonica, S. dimorphotricha, S. schomburgkii, S. henryi* and *P. japonica*. One outlier can be seen in *S. cusia*. This result is not entirely consistent with the results of other taxonomic techniques. This may be due to the confusing taxonomic results of the genus *Strobilanthes*. For example, Bremekamp divided *Strobilanthes* and its allies into over 54 genera arranged in 27 informal groups [42]. Terao recognized a broadly circumscribed *Strobilanthes* comprising all species of Strobilanthinae [43]. The results of recent molecular studies, statistical analysis and pollen and gross morphology showed that these results are problematic [44–47]. Our results provide a new explanation.

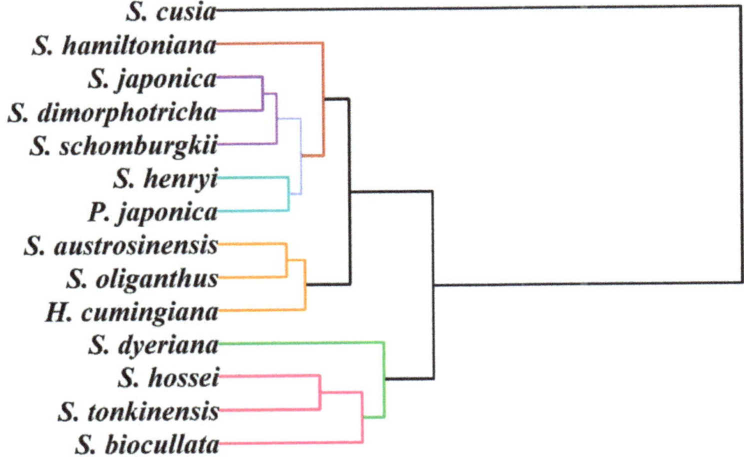

Figure 6. Dendrogram of *S. hossei, S. japonica, S. dimorphotricha, S. cusia, S. biocullata, S. oliganthus, S. hamiltoniana, S. austrosinensis, S. henryi, S. tonkinensis, S. schomburgkii, S. dyeriana, S. hamiltoniana, H. cumingiana* and *P. japonica* based on electrochemical fingerprints.

4. Conclusions

In this work, we provide an electrochemical method for potential identifying species of the dye plant indigo by using the fingerprints of electrochemically active molecules in plant tissues. Two different conditions were combined using solvents and buffer solutions for the recording of electrochemical fingerprints. The same species exhibit different fingerprint profiles under different conditions because different electrochemically active molecules were extracted and were involved in electrochemical oxidation under different pH conditions. The fingerprint profiles of some species showed similarity under one condition, but went very differently under another condition. Therefore, combining two sets of fingerprint profiles can be used to make a scatter pattern and heatmap for the identification of species. In these two pattern modes, the species were easier to identify than the DPV curves directly. The electrochemical fingerprinting presents information that can be linked to their genetic level. The dendrogram indicated that the 14 species were divided into three main clades. An outlier of *S. cusia* was observed.

Author Contributions: Conceptualization, L.F. and C.-T.L.; methodology, L.F. and Q.W.; software, W.W.; validation, B.F. and Q.W.; formal analysis, B.F. and Q.Z.; investigation, Q.W. and D.L.; data curation, Z.X. and J.Z.; writing—original draft preparation, B.F. and L.F.; writing—review and editing, C.-T.L. and H.K.-M.; supervision, W.W.; project administration, L.F.; funding acquisition, L.F. All authors have read and agreed to the published version of the manuscript.

Funding: This research was funded by the National Natural Science Foundation of China (22004026)

Institutional Review Board Statement: Not applicable.

Informed Consent Statement: Not applicable.

Data Availability Statement: Data sharing not applicable.

Conflicts of Interest: The authors declare no conflict of interest.

References

1. Fabara, A.N.; Fraaije, M.W. An overview of microbial indigo-forming enzymes. *Appl. Microbiol. Biotechnol.* **2020**, *104*, 925–933. [CrossRef] [PubMed]
2. Dutta, S.; Roychoudhary, S.; Sarangi, B.K. Effect of different physico-chemical parameters for natural indigo production during fermentation of *Indigofera* plant biomass. *3 Biotech.* **2017**, *7*, 322. [CrossRef] [PubMed]
3. Khoramdel, S.; Rezvani, P.; Hooshmand, M.; Moalem, F. Effects of cow manure levels and plant densities on yield and seed yield components, leaf and indigo yields of true indigo. *J. Plant. Prod. Res.* **2017**, *23*, 117–143.
4. Prabha, C.; Sharma, S. Extraction, Characterization and Tissue Culture of Plant Derived Indigo From *Indigofera* tinctoria-Preliminary Studies. *Int. J. Agrochem.* **2018**, *4*, 53–58.
5. Li, S.; Cunningham, A.B.; Fan, R.; Wang, Y. Identity blues: The ethnobotany of the indigo dyeing by Landian Yao (Iu Mien) in Yunnan, Southwest China. *J. Ethnobiol. Ethnomedicine* **2019**, *15*, 13. [CrossRef]
6. Zhang, L.; Wang, L.; Cunningham, A.B.; Shi, Y.; Wang, Y. Island blues: Indigenous knowledge of indigo-yielding plant species used by Hainan Miao and Li dyers on Hainan Island, China. *J. Ethnobiol. Ethnomedicine* **2019**, *15*, 31. [CrossRef]
7. Tsuji, H.; Kondo, M.; Odani, W.; Takino, T.; Takeda, R.; Sakai, T. Treatment with indigo plant (*Polygonum tinctorium* Lour) improves serum lipid profiles in Wistar rats fed a high-fat diet. *J. Med Investig.* **2020**, *67*, 158–162. [CrossRef]
8. Pattanaik, L.; Duraivadivel, P.; Hariprasad, P.; Naik, S.N. Utilization and re-use of solid and liquid waste generated from the natural indigo dye production process—A zero waste approach. *Bioresour. Technol.* **2020**, *301*, 122721. [CrossRef]
9. Nakai, A.; Tanaka, A.; Yoshihara, H.; Murai, K.; Watanabe, T.; Miyawaki, K. Blue LED light promotes indican accumulation and flowering in indigo plant, *Polygonum tinctorium. Ind. Crop. Prod.* **2020**, *155*, 112774. [CrossRef]
10. Begum, K.; Motobayashi, T.; Hasan, N.; Appiah, K.S.; Shammi, M.; Fujii, Y. Indigo as a Plant Growth Inhibitory Chemical from the Fruit Pulp of Couroupita guianensis Aubl. *Agronomy* **2020**, *10*, 1388. [CrossRef]
11. Prasad, R. Indigo—The Crop that Created History and then Itself Became History. *Indian J. Hist. Sci.* **2018**, *53*, 296–301. [CrossRef]
12. Pattanaik, L.; Padhi, S.K.; Hariprasad, P.; Naik, S.N. Life cycle cost analysis of natural indigo dye production from *Indigofera* tinctoria L. plant biomass: A case study of India. *Clean Technol. Environ. Policy* **2020**, *22*, 1639–1654. [CrossRef]
13. Lee, C.-L.; Wang, C.-M.; Hu, H.-C.; Yen, H.-R.; Song, Y.-C.; Yu, S.-J.; Chen, C.-J.; Li, W.-C.; Wu, Y.-C. Indole alkaloids indigodoles A–C from aerial parts of *Strobilanthes* cusia in the traditional Chinese medicine Qing Dai have anti-IL-17 properties. *Phytochemistry* **2019**, *162*, 39–46. [CrossRef] [PubMed]
14. Yu, H.; Li, T.; Ran, Q.; Huang, Q.; Wang, J. *Strobilanthes* cusia (Nees) Kuntze, a multifunctional traditional Chinese medicinal plant, and its herbal medicines: A comprehensive review. *J. Ethnopharmacol.* **2021**, *265*, 113325. [CrossRef]

15. Barré, P.; Stöver, B.C.; Müller, K.F.; Steinhage, V. LeafNet: A computer vision system for automatic plant species identification. *Ecol. Inform.* **2017**, *40*, 50–56. [CrossRef]

16. Wäldchen, J.; Rzanny, M.; Seeland, M.; Mäder, P. Automated plant species identification—Trends and future directions. *Plos Comput. Biol.* **2018**, *14*, e1005993. [CrossRef]

17. Wäldchen, J.; Mäder, P. Machine learning for image based species identification. *Methods Ecol. Evol.* **2018**, *9*, 2216–2225. [CrossRef]

18. Depciuch, J.; Kasprzyk, I.; Drzymała, E.; Parlinska-Wojtan, M. Identification of birch pollen species using FTIR spectroscopy. *Aerobiologia* **2018**, *34*, 525–538. [CrossRef]

19. Kendel, A.; Zimmermann, B. Chemical analysis of pollen by FT-Raman and FTIR spectroscopies. *Front. Plant Sci.* **2020**, *11*, 352. [CrossRef]

20. Sithara, N.; Komathi, S.; Rajalakshmi, G. Identification of bioactive compounds using different solvents through FTIR studies and GCMS analysis. *J. Med. Plants Stud.* **2017**, *5*, 192–194.

21. Zheng, Y.; Zhu, J.; Fu, L.; Liu, Q. Phylogenetic Investigation of Yellow Camellias Based on Electrochemical Voltammetric Fingerprints. *Int. J. Electrochem. Sci* **2020**, *15*, 9622–9630. [CrossRef]

22. Zhang, X.; Yang, R.; Li, Z.; Zhang, M.; Wang, Q.; Xu, Y.; Fu, L.; Du, J.; Zheng, Y.; Zhu, J. Electroanalytical study of infrageneric relationship of *Lagerstroemia* using glassy carbon electrode recorded voltammograms. *Rev. Mex. De Ing. Química* **2020**, *19*, 281–291. [CrossRef]

23. Fu, L.; Zheng, Y.; Zhang, P.; Zhang, H.; Wu, M.; Zhang, H.; Wang, A.; Su, W.; Chen, F.; Yu, J.; et al. An electrochemical method for plant species determination and classification based on fingerprinting petal tissue. *Bioelectrochemistry* **2019**, *129*, 199–205. [CrossRef]

24. Zhou, J.; Zheng, Y.; Zhang, J.; Karimi-Maleh, H.; Xu, Y.; Zhou, Q.; Fu, L.; Wu, W. Characterization of the Electrochemical Profiles of Lycoris Seeds for Species Identification and Infrageneric Relationships. *Anal. Lett.* **2020**, *53*, 2517–2528. [CrossRef]

25. Fu, L.; Zheng, Y.; Zhang, P.; Zhang, H.; Xu, Y.; Zhou, J.; Zhang, H.; Karimi-Maleh, H.; Lai, G.; Zhao, S.; et al. Development of an electrochemical biosensor for phylogenetic analysis of *Amaryllidaceae* based on the enhanced electrochemical fingerprint recorded from plant tissue. *Biosens. Bioelectron.* **2020**, *159*, 112212. [CrossRef] [PubMed]

26. Fu, L.; Wang, Q.; Zhang, M.; Zheng, Y.; Wu, M.; Lan, Z.; Pu, J.; Zhang, H.; Chen, F.; Su, W. Electrochemical sex determination of dioecious plants using polydopamine-functionalized graphene sheets. *Front. Chem.* **2020**, *8*, 92. [CrossRef]

27. Xu, Y.; Lu, Y.; Zhang, P.; Wang, Y.; Zheng, Y.; Fu, L.; Zhang, H.; Lin, C.-T.; Yu, A. Infrageneric phylogenetics investigation of *Chimonanthus* based on electroactive compound profiles. *Bioelectrochemistry* **2020**, *133*, 107455. [CrossRef]

28. Zhang, M.; Pan, B.; Wang, Y.; Du, X.; Fu, L.; Zheng, Y.; Chen, F.; Wu, W.; Zhou, Q.; Ding, S. Recording the Electrochemical Profile of Pueraria Leaves for Polyphyly Analysis. *ChemistrySelect* **2020**, *5*, 5035–5040. [CrossRef]

29. Li, H.; Zhang, Z.; Duan, J.; Li, N.; Li, B.; Song, T.; Sardar, M.F.; Lv, X.; Zhu, C. Electrochemical disinfection of secondary effluent from a wastewater treatment plant: Removal efficiency of ARGs and variation of antibiotic resistance in surviving bacteria. *Chem. Eng. J.* **2020**, *392*, 123674. [CrossRef]

30. Durán, F.E.; de Araújo, D.M.; do Nascimento Brito, C.; Santos, E.V.; Ganiyu, S.O.; Martínez-Huitle, C.A. Electrochemical technology for the treatment of real washing machine effluent at pre-pilot plant scale by using active and non-active anodes. *J. Electroanal. Chem.* **2018**, *818*, 216–222. [CrossRef]

31. Ehsani, A.; Mahjani, M.G.; Hosseini, M.; Safari, R.; Moshrefi, R.; Mohammad Shiri, H. Evaluation of *Thymus vulgaris* plant extract as an eco-friendly corrosion inhibitor for stainless steel 304 in acidic solution by means of electrochemical impedance spectroscopy, electrochemical noise analysis and density functional theory. *J. Colloid Interface Sci.* **2017**, *490*, 444–451. [CrossRef]

32. Szczepaniak, O.M.; Ligaj, M.; Kobus-Cisowska, J.; Maciejewska, P.; Tichoniuk, M.; Szulc, P. Application for novel electrochemical screening of antioxidant potential and phytochemicals in *Cornus mas* extracts. *Cyta-J. Food* **2019**, *17*, 781–789. [CrossRef]

33. Marsoul, A.; Ijjaali, M.; Elhajjaji, F.; Taleb, M.; Salim, R.; Boukir, A. Phytochemical screening, total phenolic and flavonoid methanolic extract of pomegranate bark (*Punica granatum* L): Evaluation of the inhibitory effect in acidic medium 1 M HCl. *Mater. Today Proc.* **2020**, *27*, 3193–3198. [CrossRef]

34. Reddy, Y.M.; Kumar, S.; Saritha, K.; Gopal, P.; Reddy, T.M.; Simal-Gandara, J. Phytochemical Profiling of Methanolic Fruit Extract of *Gardenia latifolia* Ait. by LC-MS/MS Analysis and Evaluation of Its Antioxidant and Antimicrobial Activity. *Plants* **2021**, *10*, 545. [CrossRef]

35. Lu, Y.; Xu, Y.; Shi, H.; Zhang, P.Z.H.; Fu, L. Feasibility of electrochemical fingerprinting for plant phylogeography study: A case of *Chimonanthus praecox. Int. J. Electrochem. Sci.* **2020**, *15*, 758–764. [CrossRef]

36. Vijay, A.; Chhabra, M.; Vincent, T. Microbial community modulates electrochemical performance and denitrification rate in a biocathodic autotrophic and heterotrophic denitrifying microbial fuel cell. *Bioresour. Technol.* **2019**, *272*, 217–225. [CrossRef] [PubMed]

37. Liu, J.; Shi, S.; Ji, X.; Jiang, B.; Xue, L.; Li, M.; Tan, L. Performance and microbial community dynamics of electricity-assisted sequencing batch reactor (SBR) for treatment of saline petrochemical wastewater. *Environ. Sci. Pollut. Res.* **2017**, *24*, 17556–17565. [CrossRef]

38. Wood, J.; Scotland, R. New and little-known species of *Strobilanthes* (*Acanthaceae*) from India and South East Asia. *Kew Bull.* **2009**, *64*, 3–47. [CrossRef]

39. Somprasong, W.; Vjarodaya, S.; Chayamarit, K. Taxonomic Study of the Family *Acanthaceae* used as traditional medicinal plants for ethnic groups in North, Central and Northeastern Thailand. *Thai Agric. Res. J.* **2014**, *32*, 77–88.

40. Sultana, B.; Anwar, F.; Ashraf, M. Effect of extraction solvent/technique on the antioxidant activity of selected medicinal plant extracts. *Molecules* **2009**, *14*, 2167–2180. [CrossRef]
41. Karimi-Maleh, H.; Ayati, A.; Davoodi, R.; Tanhaei, B.; Karimi, F.; Malekmohammadi, S.; Orooji, Y.; Fu, L.; Sillanpää, M. Recent advances in using of chitosan-based adsorbents for removal of pharmaceutical contaminants: A review. *J. Clean. Prod.* **2021**, *291*, 125880. [CrossRef]
42. Bremekamp, C.E.B. *Materials for a Monograph of the Strobilanthinae (Acanthaceae)*; NV Noord-Hollandsche Uitgevers Maatschappij: Amsterdam, The Netherlands, 1944.
43. Terao, H. Taxonomic study of the genus *Strobilanthes* Bl. (*Acanthaceae*): Generic delimitation and infrageneric classification. Ph.D. Thesis, Kyoto University, Kyoto, Japan, 1983.
44. Carine, M.A.; Scotland, R.W. Classification of Strobilanthinae (*Acanthaceae*): Trying to classify the unclassifiable? *Taxon* **2002**, *51*, 259–279. [CrossRef]
45. Wood, J. Notes on *Strobilanthes* (*Acanthaceae*) for the flora of Ceylon. *Kew Bull.* **1995**, *50*, 1–24. [CrossRef]
46. Moylan, E.C.; Bennett, J.R.; Carine, M.A.; Olmstead, R.G.; Scotland, R.W. Phylogenetic relationships among *Strobilanthes* s.l. (*Acanthaceae*): Evidence from ITS nrDNA, trnL-F cpDNA, and morphology. *Am. J. Bot.* **2004**, *91*, 724–735. [CrossRef] [PubMed]
47. Wood, J. Notes relating to the flora of Bhutan: XXIX. *Acanthaceae*, with special reference to *Strobilanthes*. *Edinb. J. Bot.* **1994**, *51*, 175–273. [CrossRef]

MDPI

St. Alban-Anlage 66

4052 Basel

Switzerland

Tel. +41 61 683 77 34

Fax +41 61 302 89 18

www.mdpi.com

Biosensors Editorial Office

E-mail: biosensors@mdpi.com

www.mdpi.com/journal/biosensors